GENETICS AND BREEDING FOR CROP QUALITY AND RESISTANCE

Developments in Plant Breeding

VOLUME 8

Genetics and Breeding for Crop Quality and Resistance

Proceedings of the XV EUCARPIA Congress, Viterbo, Italy, September 20-25, 1998

Edited by

G.T. SCARASCIA MUGNOZZA

E. PORCEDDU

and

M.A. PAGNOTTA
University of Tuscia,
Viterbo, Italy

SPRINGER-SCIENCE+BUSINESS MEDIA, B.V.

A C.I.P. Catalogue record for this book is available from the Library of Congress.

ISBN 978-0-7923-5844-2 ISBN 978-94-011-4475-9 (eBook)
DOI 10.1007/978-94-011-4475-9

Printed on acid-free paper

Contents

Acknowledgements

The XV Eucarpia General Congress has been organized under the high patronage of the President of the Italian Republic and under the auspices and contributions of the Ministry of Agricultural Policies, Ministry of University and Scientific and Technological Research, National Research Council, CariVit foundation, Latium Region, Municipality and Province of Viterbo and of the University of Tuscia.

This volume has been printed with the financial contribution of:
the Italian Ministry of Agricultural Policies

G.T. Scarascia Mugnozza, E. Porceddu & M.A. Pagnotta (Eds.)
Genetics and Breeding for Crop Quality and Resistance, xi-xii, 1999

Preface

The main theme of the XV EUCARPIA General Congress was "Genetics and Breeding for Crop Quality and Resistance". This theme was addressed in six different sessions, corresponding to the first six chapters of the Proceedings. Each session was introduced by a key-note speaker, and included different oral presentations. The Session on Quality, which was the largest one, was assigned two key-note lectures, respectively on agro-food quality and on the quality of wood plants. There were respectively 36 oral presentations and 183 posters exhibited. The seventh chapter of the Proceedings includes the special lectures which opened and closed the Congress.

The speakers let us know about the state of the art of the research, on the application and results of the biotechnology methods, i.e. the facts and perspectives of the genetic engineering in the improvement of crops, in their quantitative, qualitative and nutritional characters, in their adaptation to the environment, to the ecosystems, in their resistance to pests and diseases. This corpus of research and innovation will have a direct consequence in terms of reduction of environment and food pollution, of a slower erosion of natural resources (soil, water, atmosphere, biodiversity), and, ultimately, of a reduced consumption of energy. All with a final aim: the sustainability of a modern agriculture.

The Congress was attended by over 300 scientists. More than 70% of the participants were foreigners which came not only from Europe but also from Japan, South Africa, Americas, North Africa, thus demonstrating the great interest of the Congress theme and the prestige of EUCARPIA also in extra-European countries. Mention is to be made, in this respect, of the financial support awarded by EUCARPIA to numerous scientists travelling from Eastern Europe.

The XVth General Congress has been held in Viterbo, Italy, at the Teatro dell'Unione, September 20-25, 1998. The theatre, built in 1846, is a neo-classic building located in downtown Viterbo. The posters' exhibition was held in a historical church (Almadiani) of the 16th Century. The city of Viterbo is situated in a hilly area 80 km north of Rome, with pleasant surroundings. Of Etruscan origin, it was then romanized (Vicus Elbii). It was handed over to the Church in medieval times and was the scene of disputes between the Empire and the Papacy. The historical centre has an interesting structure, is still surrounded by turreted walls (11th - 13th century) and preserves much of its medieval appearance. Viterbo is also the chief seat of the University of Tuscia, founded in 1979.

The inaugural ceremony was attended by: Senator Professor Tullia Carettoni, president of the Italian UNESCO Commission; Senator Professor Edoardo Vesentini, president of the Academy of the Lincei; the Mayor of Viterbo; and the representatives of the Minister for the Agricultural Policies, the Minister of the Universities and Research, the Latium Region and the National Research Council.

During the Congress, a book was presented on the "Italian Contribution to Plant Genetics and Breeding". This book was realised awarding a proposal advanced by the undersigned and supported by the SIGA and realised by the engagement of 100 Authors. This compendium, including several chapters, has a rich table of content and in its about 900 pages quotation is made of more than 2,000 scientists. The period of studies, research and experiments considered covers mainly the last 40-50 years, and reports the increases in productivity and qualitative progresses achieved with various crops, which have been obtained in virtue of research in genetics and breeding, with specific reference to the Italian agriculture.

In conclusion, the XV Congress has confirmed two traditional characteristics of the previous EUCARPIA Congress; an important opportunity for debates and focalizations, not only on scientific themes and their progresses, but also on the interactions between agricultural genetics and biotechnological innovations and world-ranging, global problems, as: feeding increasing populations in developing countries; fight against poverty, and increase of rural classes income, particularly in developing countries; erosion of natural resources; modernisation of agrosystems in a global frame of sustainable, fair and solidaristic development of mankind.

Furthermore, the capability of the Eucarpia Association for Plant Breeding, in the scientific and technical world, in the production and impact on the modern agricultural and food systems, cross over the boundaries of Europe to reach global size, interests, attention and consensus, has been once more underlined. The wealth of human resources, scientific knowledge, experimental materials and innovations, developed within EUCARPIA, invites us - members of this association - to move toward the extension and improvement, both in Europe and elsewhere, of the co-operation with scientists and technologists working in other institutions, in professional organisations, in private enterprises and farmers associations. In particular, it is our duty to improve our partnership with our colleagues in the developing countries.

During the General Assembly, the EUCARPIA members have accomplished an act of fundamental importance for the life of our association: due to the resignation of the Vice-President, Dr M.D. Hayward, Dr G.R. Mackay (UK) was unanimously elected as new President. In his presidential address, he kindly showed the members the venue of the future XVI EUCARPIA Congress in Edinburgh. Dr. P. Ruckenbauer was elected as Vice-President.

As Past President, I wish to convey to Dr Hayward the warm gratitude of all EUCARPIA members, and to express to the new President our best wishes for his work for the future of EUCARPIA.

Viterbo, Italy
G.T. Scarascia Mugnozza

SESSION 1.

RESISTANCE TO FUNGI

G.T. Scarascia Mugnozza, E. Porceddu & M.A. Pagnotta (Eds.)
Genetics and Breeding for Crop Quality and Resistance, 3-14, 1999

Avirulence and pathogenicity genes of *Cladosporium fulvum*

P.J.G.M. De Wit
Laboratory of Phytopathology, Wageningen Agricultural University, P.O. Box 8025, 6700 EE Wageningen, The Netherlands

Abstract: The pathosystem *Cladosporium fulvum*-tomato has become a model system in molecular plant pathology as both fungus and host plant are amenable to be studied by molecular methods. Many resistance genes (*Cf*) are available in near-isogenic lines of tomato, while many races of *C. fulvum* exist that can overcome one or more *Cf* genes, giving rise to a gene-for-gene relationship. *C. fulvum* is a biotrophic pathogen that penetrates tomato leaves through stomata and colonizes the intercellular space leaf mesophyll cells, where it stays extracellular during its entire life cycle. In compatible interactions the fungus produces various extracellular proteins (ECPs), some of which represent crucial virulence factors. In incompatible interactions fungal growth is arrested very soon after penetration. Resistance is associated with the hypersensitive response (HR) and accumulation of pathogenesis-related proteins. HR-based resistance is induced in plants with *Cf* genes, by matching race-specific elicitors secreted by *C. fulvum* immediately after stomatal penetration. The race-specific elicitors AVR4 and AVR9 have been characterized and their encoding genes *Avr4* and *Avr9* have been cloned. Both elicitors are cystine-rich peptides of which AVR9 belongs to the family of cystine-knotted peptides. The resistance genes *Cf-2*, *Cf-4*, *Cf-5* and *Cf-9* have been cloned. They encode proteins that belong to a superfamily of leucine-rich repeat (LRR) proteins. The major part of the Cf proteins contains 25-38 LRRs which are extracellular, while a short C-terminal domain is cytoplasmic. Two major players in the gene-for-gene system, i.e. the *Avr* gene and the corresponding *Cf* gene, have been cloned, but so far there is no evidence that their products interact directly. Membranes of both *Cf-9*-plus and *Cf-9*-minus plants contain a similar high affinity binding site for the AVR9 peptide. To determine whether Cf-9 represents a low affinity binding site, binding studies of AVR9 to *in vitro*-produced Cf-9 protein are required. Tomato lines have been found that respond with an HR to the virulence factor ECP2, which indicates that it can also function as an avirulence factor. The resistance gene that recognizes ECP2 has been designated *Cf-ECP2* and most probably represents a durable resistance gene.

1. Introduction

Cladosporium fulvum Cooke (syn. *Fulvia fulva*) causes leaf mould of tomato. The fungus is a biotrophic pathogen that occurs world-wide and causes disease only on the genus *Lycopersicon*. Tomato (*Lycopersicon esculentum* Mill.) is susceptible to the fungus, but other species of the genus are often resistant. The fungus penetrates tomato leaves through stomata and colonization is confined to the intercellular space of tomato leaves where no specialized feeding structures such as haustoria are formed. During the 1930s losses in tomato crops caused by *C. fulvum* were high. At that time no resistance genes were present in commercially grown cultivars. Since then, many *Cf*-resistance genes originating from related wild species of tomato such as *L. chilense, L. hirsutum, L. peruvianum*, and *L. pimpinellifolium*, have been transferred to cultivated tomato in breeding programmes (Stevens & Rick, 1988). Several of these resistance genes (e.g. *Cf-2, Cf-3, Cf-4, Cf-5,* and *Cf-9*) are available in near-isogenic lines of tomato cultivar MoneyMaker. Presently, tomato crops grown in glasshouses and outdoors generally carry one or more *Cf* genes. However, after introduction of commercial cultivars carrying *Cf* genes, new races of *C. fulvum* appeared which could often overcome introgressed monogenic resistances. The availability of many near-isogenic lines of tomato that carry *Cf* genes and which respond differentially (resistant or susceptible) to the various races of *C. fulvum* has made the *C. fulvum*-tomato interaction a model system in molecular plant pathology to study gene-for-gene relationships (De Wit, 1992, 1995, 1997; Hammond-Kosack & Jones, 1996). Basic questions on pathogenicity, virulence, avirulence, resistance and specificity of resistance and accompanying defence responses have been actively pursued.

2. Compatible interaction

Conidia germinate on the lower surface of the leaf at high relative humidity and form runner hyphae. A runner hypha grows until its encounters an open stoma, which is subsequently penetrated. Once inside the leaf, the diameter of the hyphae enlarges at least two-fold. The fungus colonizes the intercellular space of leaves of susceptible genotypes without visible induction of defence responses. Hyphae grow in close contact with mesophyll cells. It is not known which factors are crucial to enable and sustain growth of the fungus inside the intercellular space. Extracellular growth of the fungus allows isolation of apoplastic fluid (AF) from infected leaves that contains compounds originating from the plant-fungus interface. AF, which is obtained by *in vacuo* infiltration of infected tomato leaves with water, followed by low speed centrifugation (De Wit & Spikman, 1982), contains compounds that are constitutively produced by the plant or fungus in addition to compounds that are produced as a result of the interaction between host plant and fungus (De Wit et al., 1986; Joosten et al., 1990).

2.1 Virulence factors

The fungus does not seem to require classical pathogenicity and virulence factors such as cell wall-degrading enzymes and toxins to colonize its host (Agrios, 1997). In order to search for putative proteinaceous pathogenicity or virulence factors that enable *C. fulvum* to colonize and reproduce in the leaves of tomato, AF isolated from compatible interactions between tomato and *C. fulvum*, has been analysed. In this way various compatible interaction-specific proteins of *C. fulvum* were identified (De Wit et al., 1986).

Two of these extracellular proteins, ECP1 and ECP2, have been purified (Joosten & De Wit, 1988; Wubben et al., 1994b) and the encoding genes, *Ecp*1 and *Ecp*2, isolated (Van den Ackerveken et al., 1993b). ECP1 and ECP2, predominantly produced by *C. fulvum in planta*, are associated with the matrix between fungal hyphae and host cell walls (Wubben et al., 1994b). Whether these proteins are essential for the adherence of *C. fulvum* to host cell walls is not known. In order to investigate whether *Ecp*1 and *Ecp*2 are required for pathogenicity or virulence of *C. fulvum*, both genes were disrupted by homologous recombination. The *Ecp*2 gene appeared not to be essential for pathogenicity or virulence of *C. fulvum* on young tomato seedlings (Marmeisse et al., 1994). However, more recent experiments have shown that on mature tomato plants the *Ecp*2 gene is essential for full virulence (Laugé et al., 1997). In the same study it was also shown that the *Ecp*1 gene encodes a virulence factor. In both cases gene replacement had a significant effect on sporulation of *C. fulvum* on the host (Laugé et al., 1997). On mature plants the *Ecp*1-minus strain of *C. fulvum* sporulated significantly less as compared to the wild type, while the *Ecp*2-minus strain hardly sporulated. It is not known whether *Ecp*1 and *Ecp*2 play a role in uptake or metabolism of nutrients. The decrease in virulence of both disruptants was associated with strong accumulation of PR proteins. This suggests that both *Ecp*1 and *Ecp*2 play a role in pathogenesis possibly by suppressing active plant defence responses during colonization. In addition a double disruptant which lacks both *Ecp*1 and *Ecp*2, showed a phenotype very similar to that of the *Ecp*2-minus strain.

3. Incompatible interaction

In incompatible interactions growth of *C. fulvum* is arrested soon after penetration of the leaf of a resistant genotype. Fungal growth is restricted to the area of a few mesophyll cells, while the cells contacted by hyphae or in advance of the growing hyphae often die. Fairly accurate measurements of fungal biomass were achieved by inoculating tomato with a transformant of *C. fulvum* constitutively expressing β-glucuronidase (GUS) (Oliver et al., 1993). By inoculating near-isogenic lines of tomato homozygous or heterozygous for different *Cf* genes with race 4 of *C. fulvum*, expressing GUS, it was found that the different *Cf* genes confer distinct abilities to restrict *C. fulvum* infection (Hammond-Kosack & Jones, 1994). However, one should keep in mind that the effectiveness of a given *Cf* gene is also dependent on the amount and stability of the race-specific elicitor produced by the matching *Avr* gene. In addition, it was found that all *Cf* genes show incomplete dominance. Significantly less growth of *C. fulvum* occurred in plants that are homozygous, compared to plants that are heterozygous for a given *Cf* gene.

3.1 Hypersensitive response

In incompatible interactions, after penetrating stomata, hyphae that are in contact with mesophyll cells induce necrosis which is considered a hypersensitive response (HR). In incompatible interactions, involving different *Cf*-genes, necrosis was sometimes observed rather late (Hammond-Kosack & Jones, 1994). AF isolated from leaves that are fully colonized by virulent races of *C. fulvum*, carrying different genes for avirulence, induced HR after injection in *Cf* lines that contain a matching resistance gene. HR that occurs in the natural plant-pathogen interaction is very difficult to quantify. The outcome of a particular race-genotype interaction depends on the stability and activity of both the race specific elicitor and the receptor. Plants that are homozygous for a particular *Cf* gene respond with HR to a two-fold lower race-specific elicitor concentration as compared with with heterozygous plants. This indicates that the receptor concentration might be a limiting factor in inducing HR (Hammond-Kosack & Jones, 1994).

3.2 Accumulation of pathogenesis related proteins

Early accumulation of several host-encoded pathogenesis-related (PR) proteins in the apoplast is characteristic for incompatible interactions (De Wit & Van der Meer, 1986; De Wit et al., 1986; Joosten & De Wit, 1989). Biochemical characterization revealed that many of these proteins are 1,3-ß-glucanases and chitinases which are hydrolytic enzymes potentially able to degrade hyphal walls that contain 1,3-ß-glucans and chitin (Joosten & De Wit, 1989). The cDNAs encoding the various basic and acidic 1,3-ß-glucanases and chitinases have been cloned (Van Kan et al., 1992) and the expression of the genes (Van Kan et al., 1992; Ashfield et al., 1994; Wubben et al., 1994a) and localization of the transcripts and the encoded proteins have been studied (Wubben et al., 1992, 1994b). Although the early accumulation of hydrolytic enzymes in the incompatible interaction coincides with the expression of HR and arrest of fungal growth, it is not clear whether the induced PR proteins play a decisive role in resistance of tomato against *C. fulvum* (Joosten et al., 1995).

4. Race-specific elicitors and their encoding genes

4.1 Isolation and cloning of avirulence genes

A major breakthrough in research on the *C. fulvum*-tomato interaction was the discovery of race-specific elicitors, the inducers of HR present in AF of *C. fulvum*-colonized tomato leaves (De Wit & Spikman, 1982; De Wit et al., 1985; Higgins & De Wit, 1985). Injection of such AF into healthy leaves of near-isogenic lines containing different *Cf* genes, resulted in the differential induction of HR. Proteinaceous compounds which induced HR on resistant tomato cultivars were correlated with the presence of avirulence genes in the races of *C. fulvum* used for inoculations. The first race-specific peptide elicitor purified from AF of *C. fulvum*-infected tomato leaves was a peptide of 28 amino acids (Scholtens-Toma & De Wit, 1988; De Wit, 1992). The purified peptide elicitor specifically induced HR on tomato genotypes that carried the matching *Cf-9* resistance gene. Races that are virulent on tomato genotype *Cf-9* do not produce the elicitor (Van Kan et al., 1991). In a similar way, the race-specific elicitor AVR4, which induces HR on Cf4 tomato genotype, has been isolated (Joosten et al., 1994). From both proteinaceous elicitors the amino acid sequence has been determined and by reverse genetics the encoding avirulence (*Avr*) genes *Avr*9 and *Avr*4, have been cloned (Van Kan et al., 1991; Van den Ackerveken et al., 1992; Joosten et al., 1994). The cloned genes revealed that both elicitors present in *C. fulvum*-infected tomato plants are post-translationally processed. The *Avr9* gene encodes a protein of 63 amino acids including a signal sequence of 23 amino acids. In culture filtrates of transformants of *C. fulvum* that constitutively produce the AVR9 elicitor, predominantly N-terminally processed peptides of 32, 33 or 34 amino acids are present (Van den Ackerveken et al., 1993a). When these latter peptides were incubated with AF from healthy tomato leaves, they were further processed into the mature elicitor of 28 amino acids, indicating that plant proteases are required for the final processing (Van den Ackerveken et al., 1993a). The *Avr4* gene encodes a protein of 135 amino acids, including a signal peptide of 18 amino acids (Joosten et al., 1994). Here, the mature peptide elicitor is processed at both the N- and C-terminus leaving a mature peptide elicitor of 86 amino acids (Joosten et al., 1997). Although there is clear evidence for the presence of proteinaceous elicitors in AF of infected leaves that induce HR on *Cf-2* and *Cf-5* genotypes of tomato, the matching AVR2 and AVR5 elicitors have not yet been characterized.

4.2 Regulation of avirulence genes

In vitro-grown *C. fulvum* hardly produces AVR4 or AVR9 elicitors. RNA gel blot analysis indicated that expression of the *Avr4* and *Avr9* genes of *C. fulvum* is specifically induced *in planta*. Accumulation of mRNAs encoding the race-specific elicitors correlates with an increase in fungal biomass in tomato leaves during pathogenesis in compatible interactions (Van Kan et al., 1991; Joosten et al., 1994). Studies on the expression of the *Avr4* and *Avr9* genes *in planta*, using transformants of *C. fulvum* carrying *Avr* promoter-GUS fusions, showed activation of the promoter in hyphae immediately after penetration of stomata, with expression levels particularly high in mycelia growing in the vicinity of the vascular tissue (Van den Ackerveken et al., 1994, Joosten et al., 1997).

The promoter of the *Avr4* gene does not contain motifs homologous to sequences known for binding of regulatory proteins. So far, the *Avr4* gene can hardly be induced under different growth conditions *in vitro*. In contrast, the *Avr9* gene could be induced under limiting concentrations of nitrogen when grown in liquid medium (Van den Ackerveken et al., 1994; Sandor Snoeijers, Wageningen Agricultural University, personal communication). Analysis of the *Avr9* promoter sequence revealed six copies of the hexanucleotide TAGATA (Van den Ackerveken et al., 1994), which has been identified as the recognition site of the NIT2 protein, a transcription factor which positively regulates gene expression under nitrogen-limiting conditions in *Neurospora crassa* and many other filamentous fungi. Therefore, the expression of the *Avr9* gene is possibly regulated in a similar way, by a *C. fulvum*-homologue of the NIT2 protein. Indeed deleting a number of the TAGATA sequences abolished the induction of the *Avr9* gene under conditions of low nitrogen concentrations *in vitro* (Sandor Snoeijers, Wageningen Agricultural University, personal communication). Two of the six TAGATA sequences are essential for *Avr9* regulation under nitrogen-limiting conditions.

4.3 Alleles of avirulence genes

Genomic DNA gel blot analysis, using *Avr4* cDNA as a probe, did not reveal any differences between races of *C. fulvum* avirulent or virulent on tomato genotype Cf4. All races contain a homologous, single copy gene, not displaying any restriction fragment length polymorphism (Joosten et al., 1994). Although none of the virulent races produce biologically active AVR4 elicitor, they produce transcripts that hybridize to an *Avr4* cDNA probe, proving that those races contain alternative alleles of *Avr4* (Joosten et al., 1994, 1997). Sequencing these alleles, showed single base pair mutations in the open reading frame (ORF) encoding the mature AVR4 protein, resulting in single amino acid changes in the AVR4 elicitor. In one case a frame shift mutation was observed leaving only 13 amino acids at the N-terminus of the AVR4 peptide intact (Joosten et al., 1994, 1997). The single amino acid changes in the AVR4 protein cause the peptide to become unstable and probably more sensitive to proteolytic degradation which would prevent it reaching a matching receptor in sufficiently high concentration. Expression of the mutant alleles in the PVX expression system indeed revealed that most still encode an active elicitor molecule which, however, has a much shorter active half life than the peptide elicitor produced by the functional avirulence allele (Joosten et al., 1997).

4.4 Structure function studies on avirulence genes

Genomic DNA gel blot analysis of races of *C. fulvum* virulent on tomato genotype Cf9 revealed that in all these races the *Avr9* gene is absent. Thus, there is always a strict correlation between virulence of races on Cf9 genotypes of tomato and absence of the *Avr9* gene (Van Kan et al., 1991). Avirulence genes *Avr4* and *Avr9* both encode relatively small globular peptide elicitors which contain 8 and 6 cysteine residues, respectively. The structure of the AVR9 peptide has been extensively studied by ^{1}H-NMR (Vervoort et al., 1997). All cysteines form disulfide bridges, which are required for elicitor activity. The AVR9 peptide consists of three anti-parallel β-strands forming a rigid region of β-sheet. AVR9 is a member of the family of cystine-knotted peptides, in which the 6 cysteine residues form a typical cystine knot found in several small proteins such as proteinase inhibitors, ion channel blockers and growth factors (Pallaghy et al., 1994). However, structural homology most probably does not represent functional homology. In contrast to the *Avr9* gene, which is absent in all races of *C. fulvum* virulent on tomato genotype Cf9, the presence of mutated *Avr4* alleles in virulent races of *C. fulvum* might suggest an essential role for its product in virulence. However, one isolate with a frame-shift mutation in the ORF encoding an AVR4 homologue of only 13 amino acids showed normal virulence, indicating that the AVR4 protein is probably dispensable (Joosten et al., 1997). The observation that transformants of *C. fulvum* in which the *Avr9* gene was replaced by a selection marker, did not show impaired virulence on non *Cf9*-containing tomato genotypes, and that in nature avoidance of *Cf-9*-specific resistance is achieved by complete deletion of the *Avr9* gene, suggests that the *Avr9* gene is dispensable for growth and virulence of *C. fulvum* (Marmeisse et al., 1993). However, the *Cf-9* resistance gene, which is present in tomato breeding lines since 1979, still provides good protection toward *C. fulvum* in commercial tomato crops, indicating that loss of the *Avr9* gene is not sufficient to overcome the *Cf-9* locus. It was found that isogenic strains of *C. fulvum* in which the *Avr9* gene had been replaced by a selection marker are only weakly virulent on *Cf-9* genotypes of tomato (Laugé et al., 1998a). This suggests that *Cf-9* genotypes of tomato contain functional *Cf-9* homologs which recognize an elicitor other than AVR9. Infection of *Cf-9* genotypes with *Avr9*-minus strains of *C. fulvum* was always associated with strong accumulation of PR proteins (Parniske et al., 1997; Laugé et al., 1998a). most probably homologues of the *Cf-9* gene provide this protection as will be discussed later.

4.5 *Virulence factors with avirulence properties*

As discussed before, *C. fulvum* secretes avirulence as well as virulence factors into the intercellular space while colonizing tomato leaves (Van den Ackerveken et al., 1992; Joosten et al., 1994; Laugé et al., 1997). The PVX expression system allows, within the host range of PVX, to search for plants that respond with an HR to virulence factors such as the *Ecp*2 gene product. A recombinant PVX construct expressing *Ecp*2 was inoculated onto various lines of species of *Lycopersicon.* In this way plants were found that responded with HR. All responding plants appeared to originate from the same ancestor. The corresponding resistance gene has been designated *Cf-ECP2* (Laugé et al., 1998b). The *Cf-ECP2* gene is anticipated to be a durable resistance gene as it recognizes a crucial virulence factor of *C. fulvum.* All races of *C. fulvum* that have been tested so far, contain an *Ecp*2 gene that induces HR on plants containing the *Cf-ECP2* gene. This finding illustrates that tomato has an efficient surveillance system that can recognize not only 'classical' avirulence factors but also crucial virulence factors.

5. Resistance genes in tomato against *Cladosporium fulvum*

Four genes for resistance against *C. fulvum* (Cf genes) have been mapped at two complex loci. The Cf-9 and Cf-4 genes have been mapped on the short arm of chromosome 1 (Jones et al., 1993), and the Cf-2 and Cf-5 genes are located on the short arm of chromosome 6 (Jones et al., 1993). The Cf-2, Cf-4, Cf-5 and Cf-9 genes have been cloned (Jones et al., 1994; Dixon et al., 1996, 1998; Hammond-Kosack & Jones, 1997; Thomas et al., 1997).

The Cf-9 gene encodes a 863-amino acid membrane-anchored, predominantly extracytoplasmic glycoprotein containing 27 leucine-rich repeats (LRRs) with an average length of 24 amino acids. These motifs occur in many plant resistance genes (Jones & Jones, 1996; Hammond-Kosack & Jones, 1997). In the Cf-9 protein seven domains (A to G) have been designated. The N-terminal domain A of 23 amino acids is consistent with a signal peptide; domain B of 68 amino acids is cysteine-rich; domain C contains 27 imperfect LRRs; domain D contains 28 amino acids; domain E contains 18 amino acids and is very acidic; domain F contains 37 amino acids and is the presumed trans-membrane domain; C-terminal domain G contains 21 amino acids, is very basic and concludes with the amino acids KKRY. Twenty-two potential N-glycosylation sites are distributed between domains B, C and D. Thus the Cf-9 gene encodes a LRR protein of which the major part (A-F) is extracellular and the C-terminal part (G) is cytoplasmic.

The LRR domain C in the Cf-9 protein is interrupted by a short region originally designated as LRR 24, which has only minimal LRR homology. This domain, now designated C2 is also present in the Cf-2, Cf-4, and Cf-5 genes (Jones & Jones, 1996; Hammond-Kosack & Jones, 1997). As a result domain C has now been divided in domains C1 (the N-terminal LRRs), C2 (with minor LRR consensus) and domain C3 (the C-terminal LRRs) (Jones & Jones, 1996; Hammond-Kosack & Jones, 1997). The LRRs match the extracytoplasmatic LRR consensus LxxLxxLxxLxLxxNxLxGxIPxx (Jones & Jones, 1996). LRR regions might be involved in various types of protein-protein interaction (Kobe & Deisenhofer, 1993).

The Cf-4 gene is very homologous to the Cf-9. The proteins have >91% identical amino acids (Thomas et al., 1997). The Cf-4 gene encodes an 806-amino acid protein with 25 LRRs. In Cf-4 two complete LRRs are deleted relative to the Cf-9 gene. DNA sequence analysis suggests that Cf-4 and Cf-9 are derived from a common gene. The amino acids that distinguish Cf-4 from Cf-9 are located at the N-terminal half of the protein. The C-terminal halves of both genes are almost identical.

The Cf-2 locus contains two functional genes that each independently confer resistance to races of *C. fulvum* carrying the Avr2 gene (Dixon et al., 1996). Each gene encodes a nearly identical 1112-amino acid protein (three amino acids are different) that is structurally very similar to the Cf-4 and Cf-9 proteins. The Cf-2 protein possesses 37 LRRs. The LRRs of Cf-2 are nearly all 24 amino acids in length and are also interrupted by a short C2 domain which divides the LRRs in a C-terminal block of 33 LRRs and a N-terminal block of 4 LRRs. The highest homology between the Cf-2, and the Cf-4 and Cf-9 proteins resides in the C-terminal part (Hammond-Kosack & Jones, 1997; Thomas et al., 1997).

The Cf-5 gene is closely linked to the Cf-2 gene and encodes a 968-amino acid protein that is very similar to the protein encoded by Cf-2 and contains 31 LRRs (Hammond-Kosack & Jones, 1997). The Cf-5 and Cf-2 proteins differ by an exact deletion of six LRRs in Cf-5. The C-terminal part of Cf-2 and Cf-5 is also very conserved.

5.1 LRR motifs in Cf proteins and their potential function

The most simple interpretation of Flor's gene-for-gene hypothesis (Flor, 1971) would be that products of avirulence genes and resistance genes interact and activate downstream signaling events eventually leading to resistance. The best candidates for binding of AVR elicitors would be the LRRs present in the Cf proteins. The N-terminal parts of the four Cf proteins are variable and could determine specificity in recognizing race-specific elicitors, while the C-terminal parts are conserved and could function as an activation domain potentially interacting with other proteins present in the plasma membrane. For the Cf-4 and Cf-9 protein, the most appropriate ligands would be the AVR4 and AVR9 elicitors, respectively. However, proof for this hypothesis has not yet been found (Kooman-Gersmann et al., 1996, 1997, 1998; De Wit 1997). Kooman-Gersmann et al. (1996) found that the AVR9 elicitor molecule binds to plasma membranes of Cf0 and Cf9 genotypes equally well (Kd=70 pM). In order to get more insight in the region of the AVR9 molecule which interacts with the high affinity binding site, *in vitro* mutagenesis of *Avr9* was performed in the PVX expression system to obtain AVR9 mutants with altered biological activities. All amino acid residues present in AVR9 were exchanged one by one with alanine (alanine scan). In this way AVR9 peptides were obtained with higher, equal or lower HR-inducing activitiy compared to wild type AVR9 (Kooman-Gersmann et al., 1997). Mutants of AVR9 with decreased HR-inducing activity on Cf genotypes of tomato also showed lower binding affinity to plasma membranes (Kooman-Gersmann et al., 1998). However, binding was decreased to plasma membranes of both Cf9 and Cf0 genotypes of tomato. It is therefore unlikely that the LRRs of the Cf-9 protein bind the AVR9 peptide directly. In case AVR9 binds to the Cf-9 protein, the affinity is too low to detect with the presently available techniques (Kooman-Gersmann et al., 1998). Binding of AVR9 to *in vitro* produced Cf-9 protein is required to detect such a low affinity binding site. LRR domains of Cf proteins could also facilitate multiple interactions with other proteins which are involved in the signal transduction pathway as has been found for other LRR-containing proteins (Jones & Jones, 1997).

Similar studies are now being carried out with the AVR4 peptide. Preliminar mutational analysis of the *Avr4* gene in the PVX expression system showed that all eight cysteine residues present in the AVR4 elicitor are involved in disulfide bridges (M.H.A.J. Joosten, Wageningen Agricultural University, personal communication). For the future, binding studies with the AVR4 elicitor to membranes of Cf4 and Cf0 genotypes and *in vitro* produced Cf-4 protein are planned.

5.2 Homologues of Cf resistance genes in tomato

When the *Cf* genes are used as a probe to a Southern blot of isogenic lines of tomato, many hybridizing bands can be observed (Jones et al., 1994; Dixon et al., 1996; Hammond-Kosack & Jones, 1997; Thomas et al., 1997; Parniske et al., 1997). All *Cf* genes appear to be members of a multigene family. The *Cf-2* locus contains two nearly identical genes (Dixon et al., 1996). Both genes are functional, but it is possible that they represent different specificities. The *Cf-4/Cf-9* locus contains ten homologues (Thomas et al., 1997; Parniske et al., 1997) including the functional *Cf-4* and *Cf-9* genes. Some of the homologues give (partial) resistance to races of *C. fulvum* for which the matching avirulence factors have not yet been identified. The *Cf-9* cluster contains two *Cf-9* homologues that confer resistance independent of AVR9 activation, while the *Cf-4* cluster contains one *Cf-4* homologue conferring resistance that is independent of AVR4 activation (Takken et al., 1998; Joosten, Wageningen Agricutural University, personal communication).

6. Exploitation of avirulence and resistance genes

Cloned avirulence and resistance genes are valuable tools to study the molecular mechanisms that determine race-specific resistance, yielding results that can be applied in molecular resistance breeding. Combining an avirulence gene and the complementary resistance gene in host plants under the control of a pathogen-inducible promoter could possibly give new horizons to molecular resistance breeding (two-component gene cassette; De Wit, 1992, 1995). Activation of the gene cassette by pathogens should occur both quickly and locally, resulting in the production of a race-specific elicitor that interacts directly or indirectly with the resistance gene product. As a result of this interaction, localized HR will be induced which will prevent further spread of an invading pathogen that is inhibited by an HR. With this system it should be possible, to use a highly specific resistance gene-avirulence gene combination (for example *Cf-9*/*Avr9* or *Cf-4*/*Avr4*) to obtain plants resistant against a wide variety of pathogens. Preliminary results with tomato plants containing the two-component gene cassette look promising (G. Honée, Wageningen Agricultural University, personal communication). For applications in plants other than tomato, the gene cassette containing both the *Cf* and *Avr* gene should be transferred to those plants. The *Cf-9* and *Avr9* gene are active in both potato and tobacco (Honée,Wageningen Agriculrural University, personal communication).

7. References

Agrios G.N. (1997) Plant Pathology, Academic Press, London pp. 63-82.

Ashfield T., Hammond-Kosack K.E., Harrison K. & Jones J.D.G. (1994) *Cf* gene-dependent induction of a ß-1,3 glucanase promotor in tomato plants infected with *Cladosporium fulvum*. Mol Plant Microbe Interact 7: 645-657.

De Wit P.J.G.M. (1992) Molecular characterization of gene-for-gene systems in plant-fungus interactions and the application of avirulence genes in control of plant pathogens. Annu Rev Phytopathol 30: 391-418.

De Wit P.J.G.M. (1995) Fungal avirulence genes and plant resistance genes: unravelling the molecular basis of gene-for-gene resistance. Adv Bot Res 21: 147-185.

De Wit P.J.G.M. (1997) Pathogen avirulence and plant resistance: a key role for recognition. Trends Plant Sci 2: 452-458.

De Wit P.J.G.M. & Spikman G. (1982) Evidence for the occurrence of race- and cultivar-specific elicitors of necrosis in intercellular fluids of compatible interactions between *Cladosporium fulvum* and tomato. Physiol Plant Pathol 21: 1-11.

De Wit P.J.G.M. & Van der Meer F.E. (1986) Accumulation of the pathogenesis-related tomato leaf protein P14 as an early indicator of incompatibility in the interaction between *Cladosporium fulvum* (syn. *Fulvia fulva*) and tomato. Physiol Mol Plant Pathol 28: 203-214.

De Wit P.J.G.M., Hofman J.E., Velthuis G.C.M. & Kuc J.A. (1985) Isolation and characterization of an elicitor ofnecrosis isolated from intercellular fluids of compatible interactions of *Cladosporium fulvum* (syn. *Fulvia fulva*) and tomato. Plant Physiol 77: 642-647.

De Wit P.J.G.M., Buurlage M.B. & Hammond K.E. (1986) The occurrence of host, pathogen and interaction -specific proteins in the apoplast of *Cladosporium fulvum* (syn. *Fulvia fulva*) infected tomato leaves. Physiol Mol Plant Pathol 29: 159-172.

Dixon M.S., Jones D.A., Keddle J.S., Thomas C.M., Harrison K. & Jones J.D.P. (1996) The tomato *Cf-2* diseaseresistance locus comprises 2 functional genes encoding leucine-rich repeat proteins. Cell 84: 451-459.

Dixon M.S., Hatzixanthis K., Jones D.A., Harrison K. & Jones J.D.G. (1998) The tomato *Cf-5* resistance gene and six homologs show pronounced allelic variation in leucine rich repeat copy number. Plant Cell 10: 1915-1925.

Flor H.H. (1971) Current status of the gene-for-gene concept. Annu Rev Phytopathol 9: 275-296

Hammond-Kosack K.E. & Jones J.D.G. (1994) Incomplete dominance of tomato *Cf* genes for resistance to *Cladosporium fulvum*. Mol Plant-Microbe Interact 7: 58-70.

Hammond-Kosack K.E. & Jones J.D.G. (1996) Resistance gene-dependent plant defense responses. Plant Cell 8: 1773-1791.

Hammond-Kosack K.E. & Jones J.D.G. (1997) Plant desease resistance genes. Annu Rev Plant Physiol Plant Mol Biol 48: 575-607.

Higgins V.J. & De Wit P.J.G.M. (1985) Use of race- and cultivar-specific elicitors from intercellular fluids for characterizing races of *Cladosporium fulvum* and resistant tomato cultivars. Phytopathol 75: 695-699.

Jones D.A. & Jones J.D.G. (1997) The role of leucine-rich repeat proteins in plant defences. Adv Bot Res 24: 89-167.

Jones D.A., Dickinson M.J., Balint-Kurti P.J., Dixon M.S. & Jones J.D.G. (1993) Two complex resistance loci revealed in tomato by classical and RFLP mapping of the *Cf-2*, *Cf-4*, *Cf-5*, and *Cf-9* genes for resistance to *Cladosporium fulvum*. Mol Plant-Microbe Interact 6: 348-357.

Jones D.A., Thomas C.M., Hammond-Kosack K.E., Balint-Kurti P.J. & Jones J.D.G. (1994) Isolation of the tomato *Cf-9* gene for resistance to *Cladosporium fulvum* by transposon tagging. Science 266: 789-793.

Joosten M.H.A.J. & De Wit P.J.G.M. (1988) Isolation, purification and preliminary characterization of a protein specific for compatible *Cladosporium fulvum* (syn. *Fulvia fulva*)-tomato interactions. Physiol Mol Plant Pathol 33:241-253.

Joosten M.H.A.J. & De Wit P.J.G.M. (1989) Identification of several pathogenesis-related proteins in tomato leaves inoculated with *Cladosporium fulvum* (syn. *Fulvia fulva*) as 1,3-ß-glucanases and chitinases. Plant Physiol 89: 945-951.

Joosten M.H.A.J., Hendrickx L.J.M. & De Wit P.J.G.M. (1990) Carbohydrate composition of apoplastic fluids isolated from tomato leaves inoculated with virulent or avirulent races of *Cladosporium fulvum* (syn. *Fulvia fulva*) Neth J Plant Pathol 96: 103-112.

Joosten M.H.A.J., Cozijnsen A.J. & De Wit P.J.G.M. (1994) Host resistance to a fungal tomato pathogen lost by a single base-pair change in an avirulence gene. Nature 367: 384-387.

Joosten M.H.A.J., Verbakel M., Nettekoven M.E., van Leeuwen J., Van der Vossen R.T.M. & De Wit P.J.G.M. (1995) The phytopathogenic fungus *Cladosporium fulvum* is not sensitive to the chitinase and 1,3-ß-glucanase defence proteins of its host tomato. Physiol and Mol Plant Pathol 46: 45-59.

Joosten M.H.A.J., Vogelsang R., Cozijnsen T.J., Verberne M.C. & De Wit P.J.G.M. (1997) The biotrophic fungus *Cladosporium fulvum* circumvents *Cf-4*-mediated resistance by producing instable AVR4 elicitors. Plant Cell 9: 1-13.

Kobe B. & Deisenhofer J. (1993) Crystal structure of porcine ribonuclease inhibitor, a protein with leucine -rich repeats. Nature 366: 751-756.

Kooman-Gersmann M., Honée G., Bonnema G. & De Wit P.J.G.M. (1996) A high-affinity binding site for the AVR9 peptide elicitor of *Cladosporium fulvum* is present on plasma membranes of tomato and other solanaceous plants. Plant Cell 8: 929-938.

Kooman-Gersmann M., Vogelsang R., Hoogendijk E.C.M. & De Wit P.J.G.M. (1997) Assignment of amino acid residues of the AVR9 peptide of *Cladosporium fulvum* that determine elicitor activity. MPMI 10: 821-829.

Kooman-Gersmann M., Vogelsang R., Vossen P., Van den Hooven H., Mahe E, Honée G. & De Wit P.J.G.M. (1998) Correlation between binding affinity and necrosis-inducing activity of mutant AVR9 elicitors. Plant Physiol 117:609-618.

Laugé R., Joosten M.H.A.J., Van den Ackerveken G.F.J.M., Van den Broek H.W.J. & De Wit P.J.G.M. (1997) The *in planta*-produced extracellular proteins ECP1 and ECP2 of *Cladosporium fulvum* are virulence factors. MPMI 10: 725-734.

Laugé R., Dmitriev A.P., Joosten M.H.A.J. & De Wit P.J.G.M. (1998a) additional resistance gene(s) against *Cladosporium fulvum* present on the *Cf-9* introgression segment are associated with strong PR protein accumulation. MPMI 11: 301-308.

Laugé R., Joosten M.H.A.J., Haanstra J.P.W., Goodwin, P.H., Lindhout W.H. & De Wit P.J.G.M. (1998b) Successful search for a resistance gene in tomato targeted against a virulence factor of a fungal pathogen. Proc Natl Acad Sci USA 95: 9014-9018.

Marmeisse R., Van den Ackerveken G.F.J.M., Goosen T., De Wit P.J.G.M. & Van den Broek H.W.J. (1993) Disruption of the avirulence gene *avr9* in two races of the tomato pathogen *Cladosporium fulvum* causes virulence on tomato genotypes with the complementary resistance gene *Cf-9*. Mol Plant-Microbe Interact 6: 412-417.

Marmeisse R., Van den Ackerveken G.F.J.M., Goosen T., De Wit P.J.G.M. & Van den Broek H.W.J. (1994) The *in-planta* induced *ecp2* gene of the tomato pathogen *Cadosporium fulvum* is not essential for pathogenicity. Curr Genet 26: 245-250.

Oliver R.P., Farman M.L., Jones J.D.G. & Hammond-Kosack K.E. (1993) Use of fungal transformants expressing ß-glucoronidase activity to detect infection and measure hyphal biomass in infected plant tissue. Mol Plant Microbe Interact 6: 521-525.

Pallaghy P.K., Nielsen K.J., Craick D.J. & Norton R.S. (1994) A common structural motif incorporating a cystine knot and a triple-stranded beta-sheet in toxic and inhibitory polypeptides. Protein Sci 3: 1833-1839.

Parniske M., Hammond-Kosack K.E., Golstein C., Thomas C.M., Jones D.A., Harrison K., Wulff B.B.H. & Jones J.D.G. (1997) Novel disease resistance specificities result from sequence exchange between tandemly repeated genes at the *Cf-4/9* locus of tomato. Cell 91: 1-20.

Scholtens-Toma I.M.J. & De Wit P.J.G.M. (1988) Purification and primary structure of a necrosis-inducing peptide from the apoplastic fluids of tomato infected with *Cladosporium fulvum* (syn. *Fulvia fulva*). Physiol Mol Plant Pathol 33: 59-67.

Stevens M.A. & Rick C.M. (1988) Genetics and Breeding. In The Tomato Crop, Atherton JG and Rudich J eds, Chapman and Hall, London 35-109.

Takken F.L.W., Schipper D., Nijkamp H.J.J. & Hille J. (1998) Identification and *Ds*-tagged isolation of a new gene at the *Cf-4* locus of tomato involved in disease resistance to *Cladosporium fulvum* race 5. Plant J 14: 401-411.

Thomas C.M., Jones D.A., Pamiske M., Harrison K., Balint-Kurti P.J., Hatzixanthis K. & Jones J.D.G. (1997) Charactarisation of the tomato *Cf-4* gene for resistance to *Cladosporium fulvum* identified sequences which determine recognitional specificity in Cf-4 and Cf-9. Plant Cell 9: 1-12.

Van den Ackerveken G.F.J.M., Van Kan J.A.L. & De Wit P.J.G.M. (1992) Molecular analysis of the avirulence gene *avr9* of the fungal tomato pathogen *Cladosporium fulvum* fully supports the gene-for-gene hypothesis. Plant J 2: 359-366.

Van den Ackerveken G.F.J.M., Vossen J.P.M.J. & De Wit P.J.G.M. (1993a). The AVR9 race-specific elicitor of *Cladosporium fulvum* is processed by endogenous and plant proteases. Plant Physiol 103: 91-96.

Van den Ackerveken G.F.J.M., Van Kan J.A.L., Joosten M.H.A.J., Muisers J.M., Verbakel, H.M. and De Wit P.J.G.M. (1993b) Characterization of two putative pathogenicity genes of the fungal tomato pathogen *Cladosporium fulvum*. Mol Plant-Microbe Interact 6: 210-215.

Van den Ackerveken G.F.J.M., Dunn R.M., Cozijnsen A.J., Vossen J.P.M.J., Van den Broek H.W.J. & De Wit P.J.G.M. (1994) Nitrogen limitation induces expression of the avirulence gene *avr9* in the tomato pathogen *Cladosporium fulvum*. Mol Gen Genet 243: 277-285.

Van Kan J.A.L., Van den Ackerveken G.F.J.M. & De Wit P.J.G.M. (1991) Cloning and characterization of cDNA of avirulence gene *avr9* of the fungal tomato pathogen *Cladosporium fulvum*, causal agent of tomato leaf mold. Mol Plant-Microbe Interact 4: 52-59.

Van Kan J.A.L., Joosten M.H.A.J., Wagemakers C.A.M., Van den Berg-Velthuis G.C.M. & De Wit P.J.G.M. (1992) Differential accumulation of mRNAs encoding extracellular and intracellular PR proteins in tomato induced by virulent and avirulent races of *Cladosporium fulvum*. Plant Mol Biol 20: 513-527.

Vervoort J., Van den Hooven H.W., Berg A., Vossen P., Vogelsang R., Joosten, M.H.A.J. & De Wit P.J.G.M. (1997) The race-specific elicitor AVR9 of the tomato pathogen *Cladosporium fulvum*: a cystine-knot protein. Sequence-specific ^{1}H NMR assignments, secondary structure and global fold of the protein. FEBS Lett 404: 153-158.

Wubben J.P., Eijkelboom C.A. & De Wit P.J.G.M. (1994a) Accumulation of pathogenesis-related proteins in the epidermis of tomato leaves infected by *Cladosporium fulvum*. Neth J of Plant Pathol 99: 231-239.

Wubben J.P., Joosten M.H.A.J. & De Wit P.J.G.M. (1994b) Expression and localization of two *in planta* induced extracellular proteins of the fungal tomato pathogen *Cladosporium fulvum*. Mol Plant-Microbe Interact 7: 516-524.

Wubben J.P., Joosten M.H.A.J., Van Kan J.A.L. & De Wit P.J.G.M. (1992) Subcellular localization of plantc chitinases and 1,3-ß-glucanases in *Cladosporium fulvum* (syn. *Fulvia fulva*)-infected tomato leaves. Physiol Mol Plant Pathol 41: 23-32.

G.T. Scarascia Mugnozza, E. Porceddu & M.A. Pagnotta (Eds.)
Genetics and Breeding for Crop Quality and Resistance, 15-24, 1999

Pathogenesis-related proteins for the control of fungal diseases of tomato

P. Veronese[1], P. Crinó, M. Tucci[2], F. Colucci[1], D.J. Yun[3], M.P. Hasegawa[3], R.A. Bressan[3] & F. Saccardo[2]
[1]ENEA C.R. Casaccia, Biotechnology & Agriculture Division, Rome, Italy; [2]Tuscia University, Plant Production Department, Viterbo, Italy; [3]Purdue University, Horticulture Department, West Lafayette, IN, USA

Abstract: Constitutive expression of genes encoding pathogenesis related (PR) proteins is one of the strategies proposed to obtain a broad and durable level of resistance to different phytopathogenic fungi. In view of this, we analyzed the response to fungal infections of transgenic tomato plants overexpressing tobacco PR-5, PR-1 and chitinase genes. Constitutive expression of the PR-5 protein osmotin was correlated with increased resistance to grey mold, powdery mildew and late blight, confirmed up to the T_3 generation. Co-expression of more than one PR-gene in the same genome might represent a further advantage. We have constructed plant expression vectors of pUC 19 derivatives that can be inserted into the cloning sites of plant transformation vectors for co-expression of up to three genes. To evaluate the use of the vectors, three target gene (PR-1, chitinase and osmotin) cassettes were constructed. Preliminary results from molecular analyses and infections with pathogens of transgenic plant material, indicate that these vectors can be used for co-transformation of multiple target genes. Additional improvements would involve the use of different promoters and/or genes encoding proteins with higher synergistic antifungal activity.

1. Introduction

Plants respond to pathogen attack by synthesizing pathogenesis-related (PR) proteins. Many of these proteins, including osmotin, a basic 24 kD PR-5 protein originally isolated from salt adapted tobacco cells, have been shown to have *in vitro* and *in vivo* antifungal activity against a large number of economically important plant pathogens (Melcher et al., 1994; Jongedijk et al., 1995; Yun et al., 1997). The antifungal action of osmotin (Abad et al., 1996) was directly correlated with its ability to increase the permeability of the fungal membrane to protons (Vigers et al., 1991, 1992). In transgenic potato plants, constitutive overexpression of osmotin was correlated with a delayed development of symptoms of *Phytophtora infestans* (Liu et al., 1994). An osmotin gene construct, with a 20 amino acid truncation triggering extracellular targeting, was particularly effective in enhancing resistance to late blight (Liu et al., 1996).

A strategy based on the use of multigene tolerance could be important to the durability as well as the degree of resistance. Co-expression of two or more antifungal proteins such as chitinase and Β-1,3 glucanase, with synergistic interactions, was able to provide increased protection against phytopathogenic fungi in transgenic plants (Leah et al., 1991; Zhu et al., 1994; Jach et al., 1995; Jongedijk et al., 1995).

Our work was aimed at testing the level of resistance towards different fungal pathogens in tomato plants constitutively overexpressing the tobacco osmotin gene and providing a plant transformation vector suitable for co-expression of multiple genes encoding antifungal proteins.

2. Materials and Methods

The plants used for testing the *in planta* antifungal activity of the transgene(s) were obtained by genetic transformation of the fresh market tomato line Cordeok, which is partially resistant to bacterial canker and was kindly provided by INRA Monfavet (Veronese et al., 1998). Transformation of cotyledonary explants was carried out using the *Agrobaterium tumefaciens* strain LBA4404 for the binary vector pKOLC3 and EHA105 for pOPC. Vector pKOLC3 is a derivative of pKYLX71 carrying the tobacco osmotin gene under the control of the 35S CaMV promoter (Liu et al., 1996). For the construction of co-expression cassettes, a plasmid which contains the cauliflower mosaic virus (CaMV) 35S promoter in *Eco*RI/KpnI sites, alpha amylase inhibitor cDNA in a Sal I site and polyadenylation signal of octopine synthase gene in *Sph*1/*Hind*III sites of pUC19 plasmid was kindly provided by M.J. Chrispeels (University of California, San Diego). To construct intermediate vectors, the *Sal*I fragment of alpha amylase inhibitor cDNA insert was cut out and self ligated to produce pTEX (Figure 1).

Upon digestion with a *Hind*III site of pTEX, the fragments were filled in with Klenow, ligated to *Eco*RI linker and self-ligated. The resulting vector being referred to as pTEX (E) (Figure 1A). Alternatively, for constructing pTEX (H), a *EcoR*I site of pTEX was replaced with a *Hind*III site by a similar procedure (Figure 1A). To construct plant expression vectors with multiple cloning sites, the *Eco*RI/*Hind*III fragment of pBI121 (Stratagene) was replaced with the *EcoR*I/*Hind*III fragment of pTEX, the resulting binary vector being referred to as pBTEX (Figure 1B). cDNA clones encoding a basic PR-1 gene and a class Ia chitinase were obtained by RT-PCR using NaCl-adapted tobacco cell total RNA as template (Ausubel et al., 1988). The cDNAs were cloned in pT7-Blue vector (Stratagene), giving rise to pT7-PR1 and pT7-chi, respectively. The *Sma*I/*Bam* HI fragment of the osmotin gene insert was isolated from pGEM-OSM (Liu et al., 1994) and subcloned in plant expression vector pBTEX to produce pBTEX-osm. The *Bam*HI/*Xba*I fragment of the PR-1 cDNA insert in the pT7-PR1 was subcloned into intermediate vector pTEX (E) between CaMV 35S promoter and octopine synthase. The chimerical 35S-PR1 gene insert was then excised as a *EcoR*I fragment and ligated to the *EcoR*I site of pBTEX-osm, the resulting binary vector being referred to as pTEX-PR1 (Figure 2). For constructing the binary vector pOPC, the *Sma*I/*Sal*I fragment of the chitinase cDNA insert in the pT/-Chi was subcloned into intermediate vector pTEX (E) between CaMV 35S promoter and Ocs-ter to produce pTEX (E)-Chi. The chimerical 35-Chi gene insert was then excised as a *EcoR*I fragment and ligated to a *Eco*RI site of pOP produced by partial digestion (Figure 2). The integrity of the OPC insert in the progenies of transgenic plants was detected by PCR utilizing primers specific for amplification of 1.2 kb (osmotin detection) and 2.8 kb (PR-1 and osmotin detection) fragments.

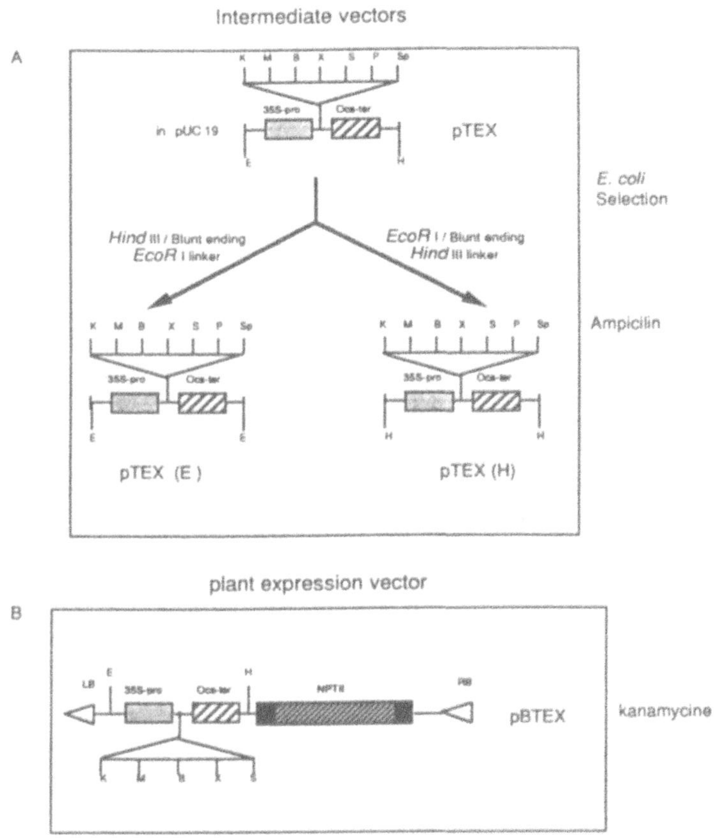

Figure 1. Construction of co-expression cassettes. A. Construction of intermediate vectors. B. Construction of plant expression vector PBTEX. RB=T-DNA right border; LB=T-DNA left border; OCS-ter=octopine synthase terminator; 35S-pro=CaMV 35S promoter; NPT-II=neomycin phosphotransferase gene; restriction endonuclease sites B=*Bam*HI; K=*Kpn*I; M=*Sma*I; P=*Pst*I; S=*Sal*I; Sp=*Sph*I; X=*Xba*I.

In both constructs, the PR gene sequences were deleted of the vacuolar-targeting signal in order to drive the accumulation of the encoded proteins in the apoplast.

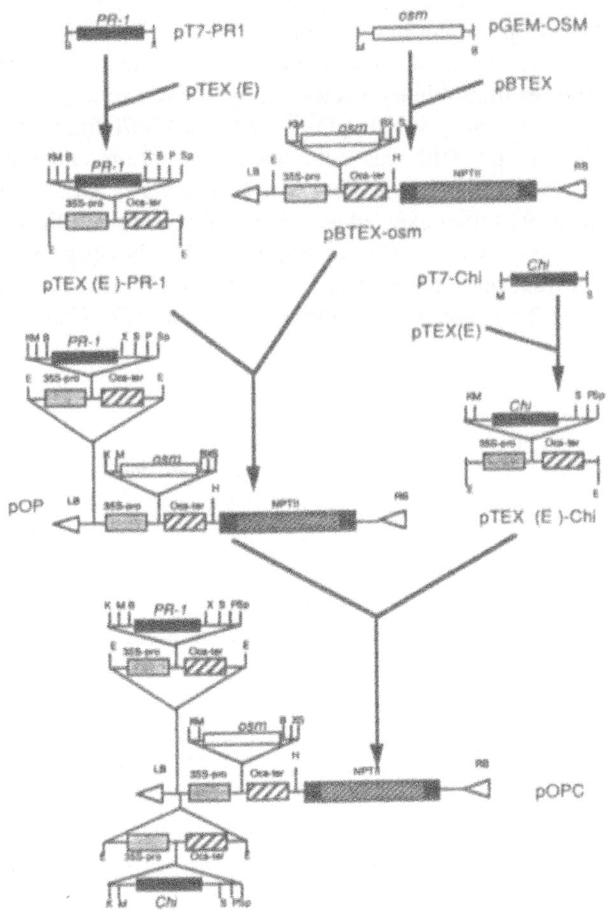

Figure 2. Construction of binary vector pOPC. In the T-DNA region flanked by right border (RB) and left border (LB) sequences, a NPTII gene as selectable marker, a PR-1, a chitinase and an osmotin cDNA.

Ten diploid primary transformants (T_0), five for the single-gene (OSM) and five for the triple-gene (OPC) construct were characterized and then tested for disease resistance to fungal pathogens. The number of T-DNA loci was determined in OSM transformants by Southern-blotting analysis of 10 µg genomic DNA digested with *Eco*RV and probed with a 1 kb DNA fragment containing the complete tobacco osmotin ORF according to standard procedures (Sambrook et al., 1989). This probe detected both tomato endogenous and tobacco osmotin. Segregation for kanamycin resistance was followed in T_1 progenies using a non destructive spraying assay, according to Weide et al. (1989). The level of expression of transgenes was determined by SDS/PAGE immunoblots of leaf proteins as described by La Rosa et al. (1992). Proteins were visualized on nitrocellulose membranes by alkaline phosphatase conjugated polyclonal antibodies.

Inoculation with the fungal pathogens *Botrytis cinerea* (causal agent of grey mold), *Phytophthora infestans* (causal agent of late blight), *Oidium lycopersicum* and *Leveillula taurica* (both causal agents of powdery mildew) were carried out on leaves of similar size (30±2 cm^2) detached from 5-7 week-old plants. *Botrytis* and *Phytophtora* isolates were provided respectively by Tuscia University (Italy) and INRA Montfavet (France), while both causal agents of powdery mildew were collected from tomato plants naturally infected in the greenhouse.

Leaflets were inoculated by placing, at the center of the leaf, 30 μl of *Botrytis* or *Phytophtora* spore suspension at a concentration of $3x10^5$ and $2x10^4$ sp. ml^{-1} respectively, or leaf disks with active sporulating powdery mildew lesions from infected tomato plants. Inoculated material was kept in Petri dishes in a growth chamber at 100% humidity, 20±2°C and 16-hrs photoperiod. Data on the development of symptoms such as infection degree (n° of infected leaf areas total^{-1} n° inoculum drops), infected leaf areas (cm^2), and hyphal growth (0-3 scale: 0= no growth, 3=high hyphal growth) were collected when 100% of leaves from non transformed plants were infected. Statistical analysis was performed according to ANOVA test. Significant differences among means were determined by Duncan's Multiple Range test (P=0.05).

3. Results

3.1 Molecular analyses of plants transgenic for osmotin

Southern blotting analyses on independent primary transformants for osmotin (OSM), revealed single and multiple insertions of the transgene (Figure 3).

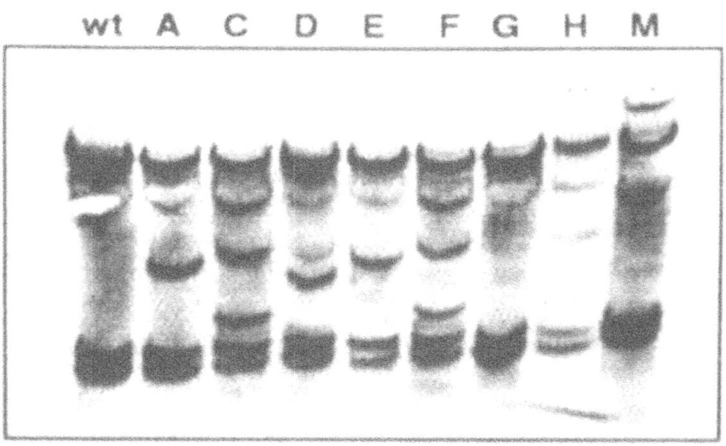

Figure 3. Southern blot analysis on T_0 tomato transformants with pKolc3 construct.

After selection of plants that were diploid and normal-looking (OSM A, E, F, H and M), segregation in T_3 progeny of these plants was followed by both Southern blotting and kanamycin resistance assay. The segregation analyses confirmed the presence of a single insertion in OSM A and E, multiple insertions at different loci in OSM F and H, and two co-segregating copies of the transgene in OSM M. Western blot analyses of leaf protein extracts revealed that the level of osmotin in all OSM transformants ranged from 0.5 to 2% of total proteins (Figure 4). These levels were maintained up to the T, generation (Figure 5).

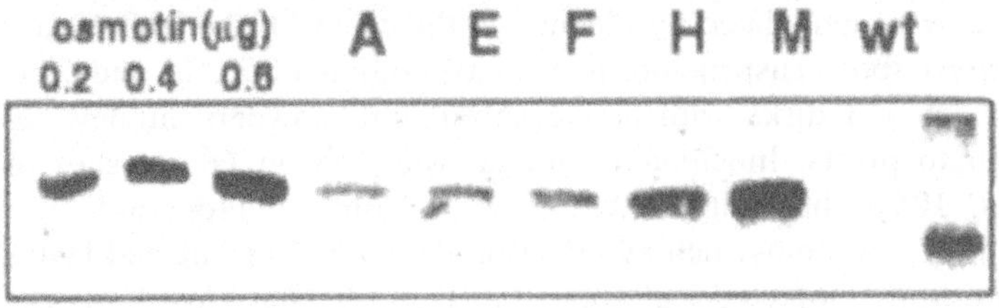

Figure 4. Western blot analysis of T_0 transformants for tobacco osmotin.

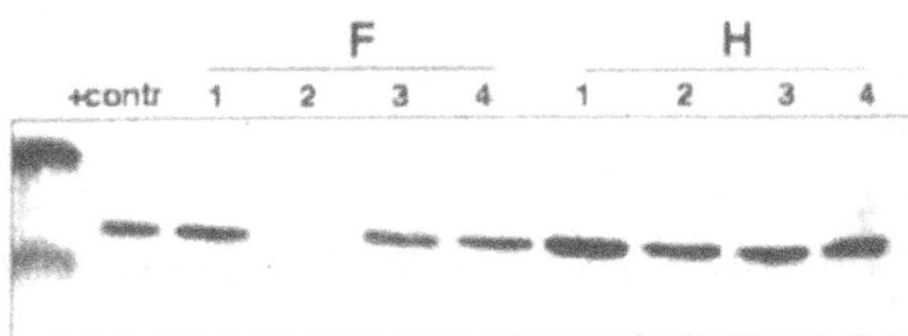

Figure 5. Western blot analysis of T3 segregating transformants for tobacco osmotin.

3.2 *Molecular analyses of plants transgenic for osmotin, chitinase and pr-1*

Spraying 200 mg l^{-1} kanamycin solution on 3 week-old plantlets allowed the selection of two independent pOPC transgenic lines, each with a single insertion. Osmotin was estimated to be 1-2% of total leaf proteins, whereas chitinase and PR-1 were detected at a low level (less than 0.1%) but higher than observed in non-transformed plants.

3.3 *Disease responses*

After inoculation with *B. cinerea*, *P. infestans*, *O. lycopersicum* and *L. taurica*, primary OSM transformants F, H and M displayed significantly less infected leaf area compared to the wild type. These genotypes also had the highest level of expression of tobacco osmotin. OSM A and E plants exhibited enhanced resistance only when challenged with late blight. Six days after infection, significant reductions in lesion development caused by grey mold (37-56%), late blight (58-92%) and powdery mildew (40-80%) were observed (Figure 6).

Frequency of infection developing from inoculation points was not affected by transgene expression, indicating that the effect of transgenes occurred some time after infection initiation.

The evaluation of sporulation index in powdery mildew infections (n° of spores released in 1 ml H_2O by a disk of infected leaf surface) and of mycelium development in late blight (stereomicroscope evaluation according to a 0-3 scale) were used as additional parameters to characterize the transformed lines. Transformants displayed 20-80% reduction in sporulation and 50-75% reduction in mycelium development for powdery mildew and late blight, respectively.

Inoculations with *B. cinerea* and *O. lycopersicum* were also made on T_2 progenies of the most interesting OSM lines. Inhibition of invasion of leaf tissues by *B. cinerea* (40-70%) and *O. lycopersicum* (50-80%) was observed in the OSM F, H and M leaves (data not shown).

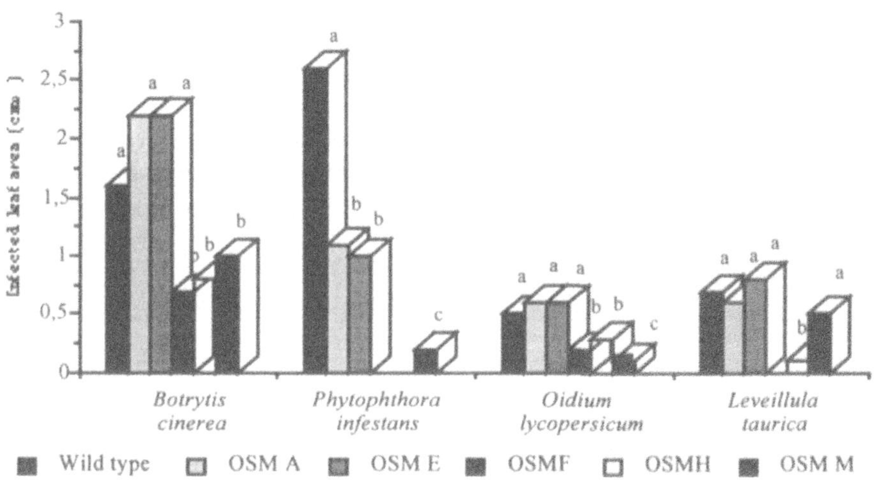

Figure 6. Response of primary transformants with osmotin gene to fungal infections on detached leaves. According to Duncan's Multiple Range test, means with the same letters are not significantly different per P=0.05.

The T_1 generation of two tomato lines transformed with the triple-gene construct was tested for resistance to both *B. cinerea* and *O. lycopersicum*. After ten days, a 74-77% reduction in the areas of necrotic lesions due to grey mold and 89-95% reduction of leaf surface covered by white powdery mildew were observed (Figure 7).

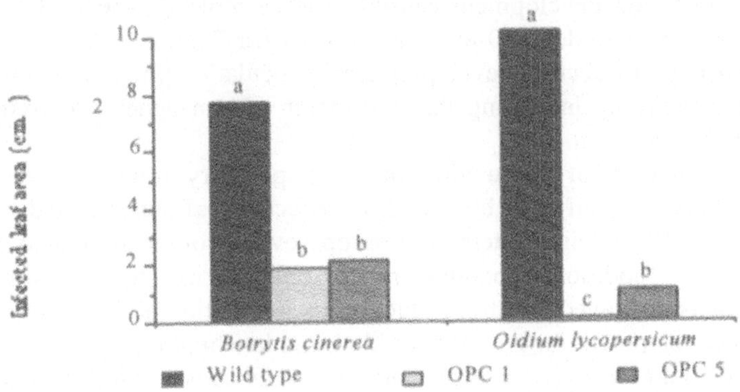

Figure 7. Response of T_1 OPC progenies of tomato to fungal infections on detached leaves. According to Duncan's Multiple Range test, means with the same letters are not significantly different per P=0.05.

Homozygous T_3 OSM H, OSM M and T_2 OPC 1 plants were inoculated *with B. cinerea* and *P. infestans* to compare the levels of disease resistance exhibited by the single and triple-gene constructs.

The percentage of inhibition of fungal infection for both pathogens confirmed the inheritance in T_3 and T_2 generations of improved resistance. However, no significant differences in resistance between transformants carrying the single or triple gene constructs was observed (Figure 8).

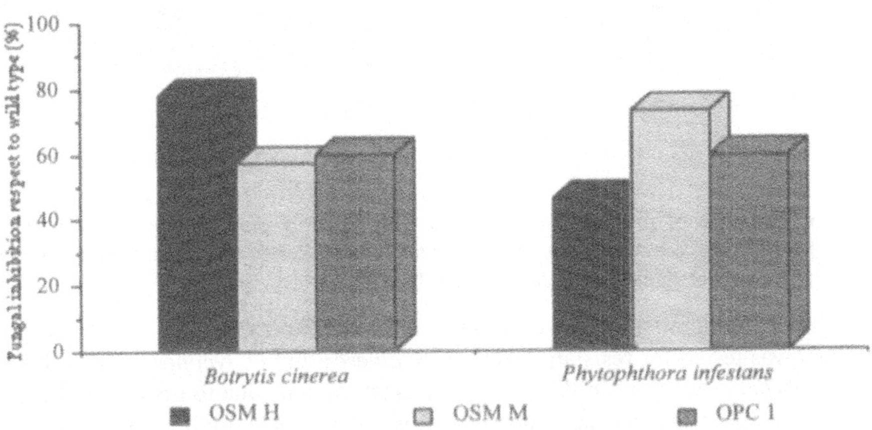

Figure 8. Inhibition of fungal infection (%) on detached leaves of selected T_3 OSM and T_2 OPC tomato transformants as compared to wild type.

4. Discussion

Co-expression of more than one gene encoding antifungal proteins such as rice chitinase and alfalfa acidic glucanase in tobacco (Zhu et al., 1994), bacterial exo-chitinase and barley RIP in tobacco (Jach et al., 1995), tobacco chitinase and glucanase in tomato (Jongedijk et al., 1995) were effective to provide synergistic protection against fungal infections. The multigene strategy has been widely applied in recent breeding programs to obtain durable and broad-range disease resistance. Grey mold, powdery mildew and late blight represent some of the most economically important fungal diseases that account for significant worldwide production losses of tomato. In the absence of available genes within the tomato gene pool for effective resistance to these pathogens, the approach of genetic transformation for simultaneous expression of genes encoding antifungal proteins may prove valuable in increasing crop protection against these fungal diseases.

Results presented in this paper provide further evidence that overexpression of PR-proteins, especially osmotin, enhances resistance of transgenic tomato plants to fungal diseases. Our data demonstrated that ectopic expression of both the single-gene and the triple-genes constructs confers resistance to very divergent plant pathogens such *as B. cinerea, P. infestans, O. lycopersicum* and *L. taurica*, against which no sources of resistance are known in the available tomato germplasm.

However, there were no clear differences between the two constructs, pOSM and pOPC, in controlling fungal infections, possibly due to the low expression levels of both PR-1 and chitinase genes. These data confirm that the effectiveness of co-expression of different antifungal proteins for further improvement of disease resistance depends on the combination of transgenes used and their expression levels. In this respect, efforts are being made to obtain new constructs with different promoters, to increase expression and avoid co-suppression, and to determine using *in vitro* assays what different combinations of proteins provide a high synergistic antifungal activity.

Aknowledgements: This work was supported by the national programme on "Plant Biotechnologies" of Italian Ministry for the Agricutural Policies. The authors thank Prof. E. Rugini for his precious contribution to obtain tomato transgenic plants, Dr. G. Chilosi from Tuscia University (Italy) and Dr. H. Laterrot from INRA Montfavet (France) for providing *Botrytis* and *Phytophthora* isolates, respectively.

5. References

Abad, L.R., Paino D'Urzo, M., Liu, D., Narasimhan, M.L., Reuveni, M., Zhu, J.K., Singh, N.K., Hasegawa, P.M. & Bressan, R.A. (1996) Antifungal activity of tobacco osmotin has specificity and involves plasma membrane permeabilization, *Plant Science* **118**, 11-23.

Ausubel, F.M., Brent, R., Kingstone, R.E., Moore, D.D., Seidman, J.G., Smith, J.A. & Struhl, K. (1988) In *Current Protocols in Molecular Biology*, John Wiley & Sons, New York, Sec. 4.3.1.-4.3.

Jach, G., Gornhardt, B., Mundy, J., Logemann, J., Pinsdor, F.E., Leah, R., Schell, J. & Maas, C. (1995) Enhanced quantitative resistance against fungal disease by combinatorial expression of different barley antifungal proteins in transgenic tobacco, *Plant J.* **8** (1), 97-109.

Jongedijk, E., Tigelaar, H., van Roekel, J.S.C., Bres-Vloemans, S.A., Dekker, I., van den Elzen, P.J.M., Cornelissen, B.J.C. & Melchers, L.S. (1995) Synergistic activity of chitinases and ß-1,3-glucanases enhances fungal resistance in transgenic tomato plants, *Euphytica* **85**, 173-180.

La Rosa, P.C., Chen, Z., Nelson, D.E., Singh, N.K., Hasegawa, P.M. & Bressan, R.A. (1992) Osmotin gene regulation is posttrascriptionally regulated, *Plant Physiol.* **100**, 409-415.

Leah,R., Tommerup, H., Svendsen, I. & Mundy, J. (1991) Biochemical and molecular characterization of three barley seed proteins with antifungal properties, *J. Biol. Chem.* **266**, 1464-1573.

Liu, D., Raghothama, K.G., Hasegawa P.M. & Bressan, R.A. (1994) Osmotin overexpression in potato delays development of disease symptoms, *Proc. Natl. Acad. Sci. USA* **91**, 1888-1892.

Liu, D., Rhodes, D., Paino D'Urzo, M., Yi Xu, Narasimhan, M.L., Hasegawa, P.M., Bressan, R.A. & Abad, L. (1996) In vivo and in vitro activity of truncated osmotin that is secreted into the extracellular matrix. *Plant Science* **121**, 123-131.

Melchers, L.S., Apotheker-de Groot, M., van der Knap, J., Ponstein, A.S., Sela-Buurlage, M., Bol, J.F., Cornelissen, B.J.C., van den Elzen, P.J.M. & Linthorst, H.J.M. (1994) A new class of tobacco chitinases homologous to bacterial exo-chitinases displays antifungal activity, *Plant J.* **5**, 469-480.

Sambrook, J., Fritsch, E.F. & Maniatis, T. (1989) Molecular cloning: A laboratory manual, 2nd Edn. Cold Spring Harbor, NY: Cold Spring Harbor Laboratory Press.

Veronese, P., Tucci, M., Crinó P., Rugini, E., Hasegawa, M.P., Bressan, R.A. & Saccardo, F. (1998) Resistance to phytopathogenic fungi in plants of tomato overexpressing an osmotin gene, in *Proceed. IX International Congress on Plant Tissue and Cell Culture,* Jerusalem, Israel, June. 14-19, 1998, p. 182 (abstract).

Vigers, A.J., Roberts, W.K. & Selitrennikoff, C.P. (1991) A new family of plant antifungal proteins, *Mol. Plant-Microbe Interact.* **4**, 315-323.

Vigers, A.J., Wiedemann, S., Roberts, W.K., Legrand, M., Selitrennikoff, C.P. & Fritig, B. (1992) Thaumatin-like pathogenesis-related proteins are antifungal, *Plant Sci.* **83**, 155-161.

Yun, D.J., Bressan, R.A. & Hasegawa, P.M. (1997) Plant antifungal proteins, *Horticultural Reviews* **14**, 39-88. John Wiley and Sons, Inc.

Weide, R., Koorneef, M. & Zabel, P. (1989) A simple, nondestructive spraying assay for the detection of an active kanamycin resistance gene in transgenic tomato plants. *Theor. Appl. Genet.* **78**, 169-172.

Zhu, Q., Maher, E.A., Masaud, S., Dixon, R.A. & Lamb, C. (1994) Enhanced protection against fungal attack by constitutive co-expression of chitinase and glucanase genes in transgenic tobacco, *Bio/Technology* **12**, 807-812.

G.T. Scarascia Mugnozza, E. Porceddu & M.A. Pagnotta (Eds.)
Genetics and Breeding for Crop Quality and Resistance, 25-32, 1999

Introgression of late blight resistance into *Solanum tuberosum*

G. Ramsay, H.E. Stewart, W. De Jong, J.E. Bradshaw & G.R. Mackay
Scottish Crop Research Institute, Invergowrie, Dundee DD2 5DA, UK

Abstract: Among 227 accessions of the Commonwealth Potato Collection, 57 resistant or very resistant accessions from 24 species were found. These species came from 7 different taxonomic series. Mexico was the source of the majority of resistant accessions, but several were also found which derived from material collected in Bolivia and Argentina. The wide taxonomic and geographical spread of resistant accessions indicates that there is likely to be a wealth of new types of resistance available in germplasm collections for breeding for resistance to late blight. Methods for accelerating the introgression of late blight resistance into cultivated lines are being explored. *Solanum papita* has been chosen as this species has an unusually high proportion of resistant accessions. An F_1 hybrid was created at the 4x level to improve the prospects for fertility and for recombination between *S. papita* and *S. tuberosum* genomes. This hybrid was backcrossed to *S. tuberosum* and segregation for late blight resistance was noted. Comparison of the parental values and those of the F_1 indicated that the resistance is additive. An approach using AFLPs to improve the efficiency of backcrossing by selecting individuals which carry the least markers from *S. papita* was outlined and is being tested.

1. Introduction

The cultivated potato *S. tuberosum* has an unusually large number of cross-compatible wild relatives for a crop plant. These wild tuber-bearing *Solanum* species contain a wide array of useful traits for the improvement of cultivated potatoes. The challenges in exploiting this wealth of useful genes are in the identification of the best forms of the trait, so that the time and resources required to move it into a cultivated background may be spent effectively, and in the development of efficient strategies for this exploitation.

Late blight, caused by the oomycete fungus *Phytophthora infestans* (Mont.) de Bary, is the most important biotic constraint worldwide for the cultivated potato. Up to 25 fungicide sprays can be required per season, with a total cost estimated to be about $1.8 billion, and crop losses in the developing countries of about $2.75 billion. In addition resistance to the fungicide metalaxyl is becoming widespread.

Further concerns are raised by the spread of a second mating strain to many parts of the world potentially permitting both the generation of increased variability in the pathogen and an ability to infect earlier in the season from resting sexual spores. In the light of these concerns it is not surprising that more effective forms of late blight resistance are eagerly sought by researchers and breeders.

Limited resistance in cultivated types prompted the use of wild germplasm as far back as 1911 (reviewed by Umaerus & Umaerus, 1994). Most of the resistance in cultivars has come from one species, *S. demissum* (Black, 1970). R-gene mediated resistance to late blight from *S. demissum* has not been stable and breeders have turned instead to searching for horizontal resistance. Although such resistance has come from *S. demissum* and other species (Black, 1970), there are many more sources of such genes which could be used to increase the range of genes and mechanisms of resistance to this pathogen. This paper reports progress evaluating, selecting, and exploiting new sources of late blight resistance with the aim of developing more effective methods of transferring such resistance into adapted cultivated backgrounds.

2. Materials and Methods

Seeds of 227 accessions from the Commonwealth Potato Collection were sown and seedlings were tested for foliage resistance to late blight according to Malcolmson & Killick (1980) with minor modifications. Two pots of 25 seedlings each were used in a randomised design. An inoculum of 5 x 10^4 zoospores ml^{-1} of a complex race (1, 2, 3, 4, 6, and 7) of *Phytophthora infestans* was sprayed onto the seedlings. They were incubated for 24 hours in a high humidity cabinet then transferred to an air-conditioned glasshouse at 15°C. Pots were sorted into five classes according to the standard scale agreed by the Association of Potato Inter-genebank Collaboration (VS – very susceptible; S – susceptible; M – medial; R – resistant; VR – very resistant). To determine the levels of resistance in segregating populations a detached leaf test on seedlings and a whole plant test on the subsequent tuber-derived plants were made. Detached leaflet tests were performed on two leaflets taken from the first fully expanded leaf of each plant. A 50 μl droplet of inoculum of 5 x 10^3 zoospores ml^{-1} was placed centrally on the abaxial surface of the leaf and leaves maintained on damp tissue in a clear plastic box at 15°C and 16 h daylength for 6 days. Lesion diameter was measured and the presence or absence of sporulation was noted. Whole plant tests used 2 plants of each genotype in a randomised design. Plants were inoculated as for seedlings and each plant given a score 7 days after inoculation using Malcolmson's scale of resistance (Cruikshank et al., 1982).

Two blight resistant clones of *S. papita* from CPC2639 and CPC2640 were used in crosses with several *S. tuberosum* cultivars in the glasshouse. Emasculated flowers of *S. papita* and *S. tuberosum* were pollinated 1-2 day later with pollen of the other species, and pollen of *S. phureja* IVP48 or IVP35 was applied to the same stigma 1-2 days later. Berries from these pollinations were collected after 18-37 days, surface-sterilised and dissected to remove developing seeds. Seeds were opened and embryos not visibly expressing the embryo spot marker from the *S. phureja* mentor pollen donor IVP48 or IVP 35 were placed on Murashige and Skoog culture medium supplemented with either 2% or 6% sucrose, 0.1 mg l^{-1} IAA, 0.01 mg l^{-1} kinetin, and 0.001 mg l^{-1} adenine sulphate and solidified with 7 g l^{-1} agar (Singsit & Hanneman, 1991). Seedlings were grown on in the same medium and transferred to the glasshouse. Plant morphology, the nodal spot marker from *S. phureja*, and chromosome counts were used to provisionally assign plants to one of the possible hybrid classes. Anchored simple sequence repeat PCR (Charters et al., 1996) was used to confirm the identity of the resulting hybrids. AFLPs were determined on the BCF_1 population between one *S. papita* x *S. tuberosum* cv. Cara F_1 hybrid and *S. tuberosum* cv. Désirée using the restriction enzymes *Mse*I and *Pst*I.

3. Results and Discussion

3.1 Genebank evaluations

Screening accessions from the Commonwealth Potato Collection using a seedling test for late blight resistance revealed 28 accessions classified as resistant and 29 very resistant from 24 of the 47 species tested (Table 1). These accessions come from seven different series from the tuber-bearing part of the genus. The results are presented here separately for the Central American and Southern American accessions. The Central American species are a particularly rich source of late blight resistance, with 44 out of 84 (52%) resistant or very resistant accessions, and only 14 out of 143 South American species (10%) in the same categories. It should be noted that the high proportion of resistant Central American accessions will be over-estimated compared to a random sample of genebank accessions from that region as a special focus was made on *S. papita*. However, even excluding *S. papita* accessions leaves a high proportion of resistant accessions in the Central American group. All of the resistant accessions from Central America in this set of evaluations were from Mexico. Mexico is widely known as the centre of diversity and probably the centre of origin of *Phytophthora infestans*. It possesses many wild species of *Solanum* carrying blight resistance (van Soest et al., 1984; Umaerus & Umaerus, 1994). Among the series of the tuber-bearing part of the genus from Central America, Demissa and Longipedicellata contain many resistant accessions. Both series have been previously reported to contain late blight resistance and species from both have been used to attempt to introduce late blight resistance into cultivated potatoes. *Solanum papita* was known from previous work at SCRI to display strong resistance to late blight, and the additional evaluations presented here confirm that most accessions of this species are resistant or very resistant. The related species *S. stoloniferum* did not have such a high proportion of resistant accessions.

Table 1. Numbers of accessions of each species in each class of evaluation score for late blight resistance in seedlings of the Commonwealth Potato Collection.

Species of	Country	Series	Nos of accessions				
			VS	S	M	R	VR
a) Central America							
agrimonifolium	GUA	Conicibaccata	1	1			
brachycarpum	MEX	Demissa		1	1	1	
demissum	MEX	Demissa			2	1	2
hougasii	MEX	Demissa			1		2
iopetalum	MEX	Demissa					2
semidemissum	MEX	Demissa					1
fendleri	MEX/USA	Longipedicellata	3	4	3	1	1
hjertingii	MEX	Longipedicellata	1			1	
polytrichon	MEX	Longipedicellata	1	2	1	4	
papita	MEX	Longipedicellata	2		1		10
stoloniferum	MEX	Longipedicellata	2	5	3	8	6
cardiophyllum	MEX	Pinnatisecta	1				1
jamesii	USA	Pinnatisecta		1			
michoacanum	MEX	Pinnatisecta		1			
trifidum	MEX	Pinnatisecta	1	1			
polyadenium	MEX	Polyadenia				1	
verrucosum	MEX	Tuberosa				2	
b) South America							
acaule	ARG/PER	Acaulia	6	5	3		
albicans	PER	Acaulia		1			
commersonii	URU/BRA/ARG	Commersoniana		1	2		1
paucijugum	ECU	Conicibaccata			1		
infundibuliforme	ARG	Cuneoalata	1				
brevidens	ARG	Etuberosa	1	1	2	1	
boliviense	BOL	Megistacroloba	1	1			
megistacrolobum	ARG/BOL	Megistacroloba	2				
raphanifolium	PER	Megistacroloba		1			
sanctae-rosae	ARG	Megistacroloba	2				
toralapanum	BOL	Megistacroloba	1	2	1		
alandiae	BOL	Tuberosa	1				1
berthaultii	BOL	Tuberosa				1	
brevicaule	BOL	Tuberosa	1				
bukasovii	PER	Tuberosa	3	2			
canasense	PER	Tuberosa	3		1		
doddsii	BOL	Tuberosa			1		
gourlayi	BOL	Tuberosa	1				
kurtzianum	ARG	Tuberosa	2		4		
leptophyes	BOL	Tuberosa	1				
microdontum	ARG	Tuberosa	2	2	2		1
okadae	ARG	Tuberosa			2	1	
oplocense	ARG/BOL	Tuberosa	2	1			
phureja	COL/ECU/VEN	Tuberosa	9	5	4		
sparsipilum	BOL	Tuberosa		3	1		
stenotomum	PER/BOL	Tuberosa	5	3	1		
tuberosum	CHL	Tuberosa	2	1			
tub. ssp. *andigena*	COL/PER/BOL	Tuberosa	15	3	1	2	
venturi	ARG	Tuberosa				1	
vernei	ARG	Tuberosa				1	
chacoense	ARG/PAR	Yungasensa	5	3		2	2
tarijense	BOL/ARG	Yungasensa			2		

Among the South American species investigated, 14 accessions resistant or very resistant were found in four series, Commersoniana, Etuberosa, Tuberosa and Yungasensia. The resistant accessions were all from Bolivia and Argentina, already identified as a secondary centre for late blight resistance by van Soest et al. (1984). The wide taxonomic base of the accessions identified as resistant gives some promise that there may be different types of resistance genes available in wild *Solanum* species, and that exploiting these in breeding will be a worthwhile exercise.

3.2 *Exploitation of resistance in wild species*

S. papita was chosen to develop improved methods for the introgression of late blight resistance into cultivated potato. Previous experience at SCRI indicated that the resistance in *S. papita* was both potent and repeatable. In previous tests all accessions of *S. papita* tested were found to be resistant, indicating that there may be something unusual about the resistance in *S. papita*. Subsequently two susceptible accessions were identified but the unusually high proportion of resistant accessions remains a feature of this species.

All series Longipedicellata species, including *S. papita,* are tetraploid, disomic, and have an endosperm balance number of 2 (EBN 2), making crosses with 4x *S. tuberosum* (EBN 4) difficult. Hybrids can be made between *S. papita* and diploid potatoes such as *S. phureja* (EBN 2) with relative ease. However, such hybrids would be triploid and sterile. Restoring fertility by doubling the chromosome complement will reduce the prospects for inter-genome recombination and therefore efficient recombination, so this approach was not taken. We chose instead to make the cross at the tetraploid level, even though this was the most difficult route as predicted from EBN numbers. This approach leaves the two genomes from the allotetraploid *S. papita* present in the monosomic state in the F_1 and therefore more liable to recombine with *S. tuberosum* chromosomes. Crosses were performed in both directions and 115 berries grew to a size worth culturing, most of them with *S. papita* as the female. Between 1 and 64 embryos were dissected from each berry and cultured. Among the surviving seedlings, only one tetraploid F_1 individual was found, from a cross between *S. papita* CPC 2639 as female and *S. tuberosum* cv. Cara. The identity of this F_1 was confirmed using anchored-SSR PCR (Figure 1, left).

BCF_1 populations between the F_1 and *S. tuberosum* cultivars Désirée and Pentland Ace were generated. They were assessed for late blight resistance along with the parents and F_1s using a detached leaf test and a whole plant test. A detached leaf test could permit screening of seedlings of a segregating population prior to flowering and hence allow a decision on which individuals from the population to use for a further round of backcrossing. The results comparing the two ways of assessing late blight resistance are presented in Figure 2. The two sets of scores for the backcross population are weakly correlated ($r = -0.3$, $p = 0.18$) indicating that this detached leaf test is a relatively poor predictor of whole plant performance and that the selection of the best individuals to take into the next generation should not be done until tubers are available for whole plant tests. The absence of any clear segregation into discrete groups in these tests suggests that the resistance may not be controlled by single major genes. A further assurance that the blight resistance in *S. papita* is of the oligogenic type comes from the intermediate position of the F_1 plants between the two parents on the graph in Figure 2. R-gene mediated resistance would be reflected in an F_1 closer to the resistant parent on this graph. The resistance present in *S. papita* is therefore attractive for breeding, being relatively potent, yet apparently not mediated by R genes. Further studies are required to determine genetics of this resistance as different strategies for utilising the resistance will be required depending on the number and dominance relationships of the loci involved.

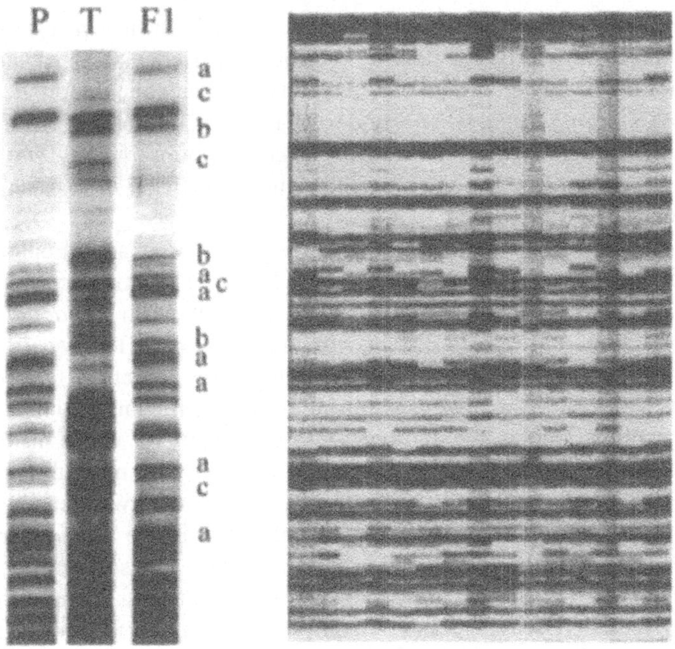

Figure 1. Anchored simple sequence repeat PCR of *S. papita* (P), *S. tuberosum* cv. Cara (T) and their F_1. a – *S. papita*-specific band, b – Cara band transmitted to the F_1, c – Cara band not transmitted to the F_1 due to heterozygosity (left). AFLPs of the segregating BCF_1 population (*S. papita* x Cara) x Désirée (right).

The efficient transfer of resistance genes from exotic to cultivated backgrounds can proceed in different ways. The traditional route is to repeatedly backcross to a cultivar while simultaneously selecting against wild traits and selecting for the trait of interest. This has been highly successful but may take decades to achieve in potatoes. The use of molecular tools can increase the efficiency of selection during backcrossing by tagging the QTLs involved and then positively selecting for markers closely linked to them. In crops such as potato, where both heterozygosity and tetrasomic inheritance make QTL analysis complex, this is currently very difficult to achieve. A simpler approach may be to use markers to aid the efficiency of selection against background genome from the wild parent while maintaining selection for reasonable expression of the trait of interest. This approach may be worth taking where there are a relatively small number of important QTLs, where the trait being introduced has high heritability and can be assessed with reasonable accuracy. We are exploring this approach as a possible general means of more rapidly introgressing traits from wild species where mapping and QTL anaylsis is difficult.

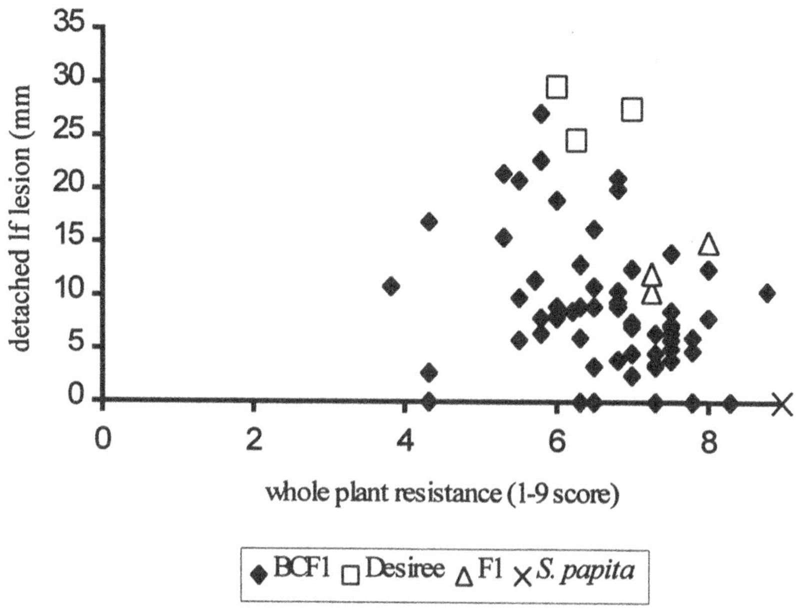

Figure 2. Segregation for late blight resistance in a BCF_1 population of *S. papita* x *S. tuberosum* cv. Cara x *S. tuberosum* cv. Désirée with the F_1 and its two parents, showing the relationship between detached leaf and whole plant scores.

AFLPs were performed on the BCF_1 population of (*S. papita* x Cara) x Désirée. Bands were scored which were unique to the *S. papita* parent and which segregated in the BCF_1 (Figure 1, right). Each individual from the BCF_1 was plotted against the whole plant resistance score (Figure 3). There was no correlation between the apparent amount f the *S. papita* genome carried by the individual and its late blight resistance score. It can be seen that there is a wide variation about the mean for the number of *S. papita*-specific bands occurring in each individual in the population. This indicates that there is the potential to make selections from such segregating inter-specific hybrid-derived populations to alter the balance of the overall content of the genome in the individuals selected for further crossing. Whether or not such selection can have an effect detectable at the agronomic level remains unknown. We are currently testing this strategy in this population to assess the feasibility of this approach.

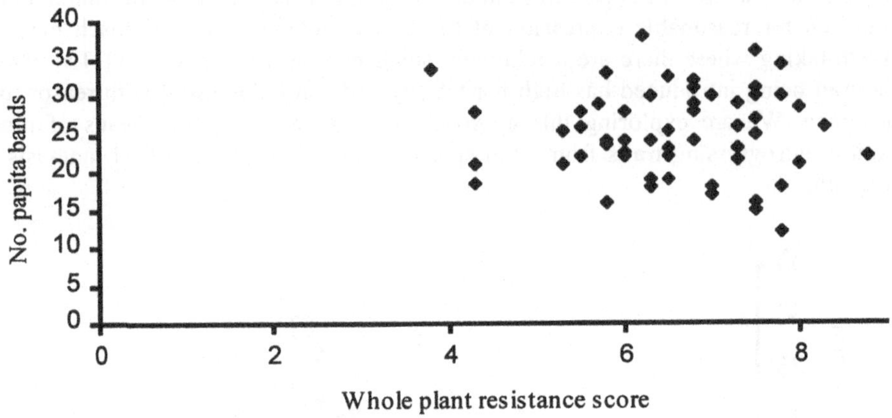

Figure 3. Scatter plot of no. of *S. papita*-specific AFLP bands against late blight resistance as measured by whole plant tests.

Acknowledgements: The financial support of the Scottish Office Agriculture, Environment and Fisheries Department and the European Commission (RESGEN CT95-34) is gratefully acknowledged. We also thank Jane Davidson, Sharon Neilson and Gail Simpson for their assistance, and are particularly indebted to Norma Dow for her contribution to this work.

4. References

Black, W. (1970) The nature and inheritance of field resistance to late blight (*Phytophthora infestans)* in potatoes. *Amer. Potato J.*, **47**, 270-288.

Charters, Y.M., Robertson, A., Wilkinson, M.J. & Ramsay, G. (1996) PCR analysis of oilseed rape cultivars (*Brassica napus* L. spp. *oleifera*) using 5' anchored simple sequence repeat (SSR) primers. *Theoretical and Applied Genetics* **92**, 442-447.

Cruikshank, G., Stewart, H.E. & Wastie, R.L. (1982) An illustrated assessment key for foliage blight of potatoes. *Potato Research* **25**, 213-214.

Malcolmson, J.F. & Killick, R.J. (1980) The breeding value of potato parents for field resistance to late blight measured by whole seedlings. *Euphytica* **29**, 489-495.

Singsit, C. & Hanneman, R.E. (1991) Rescuing abortive inter-EBN potato hybrids through double pollination and embryo culture. *Plant Cell Reports* **9**, 475-478.

Umaerus, V. & Umaerus, M. (1994) Inheritance of resistance to late blight. In*: Potato Genetics.* Eds. J.E. Bradshaw and G.R. Mackay. CAB International, Oxford.

Van Soest, L.J.M., Schoeber, B. & Tazelaar, M.F. (1984) Resistance to *Phytophthora infestans* in tuber-bearing species of *Solanum* and its geographical distribution. Potato Research **27**, 393-411.

G.T. Scarascia Mugnozza, E. Porceddu & M.A. Pagnotta (Eds.)
Genetics and Breeding for Crop Quality and Resistance, 33-42, 1999

Crown rust resistance in ryegrasses

E. Adams[1], M. De Loose[1], J. Baert[1], A. Ghesquiere[1], A. Depicker[2], E. Van Bockstaele[1,3] & I. Roldán-Ruiz[1]
[1]Department for Plant Genetics and Breeding, Centre for Agricultural Research-Gent. Caritasstraat 21, B-9090 Melle, Belgium. E-mail: M.DeLoose@CLO.FGOV.BE [2]Department of Genetics, VIB, University of Gent, KL Ledeganckstraat 35, B-9000 Gent, Belgium. [3]Depart. of Plant Production, University of Gent. Coupure Links 653 B-9000 Gent, Belgium

Key words: Crown rust, artificial infection, ryegrass, genetic control, Marker Assisted Selection

Abstract: Reciprocal crosses between *Lolium multiflorum* plants and between *L. multiflorum* and *L. perenne* plants, which were either resistant or susceptible to crown rust were performed and the mode of inheritance of the resistance was analysed. In the crosses in which the resistant plants Axis-1 or Axis-2 were involved, a maternal type of inheritance was observed. Further confirmation was obtained from a cross between a resistant second generation descendant of Axis-2 and a moderately susceptible *L. perenne* plant. In one of the crosses between one resistant and one susceptible *L. multiflorum* plant, a nuclear type of inheritance was observed. Depending on the definition of resistance, the action of a dominant resistance gene or the presence of a polygenic trait could be concluded. Based on the results obtained, it is suggested to combine BSA and QTL analysis to identify markers linked with crown rust resistance in ryegrass.

1. Introduction

The ryegrasses (Lolium ssp.) are composed of some highly productive, nutritious, persistent and well-adapted grasses which are widely used for agricultural and recreational purposes, and for stabilising soils (Jaghar, 1993). Crown rust (*Puccinia coronata* Corda) is a common and serious fungal disease of ryegrasses in parts of Europe and has been reported in most of the temperate regions of the world (Potter et al., 1990). The fungus parasites host plants and depletes their carbohydrate reserves (Simons, 1970). Severe infestations render turf unsightly, detract from growth and recuperative potential (Kopec et al., 1983).

In sustainable agriculture chemical control of pathogens is not desirable. The use of resistant cultivars and sound management practices offer the best method of disease control in grasslands (Thomas, 1991). Therefore improving genetic resistance to fungal infections is a major goal in most ryegrass breeding programs (Posselt, 1994). Pair crossing followed by recurrent selection using the polycross method to combine favourable genes is the main means of improving perennial ryegrass populations. Ideally, selection should be imposed on all important traits at each generation of selection to avoid undesirable changes resulting from pleiotropy and linkage (Wilkins, 1991). This implies that throughout the whole breeding process screening for crown rust must be performed, either by natural or by artificial infections. But natural infections do not appear every year, and artificial infections are time consuming. Genetic mapping of disease resistance genes could help improving the efficiency of plant breeding and could lead to a better understanding of the molecular basis of resistance (Lefebvre & Chèvre, 1995).

Marker assisted selection (MAS) is a selection system in which plants are screened for markers linked with the trait of interest, not for the trait itself. If DNA-markers are used, selection can be performed at seedling stage and there is no need for measurements of the trait itself. Once a gene has been found to be linked to codominant markers, plants containing the gene of interest can be easily identified; it is therefore possible to introgress recessive or dominant disease resistance genes in a minimum number of generations (Tanksley et al., 1989).

A bulk segregant analysis (BSA) strategy can be followed to identify DNA markers tightly linked with the trait of interest (Michelmore et al., 1991). BSA has recently been used for the accumulation of markers linked to a locus for beet cyst nematode resistance in sugar beet (Halden et al., 1997). Alternatively, a QTL-approach can be followed to identify genome regions involved in disease resistance. A QTL for resistance which accounts for 67% of the variation for rhizomania resistance in sugarbeet was identified by Pelsy & Merdinoglu (1996).

The main objective of this study is to identify DNA-markers linked with crown rust resistance which could be used in MAS. Taking into account that interspecific crosses are common practice during the breeding process of ryegrasses, the identification of markers of general use which are effective in different *Lolium* species would represent a very important advance. Results of previous experiments suggested a maternal mode of inheritance of the resistance in a number of crosses involving *L. multiflorum* plants (Adams et al., submitted). It is therefore of paramount importance to check the mode of inheritance of the resistance in each particular cross before it is used in a BSA analysis. In the work reported here, inheritance of crown rust resistance was checked in crosses involving *L. multiflorum* and *L. perenne* plants. Based on the results obtained a strategy has been developed for the identification of AFLP-markers linked to crown rust resistance genes in ryegrass.

2. Material and Methods

2.1 *Plant Material*

From a selection field with individual *L. multiflorum* and *L. perenne* plants, one resistant *L. perenne* plant (LpRR), and four resistant *L. multiflorum* cultivar Axis plants (Axis-1, Axis-2, Axis-3 and Axis-13) were selected. As susceptible crossing partners, one severely infected *L. perenne* plant (LpRS) and three *L. multiflorum* plants (P_1G_3-1, B-1, b90-1 and Axis-12) were selected. These plants were used to establish seven pair crosses between a susceptible and a tolerant plant, as summarised in Table 1. Seed was harvested separately on each crossing partner.

Finally, a cross was performed between a F_2 plant derived from Axis-2 (xP_1G_3-1), which showed no signs of infection in five infection rounds (LmR), and a moderately susceptible *L. perenne* plant (LpS) (see Table 1).

Table 1. Scheme of the reciprocal crosses analysed.

			SUSCEPTIBLE					
			L. multiflorum				*L. perenne*	
			P_1G_3-1	B-1	b90-1	Axis-12	LpRS	LpS
	L. perenne	LpRR				X		
		Axis-1	X	X				
		Axis-2	X	X				
RESISTANT	*L. multiflorum*	Axis-3			X			
		Axis-13					X	
		LmR						X

X: cross has been performed

2.2 Nomenclature

The following convention was used to design the different seed stocks: 'F1 Axis-1 (xP_1G_3-1)' designates the F_1 generation of the cross between Axis-1 and P_1G_3-1 harvested on Axis-1; 'F1 P_1G_3-1 (xAxis-1)' designates the F_1 generation of the same pair cross, but harvested on P_1G_3-1. 'F1 Axis-1xP1G3-1' designates the F_1 generation of the cross between Axis-1 and P_1G_3-1 harvested on both parents.

2.3 Plant Growth Conditions

Plants were grown in trays of 96 pots of 4 x 4 x 7 cm filled with common soil. They were kept in the greenhouse at 20-25°C during the day and at 15°C during the night. When necessary, illumination was supplemented for a total photoperiod of 14 hours. The first infection was carried out when the plants were 6 weeks old and was repeated two (crosses between *L. multiflorum* plants) or three (crosses between *L. perenne* and *L. multiflorum* plants) times at intervals of one month. The F_1 plants of the cross LmR x LpS were infected once.

2.4 Inoculum and Infection Conditions

Inoculum was harvested in the fields of the DvP in 1994 and in 1997 on different ryegrass cultivars as described in Adams et al. (submitted). The uredospores were preserved at -80°C.

The infection of all crosses but LmRxLpS were performed as described hereafter. The germination of the uredospores was induced by incubation during two minutes in a water bath at 45°C. The uredospores were then diluted in ten times their own volume of talc powder. The mixture was stroked with a paint brush on all the leaves of each plant individually at a density of 40 mg uredospores per 100 plants. The plants were sprayed with water and subsequently covered with a transparent plastic foil. They were kept for 36 hour at 100% relative humidity after infection. After breaking down the 100% relative humidity, the plants were kept as described in 'plant growth conditions'.

The cross LmRxLpS was infected, as described in Reheul & Ghesquire (1996). The main differences with the method described above is that in this case the inoculum was sprayed on the plants as an aqueous suspension of spores.

Fourteen days after inoculation, the plants were scored for susceptibility using the scoring scale of Birckenstaedt (1990), a 1-9 scale of increasing susceptibility. To avoid biases during the scoring process, the trays were randomly distributed and the investigator who assigned the scores did not know the origin of the plants.

2.5 *Statistical Analyses*

The mean infection score for each individual plant was calculated by averaging the scores got for this plant over the different infection rounds. Mann-Whitney U tests were performed to check for significant differences in mean infection scores between seed stocks. All calculations were carried out using the package SPSS 6.0 (Norusis, 1993).

3. Results

Hundred F_1 plants from each seed stock of the five crosses between *L. multiflorum* plants (Table 1) were subjected to infection (April-May '97). In the four crosses where the plants Axis-1 or Axis-2 were involved, the F_1 harvested on the tolerant crossing partner displayed a significantly lower mean infection score than the offspring harvested on the susceptible crossing partner (Figure 1 and Adams et al., submitted).

In Figure 1 is given an example of the results obtained after one infection round for the descendants of the cross Axis-2xP1G3-1. Figure 1a represents the results of the infection of seed harvested on the resistant crossing partner Axis-2, Figure 1b represents the results of the infection of seed harvested on the susceptible crossing partner P1G3-1. The second infection round gave a similar pattern (results not shown). These results could be explained by the presence of resistance factors in the organelle DNA, which is normally maternally transmitted in the angiosperms. When this maternal factor is present, completely different frequency distributions are found for the progenies harvested on each crossing partner (compare Figures 1a and 1b).

If this hypothesis is true, all seed harvested on a resistant genotype where these cytoplasmic resistance factor(s) is (are) present should carry it. It was therefore investigated whether this mode of inheritance was stable through several generations and even in crosses where plants of a different species were involved. A cross was performed between LmR and LpS (see Table 1). All seed harvested was grown and the plants were infected and scored once by a different person in order to increase the objectivity. The results presented in Table 2 demonstrate a maternal mode of transmission of the resistance, as the mean score for the plants harvested on the resistant crossing partner was significantly lower ($p<0.001$, Mann-Whitney U test) than that for the plants harvested on the susceptible crossing partner. These results reveal the presence of a very stable resistance factor in these plants.

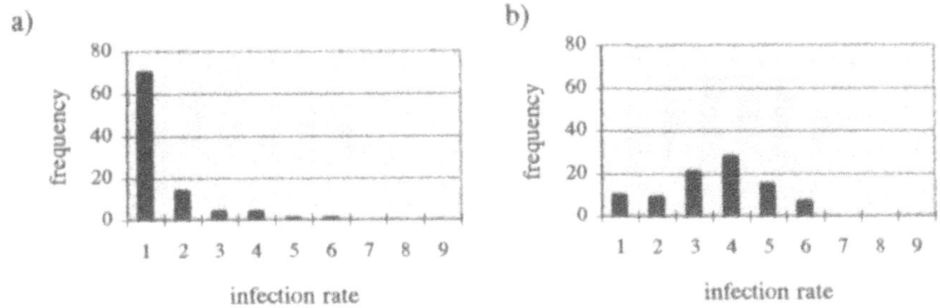

Figure 1. Example of the results obtained after infection of the F_1 offspring from the cross Axis-2xP_1G_3-1. a) F1 Axis-2(xP_1G_3-1); b) F1 P_1G_3-1 (xAxis-2).

Table 2. Summary of the crown rust scores obtained for the F_1 plants of the cross LmR x LpS.

	mean score	standard deviation	# plants analysed	% plants with score 1
F1 LmR(xLpS)	1.2	0.6	87	88
F1 LpS(xLmR)	3.2	1.7	27	26

In the cross where Axis-3 was involved (Axis-3xb90-1), no maternal effects were observed. The results of two infection rounds on 100 plants revealed no differences in average resistance score between the seed stocks (data not shown). In other words, the offspring harvested on the resistant crossing partner displayed average scores similar to the offspring harvested on the susceptible crossing partner (3.4 against 3.7). These results indicate the absence of any maternal effect in the plants used as parents in this cross (Adams et al., submitted). As no maternal effects are found, the results of the plants harvested on both crossing partners can be pooled (Figures 2a and 2b). As seen in Figure 2, the resistance is distributed near-normally.

The results were confirmed by a second analysis (June '98), in which 200 plants from each F1 seed stock were grown and screened once for crown rust (Figure 2c). The high degree of similarity between Figures 2a and 2c, which represent the crown rust scores for two independent samples of the F_1 seed stocks of Axis-3xb90-1 confirm the reproducibility of the infection technique used and the way in which the plants were scored. Also this frequency distribution graph approximates a gaussian distribution. Nevertheless, different authors have offered different definitions for resistance. Wilkins (1975) and Schmidt (1980), considered only those ryegrass plants which did not display any symptom at all as resistant. In Bush & Wise (1996), also plants with some uredinia were considered as resistant against crown rust in cultivated oat. The definition of resistance chosen influences greatly the interpretation of the results. If just those plants with no symptoms are considered resistant, only the plants which got score 1 in our study can be considered as being resistant. If some symptoms are allowed in the resistant plants, plants with score 1 or 2 (or plants with score 1, 2 or 3) should be considered as being resistant.

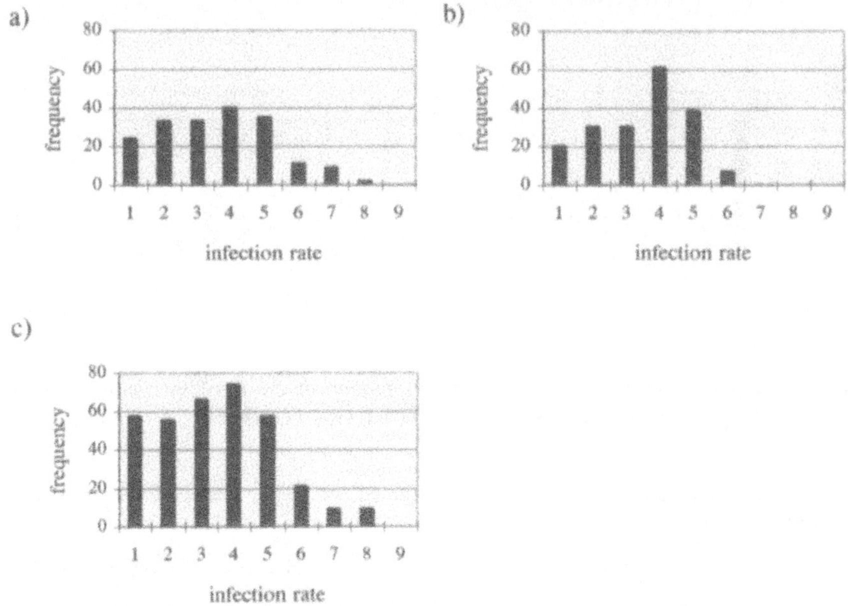

Figure 2. Infection scores of the F_1 progeny of the cross Axis-3xb90-1. a) April '97; b) May '97; c) June '98.

Table 3. Ratio between resistant and susceptible plants if different definitions of resistance are assumed.

	97/04	97/06	98/06
R(1):S(2-9)	24:163(~1:7)	20:166 (~1:8)	57:291 (~1:5)
R(1-2):S(3-9)	57:130 (~1:2)	50:136 (~1:3)	112:236 (~1:2)
R(1-3):S(4-9)	90:97 (~1:1)	80:106 (~1:1)	178:170 (~1:1)

Table 3 gives the ratio between resistant (R) and susceptible (S) plants if no infection at all is accepted (R(1):S(2-9)) and if some infection symptoms are allowed (R(1-2):S(3-9) and R(1-3):S(4-9)). From this Table it is evident that if plants with score 1, 2 or 3 are considered as being resistant, a 1:1 ratio between resistant and susceptible plants is obtained. This could be interpreted as the inheritance of a dominant resistance gene, present in heterozygous state in the resistant crossing partner.

Two hundred F_1 plants from each seed stock from the *L. multiflorum* x *L. perenne* crosses (see Table 1) were subjected to infection. The results of the infection rounds are summarised in Table 4. Unfortunately, a high mortality affected the plants after the first infection round (fifth column of Table 4), what makes the results of the second and third infection rounds less reliable. It is difficult to find an explanation to this high mortality, but probably it was not caused by the infection by crown rust because not infected as well as severely infected plants died. Only the data of the first infection round were included in the statistical analyses (Figure 3).

Table 4. Summary of the crown rust scores obtained for the F_1 seed stocks of the crosses between Axis12 and LpRR and between Axis13 and LpRS. The crosses in which the average crown rust score recorded for the two seed stocks were significantly different with p<0.05 (*) and with p<0.001 (**) in a Mann-Whitney U test are indicated.

Seed stock	infection	mean score	standard deviation	# plants analysed	% plants with score 1
F1 Axis12(xLpRR)	first	2.2*	1.1	170	38
	second	2.0	0.8	167	39
	third	1.8	0.8	165	50
	average	2.0	0.9		
F1 LpRR(xAxis12)	first	1.9*	0.8	148	47
	second	1.7	0.8	80	57
	third	1.5	0.6	51	70
	average	1.7	0.8		
F1 Axis13(xLpRS)	first	4.0**	1.1	175	1.1
	second	3.6	0.7	174	0.6
	third	3.0	0.7	171	3.5
	average	3.5	0.9		
F1 LpRS(xAxis13)	first	3.0**	1.2	153	17
	second	2.0	1.1	80	53
	third	3.0	1.4	52	29
	average	2.7	1.3		

The data of the cross Axis-12xLpRR indicate a maternal mode of transmission of the resistance, as the average score for the plants harvested on the resistant crossing partner was significantly lower (p<0.05, Mann-Whitney U test) than that for the plants harvested on the susceptible crossing partner. This could mean that also in this *L. perenne* plant a maternally transmitted factor could be present, although less pronounced than the factor found in *L. multiflorum*. Nevertheless, the frequency distributions shown in Figures 3a and 3b are not those typically found when a maternal factor is present (Figures 1a and 1b). This, together with the fact that the results are based on only one infection round, lead us to accept these conclusions with reserves.

In the F_1 progeny of the cross Axis-13xLpRS, no maternal inheritance of the resistance was observed. On the contrary, the offspring harvested on the resistant crossing partner Axis-13 were significantly (p<0.001, Mann-Whitney U test) more susceptible than the offspring harvested on the susceptible crossing partner (LpRS). The histograms displayed in Figures 3c and 3d are similar, and approximate normal distributions, what would suggest a polygenic control of the trait.

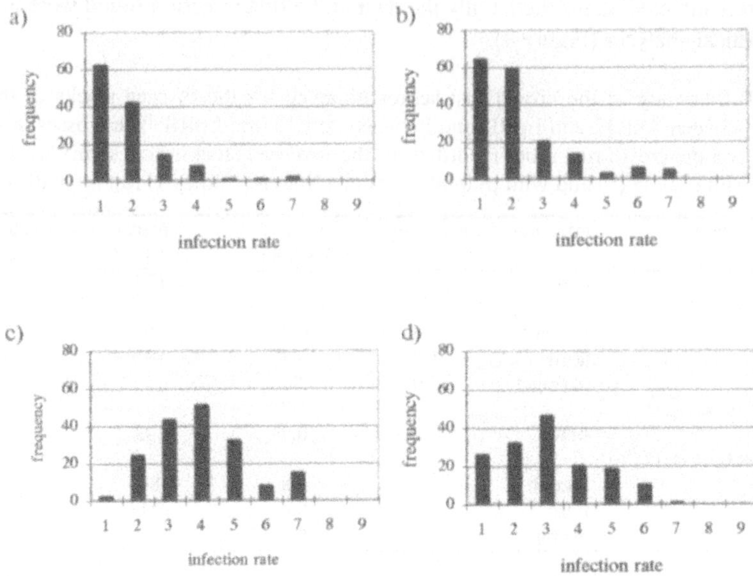

Figure 3. Histograms of the infections performed on a) F_1 LpRR (x Axis-12); b) F_1 Axis-12 (xLpRR); c) F_1 Axis-13 (xLpRS); d) F_1 LpRS (xAxis-13).

4. Discussion

From the data presented it can be concluded that in some resistant plants a cytoplasmic factor is present and that it is transmitted to the progeny in a maternal way. In Adams et al. (submitted) it was hypothesised that if the factor identified was really cytoplasmic and maternally transmitted, it should still be present in the offspring derived from a resistant plant carrying the factor, after several generations. A confirmation of this hypothesis is provided by the analysis of the cross LmRxLpS. LmR, a plant derived after two generations from a cross in which cytoplasmic factors were supposedly present (Adams et al., submitted) is still able of transmitting the resistance to its progeny following a maternal pattern. This proves that the maternal resistance factors present in these plants are stable and transmitted to the progeny in a very predictable way. These results contradict the conclusions reached by Schmidt (1980), who could not find any maternal effect when several crosses were analysed for crown rust resistance. Nevertheless, it is also possible that none of the plants used in her study had got this kind of cytoplasmic factors.

At this moment it is impossible to estimate the frequency of occurrence of cytoplasmic crown rust resistance factors in ryegrass. The high mortality that affected the plants of the cross Axis-12xLpRR does not allow to conclude nor to exclude a maternal transmission of the resistance. If a second analysis of the F_1 progeny of this cross would reveal without any doubt a significantly higher resistance on the progeny harvested on the tolerant crossing partner it could be concluded the cytoplasmic resistance factors to be a more general phenomenon than what was previously suggested in the literature. Strikingly the mortality frequency was higher among the plants harvested on LpRR. This could be related to a higher frequency of self-pollination in this seed stock than in the seed stock harvested on the Axis-12.

As it was mentioned in the introduction, our main objective is to identify molecular markers that could be used in MAS. The results discussed here suggest that in some crosses the search for resistance factors should be performed at organelle DNA level. We are currently optimising an isolation technique for organelle DNA that will allow to apply the AFLP protocol to organelle DNA (Adams et al., 1998).

The fact that the plants harvested on LpRS were more resistant than those harvested on the resistant crossing partner (Axis-13) was unexpected. A possible explanation is that Axis-13 was not resistant for the crown rust race(s) used for infection. Plants were selected in 1993, and only one year later the crown rust spores were harvested for the first time. These results require further investigation.

Complex disease resistance's are generally assumed to be under oligogenic or polygenic control (Mather & Jinks, 1971). Up to now different hypotheses have been offered to explain the mode of inheritance of crown rust resistance in ryegrass. Hayward (1977) could not find clear cut qualitative differences in disease reaction, and considered more appropriate to analyse the data as a quantitative character, although this would not preclude the possibility that the control of the trait is done by a limited number of loci. In Potter et al. (1990) resistance was described as being under polygene or major gene control. The hypothesis of Schmidt (1980) is that resistance in annual ryegrass is conditioned by a dominant gene, which action is inhibited by a lot of complementary and additive minor genes. On the other hand, Wilkins (1975) concluded that in some cases the resistance could be determined by a single major dominant gene and in other cases could be controlled by several minor genes with an additive effect. In the crosses analysed here where no maternal factors were present, the histograms of infection scores follow a Gaussian distribution. Nevertheless, we have also shown that the interpretation of our results could be completely different if a different definition of resistance is applied.

To use a QTL analysis the trait should follow a Gaussian distribution, BSA, on the other hand, is limited to monogenic traits (Lefebvre & Chèvre, 1995) or to major genes of which the action is modified by minor genes. These two approaches could be complementary in the search for molecular markers linked to nuclear factors in ryegrass. Three scenarios are possible: (i) we are dealing with a monogenic trait: a BSA approach should be appropriate to identify molecular markers linked to the gene responsible for the resistance; (ii) we are dealing with trait controlled by a major gene modified by numerous minor genes: a BSA approach where only the plants with the most extreme scores are included could help identify markers linked to major resistance genes. A QTL approach could help identify the chromosome regions where the minor genes are located; (iii) in the case we are dealing with a pure polygenic trait, a BSA approach should be unsuccessful and only a QTL analysis should help identify the location of the genes involved.

Acknowledgements: This work is supported by the 'European GRamineae Mapping Project (EGRAM)'. E.A. was supported by the Flemish Institute for the promotion of the scientific and technologic research in the industry (I.W.T.). The authors thank I. Verelst for technical assistance, the breeders D. Reheul, A. Van Wijk and J. Hintzen for kindly providing plant material and R. Peerbolte for the fruitful discussions.

5. References

Adams, E., Roldán-Ruiz, I., Depicker, A., Van Bockstaele, E. & De Loose, M. (1998) A maternal factor may affect crown rust tolerance in *Lolium multiflorum* Lam. Species. submitted.

Adams, E., Roldán-Ruiz, I., Verelst, I., Van Bockstaele, E. & De Loose, M. (1998) Crown rust tolerance in ryegrasses: a maternal factor may play an important role. Med. Fac. Landbouw. Univ. Gent, in press.

Birckenstaedt, E. (1990) Entwicklung von Methoden fur die Selektion auf kronenrost-resistenz bei *Lolium* spp. Aus phytopathologischer Sight. *PhD thesis*, Bonn.

Bush, A.L. & Wise, R.P. (1996) Crown rust resistance loci on linkage groups 4 and 13 in cultivated oat. *Journal of Heredity* 87, 427-432.

Jaughar, P.P. (1993) Cytogenetics of the *Festuca-Lolium* complex - Relevance to breeding. *Monographs of theoretical and applied genetics*, volume 18, Springer-Verlag, Berlin.

Halden, C., Sall, T., Olsson, K., Nilsson, N.O. & Hjerdin, A. (1997) The use of bulked segregant analysis to accumulate RAPD markers near a locus for beet cyst nematode resistance in *Beta vulgaris*. *Plant Breeding* 116, 18-22.

Hayward, M.D. (1977) Genetic control of resistance to crown rust (*Puccinia coronata* Corda) in *Lolium perenne* L. and its implications in breeding. *Theor Appl Genet* 51, 49-53.

Kopec, D.M., Funk, C.R. & Halisky, P.M. (1983) Sources and distribution of resistance to crown rust within perennial ryegrass. *Plant disease* 67, 98-100.

Lefebvre, V. & Chèvre, A.M. (1995) Tools for marking plant disease and pest resistance genes: a review. *Agronomie* 15, 3-19.

Mather, K. & Jinks, J.L. (1971) *Biomethrical Genetics.* Chapman and Hall, 2nd Edition, London.

Michelmore, R. W., Paran, I. & Kesseli, R. V. (1991) Identification of markers linked to disease-resistance genes by bulked segregant analysis: A rapid method to detect markers in specific genomic regions by using segregating populations. *Proc. Natl. Acad Sci. USA* 88, 9828-9832.

Norusis, N.J. (1993) SPSS for windows: base system. User's Guide, Release 6.0.

Pelsy, F. & Merdinoglu, D. (1996) Identification and mapping of raandom amplified polymprphic DNA markers linked to a rhizomania resistance gene in sugar beet (*Beta vulgaris* L.) by bulked segregant analysis. *Plant Breeding* 115, 371-377.

Potter, L.R., Cagas, B., Paul, V.H. & Birckenstaedt, E. (1990) Pathogenecity of some European collections of crown rust (*Puccinia coronata* Corda) on cultivars of perennial ryegrass. *Journal of Phytopathology* 130, 119-26.

Posselt UK (1994) Genetic aspects of crown rust resistance in the ryegrasses. In: Krohn K, Paul VH, and Thomas J (eds.), *International Conference on Harmful and Beneficial Micro-organisms in Grassland, Pastures and Turf, 17th edition.* IOBC/WPRS, Montfavet, pp. 229-235.

Reheul, D. & Ghesquire, A. (1996) Breeding perennial ryegrasses with better crown rust resistance. *Plant breeding* 115, 465-469.

Schmidt, D. (1980) La sélection du ray-grass d'Italie pour la résistance à la rouille couronnée. *Recherche agronomique en Suisse*, 19 (1, 2), 71-84.

Simons, M.D. (1970) Crown rust of oats and grasses. Monogr. 5. American Phytopathological Society. St Paul, MN. 47 pp.

Tanksley, S.D., Young, N.D., Paterson, A.H. & Bonierbale, M.W. (1989) RFLP mapping in plant breeding: new tools for an old science. *Bio/Technology* 7, 257-264.

Thomas, J. (1991) Diseases of established grassland. In *Strategies for Weed, Disease & Pest Control in Grassland. Proceedings of the BGS Conference, February 1991.* British Grassland Society, Hurley, UK, pp. 3.1-3.12.

Wilkins, P.W. (1975) Inheritance of resistance to *Puccinia coronata* Cda. and *Rhynchosporium orthospurum* Caldwell in Italian ryegrass. *Euphytica* 24, 191-196.

Wilkins, P.W. (1991) Breeding perennial ryegrass for agriculture. *Euphytica* 52, 201-214.

G.T. Scarascia Mugnozza, E. Porceddu & M.A. Pagnotta (Eds.)
Genetics and Breeding for Crop Quality and Resistance, 43-50, 1999

Results of resistance evaluations in Brassicaceae with *Plasmodiophora brassicae*, *Alternaria* and *Phoma lingam*

P. Scholze[1] & K. Hammer[2]
[1]*Federal Centre for Breeding Research on Cultivated Plants; Institute for Breeding Vegetables, Medicinal and Aromatic Plants, Neuer Weg 22/23, D-06484 Quedlinburg, Germany;* [2]*Institute of Plant Genetics and Crop Plant Research, IPK - Genebank -, Corrensstraße 3, D-06466 Gatersleben, Germany*

Abstract: Accessions of *Brassica oleracea*, *B. rapa*, *Raphanus sativus*, *Sinapis alba*, wild relatives and other species of the Brassicaceae, especially provided by the genbank at the IPK Gatersleben, have been tested for resistance to clubroot (*Plasmodiophora brassicae*; two isolations), *Alternaria* leaf blight, and *Phoma*, causing leaf spots. Most of the material reacted highly susceptible to the pathogens, but single plants/species with resistance against clubroot and *Phoma* could be found in savoy, kale, turnips, rutabagas and, especially, in *Raphanus sativus*. Wild relatives, ornamentals and other species showed a broad variability in resistance manifestation. Multiple resistance to the pathogens could be observed in several pathogen combinations, *Alternaria brassicicola*, a very aggressive parasite, inclusively. Results of a test with 10 race populations of clubroot indicated that resistance reactions seem to be based on qualitative as well as quantitative types of resistance. Some problems of stabilizing resistance manifestation are discussed.

1. Introduction

Clubroot (*Plasmodiophora brassicae* Wor.), black leaf spot (*Alternaria brassicicola* [Schw.] Wilt.; *A. brassicae* [Berk.] Sacc.), and blackleg (*Phoma lingam* [Tode ex Fr.] teleomorph: *Leptosphaeria maculans* [Desm.] Ces. et de Not.) are world-wide distributed economically important pathogens causing losses in yield, quality and seed production in cultivated Brassicaceae. Therefore, supplying of varieties with biotic resistance to these pathogens is an important aim for breeding new varieties. At present, concerning cruciferous vegetables, there are only few chances for satisfactory solving this problem since crossing partners which possess efficient absolute or partial resistance transferable into the progeny in a durable manner are not yet available.

In order to find new sources of resistance, evaluations of Brassicaceae has been carried out in the past years in cooparation between the genbank of IPK Gatersleben and the Federal Centre for Breeding Research on Cultivated Plants at Quedlinburg. Results of these investigations are presented in this paper.

2. Materials and Methods

2.1 Plant Material

Most of the material was provided by the Genebank of IPK, Gatersleben (Gladis & Hammer, 1990; Hammer, 1993). Additional material, especially new varieties came from private breeding of Germany, The Netherlands and Japan.

2.2 Screenings

For screenings the following isolates were used: two race populations of clubroot (ECD 16/07/12 and ECD 16/14/31), isolated from cabbage and Chinese cabbage, respectively, one aggressive isolate each of *Alternaria brassicicola* (RA 44/5), *A. brassicae* (Q1) and *Phoma* (RE IV), isolated from *Raphanus sativus*, white cabbage and Chinese cabbage, respectively, in several regions of Germany. All tests were conducted in a greenhouse or growth room under environmental conditions favourable to the growth of host plants and parasites. More detailed informations about the methodical prerequisitions of the tests (propagation and storage of the parasite, growth medium, inoculation techniques) are presented by Scholze (1995) and Scholze & Hammer (1998).

Disease assessment basing on a 0 (without symptoms) to 9 (highly susceptible) scale for all pathogens was carried out after incubation periods of five (*A. brassicicola*), six (*A. brassicae)*, seven (*Phoma*), and 50 (clubroot) days. A disease index (DI) was calculated and a strict standard in classification of resistance applied: clubroot, *A. brassicae* and *Phoma* ≤ DI 1.0; *A. brassicicola* DI ≤ 1.0.

3. Results

3.1 Plasmodiophora

With the exception of var. *costata*, single plants with resistance could be found in all accessions of *B. oleracea*, although most of them showed high mean disease indications as well as mean disease indices. Resistances have been detected in F_1 -hybrid cultivars of cabbage and savoy as well as in out-pollinating varieties of kale after inoculating with the highly aggressive isolate ECD 16/07/12. The stem kale variety 'Ovary' (introduction BRA 1491/18), classified as immune in the test

(Scholze & Hammer, 1998), exhibited massive clubs in a six weeks-growing period after disease assessment.

Whereas all genbank-introductions of *B. rapa* var. *chinensis* (Pak choi) and *pekinensis* (Chinese cabbage) were highly susceptible (mean DI 8.7) to clubroot, the commercially grown Japanese Chinese cabbage varieties 'Chorus', 'Nemesis', 'Marquis', 'Parkin', and 'Shinki' were resistant (DI 0.0 - 0.9). Of more than 60 accessions of var. *oleifera* and *rapa,* 11 entries reacted without symptoms.

In *Raphanus sativus* the resistance to clubroot seems to be expressed more frequently and stronger than in *B. rapa* var. *rapa* and *oleifera*. Of 107 tested accessions especially grown in Europe, 26 (24,3%) were without symptoms (immune), 32 (29.9%) reacted highly resistant.

In contrast to *Raphanus*, the entire material of *Sinapis alba*, of which 23 accessions were introduced in the screening, was highly susceptible. Remarkable variability in resistance manifestation was found in a series of introductions of the family Brassicaceae, which involve some ornamentals, many regional and worldwide distributed weeds as well as some cultivated forms. Several *Brassica* species (*B. carinata*, *B. souliei*, *B. tournefortii*, *B. fructiculosa*, *B. juncea*), all *Camelina* and *Crambe* spec., *Eruca sativa*, *Cochlearia officinalis*, *Lepidium ruderale*, *Capsella grandiflora*, *Diplotaxis tenuifolia*, *Moricandia arvensis* among others were highly susceptible. Resistance reactions were exhibited in a broad spectrum ranging from moderately susceptible (*Rapistrum perenne*, *Iberis amara*, *Rhynchosinapis cheiranthos*, *Sysimbrium officinale* and others) to heavy reduced susceptible (*Barbarea verna*, *B. vulgaris*, some accessions of *Capsella bursa-pastoris*, *Thlaspi arvense*; DI < 1,0) and introductions reacting without symptoms (*Barbarea intermedia*, *Nasturtium officinale*, *Brassica oxyrrhina*). The results could be confirmed by an enlarged test in which 12 accessions of 10 species were separately inoculated with 10 race populations, isolated from *Brassica oleracea* forms in several regions of Germany and Switzerland (Table 1).

Strong differences of symptom manifestations between the host/isolate-interrelations lead to the assumption, that inheritance of resistance is directed by qualitative as well as quantitative genetic bases.

3.2 *Alternaria and Phoma*

Of all 449 accessions of *B. oleracea* tested only one variety (savoy 'Plainpalais') exhibited a remarkable resistance to *A. brassicae*. In the *Phoma* tests, however, resistant single plants, especially in var. *sabauda* and *sabellica*, showing disease indices ranging between 0.5 and 0.8, could be detected. More than 100 entries of *Brassica rapa* (var. *chinensis*, *pekinensis*, *rapa*, and *oleifera*) were screened for *Alternaria* and *Phoma* reaction. The entire material was highly susceptible to *A. brassicicola*, whereas, in tests with *Phoma*, different responses could be observed depending on the stage of the host plants and the environmental conditions under which plants were grown (Scholze & Hammer, 1998). The Chinese cabbage cultivars 'Chorus', 'Shinki', 'Marquis', and 'Asko' exhibited resistance when tested in the 3/4-leaf- stage in the greenhouse but reacted highly susceptible when cut leaf segments of mature plants grown under field conditions were used in the tests. No resistance could be observed in *Raphanus sativus*.

Table 1. Resistance reactions (Disease Index) of some accessions to ten isolates of clubroot *(Plasmodiophora brassicae)*.

Accession	Isolates									
	I	II	III	IV	V	VI	VII	VIII	IX	X
Barbarea intermedia BAR 2/78	0.0	0.0	0.0	0.0	0.0	0.0	0.0	0.0	0.0	0.0
Nasturtium officinale NAS 3/93	0.0	0.0	0.0	0.0	0.0	0.0	0.0	0.0	0.0	0.0
Brassica rapa var. rapa BRA 1015/79	0.0	0.0	0.0	0.2	0.0	0.5	0.5	0.0	0.0	0.0
Barbarea verna BAR 3/78	0.6	0.4	0.4	0.8	0.6	0.4	0.6	0.4	0.8	0.6
Raphanus sativus RA 309/89,	0.2	0.3	0.8	0.7	0.1	0.7	0.7	0.1	1.1	0.4
Brassica nigra 460 (FU)	0.4	0.1	0.1	5.2	0.1	4.6	0.0	0.0	0.0	1.5
Rapistrum perenne CR 1798a	1.0	2.6	0.5	1.1	1.6	1.6	1.4	2.7	1.1	1.1
Hirschfeldia incana HIR 1/81	1.0	1.3	2.1	2.5	3.0	2.6	2.8	1.3	1.9	2.5
Lepidium latifolium CR 1747	0.1	0.2	0.4	6,1	4.0	5.6	3.1	0.0	3.9	0.0
Cochlearia glastifolia COCH 8/84	6.6	6.5	7.0	4.8	6.6	5.9	6.5	6.6	6.8	0.0
Brassica nigra BRA 1165	8.2	8.2	7.4	9.0	8.6	7.9	7.5	5.8	7.5	7.4

White mustard (*Sinapis alba*) is favoured as a potential donor of resistance to *Alternaria*. Therefore, a set of 23 genebank accessions was tested to the *Alternaria* species and, additionally, to *Phoma* in two series: 1. as intact young plants in the 3/4-leaf stage, and 2. by inoculating leaves cut from the same but older plants just before flowering. A broad manifestation could be shown in symptom expression in the different growth stages of the host plant and with regard to both pathogens used in the experiments. Disease indices ranged between 0.0 and 5.8, and, in relation to *A. brassicicola*, *A. brassicae* could be shown as being the more aggressive parasite.

Because of the remarkable modifications in resistance expression only one accession was selected as a possible resistance donor that should be utilized as a combination partner in somatic hybridizations.

Table 2. Reactions of some wild relatives and other species of Brassicaceae to *Alternaria brassicicola* (Aa), *A. brassicae* (Aae), *Phoma* (Pho), and clubroot (Plas).

Introduced species	Accessions	Resistance found		combinations
	tested	Aa	Aae	Pho	Plas
Barbarea intermedia Boreau	1	s	s	r	r
Barbarea vulgaris R.Br.	1	s	r	r	r
Brassica alboglabra L.H. Bailey	3	s	s	r	s
Brassica carinata Braun	4	s	s	s	s
Brassica elongata Ehrh.	1	s	s	r	s
Brassica fructiculosa Cyr.	1	s	s	r	s
Brassica juncea (L.) Czern. & Coss.	15	s	s	r/s	s
Brassica souliei (Batt.) Batt.	1	s	r	r	s
Brassica napus L.	6	s	r	r	s
Brassica napus var. *napobrassica* (L.) Han.	13	s	s	r	s/r
Brassica nigra (L.) Koch	2	s	r	s/r	r
Brassica oxyrrhina (Coss.) Coss.	1	s	s	r	r
Brassica tournefortii Gouan	1	s	s	r	s
Camelina alyssum (Mill.) Thell.	1	s	r	r	s
Camelina microcarpa Andrz. ex DC.	1	s	r	r	s
Camelina sativa (L.) Crantz	3	r	r	r	s
Capsella bursa-pastoris (L.) Medik.	3	s	s	r	s/r
Capsella grandiflora (Fauche & Chaub.) Boiss.	3	r	r	r	s
Capsella rubella Reut.	1	r	r	r	s
Cochlearia glastifolia L.	1	s	s	r	r
Cochlearia officinalis L.	1	s	r	r	s
Crambe abyssinica Hochst. ex Fries.	6	s	r	r	s
Crambe hispanica L.	6	s	r	r	s
Crambe maritima L.	1	s	s	s	?
Eruca sativa Mill.	6	s	s	s/r	s
Hesperis matronalis L.	1	r/s	?	r	s
Hirschfeldia incana (L.) Lagr.-Foss.	1	s	r	r	s
Iberis amara L.	1	s	r	r	r
Isatis tinctoria L.	2	s	s	r	s/r
Lepidium latifolium L.	2	s	s	r	r
Lepidium ruderale L.	5	s/r	s/r	r	s
Matthiola incana (L.) Br.	3	s	s	r	r
Rapistrum perenne (L.) All.	2	s	r	r	r
Rhynchosinapis cheiranthos (Vill.) Dandy	1	s	s	r	r
Sinapis arvensis L.	2	s	s	r	s
Sysimbrium irio L.	1	s	s	r	s
Sysimbrium officinale (L.) Scop.	2	s	r	r	s/r
Thlaspi arvense L.	1	s	r	r	r

As in white mustard, resistance manifestation in wild relatives, ornamentals and other cultivated races of the Brassicaceae exhibited a wide amplitude of ranging. Strong resistance to *A. brassicicola* were only found in all accessions of *Camelina sativa*, *Capsella grandiflora* and *C. rubella,* whereas in *Hesperis matronalis* and *Lepidium ruderale* single plants with reduced susceptibility could be detected. Searches for resistance to *A. brassicae*, that was shown to be less aggressive to cruciferous vegetables than *A. brassicicola* (Scholze, unpublished), were more successful: in 19 accessions single plants with strong resistance or obviously reduced susceptibility could be observed. Concerning *Phoma* leaf spot, all entries showed resistance or strongly reduced susceptibility.

Summarized results of the tests with regard to combinations of resistance to all pathogens including clubroot are presented in Table 2. Of the 39 species tested only one, *Crambe maritima*, was susceptible to all pathogens, whereas 17 and 11 species showed multiple resistance in a double and triple combination, respectively.

4. Discussion

The vast material introduced in the evaluation has been provided by the genebank at the IPK Gatersleben in which nearly 3,500 accessions of the family Brassicaceae are collected (Gladis & Hammer, 1990). In the species *Brassica oleracea*, *B. rapa* and, especially, *Raphanus sativus* several accessions and many single plants exhibiting phenotypically different resistance symptom expressions to one aggressive race population could be found. Whereas in cultivated forms resistance to leaf spots caused by *Phoma* could be observed frequently, searches to find resistance to *Alternaria* spec. were unsuccessful. In contrast to this, wild species, ornamentals as well as other cultivated forms (e.g. *Camelina* spec.) should be potential sources of resistance to all pathogens included (Scholze & Willner, 1997). More detailed informations to the material included in the screenings, the growing and propagating of pathogens as well as the discussion of the results are presented by Scholze & Hammer (1998).

For searching donors with durable resistance more than 80 accessions of *Brassica rapa* var. *rapa*, *Raphanus sativus* and some wild relatives have been tested for reaction against 10 clubroot race populations by inoculation with either the single races or a corresponding mixture. There are good reasons to suppose that the manner of host/parasite interaction is a reliable indication for the genetic basis, involved resistance types as well as the level of durability of the resistance.

Immune reaction to all race populations exhibited by *Barbarea intermedia* and *Nasturtium officinale* indicates resistance directed by one or two strong genes, whereas a weak susceptibility to all race populations, ranging in genotypically different symptom manifestation as shown in *Raphanus sativus* (RA 309/89), *Barbarea verna*, *Rapistrum perenne*, and *Hirschfeldia incana*, seems to be directed by a quantitative type of inheritance. Stronger differentiated responses to the races as exhibited in *Brassica nigra* 460 and *Lepidium latifolium* indicate more complex relations between major gene effects and the genetic (or cytoplasmatic) background as stated by Voorrips & Kanne (1997a, 1997b) and Grandclément & Thomas (1996). As shown in the *Barbarea* accessions, the *Barbarea* genome seems to possess genetic information of different resistance expression. In *Raphanus* accessions, this could be demonstrated by genetic analysis (Scholze, unpublished data). In order to increase resistance stability, Pink et al. (1997) used donors

(Böhmerwaldkohl) that showed resistance to at least three race populations of clubroot. In our attempts to transfer resistance into cabbage, which are carried out in conventional breeding manner as well as via somatic hybridization we use molecular-genetically characterized donors with absolute resistance against ten race populations, later we hope to fall back on material in which resistance is governed by quantitatively acting genes.

In relation to clubroot, more difficult problems are relevant in transferring resistance to *Alternaria* spec. from taxonomically more distant donors to cultivated forms. Even though race forming capacity has not been reported, aggressivity of these parasites, because of their semisaprophytic life cycle, is strongly influenced by environmental and host factors causing uncalculative risks in resistance evaluation (Rotem, 1994). Host plants (e.g. White mustard) which have been raised under various growth conditions varied in resistance response ranging between immunity and moderate susceptibility (accession 41/83; Scholze & Hammer, 1998). These differences are probably reduced due to a less doses effect of the quantitative genetic base. As a rule, disease expression is more or less permanent in highly susceptible hosts and there is reason to believe that hosts expressing a high degree of statistically significant changing rates indicate the ocurring of resistance. These resistances are transferable into cultivated forms via somatic hybridization (Sjödin & Glimelius, 1988; Tewari at al., 1987; Plümper & Sacristan, 1994; Ryschka et al., 1996 a.o.). In agreement with Hansen & Earle (1997) we found that, as in the combination partners, resistance transfered in somatic hybrids is being changed considerably and, after our experience, it can get lost entirely even before flowering. Therefore, repeated tests of the regenerates during growing period are essential. Stabilizing of resistance manifestation in different growth stages as well as under various environmental conditions is an important prerequisition for successful transferring into the recurrent parent. If necessary, combinations between differential resistance donors should be initiated in order to increase and stabilize dosis effects of resistance.

From this point of view, the utilization of multiple resistance appears very difficult. At present, somatic hybrids from combinations between *Barbarea* and cabbage are tested to elucidate means of resistance expression to several pathogens. When the *Barbarea* parents are genetically analysed, further treatment of the material in combination with molecular-genetic techniques is planned to find out adequate prebreeding strategies.

5. References

Gladis, T. & Hammer, K. (1990) Die Gaterslebener *Brassica*-Kollektion - eine Einführung, Kulturpflanze 38, 121-156.

Grandclément, C. & Thomas, G. (1996) Detection and analysis of QTLs based on RAPD markers for polygenic resistance to *Plasmodiophora brassicae* Woron. in *Brassica oleracea* L., Theor. Appl. Gen. **93**, 86-90.

Hammer, K. (1993) The 50th anniversary of the Gatersleben genebank, FAO/IBPGR Plant Genetic Resources Newsl. 91/91, 1-8.

Hansen, L.N. & Earle, E.D. (1997) Somatic hybrids between *Brassica oleracea* L. and *Sinapis alba* L. with resistance to *Alternaria brassicae* (Berk.) Sacc., Theor. Appl. Gen. **94**, 1078-1085.

Pink, D.A.C., King, G.J. & Ockendon, D.J. (1997) Use of doubled haploids and molecular markers in breeding for resistance to clubroot, Abstr. ISHS Symposium 10th Crucifer Gen. Workshop, Rennes (France), 120.

Plümper, B. & Sacristan, D.M. (1994) Resistance to *Alternaria brassicae* in n = 11 Brassica species and its transfer to *Brassica napus*, Abstr. ISHS Symposium on Brassicas 9th Crucifer Gen Workshop, Lisbon (Portugal), 90.

Rotem, J. (1994) *The Genus Alternaria,* The American Phytopathological Society, St. Paul, Minnesota.

Ryschka, U., Schumann, G., Klocke, E., Scholze, P. & Neumann, M. (1996) Somatic hybridization in *Brassicaceae*, Acta Hort. **407**, 201-208.

Scholze, P. (1995) Einführung eines Prüfverfahrens zur Recherche nach Resistenz gegen *Alternaria* und *Phoma* in progenerativen Stadien von Brassicaceen, Jahresber. Bundesanst. Züchtungsforsch. an Kulturpflanzen, Quedlinburg, 106.

Scholze, P. & Hammer, K. (1998) Evaluation of resistance to *Plasmodiophora brassicae*, *Alternaria* and *Phoma* in Brassicaceae, Acta Hort. **459,** 363-369.

Scholze, P. & Willner, E. (1997) Wildformen von Brassicaceen - potentielles Reservoir für Resistenzdonoren, Angewandte Wissenschaft Heft **465**, 369-371.

Sjödin, C. & Glimelius, K. (1988) Screening for resistance to blackleg *Phoma lingam* (Tode ex Fr.)Desm. within Brassicaceae, J. Phytopathol. **123**, 322-332.

Tewari, J., Conn, K.L. & Dahiya, J. (1987) Resistance to *Alternaria brassicae* in crucifers, 7th Intern. Rapeseed Congr. Poznan, Poland, 48.

Voorrips, R. & Kanne, H.J. (1997a) Genetic analysis of resistance to clubroot (*Plasmodiophora brassicae*) in *Brassica oleracea*. I. Analysis of symptom grades, Euphytica 93, 31-39.

Voorrips, R. & Kanne, H.J. (1997b) Genetic analysis of resistance to clubroot (*Plasmodiophora brassicae*) in *Brassica oleracea*. II. Quantitative analysis of root symptom measurements, Euphytica 93, 41-48.

G.T. Scarascia Mugnozza, E. Porceddu & M.A. Pagnotta (Eds.)
Genetics and Breeding for Crop Quality and Resistance, 51-59, 1999

Breeding research on resistance to Fusarium head blight in wheay

P. Ruckenbauer[1,2], H. Bürstmayr[1], H. Grausgruber[1] & M. Lemmens[2]
[1] University of Agricultural Sciences, Dept. Plant Breeding, Gregor Mendel Str. 33, A-1180 Vienna, Austria; [2] Institute for Agrobiotechnology, Dept. Plant Biotechnology, Konrad Lorenz Str. 20, A-3430 Tulln, Austria

Abstract: Since 1989 resistance breeding research is concentrated on Fusarium diseases in wheat and maize. Research on Fusarium head blight (FHB) resistance of wheat has been focused on the following topics: i) Sources of resistance and production of breeding material: A multitude of cultivars was tested for their FHB resistance in several years' field trials. Resistance sources were identified in Chinese, Japanese, South American and European wheats and crossed with adapted Austrian cultivars. Advanced breeding lines were provided to Austrian wheat breeders for further selection. ii) Artificial inoculation methods and screening procedures: A collection of a wide range of Fusarium species has been established as single-spore isolates from naturally infested crop material. Isolates with stable high virulence were identified and are now used for artificial inoculation. Reliability of resistance screening was improved by using a mist-irrigation system for the field trials in order to provide optimal humidity, and by determination of several resistance components. iii) Genetics and inheritance of FHB resistance: Chromosomes carrying major resistance genes were identified in diverse genetic backgrounds by backcross reciprocal monosomic analysis and the study of several sets of intervarietal chromosome substitution lines. Doubled haploid populations from crosses between resistant and susceptible genotypes were established for the search of molecular markers for both FHB resistance and toxin tolerance. iv) In vitro selection methods for FHB resistance: Assays based on the germination of seeds and/or the electrolyte leakage of leaves in presence of Fusarium toxins are evaluated for their possibility to select genotypes with improved FHB resistance.

1. Introduction

Graminaceous *Fusarium* species comprise pathogens that affect all classes of wheat and cause seedling blight, crown and foot rot, leaf blotch and head blight (head scab). Major problems arise in case of head blight, as diseased grain can be contaminated with mycotoxins which represent a potential risk to animal and human health.

Austrian research activities on the resistance to Fusarium head blight (FHB) in wheat started in 1989 at the University of Agricultural Sciences Vienna. Since 1994 the activities are continued at the Institute for Agrobiotechnology (IFA) Tulln. The following topics were and are still addressed in this research programme:

- improvement of artificial inoculation techniques including single ear inoculation techniques (Grausgruber et al., 1995; Lemmens et al., 1995),
- influence of *Fusarium* species and strain on resistance (van Eeuwijk et al., 1995),
- screening of Austrian grown cultivars and breeding lines (Lemmens et al., 1993; Buerstmayr et al., 1996a),
- collection and evaluation of germplasm from worldwide co-operators (Lemmens et al., 1993; Buerstmayr et al., 1996b),
- crossing program between FHB resistant 'exotic' lines and well adapted Austrian genotypes in co-operation with Austrian wheat breeders,
- studies on the inheritance of resistance (Buerstmayr et al., 1999; Grausgruber et al., 1998) and search for molecular markers (Buerstmayr et al., 1998),
- study of the relation between FHB resistance parameters based on observation of disease symptoms on the ears and/or grains and the final mycotoxin contamination in the grains (Lemmens et al., 1997),
- investigations on the role of *Fusarium* mycotoxins in the host-pathogen interaction and their effect on plant defence mechanisms, and
- development of *in-vitro* selection methods (Lemmens et al., 1994).

The results of some of these programmes are presented in detail in the present paper.

2. Non-specifity of Fusarium head blight resistance

In 1990 research institutes from five European countries (CPRO-DLO Wageningen, the Netherlands; GKI Szeged, Hungary; INRA Rennes, France; University Hohenheim, Germany, and University of Agricultural Sciences Vienna, Austria) formed a winter wheat nursery consisting of 25 *Triticum aestivum* L. cultivars and/or breeding lines. This nursery, which covered the range from resistant to susceptible, was tested in each country over three growing periods. Seventeen local strains of *F. culmorum, F. graminearum* and *F. nivale* were used for artificial inoculation.

Infestation scores (%FHB) of the single environments were combined in a genotype by environment (25 x 59) two-way table for overall analysis. Environments were defined by the strain by location by year combinations (van Eeuwijk et al., 1995). Mean %FHB of each genotype are presented in Table 1, representing low values for resistant genotypes and high values for susceptible genotypes. It was demonstrated that resistance to FHB in winter wheat is of the horizontal type, thus non-strain and non-species specific. Hence, resistance breeding against one causal pathogen will be sufficient and any reasonable aggressive strain, a *F. culmorum* strain in the cool climates and a *F. graminearum* strain in the warmer humid areas, should be satisfactory for screening purposes.

Table 1. Infestation scores (%FHB) of the 25 winter wheat genotypes investigated in five European countries.

Genotype	Origin	% FHB
Sgv/GT-Pdj*Uhr GK	GKI Szeged, Hungary	47
85-92	GKI Szeged, Hungary	48
Bence	GKI Szeged, Hungary	47
Zombor	GKI Szeged, Hungary	65
Szöke	GKI Szeged, Hungary	48
163/81/03	LSA Hohenheim, Germany	48
204/81/03	LSA Hohenheim, Germany	58
77/82/01	LSA Hohenheim, Germany	62
25/83/02	LSA Hohenheim, Germany	64
47/83/02	LSA Hohenheim, Germany	54
SVP 75059-28	CPRO-DLO Wageningen, the Netherlands	40
SVP 75059-32	CPRO-DLO Wageningen, the Netherlands	56
SVP 72017-5-10	CPRO-DLO Wageningen, the Netherlands	31
SVP 72005-20-30-1	CPRO-DLO Wageningen, the Netherlands	67
Arina	FAP Zürich-Reckenholz, Switzerland	31
Copain	Ets. C.C. Benoist, France	45
Rescler	Agri-Obtentions, France	55
Renan (RC 103)	Agri-Obtentions, France	41
82 F3 28	INRA, France	34
81 F3 79	INRA, France	35
P 2119.89	Probstdorfer Saatzucht, Austria	58
P 4371.88	Probstdorfer Saatzucht, Austria	64
NR 172/90	Saatbau Neuhof, Austria	51
Justus (SL 8/80-28)	Saatbau Linz, Austria	54
SL 34/81-12	Saatbau Linz, Austria	50

3. Improvement of the artificial inoculation method for FHB testing

Natural epidemics of Fusarium head blight (FHB) on wheat occur sporadic in time and locations and even within a field disease incidence is not uniform (Fernando et al., 1997). Several artificial inoculation methods for wheat have been described; for review, see Miedaner (1997). Since wheat is most susceptible to FHB at anthesis, experimental inoculations are carried out at flowering. Different wheat genotypes usually do not flower at the same time. They have to be inoculated according to flowering time. Infection success after artificial inoculation, however, depends on the climatic conditions which might be changing daily. This results in considerable genotype by environment interactions.

We have compared eight different methods of artificial inoculation. The aim was to select the inoculation method which was the least depending on environmental conditions. Therefore, the nursery of 25 European winter wheats described above was sown in 1994 and 1995 in two and/or three replications, which also represented different sowing dates. Artificial inoculation was carried out at anthesis with both a

F. graminearum and a *F. culmorum* inoculum consisting of mycelium and macroconidia, respectively. Twenty millilitre of *Fusarium* suspension was sprayed on bunches of 25 flowering ears. The following inoculation methods were tested: M1, inoculation at midday and subsequent covering of the bunches with a plastic bag for 24 hours; M2, inoculation in the evening and subsequent covering of the bunches with a plastic bag for 18 hours; M3, inoculation in the evening, and M4, inoculation in the evening followed by a second inoculation two days later. All four inoculation methods were performed with and without mist irrigation. Disease severity was assessed five times by visual evaluation of the percentage of diseased spikelets per bunch. The area under the disease progress curve (AUDPC) was calculated and used for further analysis. Accuracy and repeatability for each inoculation method and inoculation method by inoculum combination were calculated. In addition, 15 ears of each bunch were harvested for mycotoxin analysis. Highly significant differences between wheat genotypes, inoculation methods and isolates were detected in both years. Both accuracy ($r = 0.62$-0.83) and repeatability ($h^2 = 0.61$-0.91) were highest for the inoculation methods without bag covering. It was demonstrated that on sunny days the temperature in the bags rises above 45°C. Under such conditions infection success decreased dramatically, reducing both accuracy and repeatability of the inoculation methods using bags to provide humidity. The *F. culmorum* strain, which was a mere macroconidia suspension, was superior to the *F. graminearum* mycelium suspension. The content of deoxynivalenol (DON) in the harvested samples is analysed and the influence of the inoculation technique on DON content is investigated.

4. Studies on the inheritance of resistance

So far, all genetic studies have indicated that resistance to FHB is under oligogenic control. Most studies on chromosomal location of resistance genes have been carried out in Chinese spring wheats (Yu, 1982, 1991; Liao & Yu, 1985; Yu et al., 1986). These studies, however, considered only head infection severity by visual evaluation of the percentage of head blight symptoms. Therefore, only resistance to initial infection and fungal spread (Schroeder & Christensen, 1963) were evaluated. Buerstmayr et al. (1999) considered in their back-cross reciprocal monosomic analyses also grain damage and deoxynivalenol contamination. Grausgruber et al. (1998) investigated resistance to head infection and yiel loss (tolerance), and *in vitro* toxin tolerance using inter-varietal single chromosome substitution lines. Considering the results of all chromosomal location studies it can be summarized that chromosomes 3B, 5A and 6B seem to carry the most effective resistance genes, since they were reported most frequently. Recently, using AFLPs and RFLPs Anderson et al. (1998) reported also genomic regions on the short arms of chromosomes 3B and 6B to be associated with FHB resistance. The number of two to three major resistance genes is also in correspondence with the quantitative resistance studies carried out by Zhou et al. (1987), Bai et al. (1989), Singh et al. (1995) and van Ginkel et al. (1996).

Molecular markers for FHB resistance offer a promising tool for more efficient breeding and may reduce the need for tedious resistance screening in the field. Therefore, we are currently developing markers (RFLP, AFLP) for this resistance trait (Buerstmayr et al., 1998).

5. Possibilities for *in vitro* selection

Trichothecenes, like deoxynivalenol (DON), are *Fusarium* mycotoxins generally produced after invasion of the wheat plant. It has been reported that plants tolerant to these toxins have also an increased FHB resistance. Mycotoxins probably play a role in the aggressiveness of the pathogen and promote disease development and colonisation by interfering with the plant's defence mechanisms which require *de nuovo* protein synthesis (Snijders & Krechting, 1992). Tests with trichothecene-deficient mutants of *F. graminearum*, however, were inconsistent over genotypes in causing less disease than their wild-type progenitor strain (Proctor et al., 1995).

Several methods are described to select wheat plants with increased level of toxin tolerance. Either crude toxic metabolite extracts or highly purified *Fusarium* toxins have been used. Selection was carried out at the level of germinating seeds and seedlings, callus, coleoptiles or microspores. High correlations between *in vitro* tests and FHB resistance have been reported by Wang & Miller (1988), Wakulinski (1989) and Guo Lijuan et al. (1991).

To measure differences in toxin sensitivity between wheat genotypes an assay was developed based on the time course of germination of seeds on toxin containing medium. The 25 European winter wheat genotypes described above were used for these investigations, since their resistance level was meticulously investigated and can be regarded as most precise. The bioassay was continuously optimised for the best correlation with the known field resistance data. Surface sterilised seeds (100 per petri-dish, 2 replications) were placed with the embryo upwards on medium containing 10 to 30 ppm DON and on control medium. Germination took place in the dark at 5°C. The cumulative number of germinated seeds was counted daily, plotted against days after imbibition and subsequently fitted to a Gompertz function. As a measure for toxin tolerance the number of germinated seeds in the toxin medium when in the control 90% of the seeds were germinated (NGT_{C90}) was used (Lemmens et al., 1994).

Reproducible and statistically significant differences in toxin tolerance between the wheat cultivars were observed. If DON contributes significantly to the aggressiveness of the fungus, differences in toxin tolerance of wheat genotypes are expected to correlate with differences in FHB resistance. Correlation coefficients between field resistance data and NGT_{C90} varied from -0.60 to -0.85 depending on the year of testing (3 growing periods). An example is illustrated in Figure 1 (r = -0.77). This result suggests that tolerance to DON is one mechanism contributing to the quantitative nature of FHB resistance. Therefore, the role of DON as an aggressiveness factor is expected to be minimal in highly toxin tolerant cultivars.

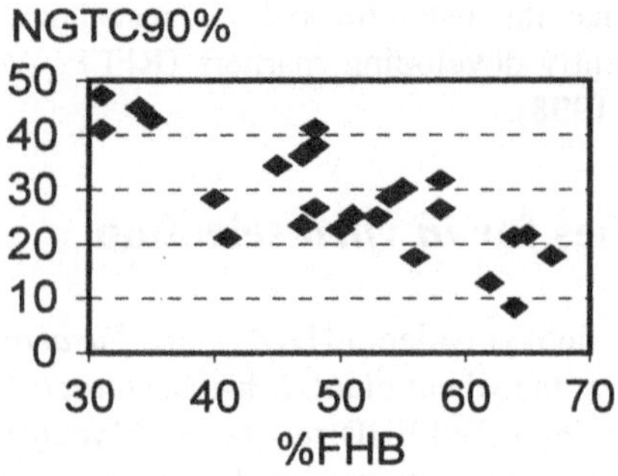

Figure 1. Relation between field resistance (%FHB) and DON tolerance (NGTC90) for 25 winter wheats.

This bioassay was evaluated with 100 different wheat genotypes, varying in their resistance level from highly resistant to very susceptible, thus covering the whole range of FHB resistance known to date (Buerstmayr et al., 1996b). A highly significant correlation coefficient between field resistance and toxin tolerance was calculated. Coefficient of determination (R^2) was about 0.36, indicating that about 36% of the observed variability in field resistance can be explained by DON tolerance. This also suggests that besides toxin tolerance other resistance mechanisms have to be present. Currently we are checking whether chitinase and glucanase activities in the plant are contributing to the quantitative resistance of wheat towards FHB.

For this bioassay, based on the germination of seeds, at least 200 seeds of a homogeneous wheat line are required. A method based on electrolyte leakage measurements, testing toxin tolerance of a single plant only, has now been established and is currently evaluated.

6. Investigations on the relationship between disease symptoms and mycotoxin content

Quality loss because of toxin contamination of the harvested grain is probably the worst consequence of FHB. The main toxins found in Austria are deoxynivalenol (DON), moniliformin (MON) and zearalenone (ZON). *Fusarium* spp. producing more toxic metabolites such as T2-toxin and fumonisins play only a minor role in Austria (Lew, 1993). The goal of any resistance breeding program against FHB is improvement of the resistance with a simultaneous reduction of the toxin content. The relationship between disease symptoms and DON content is not well understood. Mesterházy & Bartók (1992) reported that genotypes with a similar FHB resistance level can have strikingly different concentrations of DON. Toxin analyses are costly and time consuming and hence it is impossible to use them routinely for practical breeding purposes where hundreds of lines should be tested.

A parameter which correlates closely with the toxin content after artificial inoculation would be very valuable for making a preselection.

The aim of this work was to identify a disease parameter with which the mycotoxin content in kernels of artificially inoculated wheat genotypes can be predicted for breeding purposes. In total 108 winter and spring wheat genotypes, the resistance of which varied from highly resistant to very susceptible, were artificially inoculated with three different *Fusarium* spp.: (a) a *F. culmorum* isolate known to produce large amounts of DON, (b) a *F. avenaceum* isolate producing large amounts of MON and (c) a *F. graminearum* isolate capable of producing ZON. Several field and laboratory parameters (AUDPC, percentage of *Fusarium* damaged kernels (FDK), density of wheat kernels etc.) were investigated and the data were correlated with the toxin content of the harvested wheat kernels.

Correlation coefficients between disease parameters and DON-contamination varied from 0.33 to 0.81. It is concluded that for practical breeding purposes visual scores of disease symptoms on the field (*e.g.* AUDPC, r = 0.81; see also Figure 2) or on harvested grain (*e.g.* FDK, r = 0.76) give a good estimation of the DON content for a reasonable input of labour. Similar conclusions can be drawn for contamination with MON: the highest correlation coefficient (r = 0.7) was found between MON-content in the harvested wheat probes and FDK. Low correlation coefficients (r = 0.18-0.39), however, were found between ZON-content and disease parameters. Only genotypes with a very high level of FHB resistance consistently showed low ZON-contents.

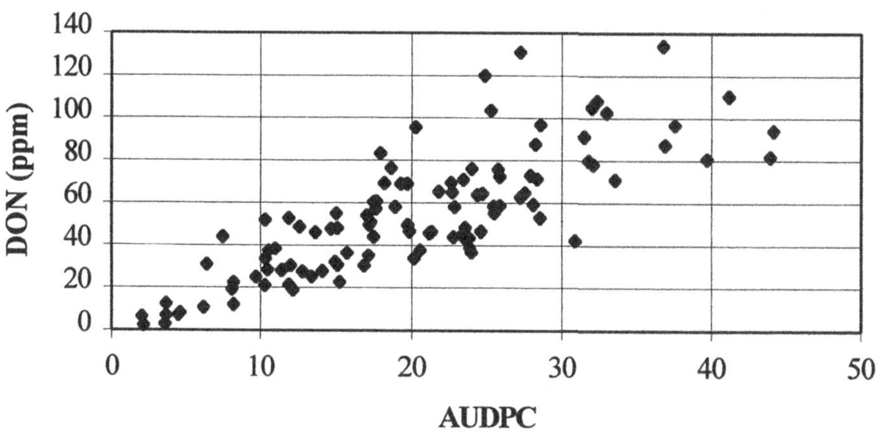

Figure 2. Relation between DON-content (ppm) and AUDPC of 108 winter and spring wheats artificially inoculated with a *F. culmorum* isolate (data represent means of two replications).

Acknowledgements: We gratefully acknowledge the financial support from Austrian Science Funds (projects P9190-BIO and P11884-GEN), Austrian National Bank (project 4906) and Ministry of Agriculture (projects L687/91 and L850/94).

7. References

Anderson, J.A., Waldron, B.L., Moreno-Sevilla, B., Stack, R.W. & Frohberg, R.C. (1998) Detection of Fusarium head blight resistance QTL in wheat using AFLPs and RFLPs, in A.E. Slinkard (ed.), *Proc. 9th Int. Wheat Genet. Symp., Vol. 1*, Univ. Extension Press, Univ. Saskatchewan, Saskatoon, pp. 135-137.

Bai, G.H., Xiao, Q.P. & Mei, J.F. (1989) Studies on the inheritance of scab resistance in six wheat varieties, *Acta Agric. Shanghai* **5**, 17-23.

Buerstmayr, H., Lemmens, M., Grausgruber, H. & Ruckenbauer, P. (1996a) Head blight (*Fusarium* spp.) resistance of wheat cultivars registered in Austria, *Bodenkultur* **47**, 183-190.

Buerstmayr, H., Lemmens, M., Grausgruber, H. & Ruckenbauer, P. (1996b) Scab resistance of international wheat germplasm, *Cereal Res. Comm.* **24**, 195-202.

Buerstmayr, H., Doldi, L., Lemmens, M., Steiner, B., Berlakovich, S. & Ruckenbauer, P. (1998b) Analysis of Fusarium head blight (scab) resistance in wheat by classical, cytogenetical and molecular tools, in A.E. Slinkard (ed.), *Proc. 9th Int. Wheat Genet. Symp., Vol. 1*, Univ. Extension Press, Univ. Saskatchewan, Saskatoon, pp. 197-199.

Buerstmayr, H., Lemmens, M., Fedak, G. & Ruckenbauer, P. (1999) Back-cross reciprocal monosomic analysis of Fusarium head blight resistance in wheat (*Triticum aestivum* L.), *Theor. Appl. Genet.* **98**: 76-85.

Fernando, W.G.D., Paulitz, T.C., Seaman, W.L., Dutilleul, P. & Miller, J.D. (1997) Head blight gradients caused by *Gibberella zeae* from area sources of inoculum in wheat field plots, *Phytopathology* **87**, 414-421.

Grausgruber, H., Lemmens, M., Bürstmayr, H. & Ruckenbauer, P. (1995a) Evaluation of inoculation methods for testing *Fusarium* head blight resistance of winter wheat on single plant basis, *Bodenkultur* **46**, 39-49.

Grausgruber, H., Lemmens, M., Bürstmayr, H. & Ruckenbauer, P. (1998) Chromosomal location of Fusarium head blight resistance and *in vitro* toxin tolerance in wheat using the Hobbit 'sib' (*Triticum macha*) chromosome substitution lines, *J. Genet. Breed.*, in press.

Guo Lijuan, Yao Qingxiao, Hu Qide, Zhang Hao, Deng Fuyou, Zheng Huaquan & Huang Wufang (1991) Studies on screening resistant mutants of wheat to *Fusarium graminearum* by tissue culture, *Genet. Manipul. Plants* **7**, 25-33.

Lemmens, M., Buerstmayr, H. & Ruckenbauer, P. (1993) Variation in *Fusarium* head blight susceptibility of international and Austrian wheat breeding material, *Bodenkultur* **44**, 65-78.

Lemmens, M., Reisinger, A., Buerstmayr, H. & Ruckenbauer, P. (1994) Breeding for head blight (*Fusarium* spp.) resistance in wheat: development of a mycotoxin-based selection method of seedlings, *Acta Horticulturae* **355**, 223-232.

Lemmens, M., Bürstmayr, H., Wimmer, E., Grausgruber, H. & Ruckenbauer, P. (1995) Ährenfusariose bei Weizen: Vergleich mehrerer künstlicher Inokulationstechniken, in *Bericht 46. Züchtertagung Vereinigung Österreichischer Pflanzenzüchter*, BAL Gumpenstein, pp. 235-237.

Lemmens, M., Josephs, R., Schuhmacher, R., Grausgruber, H., Buerstmayr, H., Ruckenbauer, P., Neuhold, G., Fidesser, M. & Krska, R. (1997) Head blight (*Fusarium* spp.) on wheat: investigations on the relationship between disease symptoms and mycotoxin content, *Cereal Res. Comm.* **25**, 459-465.

Lew, H. (1993) Die aktuelle Mykotoxinsituation in der heimischen Landwirtschaft, *Veröff. Bundesanst. Agrarbiol. Linz* **21**, 5-26.

Liao, Y.C. & Yu, Y.J. (1985) Genetic analysis of scab resistance in the local wheat variety Wang Shui Bai, *J. Huazhong Agric. Coll.* **4**, 6-14.

Mesterházy, Á. & Bartók, T. (1992) Resistance and pathogenicity influencing toxin (DON) contamination of wheat varieties following Fusarium infection, in *Proc. 3rd Europ. Fusarium Sem., Part II*, IHAR Radzików, Poland, pp. 9-16.

Miedaner, T. (1997) Breeding wheat and rye for resistance to Fusarium diseases. *Plant Breeding* **116**, 201-220.

Proctor, R.H., Hohn, T.M., and McCormick, S.P. (1995) Reduced virulence of *Gibberella zeae* caused by disruption of a trichothecene toxin biosynthetic gene, Mol. Plant Microbe Interact. 8, 593-601.

Schroeder, H.W. & Christensen, J.J. (1963) Factors affecting resistance of wheat to scab caused by *Gibberella zeae*, *Phytopathology* **53**, 831-838.

Singh, R.P., Ma, H. & Rajaram, S. (1995) Genetic analysis of resistance to scab in spring wheat cultivar Frontana, *Plant Dis.* **79**, 238-240.

Snijders, C.H.A. & van Eeuwijk, F.A. (1991) Genotype x strain interactions for resistance to Fusarium head blight caused by *Fusarium culmorum* in winter wheat, *Theor. Appl. Genet.* **81**, 239-244.

Snijders, C.H.A. & Krechting, C.F. (1992) Inhibition of deoxynivalenol translocation and fungal colonization in *Fusarium* head blight resistant wheat, *Can. J. Bot.* **70**, 1570-1576.

van Eeuwijk, F.A., Mesterházy, Á., Kling, C.I., Ruckenbauer, P., Saur, L., Buerstmayr, H., Lemmens, M., Keizer, L.C.P., Maurin, N. & Snijders, C.H.A. (1995) Assessing non-specificity of resistance of wheat to head blight caused by inoculation with European strains of *Fusarium culmorum, F. graminearum and F. nivale*, using a multiplicative model for interaction, *Theor. Appl. Genet.* **90**, 221-228.

van Ginkel, M., van der Schaar, W., Yang, Z. & Rajaram, S. (1996) Inheritance of resistance to scab in two wheat cultivars from Brazil and China, *Plant Dis.* **80**, 863-867.

Wakulinski, W. (1989) Phytotoxicity of the secondary metabolites of fungi causing wheat head fusariosis (head blight), *Acta Physiol. Plant.* **11**, 301-306.

Wang, Y.Z. & Miller, J.D. (1988) Effects of *F. graminearum* metabolites on wheat tissue in relation to *Fusarium* head blight resistance, *J. Phytopath.* **122**, 118-125.

Yu, Y.J. (1982) Monosomic analysis for scab resistance and yield components in the wheat cultivar Soo-mo 3, *Cereal Res. Comm.* **10**, 185-189.

Yu, Y.J. (1991) Genetic analysis for scab resistance in five wheat varieties, Ping Hu Jian Zi Mai, Hong Hu Da Tai Bao, Chong Yang Hong Mai, Yan Gang Fang Zhu and Wan-Nian 2, *Acta Agron. Sinica* **17**, 248-254.

Yu, Y.J., Liao, Y.C. & Li, Y.F. (1986) Monosomic analysis for scab resistance genes in wheat cultivars, in *Proc. 1st Int. Symp. Chromosome Engineering in Plants*, Xian, China, pp. 38-40.

Zhou, C.F., Xia, S.S., Qian, C.M., Yao, G.C. & Shen, J.X. (1987) On the problem of wheat breeding for scab resistance, *Scientia Agric. Sinica* **20**, 19-25.

G.T. Scarascia Mugnozza, E. Porceddu & M.A. Pagnotta (Eds.)
Genetics and Breeding for Crop Quality and Resistance, 61-66, 1999

Genetics of durabel adult plant resistance to rust diseases in wheat (*Triticum aestivum* L.)

A. Börner[1], O. Unger[2] & A. Meinel[2]
[1]Institut für Pflanzengenetik und Kulturpflanzenforschung (IPK), Corrensstrasse 3, D-06466 Gatersleben, Germany; [2]NORDSAAT Saatzuchtgesellschaft mbH, Hauptstrasse 1, D-38895 Böhnshausen, Germany

Abstract: Adult plant resistance against rusts is controlled by both specific and non specific resistance genes. Although the progress in breeding and research by the exploitation of major resistance genes is evident, the vulnerability of specific resistance by a quick adaptability of the pathogen is a principal disadvantage. In a wheat breeding program for disease resistance a few lines were selected having a stable non specific adult plant resistance against stripe rust and leaf rust. F_3 progenies of 150 - 160 F_2 plants of four crosses were analyzed within two years. For three crosses it is shown that the durable adult plant resistance against rusts in wheat is determined by major genes.

1. Introduction

Resistance to yellow rust (*Puccinia striiformis*) and leaf rust (*Puccinia recondita*) in wheat can be divided into specific and non specific resistance. Specific resistance effective at the seedling and adult plant stage is assumed to follow the gene-for-gene relationship (Floor, 1959). The combination of different specific resistance genes resulted in many cases in a remarkable progress of stability of resistance in different regions and over some years. However, the vulnerability of specific resistance by a quick adaptability of the pathogen is a principal disadvantage.

As an alternative, plant breeders are interested in non specific resistance, present in plants that are susceptible at the seedling stage but resistance at the adult plant stage. The genetics of this adult plant resistance have so far been studied only by a few research groups including, Barina & McIntosh (1995), Line et al. (1996) and Law & Worland (1997). This paper describes the identification of loci determining non specific resistance against yellow and leaf rust by using wheat lines identified from a 20 years program of resistance breeding against rusts at the breeding station at Böhnshausen. The expression of the adult plant resistance of these lines was not influenced by any change of virulences since the 1970s (Meinel & Unger 1998).

2. Material and Methods

For studying the inheritance of the adult plant resistance segregating populations were produced using two lines resistant to yellow rust and two lines resistant to leaf rust. The four resistant lines are characterized by susceptibility against the actual virulent races at the seedling stage tested in the green house indicating that no specific resistance gene is present. At the adult plant stage, however, the lines are resistant against a mixture of the most virulent races. The characterization of the resistance of the parents is given in Table 1.

Table 1. Seedlings and adult plant resistance of the parents classified as non specific resistant or susceptible to yellow or leaf rust. Coefficients of infection were calculated as described by Stubbs et al. (1986).

Disease	Parents	Seedlings test	Adult plant test (1979-1998)	Coefficient of infection Adult plant test (1979-1998)
Yellow rust	Lgst.79-74	susceptible	resistant	0-4.0
	Lgst.209-75	susceptible	resistant	0-4.0
	Winzi	susceptible	susceptible	80-100
Leaf rust	Lgst.386-72	susceptible	resistant	0-0.2
	N-POL-2	susceptible	resistant	0-4.0
	Alcedo	susceptible	susceptible	40-60

The yellow rust resistant and leaf rust resistant lines were crossed with 'Winzi' (susceptible to yellow rust) and 'Alcedo' (susceptible to leaf rust), respectively. The obtained F_1 hybrids were used to produce selfed F_2 seeds. From each cross 150 - 160 F_2 plants were grown to get F_3 families analyzed for their resistance at the seedlings and adult plant stage.

The seedlings tests were performed in the greenhouse inoculating 20 seedlings per F_3 family with three yellow rust races and three leaf rust races containing the virulences 1, 2, 3, 6, 7, 9, 17 and 1, 2a, 2b, 2c, 3bg, 3ka, 10, 11, 13, 14a, 14b, 15, 16, 17, 18, 20, 21, 23, 26, 28, 30, 32, 33, 35, 37, 38, 44, respectively. The adult plant resistance was tested in the field during the seasons 1997 or 1998 growing F_3 double rows (90 cm). The plants were inoculated with a mixture of actual virulences using spreader rows. For classification the 'Coefficient of infection' described by Stubbs et al. (1986) was calculated. Plants classified as adult plant resistant need to have a coefficient of <8.0 combined with a race non specific susceptibility at the seedling stage. To test the goodness of fit of observed ratios to theoretical expectations the χ^2 analyses were used.

3. Results and Discussion

The inheritance of the adult plant resistance to yellow rust was studied during the 1997 growing season. Both monogenic (Figure 1a) and polygenic (Figure 1b) segregation patterns were observed.

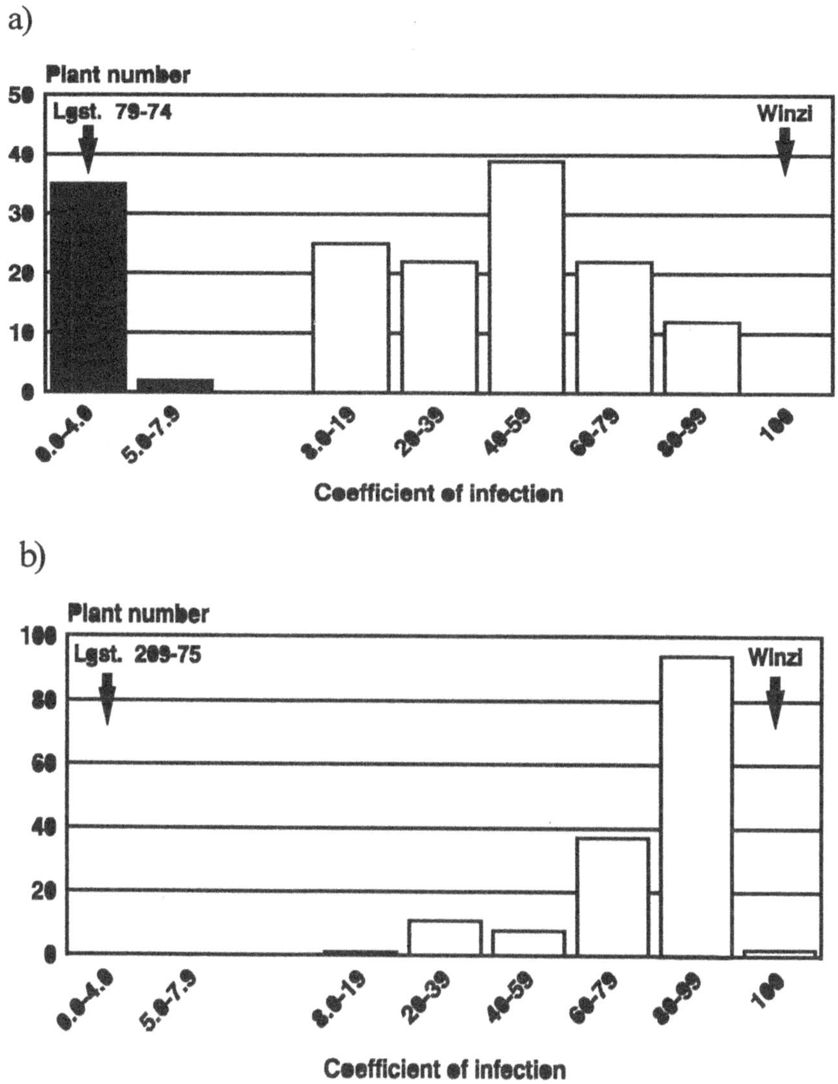

Figure 1. F_2 segregation pattern for adult plant resistance against yellow rust of the crosses 'Lgst.79-74 x Winzi' (a) and 'Lgst.209-75 x Winzi' (b) based on analysis of F_3 scoring data of 1997 field experiments. Black and white bars show number of resistant and susceptible genotypes, respectively. The means of the parents are marked by the arrows.

The F_3 progenies of the 'Lgst.79-74 x Winzi' cross could be classified into resistant and susceptible families. The segregation ratio is given in Table 2 and fitted the expected 3:1 ratio tested by the χ^2 test.

Table 2. Segregations and χ^2-test values for adult plant disease resistance.

Disease	Cross	observed segregation	expected segregation	χ^2 - value	P -value
Yellow rust	Lgst.79-74 x Winzi	37 : 120	1 : 3	0.18	P>0.60
Leaf rust	Lgst.386-72 x Alcedo	38 : 91	1 : 3	1.36	P>0.20
	N-POL-2 x Alcedo	104 : 55	3 : 1	7.80	P>0.001

In contrast to this, no resistant F_3 families were observed analysing the 'Lgst.209-75 x Winzi' cross. The susceptible progenies followed a normal distribution indicating a polygenic control of the adult plant resistance of 'Lgst.209-75'. These results support the conclusion of Van der Planck (1968) that there may exist more than one sort of horizontal resistance to a disease and that there may be both polygenic and oligogenic inheritance. Both monogenic and digenic inheritances of adult plant yellow rust resistance were described by Barina & McIntosh (1995) studying Australian wheats.

The F_3 progenies of two different crosses were tested for adult plant leaf rust resistance in 1997 or 1998. In both combinations major gene segregations were observed (Figure 2), although the mode of inheritance of the resistance was different. Whereas the resistance of Lgst.386-72 is inherited recessive, 'N-POL-2' was found to carry a dominant leaf rust resistance gene. The segregation data along with the χ^2-test values are given in Table 2. For the 'N-POL-2 x Alcedo' cross a distorted segregation was observed.

The presence of major genes for adult plant resistance to leaf rust was also described by Brammer et al. (1998). The genetic analysis of progenies from crosses between the Brazilian variety 'Torpi' and a monosomic series of 'Cappelle Desprez' indicated that two genes for adult plat resistance in 'Toropi' are located on chromosomes 1A and 4D.

The results presented here give clear evidence that major genes are available controlling non specific durable resistance against yellow and leaf rusts in wheat. The understanding of the genetics of adult plant disease resistance is necessary for their exploitation in breeding programs. Because of the qualitative inheritance it should be possible to detect molecular markers for adult plant resistance. The use of molecular markers will offer possibilities for a marker aided selection of non specific resistant genotypes.

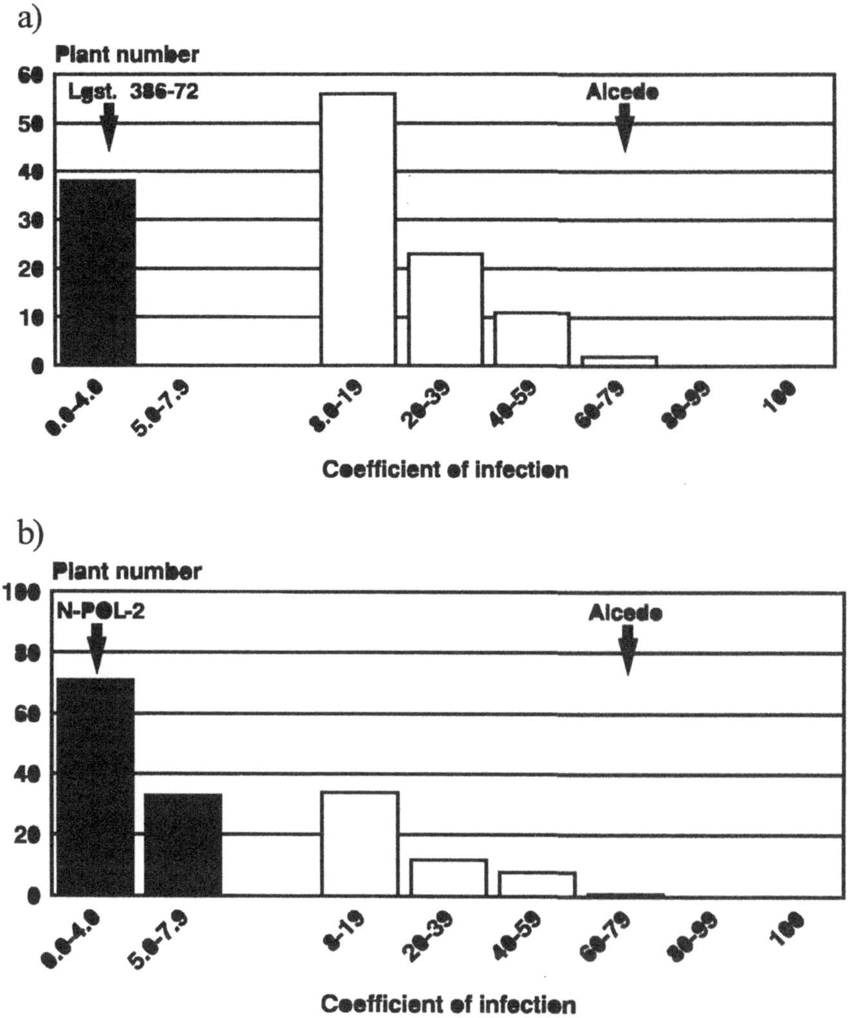

Figure 2. F_2 segregation pattern for adult plant resistance against leaf rust of the crosses 'Lgst.386-72 x Alcedo' (a) and 'N-POL-2 x Alcedo' (b) based on analysis of F_3 scoring data of 1997 (a) and 1998 (b) field experiments. Black and white bars show number of resistant and susceptible genotypes, respectively. The means of the parents are marked by the arrows.

4. References

Barina, H.S. & McIntosh R.A. (1995) Genetics of adult plant stripe rust resistance in four Australian wheats and the French cultivar 'Hybride-de-Bersee, *Plant Breed.* **114,** 485-491.

Brammer, S.P., Worland, A., Barcellos, A.L. & Fernandes, M.I.B.de M. (1998) Monosomic analysis of adult-plant resistance to leaf rust in the Brazilian wheat cultivar 'Toropi', *Proc. 9th Int. Wheat Genet. Symp. Vol. 2,* Saskatoon, Saskatchewan, Canada, 17 - 18.

Floor, H.H. (1959) Genetic controls and host parasite interactions in rust diseases, in C.S. Holton et al. (eds.), Plant pathology, Problems and Progress 1908 - 1958, Univ. of Wisconsin Press, Madison, Wisc.

Law, C.N. & Worland, A.J. (1997) The control of adult plant resistance to yellow rust by the translocated chromosome 5BS-7BS of bread wheat, Plant Breed. 116, 59-63.

Line, R.F., Chen, X.M., Gale, M.D. & Leung, H. (1996) Development of molecular markers associated with quantitative trait loci in wheat for durable resistance to Puccinia striiformis, Proc. 9th European and Mediterranean Cereals Rust and Mildew Conf., Lunteren, The Netherlands, 234.

Meinel, A. & Unger, O. (1998) Breeding aspects of partial resistance to airborn pathogens in wheat. Czech. J. Genet. Plant Breed. 34 (in press).

Stubbs, R.W., Prescott, J.M., Saari, E.E. & Dubin, H.J. (1986) Cereal disease methodology manual, Cimmyt, Mexico and IPO, Wageningen, 46 p.

Van der Planck, J.E. (1968) Disease resistance in plants, Academic Press, London, 206 p.

G.T. Scarascia Mugnozza, E. Porceddu & M.A. Pagnotta (Eds.)
Genetics and Breeding for Crop Quality and Resistance, 67-76, 1999

Assessment and genetics of host plant resistance to yellow rust in bread wheat germplasm adapted to the East African highlands

W.W. Wagoire, O. Stølen, J. Hill & R. Ortiz
The Royal Veterinary and Agricultural University, Dept. of Agricultural Sciences, 40 Thorvaldsensvej, DK-1871 Frederiksberg C, Copenhagen, Denmark, E-mail: ro@kvl.dk.

Key words: *Triticum aestivum*, *Puccinia striiformis*, breeding, combining ability, heritability

Abstract: Lack of adapted, high yielding and disease-resistant wheat cultivars constrains grain production in some locations of the East African highlands. Between 1994 and 1996 eight bread wheat cultivars were investigated, using a complete diallel crossing scheme, to determine (i) yield loss caused by yellow rust and the 'cost' of incorporating yellow rust resistance genes; (ii) the genetic basis of adult field resistance to yellow rust, and (iii) the effect of environment on the genetic parameters controlling grain yield. Comparison of F_1 and F_2 diallels sown in fungicide treated and untreated experiments showed that yield loss due to all foliar diseases was 83%, of which yellow rust accounted for 25%. Incorporation of resistance had no extra 'cost' in terms of grain yield at a disease-free location, where resistant and susceptible crosses had similar yields. Diallel analysis revealed the importance of additive genetic effects for days to heading, plant height and grain yield. Additivity, dominance, and epistasis played a significant part in the genetic control of host plant resistance to yellow rust. For grain yield additive effects interacted significantly with seasons and locations. Additive effects were significant at the location where the parental genotypes were selected for their wide response to yellow rust, suggesting that parents should be selected at the target location.

1. Introduction

Wheat (*Triticum aestivum* L.) has been grown in Uganda since the beginning of this century at high altitudes, ranging from about 1800 to 2400 m above sea level (m a.s.l.). These areas are in the South Western highlands and in the East on the slopes of Mount Elgon. In the former, it is grown mainly by small scale farmers on 0.5 to 1.0 ha plots while in the latter it is usually grown on larger farms where mechanization can be practised. In both areas, the crop may be grown twice a year depending on the wishes of the farmer. In these areas however, very little additional land is available. While efforts are being made to develop wheat cultivars adapted to mid-altitude areas ranging from 1500 to 1800 m a.s.l., where there is more unutilized land, provision of improved cultivars for the high altitude areas would be a more appropriate short term remedy.

One of the main factors limiting wheat production in these areas is the fungal disease yellow rust, caused by *Puccinia striiformis* Westend. (Wagoire et al., 1998a). Nothing is known at present about the inheritance of resistance to this disease in wheat cultivars adapted to Ugandan environments, nor is anything known about which races of the pathogen prevail. If such information were available it would speed up breeding programmes aimed at developing high yielding, yellow rust-resistant cultivars. Therefore, the aim of this research was to determine (i) yield loss caused by yellow rust and the 'cost' of incorporating yellow rust resistance genes, (ii) the genetic basis of adult field resistance to yellow rust, and (iii) the effect of environment on the genetic parameters controlling grain yield.

2. Materials and Methods

The F_1 and F_2 generations of genotypes derived from a complete diallel cross among eight bread wheat lines (Table 1) were used in this study. The F_2s were generated by selfing the corresponding F_1s in the preceding season. All the parents were selected from the Ugandan Wheat Development Project. Thus, K. Chiriku has already been released in Uganda, while others are promising candidates for national release. The experiments were planted in two cropping seasons (A and B) at two locations, Kalengyere and Buginyanya, both having a bimodal rainfall distribution. Kalengyere, a high-rust site at 2400 m a.s.l, has an Andosol with pH 5.7 and an average temperature of 16°C throughout the year. The high rainfall season B (750 mm) lasts from September to March, and the relatively low rainfall season A (480 mm) from March to August. Buginyanya is a low-rust site at 2100 m a.s.l., having an Andosol with pH 5.5 and an average temperature of 18°C. The high rainfall season B (560 mm) occurs from September to March and the relatively low rainfall season A (470 mm) from March to August.

This study was carried out from 1994 to 1996. In each season the 64 genotypes (8 selfed parents and their 56 F_1 crosses or the corresponding F_2 populations) were sown in a completely randomized block design with two replicates at each site. An extra environment was created by applying fungicide (150 ml ha^{-1} of 250 g l^{-1} of the active ingredient [alpha-tert-butyl-alpha-(para-chlorophenethyl)-1-H-1,24=q-zole-lethanol]) to control yellow rust at Kalengyere in the 1995 B season. Due to limited availability of seed, the F_1 experimental plots comprised two rows of 1.5 m length, 0.3 m between rows and 0.15 m between plants, (i.e., about 20 plants $plot^{-1}$), whereas for the F_2s there were two rows of 5 m length and 0.3 m inter-row spacing. All experiments were given nitrogen equivalent to 50 kg ha^{-1} prior to planting. Experiments were hand-weeded, while bird-scaring was practised from anthesis to harvest to minimise losses.

Table 1. Cultivars of bread wheat lines adapted to the East African highlands and their yellow rust reaction.

Code	Accession name or pedigree	Host reaction to yellow rust (%)
1	BURI CM58340-A-1Y-3Y-2M-2Y-0M	Resistant (0-2)
2	K. CHIRIKU K.TEMBO/CARPINTERO"S"	Resistant (0-2)
3	ESDA/LIRA CM78428-017M-013M-013Y-03AL-3Y-3AL-0Y	Resistant (0-5)
4	VEE"S"/JUP73/EMU"S"//GJO"S" CM74465-05AP-300AP-4AP-300AL-0AP	Moderately resistant (10-20)
5	ATTILA CM85836-4Y-0M-0Y-OPZ	Moderately susceptible (40-60)
6	CY8801	Susceptible (60)
7	F60314.76/4/CNO76/7C//KAL/BB/3/PC1"S"/5/CNO79	Susceptible (60-100)
8	Car853/Coc//VEE"S"/3/E7408/PAM"S"/HORK"S"/PF73226	Susceptible (90-100)

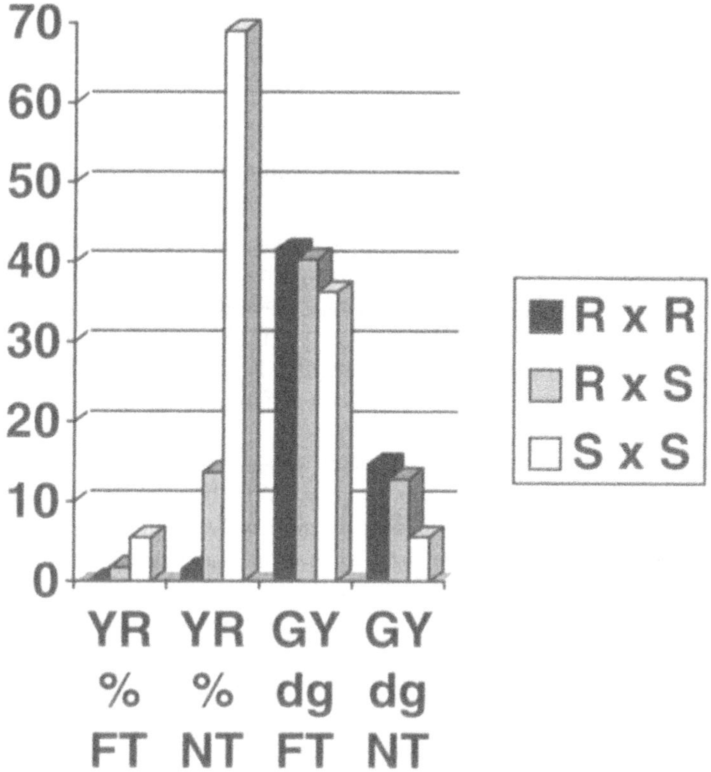

Figure 1. Yellow rust (YR, %) and grain yield (GY, dg) at Kalyengere according to cross. R and S mean resistant and susceptible parent. FT and NT mean fungicide-treated and non-treated plots.

Days to heading were recorded on individual plants in each plot and plant height was measured at physiological maturity. Yellow rust was scored on the flag leaf of individual plants when the severity on the most susceptible parent was about 100%, i.e., most of the leaf surface was covered with uredinia. The modified Cobb scale (Peterson et al., 1948) was used for scoring the percentage of possible tissue rusted (disease severity). Host response to infection was scored using "T=0.1" to indicate immunity; "R= 0.2" to indicate resistance in plants showing miniature uredinia; "MR = 0.4" to indicate moderate resistance in plants exhibiting small uredinia; "MS = 0.8" to indicate moderate susceptibility in plants with moderate sized uredinia (smaller than fully susceptible type), and "S=1" to indicate full susceptibility. Disease severity and host response scores were multiplied together to give the coefficient of infection (C.I.) for data analysis (Wagoire, 1997). Whole plots were hand-harvested, threshed, cleaned, sun-dried and the grain weighed at approximately 12% moisture content. Plot grain yield was transformed to g m^{-2}. Analyses of variance were carried out on plot means for all characters, following the methods outlined in Hill et al. (1998).

3. Results and Discussion

3.1 Assessment of yield losses caused by diseases

At Kalengyere, the difference in mean yield between the fungicide treated and untreated plots of the susceptible material (susceptible (S) parents and S x S crosses) represented the loss due to foliar diseases that can be eliminated by chemical control. Averaged across generations during the test period this loss was 83% (Figure 1). Yield loss due to foliar diseases other than yellow rust was estimated from the difference in mean yield between fungicide treated and untreated plots of the resistant material (resistant (R) parents and R x R crosses). This amounted to 58% after averaging across generations. The difference, 25%, represented the yield loss due to yellow rust.

The high losses attributed to diseases other than yellow rust indicate that breeding for yellow rust resistance alone would not necessarily increase wheat yields. There is therefore a need to develop material that incorporates resistance to other common diseases in Uganda, such as the leaf blotches caused by *Septoria* spp. (Wagoire et al., 1998a). Such material should be tested in a range of environments covering the target diseases. This study also indicates that incorporation of yellow rust resistant genes has no adverse effect on grain yield, because the results at Buginyanya, a disease-free location, disclose that resistant and susceptible crosses have a similar grain yield (Figure 2). Hence, material resulting from such a breeding programme could be utilized in disease free areas having yields comparable to Buginyanya.

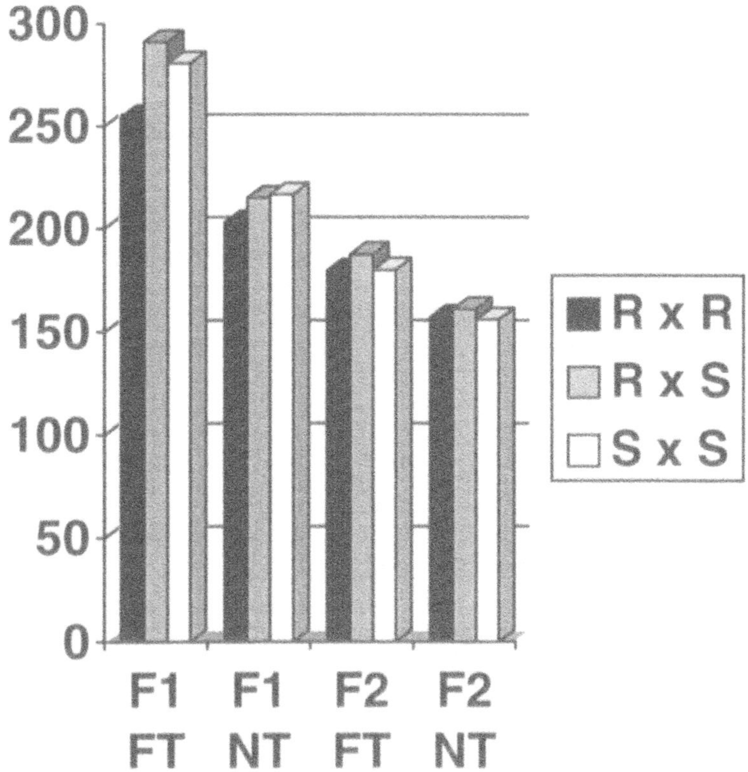

Figure 2. 'Cost' of yellow rust resistance in grain yield (g) at Buginyanya according to cross in F_1 and F_2 generations. R and S mean resistant and susceptible parent, while FT and NT mean fungicide-treated and non-treated plots.

3.2 Genetics of yellow rust resistance

Yellow rust resistance appears to be polygenically controlled. Resistance genes may be dominant or recessive. Furthermore, the expression of resistance may be influenced by the environment and genetic background (Johnson, 1988; Lewellen & Sharp, 1968). Some resistance genes may be expressed during the seedling stage; others are expressed only in adult plants. The latter is usually referred to as field resistance. Evaluation of genotypes in this study was conducted under field conditions, and yellow rust was assessed during the adult stage. Resistance to yellow rust was under the control of genes exhibiting additive and dominance effects (Table 2). Analysis of the F_1s revealed transgressive segregation suggesting the presence of epistasis between genes that enhance host plant resistance. Epistatic effects were removed by omitting two of the resistant parental lines (Table 2), a method suggested by Jinks (1954) and Jana (1975). Likewise, omission of two of the most susceptible parents removed epistatic effects in the F_2s, because offspring with higher susceptibility were derived by crossing these highly susceptible parents (data not shown). Yellow rust resistance in these lines was under the control of dominant genes, carried by the most resistant parents, acting additively. The most susceptible parental lines carry recessive genes that enhance susceptibility to yellow rust (Wagoire et al., 1998b). The high heritability values for yellow rust in the F_1 diallel (Table 2) suggests that this material can be utilized in breeding for yellow rust resistance.

Table 2. Genetic parameters for stripe rust resistance in F_1's of a full 8 x 8 diallel of bread wheat lines and its reduced form (6 x 6 diallel) after fitting the additive-dominance model in 1995.

	8x8 diallel			6x6 diallel	
Variance	1994B[x]	1995A[x]	1995B[x]	1995A[x]	1995B[x]
Additive	112.60	941.13	798.04	970.98	750.84
$Dominance_1$ [z]	53.58	1153.30	651.43	1241.47	652.48
$Dominance_2$ [z]	58.62	960.15	567.36	889.06	466.52
V_{dAD} [y]	-21.51	-148.22	20.95	-351.00	-293.34
Environment	30.24	28.30	75.78	40.31	123.80
Dominance ratio	0.69	1.11	0.90	1.13	0.93
Narrow sense heritability	0.68	0.72	0.71	0.78	0.78
Broad sense heritability	0.84	0.98	0.94	0.98	0.92

[z] When $dominance_s$ < $dominance_l$, allelic frequencies are unequal at those loci exhibiting dominance for the character; [y] a positive value indicates that dominant alleles are more frequent than recessive alleles, and vice versa when negative; [x] cropping seasons (see text).

3.3 *Genetics of agronomic characters*

Environments significantly affected days to heading, plant height, and grain yield in the F_1 generation (Figure 3). General combining ability (GCA) was also highly significant for all characters (Figure 4). Further analyses revealed the predominant role of additive and dominance gene action for grain yield, plant height and days to heading as shown by the significant GCA and specific combining ability (SCA) effects (Wagoire et al., 1998c). Significance of genotype-by-environment (GxE) interactions for these characters (data not shown), suggests that selection for specific adaptation may maximise the use of the germplasm developed in this study. This observation was supported by the results from the stability analysis for each character, which reveals that GCA was not correlated with phenotypic stability while SCA was correlated with yield stability across environments (Wagoire, 1997). Further, joint regression analysis suggested that the linear regression model did not satisfactorily explain the GE interaction for grain yield. Nevertheless, the regression method may be still of value because of our interest in genotypes showing specific adaptation to targeted environments (Becker & León, 1988).

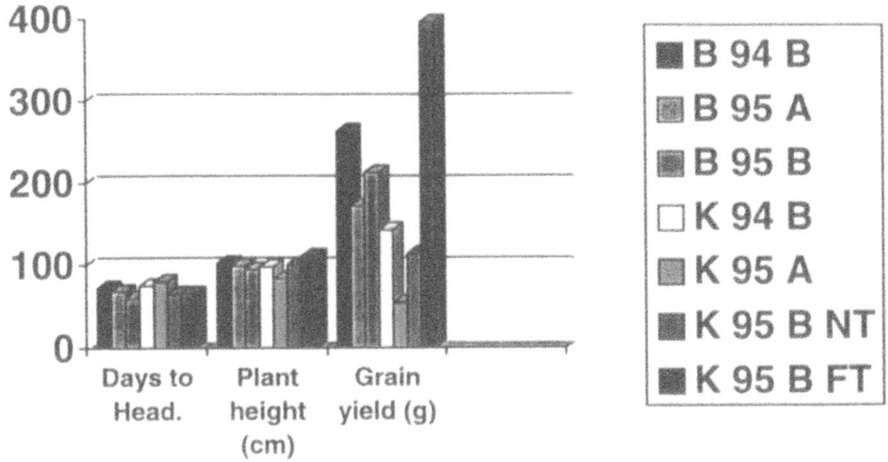

Figure 3. F_1 performance per environment for days to heading, plant height, and grain yield in seven Ugandan environments. B is Buginyanya and K is Kalyengere, while FT and NT are fungicide treated and non-treated plots. The standard errors were 0.60 for DH (days), 1.77 for PH (cm), and 13.00 for GY (g m^{-2}).

Individual analysis of the various diallel crosses for each season and location (environments), reveals homogeneity of error variances for the parents and F_1 or F_2 generations within a particular environment (Table 3). Additive effects were predominant at Kalengyere, whereas narrow-sense heritability was low at Buginyanya (Table 4), where the genotypes showed reduced variability for this character as there was no yellow rust at this location (Wagoire et al., 1999). Thus, there is no relationship between yield level and respective heritability values in this study. Similar views have been expressed by Cooper & DeLacy (1984) and by Singh & Ceccarelli (1995).

It should be noted that these parental lines have been selected at Kalengyere for their wide response to yellow rust. Consequently, they display a range of grain yields at this location. Indeed, Hill et al. (in press) found that in the unsprayed plots at Kalengyere, indirectly selecting for yield, by choosing the most rust resistant material, was more effective than direct selection for yield itself. This was attributable to the heritability of each character (Tables 2 and 4), but above all to the high additive genetic correlation between them in this broad-based spring wheat germplasm adapted to the East African highlands.

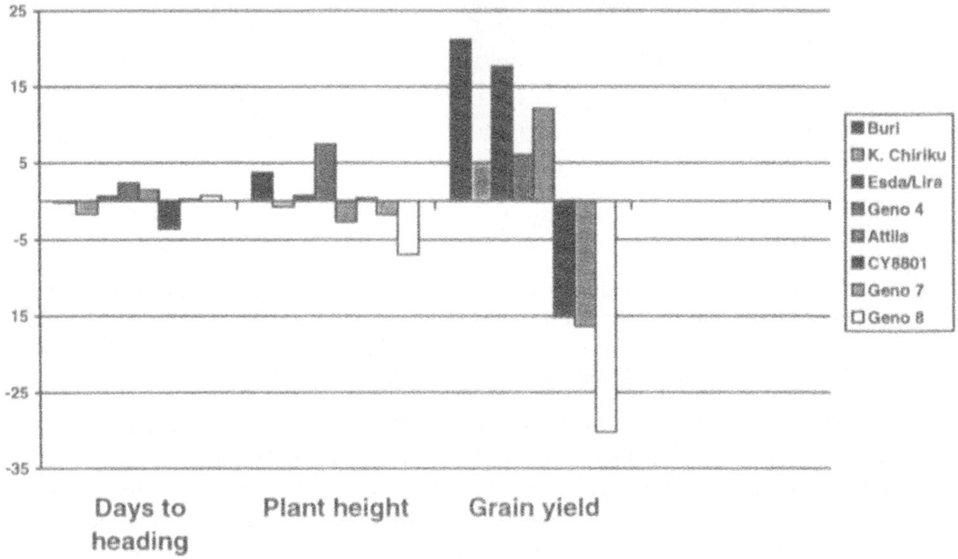

Figure 4. General combining ability (GCA) of eight wheat lines adapted to the East African highlands for days to heading (DH), plant height (PH), and grain yield (GY) across seven Ugandan environments. The standard errors of GCA were 0.24 for DH (days), 0.59 for PH (cm), and 5.24 for GY (g m^{-2}), while the standard errors for GCA differences were 0.36, 0.89, and 7.92 respectively.

Table 3. Mean parental, F_1 and F_2 grain yield (g m^{-2}) for a full diallel of bread wheat lines grown at two Ugandan locations (date of planting season indicated in brackets).

Location and season		Mean	d.f.[z]	Error	Variance ratio
F$_1$ generation					
Kalengyere,1994 (B)	Parent	139.98±16.375	7	4290.45	
	F_1	143.50±3.961	55	1757.27	2.44
Kalengyere,1995 (A)	Parent	55.28 ± 10.909	7	1904.00	
	F_1	54.50 ± 2.575	55	742.79	2.56
Kalengyere,1995 (B)	Parent	101.20 ± 8.837	7	1249.60	
	F_1	115.31 ± 4.707	55	2481.28	1.99
Buginyanya,1994 (B)	Parent	239.31 ± 10.993	7	1933.86	
	F_1	235.84 ± 4.158	55	1936.44	1.00
Buginyanya,1995 (A)	Parent	116.11 ± 9.520	7	1450.09	
	F_1	172.63 ± 4.106	55	1887.96	1.30
Buginyanya,1995 (B)	Parent	171.32 ± 13.927	7	3103.34	
	F_1	218.52 ± 5.330	55	3181.47	1.02
F$_2$ generation					
Kalengyere,1995 (A)	Parent	105.71 ± 10.731	7	1842.48	
	F_2	115.23 ± 4.569	55	2338.31	1.27
Kalengyere,1995 (B)	Parent	75.63 ± 9.376	7	1406.44	
	F_2	92.41 ± 3.232	55	1170.10	1.20
Buginyanya,1995 (A)	Parent	237.42 ± 19.255	7	5932.15	
	F_2	222.08 ± 5.072	55	2881.48	2.06
Buginyanya,1995 (B)	Parent	173.12 ± 4.102	7	269.24	
	F_2	156.84 ± 1.226	55	168.35	1.60

[z] degrees of freedom

Table 4. Genetic parameters for grain yield combined over seasons for F_1 and F_2 generations diallel of bread wheat lines grown at two Ugandan locations.

Variance	Kalengyere	Buginyanya
Additive	7023.66	1217.80
Dominance$_1$ [z]	2397.16	5349.10
Dominance$_2$ [z]	3599.61	3690.60
V_{dAD} [y]	-3845.83	1194.97
Environment	3031.65	3554.03
$/Vd_{D1}/Vd_A$	0.58	2.10
Narrow sense heritability	0.55	0.16
Broad sense heritability	0.65	0.33

[z] When dominance$_s$ < dominance$_1$, allelic frequencies are unequal at those loci exhibiting dominance for the character;[y] a positive value indicates that dominant alleles are more frequent than recessive alleles, and vice versa when negative.

4. References

Becker, H.C. & Léon, J. (1988) Stability analysis in plant breeding. *Plant Breeding* **101**, 1-23.

Cooper, M. & DeLacy, I.H. (1994) Relationships among analytical methods used to study genotypic variation and genotype-by-environment interaction in plant breeding multi-environment experiments. *Theor. and Appl. Genet.* **88**, 561-572.

Hill, J., Becker, H.C. & Tigerstedt, P.M.A. (1998) *Quantitative and Ecological Aspects of Plant Breeding*, Chapman & Hall, London.

Hill, J., Ortiz, R., Wagoire W.W. & O. Stølen, O. (in press) Effectiveness of indirect selection for wheat yield in a stress environment. *Theor. Appl. Genet.*

Jana, S. (1975) Genetic analysis by means of diallel graph. *Heredity* **35**, 1-19.

Jinks, J. L. (1954) The analysis of continuous variation in a diallel cross of *Nicotiana rustica* varieties. *Genetics* **39**, 767-788.

Johnson, R. (1988) Durable resistance to yellow (stripe) rust in wheat and its implications in plant Breeding, in N.W. Simmonds and S. Rajaram (eds.), *Breeding Strategies for Resistance to the Rusts of Wheat*, CIMMYT, Mexico, pp. 63-75.

Lewellen, R.T. & Sharp, E.L. (1968) Inheritance of minor reaction gene combinations in wheat to *Puccinia striiformis* at two temperature profiles. *Can. J. Bot.* **46**, 21-26.

Peterson, R.F., Campbell, A.B. & Hannah, A.E. (1948) A diagrammatic scale for estimating rust intensity of leaves and stem of cereals. *Can. J. Res.* Sect. C. **26**, 496-500.

Singh, M. & Ceccarelli, S. (1995) Estimation of heritability using varietal trials data from incomplete blocks. *Theor. and Appl. Genet.* **90**, 142-145.

Wagoire, W.W. (1997) *Yellow Rust Resistance of Wheat Cultivars in Uganda*, PhD Thesis, The Royal Veterinary and Agricultural Univ., Copenhagen, Denmark.

Wagoire, W.W., Stølen, O., Hill, J. & Ortiz, R. (1998a) Is there a cost for wheat cultivars with genes for resistance to yellow rust caused by *Puccinia striiformis*? *Crop Protection* **17**, 337-340.

Wagoire, W.W., Stølen, O., Hill, J. & Ortiz, R. (1998b) Inheritance of adult field resistance to yellow rust disease among broad-based hexaploid spring wheat germplasm. *Theor. Appl.Genet.* **97**, 502-506.

Wagoire, W.W., Stølen, O. & Ortiz, R. (1998c) Combining ability analysis in bread wheat adapted to the East African highlands. *Wheat Information Service* **87**, in press.

Wagoire, W.W., Hill, J., Stølen, O. & Ortiz R. (1999) Impact of genotype-environment interactions on the inheritance of wheat yield in low-yielding environments. *Euphytica.* 105(1): 17-23,

G.T. Scarascia Mugnozza, E. Porceddu & M.A. Pagnotta (Eds.)
Genetics and Breeding for Crop Quality and Resistance, 77-82, 1999

The role of b-32 protein in protecting plants against pathogens

M. Maddaloni[1], F. Forlani[1], V. Balmas[2], G. Donini[1], L. Corazza[2], H. Fang[3], S. Pincus[3] & M. Motto[1]

[1] Istituto Sperimentale per la Cerealicoltura, sezione di Bergamo, Via Stezzano 24, 24100 Bergamo, Italy- Tel. -39 035 311422; Fax -39 035 316054; [2] Istituto Sperimentale per la Patologia vegetale, Via Bertero 22, 00100 Roma, Italy; [3] Department of Microbiology, Montana State University, Bozeman, Montana, USA.

Key words: b-32, fungal tolerance, *Rhizoctonia solani*, ribosome-inactivating protein, transgenic plant.

Abstract: The maize b-32 protein is a functional ribosome-inactivating protein (RIP), inhibiting *in vitro* translation in the cell-free reticulocyte-derived system and having specific N-glycosidase activity on 28S rRNA. previous results indicated that *opaque-2* (*o2*) mutant kernels, lacking b-32, show an increased susceptibility to fungal attack and insect feeding and that ectopic expression in plants of a barley and a pokeweed RIP leads to increased tolerance to fungal and viral infection. This prompted us to test whether b-32 might function as a protectant against pathogens. The *b32.66* cDNA clone under the control of the potato *wun1* gene promoter was introduced into tobacco by *Agrobacterium tumefaciens*-mediated transformation. Out of 23 kanamycin resistant regenerated shoots, 16 contained a PCR fragment of the corrrect size spanning the boundary between the promoter used and the coding region of the *b-32* gene. Eight independently transformed tobacco lines were randomly chosen for protein analysis: all of them expressed b-32 protein. The data presented indicate that transgenic tobacco plants expressing b-32 show an increased tolerance against infection by the soil-borne fungal pathogen *Rhizoctonia solani* Kühn. The exploitation of b-32 protein in immunoconjugates directed against HIV-infected cells is also discussed.

1. Introduction

Plants have developed an array of natural defense strategies to protect themselves against infection by pathogens (Bowles, 1990; Hammond-Kosack & Jones, 1996). These include the synthesis of antimicrobial metabolites, enhanced strengthening of the cell wall, stimulation of lytic enzyme synthesis, and synthesis of pathogenesis-related proteins. plants also constitutively accumulate high levels of proteins which are either toxic to or inhibitory against pathogens and pests (Broekaert et al., 1995).

Some of the most potent plant toxins belong to the class of translation inhibitors called ribosome-inactivating proteins (RIPs). These proteins are widely distributed in higher plants and catalytically inactivate eukaryotic, and in some cases prokaryotic, ribosomes. RIPs act as N-glycosidases to remove a specific adenine residue in the large ribosomal RNA (Endo et al., 1987). This irreversible modification blocks the elongation factors EF-1 and EF-2-dependent GTPase activity and renders the ribosome unable to bind EF-2, thereby blocking translation. RIPs have been classified as type 1, composed of a single polypeptide chain, and type 2, composed of a catalytic polypeptide which is linked via a disulphide bridge to a docking lectin. This lectin is responsible for the internalization into the cell (Barbieri & Stirpe, 1982; Stirpe et al., 1992). The biological role played by these proteins is completely unknown (Stirpe et al., 1992). However, it is commonly accepted that the physiological role of RIPs is one of defence protecting the plant or seed from attack by predators or pathogens (Olsnes & Pihl, 1982; Lord et al., 1991). Transgenic plants expressing a RIP from barley and pokeweed have improved resistance to fungal and viral infection, respectively (Logemann et al., 1992; Lodge et al., 1993). Moreover, interest in RIPs is currently growing in view of their possible use as the toxic part of immunotoxins for human-cancer therapy and, hopefully, the anti-HIV properties exhibited by some RIPs (Spooner & Lord, 1990; Lee-Huang et al., 1991).

In maize endosperm a cytosolic albumin with a molecular weight of 32 kda, termed b-32, is synthesized in temporal and quantitative coordination with the deposition of storage proteins (Soave et al., 1981). Both cDNA and genomic clones encoding b-32 have been isolated and it was shown that the *b-32* genes form a small gene family (Hartings et al., 1990). The *b-32* gene, as well as the 22 kda storage protein zeins, are under the control of the seed-specific transcriptional activator *Opaque-2* (*O2*) (Lohmer et al., 1991). Although the role of *b-32* in maize endosperm remains unclear, this protein has homology with several previously characterized RIPs (Lohmer et al., 1991). b-32 is a functional RIP by the criteria of inhibition of *in vitro* translation in a cell-free rabbit reticulocytes system and specific N-glycosidase activity on 28S rRNA (Maddaloni et al., 1991; Walsh et al., 1991; Bass et al., 1992).

In order to further assess the biological properties of b-32 we have produced transgenic tobacco plants to address the question of whether the plants expressing this gene have increased levels of protection against infection by the fungus *Rhizoctonia solani*. In addition, the exploitation of the b-32 as a toxic moiety in immunotoxins directed against HIV-infected cells.

2. Results and Discussion

2.1 *Production of transgenic tobacco plants expressing b-32*

Tobacco cv. "Petite Havana SR1" leaf discs were transformed with *Agrobacterium tumefaciens* LBA4404 harbouring the *pbinwunb32* chimaeric gene as previously described (Maddaloni et al., 1997). Twenty-three independent shoots were able to root on medium containing 100 μg ml^{-1} of kanamycin. After transplantation to soil, genomic DNA extracted from young leaves of each independent shoot was analysed by polymerase chain reaction (PCR) amplification of a 470 bp region spanning the boundary between the *wun1* promoter and *b-32* coding region. Of the 23 kanamycin resistant independent transformants, 16 were shown to contain the PCR fragment of the correct size, suggesting that these plants carried *pbinwunb32* sequences. By contrast, DNA extracted from young leaves of untransformed SR1 plants and used in PCR reactions identical to that of putative transformed plants did not give rise to any detectable signal.

The transgenic tobacco plants appeared phenotypically normal. Primary transformants (T_0 generation) were grown in the soil and self-pollinated. From each T_0 plant, seeds (T_1 generation) were scored for antibiotic resistance on MS medium containing kanamycin, in order to determine the number of transgenic loci present in their genomes. The results of the genetic analysis revealed that 6 and 1 T_1 populations exhibited 3:1 and 15:1 segregation ratios for kanamycin resistance, respectively. these results indicate that either one (3:1 segregation) or two (15:1 segregation) functional kanamycin resistance loci were present in these transgenic lines. only one transformant, line Tr10, contained apparently more than two independent loci responsible for kanamycin resistance.

Eight transgenic tobacco lines were randomly chosen for protein analysis. Sampled young leaves from T_1 selfed plants and from an untransformed SR1 plant were wounded to trigger the synthesis of b-32. The extracted proteins were separated by SDS-PAGE, blotted onto nitrocellulose membrane and probed with a polyclonal antiserum raised against b-32. An immunoreactive band, migrating at the same position as the affinity purified maize b-32, appears specifically in protein extracts from transgenic lines. By contrast, no specific immunoreactive bands were detectable in the untransformed control. The concentration of b-32 in the transgenic lines was estimated to range from 130 ng (line Tr4) to 520 ng (line Tr16) per mg of total leaf extracted protein based on measuring the intensity of the immunostained bands and using purified maize b-32 as a quantitative standard. The stability of b-32 expression was monitored in the T_2 generation: all the T_2 progenies analyzed expressed the b-32 protein.

2.2 *In vitro bioassay of b-32 activity*

The phytopathogenic fungus *R. solani* is the causal agent of root rot, stem canker and damping off observed in a wide range of crops (Anderson, 1982; Ogoshi, 1987; Kataria & Verma, 1992). As the fungus infects also tobacco (Roby et al., 1990;

Broglie et al., 1991; Logemann et al., 1992), a *R. solani* / tobacco system can be used in studies aiming to develop transgenic plants expressing foreign proteins with antifungal properties. An *in vitro* bioassay performed as described (Woloshuk et al., 1991) showed that b-32 protein has an inhibitory effect on the growth of *R. solani.* Such inhibitory effect was already detectable at the lowest concentration. It was also evident that the range of concentrations used in this study is comparable to that reported by Leah et al. (1991) in a similar assay about the inhibition of fungal growth by a barley RIP. In fact the concentrations of b-32 used in the present paper range from 0.6 to 60 $\mu g\ ml^{-1}$, corresponding to 0.1 to 10 μg $microtiter^{-1}$ well. *R. solani* was also chosen because it is sensitive to cereal RIPs (Coleman & Roberts, 1982).

2.3 Evaluation of tolerance of transgenic tobacco lines against Rhizoctonia solani

To perform the phytopathological test, T_1 seeds from the 8 chosen lines were used. Sixty plants, contained in the three replications, per each independent transformant were evaluated 30 days after infection. Each plant was assigned to a class ranging from 0 to 4 as described (Maddaloni et al., 1997) according to the following damage parameters: i) the peculiar root rotting damage caused by *R. solani* (Ogoshi, 1987) and ii) the general vigour of the above-ground organs.

The mean values of each entry for root rotting and above-ground plant damage of the nontransformed control, transformants and noninfected control assayed for tolerance to *R. solani*. It must be pointed out that the T_1 plants used in the test were not genetically homogeneous in that the transgene(s) copies were segregating, thus very likely resulting in underestimation of tolerance. For both traits, transgenic plants were definitely more tolerant to *R. solani* infection than control plants. Root rotting scores for transgenic plants were on average 43% lower than for control plants (0.69 *vs* 1.20). A similar trend of response to infection, although less pronounced, was noticed for the above-ground plant damage. It was also evident that the magnitude of phenotypic variation among the different transgenic lines was appreciably larger for each of the traits considered compared to the sets of the untransformed plants. In fact, the mean of replications for the root rotting ranged from 1.03 (line Tr13) to 0.51 (line Tr9), while for the above-ground plant damage it ranged from 1.78 (line Tr14) to 0.47 (line Tr9). On considering root rotting, transgenic lines Tr1, Tr7, Tr9 and Tr16 showed an increased protection to fungal attack with respect to control. The transgenic line Tr9 approached the value of the noninfected control for root rotting and performed much better than noninfected control with respect to the above ground plant damage.

The previous data were also processed by McKinney's formula (McKinney, 1923), which generates a numeric index ($I_{McKinney}$) of the severity of the attack. The $I_{McKinney}$ values for transformed, untransformed control, and noninfected control plants indicate that transgenic tobacco lines were on average, for both root rotting and above-ground plant damage, more tolerant to infection by *R. solani* than the control plants. The transgenic line Tr9 had the highest protection against fungal attack.

Fungal infection of tobacco plants results in necrosis which in turn reduces plant fitness. It was known that, on average, infected transformed tobacco plants had a superior dry matter accumulation, an index of plant growth, as compared to infected controls. In particular, dry matter accumulation of line Tr1, approximately twice that of the controls, confirmed that transformed plants were more tolerant to damage induced by *R. solani* and grew better than infected control plants. Altogether our results showed that transgenic tobacco plants, in which the expression of b-32 gene is driven by the wun1 promoter, had increased protecion against infection of the soil-borne fungal pathogen *R. solani.*

2.4 Exploitation of b-32 protein as a toxic moiety in immunoconjugates directed against HIV-infected cells

The present project deals with the use of b-32 as a toxic moiety in immunotoxins directed against HIV-infected cells. The identification of novel toxins that are less systemically toxic, but maintain potent activity when targeted to virus infected cells or malignancies, is likely to increase the therapeutic utility of immunotoxins. Advantages of b-32 over other toxins include: i) it lacks the internalization domain which, when present, delivers the drug to a large array of cells; ii) it shows a low catalytic activity, but still exerts potent effects; iii) no genetically modified organisms are needed for products; iv) it can be produced in very large quantities at minimum costs; v) it is reasonably easy to purify; vi) it is stable during purification. Although researches are still in progress, preliminary results indicate that the monoclonal 1924-b-32 immunoconjugate selectively kills H9 HIV infected cells.

Acknowledgements: Research cited from authors' laboratory was financially supported by grants from "Ministero delle Politiche Agricole", Rome, special grant: Programma nazionale "Resistenze genetiche delle piante agrarie agli stress biotici ed abiotici".

3. References

Anderson, N.A. (1982) The genetics and pathology of *Rhizoctonia solani*, Annu. Rev. Phytopathol. 20, 327-329.

Barbieri, L. & Stirpe, F. (1982) Ribosome-inactivating proteins from plants: properties and possible uses, Cancer Surv. 1, 489-520.

Bass, H.W., Webster, C., O'Brian, G.R., Roberts, J.K.M. & Boston, R.S. (1992) A maize ribosome-inactivating protein is controlled by the transcriptional activator Opaque-2, Plant Cell 4, 225-234.

Bowles, R.J. (1990) Defense related proteins in higher plants, Annu. Rev. Biochem. 59, 873-907.

Broekaert, W.F., Terras F.R.G., Cammue B.P.A. & Osborn, R.W. (1995) Plant defensins: novel antimicrobial peptides as components of the host defense system, Plant Physiol. 108, 1353-1358.

Broglie, K., Chet, I., Holliday, M., Cressman, R., Biddle, P., Knowlton, S., Mauvais, C.J. & Broglie, R. (1991) Transgenic plants with enhanced resistance to the fungal pathogen *Rhizoctonia solani*, Science 254, 1194-1197.

Coleman, W.H. & Roberts, W.K. (1982) Inhibitors of animal cell-free protein synthesis from grains, Biochem. Biophys. acta 696, 239-244.

Endo, Y., Mitsui, K., Motizuchi, M. & Tsurugi, K. (1987) The mechanism of action of ricin and related toxic lectins on eukaryotic ribosomes, J. Biol. Chem. 262, 5908-5912.

Hammond-Kosack, K.E. & Jones, J.D.G. (1996) Plant disease resistance genes, Annu. Rev. Plant Physiol. Plant Mol. Biol. 48, 575-607.

Hartings, H., Lazzaroni, N., Ajmone Marsan, P., Aragay, A., Thompson, R., Salamini, F., Di Fonzo, N., Palau, J. & Motto, M. (1990) The b-32 protein from maize endosperm: characterization of genomic sequences encoding two alternative central domains, Plant Mol. Biol. 14, 1031-1040.

Kataria, H.R. & Verma, P.R. (1992) *Rhizoctonia solani* damping-off and root rot in oilseed rape and canola, Crop Prot. 11, 8-13.

Leah, R., Tommerup, H., Svendsen, I.B. & Mundy, J. (1991) Biochemical and molecular characterization of three barley seed proteins with antifungal properties, J. Biol. Chem. 266, 1564-1573.

Lee-Huang, S., Huang, P.L., Kung, H.F., Li, B.Q., Huang, P., Huang, H.L. & Chen, H.C. (1991) TAP29: An anti-human immunodeficiency virus protein from *Trichosanthes kirilowii* that is nontoxic to intact cells, Proc. Natl. Acad. Sci. USA 88, 6570-6574.

Lodge, J.K., Kaniewski, W.K. & Tumer, N.E. (1993) Broad-spectrum virus resistance in transgenic plants expressing pokeweed antiviral protein, Proc. Natl. Acad. Sci. USA 90, 7089-7093.

Logemann, J., Jach, G., Tommerup, H., Mundy, J. & Schell, J. (1992) Expression of a barley ribosome-inactivating protein leads to increased fungal protection in transgenic tobacco plants, Bio/technology 10, 305-308.

Lohmer, S., Maddaloni, M., Motto, M., Di Fonzo, N., Hartings, H., Salamini, F. & Thompson, R. (1991) The maize regulatory locus *Opaque-2* encodes a DNA-binding protein which activates the transcription of the b-32 gene, EMBO J. 10, 617-624.

Lord, J.M., Hartley, M.R. & Roberts, L.M. (1991) Ribosome inactivating proteins by plants, in J.M. Lord (ed.), Redirecting Natureís Toxins. Sem. Cell Biol. 2, 15-22.

Maddaloni, M, Barbieri, L., Lohmer, S., Motto, M., Salamini, F. & Thompson, R. (1991) Characterization of an endosperm-specific developmentally regulated protein synthesis inhibitor from maize seeds, J. Genet. & Breed. 45, 377-380.

Maddaloni, M., Forlani, F., Balmas, V., Donini, G., Stasse, L., Corazza, L. & Motto, M. (1997) Tolerance to the fungal pathogen Rhizoctonia solani AG4 of transgenic tobacco expressing the maize ribosome inactivating protein b-32, Transgenic Res. 6, 1-10.

McKinney, H.H. (1923) Influence of soil temperature and moisture on infection on wheat seedling by *Helmintosporium sativum*, J. Agric. res. 26, 195-217.

Ogoshi, A. (1987) Ecology and pathogenicity of anastomosis and intraspecific groups of *Rhizoctonia solani* Kühn, Annu. Rev. Phytopathol. 25, 125-143.

Olsnes, S. & Pihl, A. (1982) Toxic lectins and related proteins, in P. Cohen and S. van Heyningen (eds.), Molecular Action of Toxins and Viruses, Elsevier, New York, pp. 51-105.

Roby, D., Broglie, K., Cressman, R., Biddle, P., Chet, I. & Broglie, R. (1990) Activation of a bean chitinase promoter in transgenic tobacco plants by phytopathogenic fungi, Plant Cell 2, 999-1007.

Soave, C., Tardani, L., Di Fonzo, N. & Salamini, F. (1981) Zein level in maize endosperm depends on a protein under control of the *opaque-2* and *opaque-6* loci, Cell 27, 403-410.

Spooner, R.A. & Lord, M.J. (1990) Immunotoxins: status and prospects, Trends Biotechnol. 8, 189-193.

Stirpe, F., Barbieri, L., Battelli, M.G., Soria, M. & Lappi, D.A. (1992) Ribosome-inactivating proteins from plants: present status and future prospects, Bio/Technology 10, 405-412.

Walsh, T.A., Morgan, A.E. & Hey, T.D. (1991) Characterization and molecular cloning of a proenzyme form of a ribosome inactivating protein from maize, J. Biol. Chem. 266, 23422-23427.

Woloshuk, C.P., Meulenhoff, J.S., Sela-Burlage, M., van der Elzen, P.J.M. & Cornelissen, B.J.C. (1991) Pathogen induced proteins with inhibitory activity toward *Phytophtora infestans*, Plant Cell 3, 619-628.

G.T. Scarascia Mugnozza, E. Porceddu & M.A. Pagnotta (Eds.)
Genetics and Breeding for Crop Quality and Resistance, 83-92, 1999

Breeding for fungal resistance in *Musa*

J. B. Hartman & D. Vuylsteke
International Institute of Tropical Agriculture-Eastern and Southern Africa Regional Center, P.O. Box 7878, Kampala, Uganda

1. Introduction

Bananas and plantains are both important staple foods and cash crops for millions of people. Bananas are large perennial herbs of the genus *Musa*. Cultivated bananas are primarily triploids (*3x*) derived from intraspecific and interspecific crosses of two diploid species *Musa acuminata* Colla (*M.a.*) and *Musa balbisiana* Colla (*M.b.*) (Simmonds, 1995). In West and Central Africa, plantains provide more than 25% of the carbohydrate requirements for over 70 million people (Vuylsteke et al., 1993).In Uganda, Burundi, and Rwanda, consumption of cooking bananas exceeds 200kg per person per year.The spread of the fungal pathogen *Mycosphaerella fijiensis* Morelet into Africa has lead to serious declines in the productivity of banana and plantain based farming systems, and lead to the establishment of a plantain breeding program at the International Institute of Tropical Agriculture (IITA) in 1987 (Vuylsteke et al., 1993). Yield losses from fungal pathogens have been the main impetus for the establishment of banana breeding programs throughout the world (Buddenhagen, 1990).The first breeding program was established in 1922 at the Imperial College of Tropical Agriculture (ICTA) in Trinidad to produce dessert bananas resistant to *Fusarium oxysporum* f. sp. *cubense* (*F.o.c.*) the causal agent of Fusarium wilt or Panama disease (Simmonds, 1966). Breeding at ICTA was latter transferred to the Banana Board of Jamaica (BBJ). A second breeding program was established by United Fruit Company in Honduras to breed for resistance to yellow Sigatoka, *Mycosphaerella musicola* (*M.m.*), and Fusarium wilt and was later turned over to the Fundación Hondureña de Investigación Agricola (FHIA) (Rowe, 1984).These early breeding efforts have been reviewed by others (Simmonds, 1966; Sheperd, 1974; Stover & Buddenhagen, 1986; Rowe & Rosales, 1990; Ortiz et al., 1995).The past decade has seen the release of cultivars from breeding programs at FHIA, IITA, the Centre de Coopération Internationale en Recherche Agronomique pour le Dévelopment (CIRAD), the Centre Régional Bananiers et Plantains (CRBP), the Centro Nacional de Pesquisa de Mandioca Fruticultura of Empresa Brasileira de Pesquisas (CNPMF/EMBRAPA), the Instituto Nacional de Investigacion Viandas Tropicales (INIVIT), and the Taiwan Banana Research Institute (TBRI). These cultivars have resistance to Fusarium wilt, black

Sigatoka, or yellow Sigatoka, and many have multiple resistances (Rowe & Rosales, 1996). The methods used to produce these cultivars and the "breeding philosophies" behind them vary greatly. This paper will review the methods and philosophies of breeding for fungal resistance in *Musa* research, and discuss the future of new resistant banana and plantain cultivars and resistance breeding in *Musa*.

Once a pest or pathogen has been identified as a problem that requires a resistant cultivar(s), scientific breeding can be divided into five basic activities: i) characterization of the host/pathogen interaction, ii) development of screening procedures, iii) identifying sources of resistance, iv) studying inheritance of resistance, and v) designing and implementing a breeding program based on the results of steps i through iv.In a practical breeding program much of the research is concurrent, and steps iv and v are usually combined with breeding and selection commencing prior to a full understanding of inheritance. An ongoing breeding program is then modified to fit new information as it is obtained.

1.1 Major fungal diseases of banana

Fusarium wilt can cause complete destruction of susceptible plantations.Several reviews of the Fusarium wilt epidemic and its effect on the Gros Michel based export trade have been published (Stover, 1990a; Ploetz, 1994).The causal agent, *F.o.c.*, is a soil-borne filamentous fungus that colonizes and occludes the xylem of banana plants (Ploetz, 1994). Four races of *F.o.c.* have been identified by host differential testing (Ploetz, 1994). The emergence of subtropical race 4 on Cavendish clones and the apparent "breakdown" of resistance in 'IC-2' has raised questions about the durability of resistance to Fusarium wilt (Moore et al., 1991; Boehm et al., 1994).

The Sigatoka leaf spots are ascomycetes, very similar in their effect on susceptible cultivars. Both leaf spots reduce leaf area available for photosynthesis and cause premature senescence leading to yield losses of up to 50 % (Mobambo et al., 1993; Stover, 1990b). Black Sigatoka causes more severe damage, develops more rapidly, and attacks plantains, which are not normally attacked by yellow Sigatoka, and is regarded as a greater threat to *Musa* production (Stover, 1990b). Both Sigatoka leafspot diseases are widespread in banana growing areas (Mourichon & Fullerton, 1990). Yellow Sigatoka is generally a problem in highlands where black Sigatoka has failed to penetrate.

1.2 Screening procedures

The first Screening method for Fusarium resistance was developed for the breeding program at ICTA. Clones to be screened were planted into holes in the field from which Gros Michel plants that had succumbed to Fusarium wilt had been removed (Larter, 1947). The breeding programs at EMBRAPA and TBRI, and the germplasm screening programs at the South African Banana Board – Banana Plant Improvement Unit (BPIU) and Queensland Department of Plant Industry (QDPI) use similar "infested field" methods.At FHIA, clones are planted into holes in which chopped corm material from heavily infested plants has been placed (Rowe & Rosales, 1996). Disease severity is assessed by scoring outward symptoms (e.g.

yellowing of leaves) and evaluating the degree of discoloration of vascular bundles in the corm and pseudostem (Robinson, 1996 p. 181).

Similarly, most screening for Sigatoka leafspot resistance has been done using natural infestations in field tests. Several methods of scoring Sigatoka resistance in the field have been proposed, including youngest leaf spotted (Vakili, 1968), incubation time (Foure, 1982), disease development time (Foure, 1982), index of spotted leaves (Craenen & Ortiz, 1997), and index of non-spotted leaves (Ortiz et al., 1997). For purposes of selection for resistance, Craenen & Ortiz (1997) found that youngest leaf spotted at flowering sufficed to distinguish black Sigatoka resistant hybrids, whereas other measurements are likely more important for screening pathogen variation or characterizing the nature of host plant response (Foure et al., 1990). Experimental designs for field testing for Sigatoka leaf spot resistance have been highly developed (Nokoe & Ortiz, 1998).

Field screening for Fusarium and Sigatoka leafspots is relatively expensive in both time and resources due to the large size and long lifecycle of banana. Susceptible clones may also escape disease challenge due to edaphic variation in the field. Pathogen populations in the screening field may not resemble those in many locations where new cultivars are to be grown (*e.g.* most breeding programs are located in the tropics, whereas *F.o.c.* race 4 occurs mostly in temperate growing regions). Seedling screening and *in vitro* screening have been used with varying success. Vakili (1965) found seedlings resistant to *F.o.c.* in greenhouse flats were also resistant as mature plants in the field with the exception of progeny derived from some *M. accuminata* subspecies. Hwang & Ko (1987) also reported similar results from nursery screening and field screening for Fusarium wilt resistance. Stover (1987) recommended a two-tiered approach to Sigatoka screening in which seedlings were first screened by stapling small pieces of ascospore-bearing tissue to leaves, and then clones deemed resistant in the seedling screen were re-tested in the field. Fullerton & Olsen (1993) found that seedlings of many clones with a high level of field resistance to *M.f.* were susceptible as seedlings, thus seedling screening may be conservative, causing breeders to discard clones with field resistance. *In vitro* screening methods using mycelium, conidia, ascospores, crude extracts, and specific toxins have been used (Natural, 1990). *In vitro* screening using toxins does not always correlate well with results of *in vivo* screening for *F.o.c.* (Morpurgo et al., 1994) or *M.f.* (Harelimana et al., 1997). Thus, the utility of *in vitro* screening using toxins has been questioned (Harelimana et al., 1997).

1.3 Sources of resistance identified and characterized

Several authors have screened large numbers of clones for resistance to Fusarium wilt, yellow Sigatoka and black Sigatoka. Resistance to these pathogens is remarkably common among the wild and cultivated *Musa* species (Table 1). Although many clones with resistance to these diseases have been identified, surprisingly few have been used in breeding programs.

Inheritance studies have been conducted on only a handful of the resistant clones. Low seed-set, mixed ploidy in progeny, and non-disomic segregation complicate inheritance studies; however, several genetic models for inheritance of fungus resistance genes have been proposed (reviewed by Ortiz, 1995). Fusarium wilt resistance to race 1 derived from 'Pisang lilin' and 'Calcutta 4' is believed to be controlled by a single dominant gene (Larter, 1947; Vakili, 1965). Black sigatoka resistance derived from 'Calcutta 4' fits a model of a major recessive gene with dosage effects combined with two additive genes with dosage effects (Ortiz & Vuylsteke, 1994).

Table 1. Percentage of screened clones deemed tolerant or resistant to *Musa* fungal pathogens.

Pathogen	Clones Screened		Clones Resistant		% Resistant		Source
	cultivars	wild	cultivars	wild	cultivars	wild	
F.o.c. race 1	51	3	40	3	78	100	Waite, 1977
M. musicola	376	79	118	64	31	81	Vakili, 1968
M. fijiensis	43	4	32	3	74	75	Foureet al., 1990
	400	> 10	20	10	8	—	Vuylstekeet al., 97

Although it is known that many landraces and bred cultivars carry multiple resistances (*e.g.* to both black and yellow Sigatoka), it is not known whether or not these resistances are due to the same genes. Genetic models that fit patterns of inheritance of resistance to yellow Sigatoka were found to differ with the source of the resistance genes an indication that different genes were governing resistance in different clones (Vakili, 1968; Rowe, 1984; Shepherd, 1990). Unfortunately, complementation tests of resistance genes from different sources or for different pathogens have yet to be conducted in *Musa*.

Pedigree analysis of released cultivars identifies the primary sources of fungal resistance used in breeding programs. Fusarium and sigatoka resistance in the first bred cultivar 'IC-2' comes from wild diploid *M. a.* ssp. *malaccensis* whereas resistance in 'Bodles Altafort' is derived from the edible diploid 'Pisang lilin' (AA) believed to be derived from *M.a.* ssp. *malaccensis* (Simmonds, 1966). Pisang lilin and wild *M. a.* ssp. *burmannica* (primarily 'Calcutta 4') are the main sources of fungal resistance in IITA, EMBRAPA-CNPMF, CIRAD-FHLOR, CRBP and most FHIA released hybrids. The use of the same few sources of resistance has contributed to the high levels of relatedness between clones from the various breeding programs.

The design of an efficient breeding program requires not only an understanding of the inheritance of a single trait, but also of the relationship of that trait to other traits of interest. Several recent studies have clarified the relationship between resistance to black Sigatoka and agronomically important traits such as bunch weight. The BS_1 locus has been linked to the P_1 locus for fruit parthenocarpy (Ortiz & Vuylsteke, 1992, 1994). The net effect of the BS_1 locus on bunch weight varies from 12 to 39% (Craenen & Ortiz, 1996), while the combined effects of allele substitution at BS_1 and P_1 accounted for 88 % of the variation in bunch weight in progeny of one Plantain x Calcutta 4 cross (Ortiz et al., 1997).

The ability to use markers for indirect selection of important traits could greatly increase the efficiency of *Musa* breeding, especially when screening can only be done on mature plants (Ortiz & Vuylsteke, 1996; Crouch et al., 1998a). A molecular marker (SSRLP or microsatellite) associated with the P_1 locus has recently been reported (Crouch et al., 1998b). At present, IITA is using single marker analysis to identify candidate markers for marker assisted selection; however, the eventual mapping and integration of genetic and molecular markers will greatly enhance our understanding of the nature of fungal resistance in *Musa*

(Crouch et al., 1998a). In addition to new molecular methods, new biometrical models that allow the analysis of quantitative traits in multi-ploidy progeny from non-disomic inheritance should soon greatly enhance the ability of breeders to study the inheritance of fungal resistance and other important traits in *Musa* (Tenkouano, 1998 personal communications). When new biometrical methods are used in combination with rapid screening for euploidy and/or markers, previously impractical studies of inheritance in *Musa* will be possible (Dolezel et al., 1994; Tenkouano et al., 1998).

2. Musa improvement

The primary objective of the different *Musa* improvement programs around the world is to produce disease resistant triploid cultivars. Several strategies are being employed towards this objective. These strategies fall into three broad categories, classical breeding, mutation breeding, and transformation.

Classical breeding programs have employed various combinations of three distinct methods: i) diploid improvement, ii) polyploid improvement, and iii) "evolutionary" or "triploid breeding". The production of acceptable primary tetraploids (*i.e.* from *3x* landraces x *2x*) and triploid cultivars had been thought to be dependant solely on the production of improved diploids through diploid breeding (*i.e.* the "fixed genome theory", see Dodds, 1943). In plantains, two major factors have contributed to the production of acceptable primary tetraploids from crosses of *3x* landraces to Calcutta 4. One is the presence of favourable alleles masked by intra- and interlocus interactions, for example the presence of black Sigatoka resistance alleles in highly susceptible plantain landraces (Ortiz & Vuylsteke, 1994). The second is the occurrence of recombination during the formation of 2n megaspores (Vuylsteke et al., 1993; Crouch et al., 1998a). These discoveries opened up the possibility of polyploid improvement and led Ortiz (1997) to propose a method combining recurrent selection in both diploid and polyploid lines with selection for specific combining ability (SCA). Cycles of diploid improvement combined with cycles of polyploid improvement have lead to the release of the tertiary triploid FHIA-25 (Rowe, 1998) and to the IITA hybrids PITA-15, PITA-16, PITA-19 and PITA-20 (Ortiz et al., 1998). Stover & Buddenhagen (1986) saw the limitations of diploid breeding and proposed "evolutionary breeding" using colchicine to double improved *2x* to get *4x* then cross to other improved *2x* to obtain improved *3x* cultivars. The breeding program at CIRAD-FHLOR uses an adaptation of this technique known as "triploid breeding" and have recently released the black and yellow sigatoka resistant triploid cultivar, 'IRFA-909'(Tézenas du Montcel et al., 1996).

Hybrids from Gros Michel and 'Highgate' have never matched the quality of clones of the Cavendish subgroup (Stover & Buddenhagen, 1986). Researchers have cited the failure of traditional breeding methods to produce acceptable clones as a justification for mutation breeding. The search for somaclonal variants within the commercial standard Cavendish subgroup has produced clones resistant to *F.o.c.* race 4. (Hwang & Tang, 1996), but clones with acceptable horticultural characters have not yet been produced (de Beer, 1993). Although no commercially acceptable disease resistance clones have yet been produced, researchers continue to argue that this will be the quickest way to fungus resistant clones acceptable to farmers, shippers, and consumers (Hwang & Tang, 1996). Somaclonal and other mutation breeding methods have been employed to identify host plant resistance to Race 4 of *Fusarium* in Taiwan, Australia, and South Africa (Ortiz & Vuylsteke, 1996).

Like mutation-derived clones, transgenic clones offer the advantage of combining the desirable quality traits of current standard cultivars with resistance genes unlinked to undesirable traits (Panis et al., 1996). The past 10 years has seen the development of many of the techniques required for the transformation of *Musa* clones. Protocols for the production and regeneration of plants from cell suspension and protoplast culture have been developed (Novak et al., 1989). Methods for inserting transgenes, including electroporation (Sagi et al., 1994), particle bombardment (Sagi et al., 1995) and cocultivation with *Agrobacterium* (May et al., 1995) have been devised. Transgenes for antifungal proteins have been successfully inserted and expressed in *Musa* at Katholieke Unversiteit Leuven, Belgium (Remy et al., 1998). The effectiveness of the chitinase gene *Chi2* as an antifungal transgene is scheduled to be tested by Centro Agronómico Tropical de Investigacíon y Enseñanza in Costa Rica (Panis et al., 1996). These transgenic plants must overcome several regulatory barriers before field-testing can proceed.

2.1 Pathogen diversity and durability of resistance

2.1.1 Pathogen Diversity and Variation in Pathogenicity

Much has been gained in our knowledge of pathogen diversity and its implications for the durability of resistance in the past few years. Four races of *F.o.c.* are recognized using a host differential assay (Stover & Simmonds, 1987). Recent studies of diversity within *F.o.c.* have identified at least sixteen vegetative compatibility groups (Boehm et al., 1994; VCG). Races of *F.o.c.* are found across VCG and multiple races are found within single VCG. Within races within VCG variation in chromosome number, genome size, RFLP patterns, and genomic sequences have been identified (Boehm et al., 1994). This evidence points to convergent evolution of pathogenicity on different *Musa* clones (O'Donnell et al., 1998). Although molecular, genetic, and VCG evidence strongly supports coevolution of *Musa* and *F.o.c.* in Southeast Asia followed by the spread of the pathogen through infested planting material (Ploetz, 1994), virulence on specific clones (*e.g.* race 4 on Cavendish) appears to have arisen in several *Fusarium* lineages.

Morpurgo et al. (1994) have demonstrated an elicitor induced systemic defense response in *Musa* to a *F.o.c.*; however, it is not believed that the *Musa*/*F.o.c.* interaction is gene-for-gene (Stover & Buddenhagen, 1986). The question of the origin of *F.o.c.* host specificity has serious implications for *Fusarium* resistance breeding. If virulence on specific *Musa* clones arose convergently in different *Fusarium* lineages (O'Donnel et al., 1998), it is likely that virulence is governed by relatively few genes and currently resistant cultivars may "breakdown". If virulence on specific *Musa* clones predates mutations at *het* loci, all isolates within a VCG have the same virulence and differences in pathogenicity of VCG's over environments are due to genotype by environment interactions (S. Bentley, 1998 personal communication). Studies that examine differences in pathogenicity of the same races from different lineages in the same environment are needed. An expansion of host differential testing to include clones being used as sources of *F.o.c.* resistance in breeding programs has been recommended (Buddenhagen, 1990). Without knowledge about the frequency of isolates virulent on resistant clones, meaningful predictions about the durability of new *Fusarium* wilt resistant cultivars cannot be made.

M.f. and *M.m.* are also believed to have coevolved with *Musa* in Southeast Asia and then spread by accidental introductions on contaminated planting material to their present distribution (Stover, 1980; Mourichon & Fullerton, 1990). Analysis of DNA polymorphisms in *M.f.* and *M.m.* have revealed a great degree of both inter and intraspecific variation (Carlier et al., 1996). The DNA evidence supports origins in Southeast Asia with limited introductions into other banana growing regions. Although diversity in Africa, Latin America, and the Pacific Islands was much less than that in Southeast Asia, indicating founder effects, significant diversity was still present in areas where black Sigatoka has only been recently introduced. Although races of *M.f.* and *M.m.* have not been defined, differential screening has revealed variation in pathogenicity of *M.f.* and *M.m.* isolates on resistant cultivars (Vakili, 1968; Fullerton & Olsen, 1993). Field resistance to black Sigatoka in at least two cultivars 'Paka' and 'T8' has broken down after 8 years of exposure in the Cook Islands. It has been suggested that resistant clones be tested over several years in Southeast Asia in order to determine the frequency of virulent isolates and to predict the durability of resistance (Fullerton & Olsen, 1993; Carlier et al., 1996).

2.1.2 Multi-location and Multi-Isolate Testing of Resistance

The preceding section shows the great diversity within the pathogen populations. Locational variation in host response has also been reported in several studies. Perhaps the best known example is that of *F.o.c.* race 4 resistance in Cavendish clones. In the tropics, Fusarium wilt resistance in Cavendish clones has endured for more than 40 years, and yet race 4 is prevalent in temperate growing regions (Buddenhagen, 1990). Fusarium wilt resistance in TBRI clones did not appear to be as strong in South Africa as in Taiwan where they were selected (de Beer, 93). Resistance to Fusarium wilt in several FHIA clones has been relatively stable in tests around the world, but differences in resistance to race 4 isolates from Taiwan and Australia have been reported (Daniells et al., 1995).

As was stated above, there is a need to expand testing of resistant clones in order to make any predictions about the durability of resistance. Fullerton & Olsen (1993) identified isolates from Papua New Guinea and the Pacific Islands virulent on Calcutta 4 and Pisang lilin. These same tests found that resistance in 'Tuu Gia' and 'TU8' are relatively more stable. In the absence of any knowledge of mitigating factors (*i.e.* initial frequency of virulence alleles, degree of stabilizing selection for avirulence, mutation rates at virulence loci), stability in multi-isolate or multi-locational trials is the best estimate of durability. Two stages of multilocational testing occur in IITA's breeding program. A multilocational evaluation trial (MET) that is conducted at several sites, and an advanced Musa yield trial (AMYT) that may be planted in 20 or more locations in countries throughout Africa (Ortiz et al., 1997). Calcutta 4 and its hybrid progeny were found to exhibit stable resistance to black sigatoka in multilocational trials throughout the banana growing regions of Africa (Ortiz et al., 1997) and has held up at the original breeding station for more than a decade. However, pathogen diversity is much less in Africa than it is in Southeast Asia and the Pacific (Ploetz, 1994; Carlier et al., 1996). In response to the need for global testing of new cultivars, the International Network for The Improvement Of Banana and Plantain (INIBAP) initiated an International *Musa* Testing

Program (IMTP) to evaluate new banana and plantain hybrids for resistance to black Sigatoka and Fusarium wilt (Jones, 1994).

Resistance genes themselves may be more stable depending upon their mode of action and the particular type of resistance they control. Calcutta 4 expresses two types of resistance, a hypersensitive response (HR) which stops all development of the pathogen, and partial resistance (PR), which merely slows pathogen development down (Ortiz & Vuylsteke, 1994). It is proposed that PR will be more stable and difficult to overcome than HR, which has already failed in Papua New Guinea (Craenen & Ortiz, 1996). A HR is often associated with race specificity in the pathogen and a gene-for-gene host pathogen interaction; however this theory has not been tested in *Musa*. It has also been proposed that recessive resistance, such as that regulated by bs_1, is more stable than dominant resistance because virulence requires a rare mutation to the dominant allele of a virulence locus (Ortiz & Vuylsteke, 1994). In the meantime, efforts are underway to broaden sources of resistance at IITA.

3. Conclusions

After 76 years of breeding bananas, fungal diseases are still the primary driving force behind banana breeding. We can finally say that breeding programs are putting new fungal resistant cultivars in farmer's fields, and the final stage of breeding, farmer and consumer selection, has begun. Many more new cultivars will soon be reaching farmers. In the next, few years we will test transgenic bananas for fungal resistance, and the initiation of MAS for fungal resistance is not too far off. In the meantime, we will continue to accumulate more knowledge about *Musa*, *Musa* diseases, and *Musa* breeding that should allow us to answer important questions about the durability of fungal resistance genes. This review shows that much has been gained in our knowledge of both pathogen diversity and *Musa* genetics in the past few years. What we are still lacking is a clear understanding of the nature of the interactions between *Musa* and its major fungal pathogens. How does host plant resistance in *Musa* affect pathogen populations? The answers to this question should shape the course of future *Musa* breeding.

4. References

Boehm, E.W.A., Ploetz, R.C. & Kistler, H.C. (1994) Statistical analysis of electrophoretic karyotype variation among vegetative compatibility groups of *Fusarium oxysporum* f. sp. *cubense*. *Mol. Plant Microbe Interact.* 7, 196-207.

Buddenhagen, I.W. (1990). Banana breeding and fusarium wilt. In R.C. Ploetz (eds.) Fusarium wilt of bananas. APS Press. St. Paul. Minn. pp 107-113.

Carlier, J., Lebrun, M.H., Zapater, M.F., Dubois, C. & Mourichon, X. (1996). Genetic structure of the global population of banana black leaf streak fungus, *Mycosphaerella fijiensis*. *Mol. Ecol.* **5**, 499-510.

Craenen, K. & Ortiz, R. (1996) Effect of the black Sigatoka resistance gene bs_1 and ploidy level on fruit and bunch traits of plantain-banana hybrids, *Euphytica* **87**, 97-101.

Craenen, K. & Ortiz, R. (1997) Influence of black Sigatoka disease on the growth and yield of diploid and tetraploid hybrid plantains, *Crop Protection*. **16**,

Crouch, H.K., Crouch, J.H., Jarret, R.L., Cregan, P.B. & Ortiz, R. (1998a) Segregation of microsatellite loci from haploid and diploid gametes in *Musa*. *Crop Sci.* **38**, 211-217.

Crouch, J.H., Ortiz, R., Crouch, H.K., Ford-Lloyd, B.V., Howell, E.C., Newbury, H.J. & Jarret, R.L. 1998b. Utilization of molecular genetic techniques in support of plantain and banana improvement. *Acta Horticulturae*, in press

Daniells, J., Pegg, K., Searle, C., Whiley, T., Langdon, P., Bryde, N. & O'Hare, T. (1995) Goldfinger in Australia: a banana variety with potential. *InfoMusa* **4**, 5-6.

De Beer, Z.C. (1993) Breeding for resistance to *Fusarium* wilt in South Africa. In R.V. Valmayor, S.C. Hwang, R. Ploetz, S.W. Lee & V.N. Roa (eds.). *Proceedings: International Symposium on Recent Developments in Banana Cultivation Technology.* Chiuju, Pingtung, Taiwan, 14-18 December 1992. INIBAP/ASPNET. Los Baños, Philippines. pp. 75-83.

Dodds, K.S. (1943) The genetic system of banana varieties in relation to banana breeding. *Emp. J. Exp. Agric.* **11**, 89-98.

Dolezel, J., Dolezelova, M. & Novak, F.J. (1994) Flow cytometric estimation of nuclear DNA amount in diploid bananas (*Musa acuminata* and *M. balbisiana*), *Biologia Plantarum* **36**, 351-357.

Foure, E. (1982) Etude de la sensibilité variétale des bananiers et plantains à *Mycosphaerella fijiensis* au Gabon. *Fruits* **37**, 749-770.

Foure, E., Mouliom Pefoura, A. & Mourichon, X. (1990) Etude de la sensibilité variétale des bananiers et plantains à *Mycosphaerella fijiensis* Morelet au Cameroun: Caractérisation de la résistance au champ de bananiers appartenant à divers groupes génétiques. *Fruits* **45**, 339-345.

Fullerton, R.A. & Olsen, T.L. (1993) Pathogenic diversity in *Mycosphaerella fijiensis* Morelet. In J. Ganry (eds.) *Breeding banana and plantain for resistance to diseases and pests*: Proceedings of a conference held at Montpellier, France September 7-9, 1992 CIRAD and INIBAP. Montpellier, France, pp. 201-211.

Harelimana, G., Lepoivre, P., Jijakli, H. & Mourichon, X. (1997) Use of *Mycoshpaerella fijiensis* toxins for the selection of banana cultivars resistant to black leaf streak. *Euphytica* **96**, 125-128.

Hwang, S.C. & Ko, W.H. (1987) Somaclonal variation of bananas and screening for resistance to *Fusarium* wilt. In G.J. Persley & E.A. De Langhe (eds.), *Banana and plantain breeding strategies*: proceedings of an international workshop held at Cairns, Australia, October 13-17, 1986. ACIAR, Canberra, Australia pp. 151-156.

Hwang, S.C. & Tang, C.Y. (1996) Somaclonal variation and its use for improving Cavendish (AAA dessert) bananas in Taiwan. In E.A. Frison, J-P. Horry & D. De Waele (eds.) *New Frontiers in Resistance Breeding for Nematode, Fusarium and Sigatoka.* INIBAP, Montpellier, France pp. 173-181.

Jones, D.R. (1994) International *Musa* Testing Program. In D.R. Jones (eds.)*The Improvement and Testing of Musa: a Global Partnership.* INIBAP, Montpellier, France, pp. 12-20, 23-31.

Langdon, P.W. & Pegg, K.G. (1988) Field screening for resistance to *Fusarium* wilt: *ACIAR Banana Improvement Project for the Pacific and Asia Research Report.*

Larter, L.N.H. (1947) *Report on Banana Breeding*: *Bulletin No. 34*. Dept. Agric. Jamaica, Kingston, Jamaica, 24 p.

May, G., Afza, R., Mason, H., Wiecko, A., Novak, F. & Arntzen, C. (1995) Generation of transgenic banana (*Musa acuminata*) plants via *Agrobacterium*-mediated transformation. *Bio/Technology* **13**, 486-492.

Maynard-Smith, J. (1989) Evolutionary genetics. Chapman Hall. New York.

Mobambo, K.N., Gauhl, F., Vuylsteke, D., Ortiz, R., Pasberg-Gauhl, C. & Swennen, R. (1993). Yield loss in plantain from black Sigatoka leaf spot and field performance of resistant hybrids. *Field Crops Res.* **35**, 35-42.

Moore, N.Y., Hargreaves, P.A., Pegg, K.G. & Irwin, J.A.G. (1991) Characterization of strains of *Fusarium oxysporum* f.sp. *cubense* by production of volatiles. *Australian J. Bot.* **39**, 161-166.

Morpurgo, R., Lopato, S.V., Afza, R. & Novák, F.J. (1994) Selection parameters for resistance to *Fusarium oxysporum* f.sp. *cubense* race 1 and 4 on diploid banana (*Musa acuminata* Colla). *Euphytica* **75**, 121-129.

Mourichon, X. & Fullerton, R.A. (1990) Geographical distribution of the two species *Mycosphaerella musicola* Leach (*Cercospora musae*) and *M. fijiensis* Morelet (*C. fijiensis*), respectively agents of sigatoka disease and black leaf streak disease in bananas and plantains. *Fruits* **45**, 213-218.

Natural, M.P. (1990) An update on the development of an *in vitro* screening procedure for resistance to Sigatoka leaf diseases of banana. In RA Fullerton & RH Stover (eds.) *Sigatoka leaf spot diseases of bananas.* INIBAP, France. pp. 208-230.

Nokoe, S. & Ortiz, R. 1998. Optimal plot size for banana trials, *HortScience* **33**, 130-132.

Novak, F.J., Afza, R., van Duren, M., Perea-Dallos, M., Conger, B.V. & Xiaolang, T. (1989) Somatic embryogenesis and plant regeneration in suspension cultures of dessert (AA and AAA) and cooking (ABB) bananas (*Musa* spp.). *Biotechnology* **7**, 147-158.

O'Donnell, K.O., Corby Kistler, H., Cigelnik, E. & Ploetz, R.C. (1998) Multiple evolutionary origins of the fungus causing Panama disease of banana: Concordant evidence from nuclear and mitochondrial gene genealogies. *Proc. Natl. Acad. Sci. USA* **95**, 2044-2049.

Ortiz, R. (1995) *Musa* genetics. In S. Gowen (eds.).*Bananas and Plantains.* Chapman and Hall, London, pp 84-109.

Ortiz, R. (1997) Secondary polyploids, heterosis, and evolutionary crop breeding for further improvement of the plantain and banana (*Musa* spp. L.) genome. *Theor. Appl. Genet.* **94**, 1113-1120.

Ortiz, R., Ferris, R.S.B. & Vuylsteke, D.R. (1995) Banana and plantain breeding. In S. Gowen (eds.).*Bananas and Plantains.* Chapman and Hall, London, UK. pp 110-146.

Ortiz, R. & Vuylsteke, D.R. (1992) Inheritance of black sigatoka resistance and fruit parthenocarpy in triploid AAB plantains. *Agronomy Abstracts*, **84**, 109.

Ortiz, R. & Vuylsteke, D.R. (1994) Inheritance of black sigatoka disease resistance in plantain-banana (*Musa* spp.) hybrids. *Theor. Appl. Genet.* **89**, 146-152.

Ortiz, R. & Vuylsteke, D.R. (1996) Recent advances in *Musa* genetics, breeding and biotechnology. *Plant Breeding Abstracts* **66**, 1355-1363.

Ortiz, R., Vuylsteke, D., Ferris, R.S.B., Okoro, J.U., Guessan, A.N., Hemeng, O.B., Yeboah, D.K., Afreh-Nuamah, K., Ahiekpor, E.K.S., Foure, E., Adelaja, B.A., Ayodele, M., Arene, O.B., Ikediugwu, F.E.O., Agbor, A.N., Nwogu, A.N., Okoro, E., Kayode, G., Ipinmoye, I.K., Akele, S. & Lawrence, A. 1997. Developing new plantain varieties for Africa. Plant Varieties and Seeds 10, 39-57.

Ortiz, R., Vuylsteke, D.R., Crouch, H.K. & Crouch, J.H. (1998) TM3x: triploid black sigatoka-resistant *Musa* hybrid germplasm. *Hortscience* **33**, 362-365.

Panis, B., Côte, F., Escalant, J.V. & Sagi, L. (1996) Aspects of genetic engineering in banana. in E.A. Frison, J-P. Horry & D. De Waele (eds.) *New Frontiers in Resistance Breeding for Nematode, Fusarium and Sigatoka*. INIBAP, Montpellier, France pp. 182-198.

Ploetz, R.C. (1994) Panama disease: return of the first banana menace. *Intern. J. Pest Manag.* **40**, 326-336.

Remy, S., Francois, I., Schoofs, H., Panis, B., Cammue, B., Swennen, R. & Sagi, L. (1998) Genetic transformation as a technology to create disease resistance in banana. *Acta Horticulturae* **In press.**

Robinson, J.C. (1996) *Bananas and Plantains*. CAB International, Wallingford, UK.

Rowe, P. (1984) Breeding bananas and plantains. *Plant Breeding Reviews* **2**, 135-155.

Rowe, P.R. (1998) A breakthrough in breeding cooking bananas resistant to black sigatoka. *FHIA An. Rep.*

Rowe P.R. & Rosales, F.E. (1990) Breeding bananas and plantains with resistance to black Sigatoka. In R.A. Fullerton and R.H. Stover (eds.) *Sigatoka Leaf Spot Diseases of Bananas*. INIBAP, Montpellier, pp. 243-251.

Rowe P. & Rosales, F.E. (1996) Bananas and plantains. in J. Janick & J. Moore (eds.). *Fruit Breeding Volume I: Tree and Tropical Fruits*. John Wiley, New York, USA. pp. 167-211.

Sagi L., Remy S., Panis B., Swennen R. & Volckaert G. (1994) Transient gene expression in electroporated banana (*Musa*, cv. Bluggoe, AB group) protoplasts isolated from regenerable embryogenic cell suspensions. *Pl. Cell Rep.* **13** 262-6.

Sagi, L., Panis, B., Remy, S., Schoofs, H., De Smet, K., Swennen, R. & Cammue, B. (1995) Genetic transformation of banana (*Musa* spp.) via particle bombardment. *Bio/Technology* **13**, 481-485.

Sheperd, K. (1974) Banana research at I.C.T.A. *Trop. Agric. Trin.* **51**, 482-490.

Simmonds, N.W. (1966) *Bananas*. 2nd ed. Longman Group Ltd., Essex, UK.

Simmonds, N.W. (1995) Bananas: *Musa* (Musaceae). In J. Smartt & N.W. Simmonds (eds.) *Evolution of Crop Plants*. 2nd ed., Longman Scientific and Technical, Essex, UK pp. 370-375.

Stover, R.H. (1987) Measuring response of *Musa* cultivars to Sigatoka pathogens and proposed screening procedures. In G.J. Persley & E.A. De Langhe (eds.), *Banana and plantain breeding strategies*: proceedings of an international workshop held at Cairns, Australia, October 13-17, 1986. ACIAR, Canberra, pp. 114-118.

Stover, R.H. (1990a) *Fusarium* wilt of banana: some history and current status of the disease. In R.C. Ploetz (eds.) *Fusarium Wilt of Banana*, APS Press, St. Paul, MN, USA. pp. 1-70.

Stover, R.H. (1990b) Sigatoka leaf spots, thirty years of changing control strategies, 1959-1989. In R.A. Fullerton & RH Stover (eds) *Sigatoka leaf spot diseases of bananas*. INIBAP, Montpellier. pp. 66-74.

Stover, R.H. & Buddenhagen, I. (1986) Banana breeding: polyploidy, disease resistance and productivity. *Fruits* **41**, 175-191.

Stover, R.H. & Simmonds, N.W. (1987) *Bananas*. 3rd ed., Longman Scientific and Technical, Essex UK.

Tenkouano, A., Crouch, J.H., Crouch, H.K. & Ortiz, R. (1998). Genetic diversity, hybrid performance, and combining ability for yield in *Musa* germplasm. *Euphytica* 102(3): 281-288.

Tézenas du Montcel, H., Carreel, F. & Bakry, F. (1996) Improve the diploids: the key for banana breeding. In E.A. Frison, J-P. Horry & D. De Waele (eds.) *New Frontiers in Resistance Breeding for Nematode, Fusarium and Sigatoka*. INIBAP, Montpellier, France pp. 173-181.

Vakili, N.G. (1965) Fusarium wilt resistance in seedlings and mature plants of *Musa* species. *Phytopath* **55**, 135-140.

Vakili, N.G. (1968) Responses of *Musa acuminata* species and edible cultivars to infection by *Mycosphaerella musicola*. *Trop. Agric. Trin.* **45**, 13-22.

Vuylsteke, D., Swennen, R. & Ortiz, R. (1993) Development and performance of black sigatoka-resistant tetraploid hybrids of plantain (*Musa* spp., AAB group). *Euphytica* **65**, 33-42.

Vuylsteke, D., Ortiz, R., Ferris, S.B. & Crouch, J.H. (1997) Plantain improvement. *Plant Breeding Reviews* **14**, 267-320.

Waite, B.H. (1977) Inoculation studies and natural infection of banana varieties with races 1 and 2 of *Fusarium oxysporum* f.sp. *cubense*. *Plant Disease Reporter* **61**, 15-19.

G.T. Scarascia Mugnozza, E. Porceddu & M.A. Pagnotta (Eds.)
Genetics and Breeding for Crop Quality and Resistance, 93-100, 1999

Mapping different resistances against downy mildew in sunflower

L. Brahm, T. Röcher, R. Horn, M. Prüfe & W. Friedt
Institute of Crop Science and Plant Breeding, University Giessen; Ludwigstrasse 23, D-35390 Giessen, Germany

Abstract: Downy mildew caused by *Plasmopara halstedii* is one of the major dangers of sunflower in most of the world's growing areas. A number of major dominant resistance genes (*Pl* genes) have been either identified in cultivated sunflower or were introduced from wild *Helianthus annuus* or other wild *Helianthus* species. However, many aspects of the resistance genetics remain unclear. RAPD and AFLP analysis of near isogenic lines differing in the Pl_2 locus and bulked segregant analysis (BSA) of F_2 populations, segregating for race 7 and 9 resistances originating from *H. annuus* (Pl_2), and race 5 resistance derived from *H. annuus* (Pl_6) and *H. argophyllus* (Pl_{arg}), respectively, were used to identify molecular markers for different sources of resistance against downy mildew. On the basis of these markers, linkage maps for both populations were constructed representing the genomic regions carrying the respective resistance locus. The investigations confirmed the close relationship of Pl_2 and Pl_6, whereas no association was found for loci Pl_2 and Pl_{arg}.

1. Introduction

Downy mildew of sunflower caused by *Plasmopara halstedii* [Farl.] Berl. & de Toni is one of the main diseases in most of the sunflower growing areas of the world. The incidence of the disease in the field ranges from traces up to 90% under extreme conditions, depending on the inoculum potential, the variety used, and the occurrence and intensity of rainfall during the planting period (Viranyi, 1992). A number of major resistance genes have been either identified in cultivated sunflower or were introduced from wild *Helianthus annuus* or other wild *Helianthus* species (e.g., Miller, 1992; Korell et al., 1996).

These dominant resistance genes have been designated as *Pl* genes providing resistance either to a single race of downy mildew or against two or more races of the pathogen (Miller, 1992). Some classical segregation experiments suggested that the different *Pl* genes are inherited independently (Zimmer & Kinman, 1972; Miller & Gulya, 1991), whereas others (Mouzeyar et al., 1992) contradict these results with regard to certain *Pl* genes. Mapping of genes Pl_1, Pl_2 and Pl_6, originating from wild *H. annuus* or identified in cultivated sunflower itself, showed that these genes share the same map position (Mouzeyar et al.,1996; Reockel-Drevet et al., 1996). Moreover, it was demonstrated by Vear et al. (1997) that the "locus" Pl_6, providing resistance against all known races of sunflower downy mildew, consists of at least two genes or groups of genes conferring resistance against French races A, B and C (=race 3) and 1 and D (=race 2), respectively.

However, particular aspects of the resistance genetics remain unclear. For example, it is still not proven whether genes of different origin are inherited independently. Additionally, it is unknown, if resistance genes other than Pl_6, which are effective against more than one race of sunflower downy mildew, represent clusters of tightly linked genes or single genes.

Mapping experiments involving different sources of resistance against downy mildew of sunflower will help to answer some of the questions concerning the genetics of Plasmopara resistance as already demonstrated. This report deals with resistance against a field isolate comparable with races 7 and 9 and resistance against race 5. Resistant genotypes originate from inbred AS110Pl_2 (Pl_2), and from lines HA335 (Pl_6) and ARG1575-2 (Pl_{arg}). Since collocation of different *Pl* genes has been proven for three genes, all derived from *H. annuus* (Pl_1, Pl_2, Pl_6), it will be of special interest whether resistance genes originating from interspecific crosses to other *Helianthus* species also share the same genomic location, e.g. race 5 resistance derived from *H. argophyllus* in ARG1575-2.

2. Materials and Methods

2.1 *Sunflower genotypes*

Table 1. Resistant and susceptible inbred lines used as parents for the development of segregating populations.

F_2 population	Cross female x male	segregating for resistance against
I	HA89(cms) x AS110Pl_2(Pl_2)	races 1, 2, 7 + 9
II	HA342(cms) x ARG1575-2(Pl_{arg})	all races
III	AS110Pl_2(Pl_2) x HA335(Pl_6)	all races

An overview of the crosses which we carried out for this work is given in Table 1. Either cytoplasmic male sterile (cms) inbred lines as females and inbred restorer lines as males (I and II) or an emasculated restorer line as a female and a maintainer as a male (III) were used in the different cross combinations. In addition, two sets of near isogenic lines AS110/AS110Pl_2 and S1358/S1358Pl_2 were used for the identification of molecular markers. Both sets were originally developed by Dr. V.A. Vranceanu (Fundulea, Romania) with special reference to the Pl_2 locus, conferring resistance against downy mildew races 1, 2, 6, 7 and 9. Another two sets of near isogenic lines (NIL) differing in the *Pl6* locus were donated by Maïsadour (France). These lines, P7/PEM77 and CR2/CEM22, were screened with the markers identified for the *Pl2* gene. The resistant lines PEM77 and CEM22 were developed via 6 backcross generations using HA335 and 7 backcross generations with HA336 as a donor parent, respectively.

2.2 Resistance tests

Downy mildew resistance was tested applying the whole seedling immersion method described by Gulya et al. (1991). Resistance of F_2 plants was reinvestigated by testing 20-24 F_3 germlings per F_2 individual. Symptoms were observed two weeks after inoculation, following 72h under a saturated atmosphere. Population I was screened by using a field isolate (GG-F5), collected at our breeding station at Gross-Gerau near Frankfurt/Main. This isolate reacts similarly to American races 7 and 9, according to tests of USDA lines HA821, RHA266, RHA274, DM2, 799-2, and HA335 as differentials (Spring & Rozynek, personal communication). Populations II and III were tested with race 5 donated by Prof. F. Viranyi (Gödöllö, Hungary).

Depending on the host-pathogen-interaction, resistance was defined as absence of sporulation (population I), and absence or slight sporulation on cotyledons (populations II and III), which is known as cotyledon limited infection (Sackston, 1992). Coteledon limited infection was not observed in population I.

2.3 Molecular analysis

DNA was extracted from leaf material collected at flowering according to Doyle & Doyle (1990). Bulked segregant analysis (BSA) of populations I, II and III was performed according to Michelmore et al. (1991). For all populations, resistant and susceptible bulks were compiled consisting of 10 homozygous resistant and homozygous susceptible plants, each. Molecular markers for resistance against field isolate GG-F5 (population I, *Pl2*) were developed by analysing near isogenic lines (RAPDs) and bulks (RAPDs, AFLPs), respectively. Markers for race 5 resistance (population II, RAPDs) were identified by investigating bulks only. Near isogenic lines P7/PEM77 and CR2/CEM22 as well as bulks of population III were screened by using the RAPD markers identified for the *Pl2* locus. RAPD analysis was performed as described by Brahm & Friedt (1996) employing Operon decamer primers. AFLP markers were generated by using AFLP Analysis System I (Gibco BRL/Life Technologies). *Eco*RI specific primers were labelled with [γ-^{33}P]dATP (NEN Life Science). Amplification and fragment analysis were carried out according to the suppliers protocols.

2.4 Linkage analysis

Linkage maps for the specific *Pl* regions in all populations were constructed using Mapmaker 3.0 software (Lander et al., 1987). Map units were computed by applying the Kosambi function (Kosambi, 1944). Linkage groups were identified by utilizing the default settings. Initial orders were determined with the "compare" command following three point analysis. Additional markers were integrated into the maps using the "build" command with a strict exclusion threshold of LOD 3.0. Final map orders were tested with the "ripple" command.

3. Results

3.1 Resistance tests

Population I derived from the cross HA89(cms) x AS110*Pl₂* was segregating for resistance against the field isolate GG-F5 (similar to sunflower downy mildew races 7 and 9), whereas populations II and III were segregating for race 5 resistance. Homozygous and heterozygous F_2 individuals were distinguished following the test of F_3 families of each F_2 plant. Subsequently, segregation pattern in all F_2 populations fitted a 1:2:1 ratio as is expected for a single dominant trait (Table 2).

Table 2. Segregation ratios for resistance to downy mildew of sunflower in the F_2 generation.

Population	Cross	Segregation (RR : RS : SS) observed	tested	χ^2	P (DF = 2)
I	HA89(cms) x AS110*Pl*$_2$	43 : 63 : 35	1 : 2 : 1	2.50	0.29
II	HA342(cms) x ARG1575-2	32 : 63 : 33	1 : 2 : 1	0.04	0.97
III	AS110*Pl*$_2$ x HA335	32 : 56 : 24	1 : 2 : 1	1.14	0.57

3.2 Identification of markers

A total of 380 RAPD primers were screened using the NIL pairs AS110/AS110*Pl*$_2$ and S1358/S1358*Pl*$_2$. Only 21 primers which generated fragments polymorphic between both sets of near isogenic lines were considered for further analyses (Brahm & Friedt, 1996). Three primers identified a marker for the resistance locus (Table 3). Subsequently, the bulks of Population I (HA89(cms) x AS110*Pl*$_2$) were tested with an additional set of 500 RAPD primers. Five primers generated fragments with linkage to resistance locus *Pl*$_2$ (Table 3). Moreover, 23 Eco/Mse primer combinations used in an AFLP analysis detected two markers (E3548_3 and E4162_2) generated by primer combinations E35/M48 and E41/M62 for the resistance gene.

The RAPD markers detected before with regard to the *Pl*$_2$ locus were then screened using the near isogenic lines differing in the *Pl*$_6$ locus. Since Mouzeyar et al. (1996) showed the collocation of genes *Pl*$_1$, *Pl*$_2$ and *Pl*$_6$, this markers should be useful also when dealing with the *Pl*$_6$ locus. Primer AA11 differentiated between P7(s) and PEM77(r) as well as CR2(s) and CEM22(r), generating an additional fragment on lines P7 and CEM22 of a size similar to *Pl*$_2$ marker AA11-1008. Primers A02, C12, AA14 and AC20 only discriminated between near isogenic lines P7 and PEM77. They generated an additional fragment on the susceptible line P7 of the same size as the original markers A02_630, C12_459, AA14_750 and AC20_831 (Table 3).

Table 3. RAPD markers identified for the *Pl*$_2$ gene analysing near isogenic lines and bulks, respectively.

Primer	Fragment size (bp)	HA89(cms)	AS110*Pl*$_2$	HA335	CR2	CEM22	P7	PEM77
A02	630	-	+	-	-	-	+	-
B08	730	-	+	-	-	-	-	-
C12	459	-	+	-	-	-	+	-
O04	486	+	-	-	-	-	-	-
Z15	700	-	+	-	-	-	-	-
AA11	1008	-	+	+	-	+	+	-
AA14	750	-	+	-	-	-	+	-
AC20	831	-	+	-	-	-	+	-
AS12	280	+	-	-	-	-	-	-

+ = fragment present, - = fragment absent

All RAPD primers which detected a marker for the *Pl*$_2$ locus were also used to analyse the bulks of population III. All markers with linkage to the resistance allele of the *Pl*$_2$ locus with exception of marker AA11_1008 differentiated between bulks resistant and susceptible to race 5 of downy mildew.

Bulked Segregant Analysis of population II (HA342(cms) x ARG1575-2) was carried out with 400 RAPD primers. Fragments C02_360, R03_490, AE15_640, AI06_410 and AJ15_510 generated by the corresponding Operon primers showed linkage to *Pl*$_{arg}$ (race 5 resistance).

3.3 *Construction of linkage maps*

Markers for *Pl*$_2$, conferring resistance against downy mildew races 7 and 9, respectively, in population I were used to construct a linkage map of that particular region of the sunflower genome using 132 F_2 plants (Köhler et al., 1997). Subsequently, this partial linkage map was integrated into an AFLP linkage map which was constructed for population I with a set of 90 F_2 individuals (data not shown). The resistance locus is flanked by AFLP marker E35M48_3 and RAPD marker AA14_750 in 6.7 and 8.2 cM distance, respectively, on linkage group 1 of the same AFLP map (Figure 1A).

The RAPD markers for the *Pl*$_2$ resistance gene also distinguished between the resistant and susceptible bulks of population III, which is segregating for the *Pl*$_6$ locus (resistance to all races). Resistance against race 5 of downy mildew was estimated for the F_2 plants and this data was used to map the resistance locus.

Up to now, *Pl*$_2$ markers A02_630, Z15_700, AA14_750 and AC20_831 were involved in the mapping experiments (Figure 1B) based on 96 F_2 individuals. Marker AC20_831 shows the closest linkage to the *Pl*$_6$ gene with a distance of 6.7 cM. Originally, the marker fragments were generated in the inbred line AS110*Pl*$_2$ (Table 3) and linked to the resistance allele of the *Pl*$_2$ locus. The markers are generated in the F_2 plants of population III (AS110*Pl*$_2$ x HA335) if the susceptible allele is present.

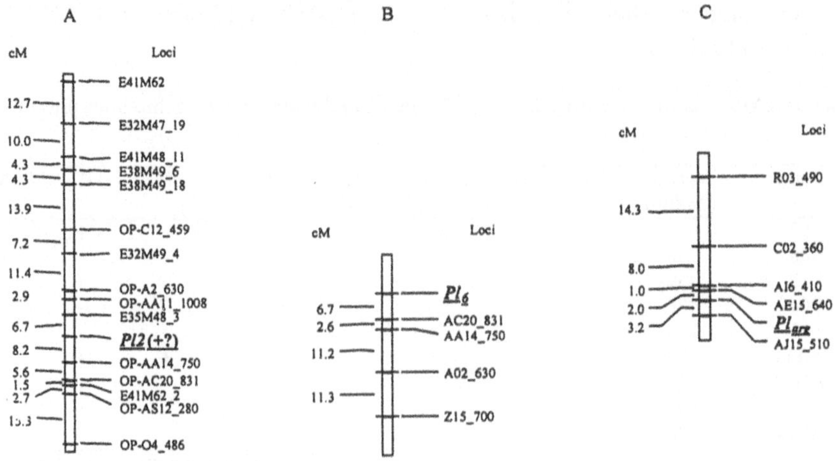

Figure 1. Linkage maps of the genomic regions carrying the investigated resistance genes. A: Linkage group 1 of the AFLP map based on 90 F_2 plants of the cross HA89(cms) x AS110*Pl*$_2$ with integrated RAPD markers. B: Linkage map of the *Pl*$_6$ region of the cross AS110*Pl*$_2$ x HA335 constructed using markers originally identified for the *Pl*$_2$ gene. C: Linkage map of the *Pl*$_{arg}$ region in the F_2 population HA342 x ARG1575-2.

For population II (HA342(cms) x ARG1575-2) segregating for *Pl*$_{arg}$ (race 5 resistance), a linkage map of the sunflower genomic region carrying the resistance locus was constructed using the four RAPD markers which were identified in bulked segregant analysis (Figure 1c). A total of 123 F_2 plants were analysed. The resistance locus is flanked by marker AJ15_530 in a distance of 3.2 cM, and by AE15_640, in a distance of 2.0 cM.

Subsequently, bulks and parental lines of population I were tested with primers AJ15 and AI06, and bulks and parents of population II were tested with primers AA11 and AA14. AJ15 and AI06, both generating markers for the *Pl*$_{arg}$ locus, gave no polymorphisms between bulks and parents of population I. Primers AA11 and AA14, both amplifying markers for the *Pl*$_2$ locus, generated no fragments polymorphic between bulks and parental lines of population II.

4. Discussion

Earlier publications concluded from segregation analysis of populations derived from crosses between different sources of resistance, that resistance against downy mildew of sunflower was controlled by independent single dominant *Pl* genes. On the contrary, the results of Mouzeyar et al. (1992) and Mouzeyar et al. (1996) indicate that the *Pl* genes are closely linked rather than independent. Mapping experiments of Mouzeyar et al. (1996) and Roeckel-Drevet et al. (1996) gave evidence for the collocation of genes *Pl*$_1$, *Pl*$_6$ and *Pl*$_2$ in the same region of the sunflower genome. These results lead to the hypothesis, that those *Pl* loci conferring resistance to more than one race of downy mildew do not represent single genes but are actually a cluster of a number of closely linked genes (Mouzeyar et al., 1996).

Subsequently, Vear et al. (1997) were able to show that the Pl_6 gene, providing resistance to all known races so far, indeed consists of at least two different loci, one providing resistance to the French races A, B and C (similar to American races 4, 3, and indistinguishable from race 3) and one to races 1 and 2 (French race D), which are linked in a distance of 0.6 cM.

The mapping experiments presented in this paper confirm the findings of Mouzeyar et al. (1996) and Roeckel-Drevet et al. (1996) with regard to genes Pl_2 and Pl_6. Both genes were originally identified in inbred lines which were developed by hybridization of wild *H. annuus* accessions with cultivated sunflower (Zimmer & Kinman, 1972; Miller & Gulya, 1988). Referring to the results of Vear et al. (1997), the collocation of the Pl_2 gene with the complex resistance locus Pl_6 indicates that these loci may have a common origin. It seems to be likely that the Pl_2 locus (resistance to races 1, 2, 7 and 9) represents a smaller part of the Pl_6 cluster (resistance to all races), which then would be the "complete" *Pl* resistance cluster of wild *H. annuus*.

On the other hand, the Pl_2 markers which we used to map the race 5 resistance of the Pl_6 locus are linked to race 7 and 9 resistances and to race 5 susceptibility of inbred AS110Pl_2 indicating that there is a allelic relationship between the Pl_2 gene of AS110Pl_2 and the Pl_6 of the line HA335. If the Pl_2 gene were a part of the Pl_6 cluster, there should be no segregation of the Pl_2 locus and the adjacent genomic region in population III, including at least the markers with the closest linkage to the Pl_2 gene. Therefore, they should be monomorphic in population III and inapplicable for mapping the Pl_6 cluster. In contrast, besides marker AA11_1008 all markers linked to the resistance allele of the Pl_2 locus are polymorphic between inbreds AS110Pl_2 and HA335 and useful to map the locus Pl_6. This is affirmed by the screening of near isogenic lines P7/PEM77 and CR2/CEM22 with the RAPD markers for the Pl_2 gene. Both race 5 resistant lines carry the Pl_6 locus. If the Pl_2 gene were a part of the Pl_6 locus, they should both have shown the marker fragments in the screening. But with exception of AA11_1008, which was generated in CEM22, they did not.

However, it may well be that *Pl* genes derived from interspecific hybridization with other *Helianthus* species like Pl_7 from *H. preacox* ssp. *runyonii* or Pl_8 from *H. argophyllus* (Miller & Gulya, 1991) do not share the same location in the sunflower genome, and therefore, may not be a part of a "common" *Plasmopara* resistance locus but may encode different mechanisms of resistance. Although, segregation analysis of a testcross between HA335(Pl_6) and HA338(Pl_7) by Mouzeyar et al. (1996) showed no segregation in a test with French race A, indicating that genes Pl_6 and Pl_7 are not independent. In a first screening bulked segregant analysis using the bulks of population I (HA89[cms] x AS110Pl_2) were tested with markers flanking the Pl_{arg} gene and bulks of population II (HA342[cms] x ARG1575-2) were tested by using markers flanking the Pl_2 locus. For both, no polymorphisms could be obtained in the RAPD pattern. This may be a first hint that there is no association between the two genetic loci.

Acknowledgements: We wish to thank Dr. V.A. Vranceanu, Fundulea (Romania), and Maïsadour (France) for the donation of the near isogenic lines and Prof. F. Viranyi, Gödöllö (Hungary) for supplying several downy mildew pathotypes. We also thank Prof. O. Spring, Hohenheim (Germany) and his coworkers for the characterization of our downy mildew field isolate. This work is supported by the Deutsche Forschungsgemeinschaft (DFG) and the Gesellschaft zur Förderung der privaten deutschen Pflanzenzüchtung (GFP), Bonn.

5. References

Brahm, L. & Friedt, W. (1996) Identifizierung von RAPD-Fragmenten mit Kopplung an die Resistenz gegen den Falschen Mehltau (Rasse 2) der Sonnenblume. Vortr Pflanzenzüchtg 32: 106-108.

Doyle, J.L. & Doyle, J.L. (1990) Isolation of plant DNA from fresh tissue. Focus 12: 13-15.

Gulya, T.J., Miller, J.F., Viranyi, F. & Sackston W.E. (1991) Proposed internationally standardized methods for race identification of *Plasmopara halstedii*. HELIA 14: 11-20.

Köhler, H., Brahm, L., Röcher, T.& Friedt, W. (1997) Anwendung molekulargenetischer Techniken in der Resistenzzüchtung bei der Sonnenblume (*Helianthus annuus*). Vortr Pflanzenzüchtg 36: 47-50.

Korell, M., Brahm, L., Horn, R. & Friedt, W. (1996) Interspecific and intergeneric hybridization in sunflower breeding, II: Specific uses of wild germplasms. Plant Breeding Abstracts 66, 1081-1091.

Kosambi, D.D. (1944) The estimation of map distances from recombination values. Ann. Eugen 12: 172-175.

Lander, E.S., Green, P., Abrahamson, J., Barlow, A,. Daly, M.Y., Lincoln, S.E. & Newburg, L. (1987) An interactive computer package for constructing primary genetic linkage maps of experimental and natural populations. Genomics 1: 174-181.

Michelmore, R.W., Paran, I. & Kesseli, R.V. (1991) Identification of markers linked to disease-resistance genes by bulked segregant analysis: A rapid method to detect markers in specific genomic regions by using segregating populations. Proc Natl Acad Sci USA 88: 9828-9832.

Miller, J.F. & Gulya, T.J. (1988) Registration of six downy mildew resistant sunflower germplasm lines. Crop Sci 28: 1040-1041.

Miller, J.F. & Gulya, T.J. (1991) Inheritance of resistance to race 4 of downy mildew derived from interspecific crosses in sunflower. Crop Sci 31: 40-43.

Miller, J.F. (1992) Update on inheritance of sunflower characteristics, in CETIOM Paris (eds.), Proc. 13th Int. Sunflower Conf., Pisa, Italy, 905-945.

Mouzeyar, S., Phillippon, J., Vear, F. & Tourvieille, D. (1992) Genetical studies of resistance to downy mildew (*Plasmopara helianthi* Novot.) in sunflowers, in CETIOM Paris (eds.), Proc. 13th Int. Sunflower Conf., Pisa, Italy, 1162-1168.

Mouzeyar, S., Roeckel-Drevet, P., Gagne, G., Philippon, J., Gentzbittel, L., Mestries, E. & Nicolas, P. (1996) Inheritance of resistance to downy mildew (*Plasmopara halstedii*) in sunflowers, in CETIOM Paris (eds.), Proc. 14th Int. Sunflower Conf. Beijing, China, 22-27.

Roeckel-Drevet, P., Gagne, G., Mouzeyar, S., Gentzbittel, L., Philippon, J., Nicolas, P., Tourvieille De Labrouhe, D. & Vear, F. (1996) Collocation of downy mildew (*Plasmopara halstedii*) resistance genes in sunflower (*Helianthus annuus* L.). Euphytica 91: 225-228.

Sackstone W. (1992) Cotyledon limited infection (CLI) and leaf disk immersion (LDI) inoculation of sunflower by downy mildew (*Plasmopara halstedii*). Proc. 13th Int. Sunflower Conf., Pisa, Italy, 840-848.

Vear, F., Gentzbittel, L., Philippon, J., Mouzeyar, S., Mestrie, E., Roeckel-Drevet, P., Tourvieille de Labrouhe, D. & Nicolas, P. (1997) The genetics of resistance to five races of downy mildew (*Plasmopara halstedii*) in Sunflower (*Helianthus annuus* L.). Theor Appl Genet 95: 584-589.

Viranyi, F. (1992) Downy mildew of sunflower, in Chaube, Kumar, Mukhopadhyay, Singh (eds.), Plant disease of international importance, Vol. II, Diseases of vegetables and oil seed crops.Prentice Hall Inc., New Jersey, USA, 328-344.

Zimmer, D.E. & Kinman, M.L. (1972) Downy mildew resistance in cultivated sunflower and its inheritance. Crop Sci 12: 749-751.

SESSION 2

RESISTANCE TO BACTERIA

G.T. Scarascia Mugnozza, E. Porceddu & M.A. Pagnotta (Eds.)
Genetics and Breeding for Crop Quality and Resistance, 103-110, 1999

Resistance to bacteria in tomato

H.E. Laterrot
Unité de Génétique et d'Amélioration des Fruits et Légumes; INRA-Avignon; BP 94. F84143, Montfavet, Cedex, France

Abstract: Tomato cultures are parasited by a large number of pathogens. The genetic control of a series of these is widely used in practice. The resistance sources are the wild species related to the cultivated tomato. At the world level four bacteria are very important. Genetic solutions are used to control two of them. The dominant gene *Pto* is very efficient against *Pseudomonas syringae* pv tomato. The genetic control of *Ralstonia solanacearum* is more complicated. New genetic hopes appear to limit the effect of *Clavibacter michiganensis* and *Xanthomonas campestris* pv *vesicatoria.*

1. Introduction

All types of pathogens can be found in tomato crops. Chemical control gives good results for some of these, such as airborne fungi and some insect pests. Chemical treatments have little effect against bacteria and some insects and have practically no effect on viruses.

Genetic resistance can be used to control a small but ever increasing number of pathogens. Most resistances actually used are monogenic and dominant, and breeders are generating F_1 hybrids cumulating several resistance genes. Wild species of *Lycopersicon* are sources of the resistances present in the modern varieties. The importance of wild species is always increasing. In practice, 13 different pathogens are now controlled by genetic resistance. The effectiveness of these resistances varies widely with respect to their level of expression and to their stability over time when confronted with an expanding virulent spectrum of pathogens (review by Laterrot, 1997).

Four of the bacterial species encountered on tomatoes are very important at the world level. The situation of their genetic control is very different.

1.1 *Pseudomonas syringae pv. tomato, causing bacterial speck*

The genetic resistance to this bacterium is a good example of chance for the breeders. In Canada, during 1978, a severe attack of *Pseudomonas tomato* in the field showed a high resistance level of several tomato lines. The common origin of these resistant lines was *L. pimpinellifolium*. This resistance is controlled by one dominant gene, *Pto,* localized on chromosome 5 (Pitblado, 1983).

In France during 1984, populations of the leafminer *Liriomyza trifolii* on tomatoes were sprayed with the insecticide Lebaycid (active substance Fenthion). Four days after the treatment, necrosis appeared only on the foliage of the plants carrying the gene *Pto* (Laterrot, 1985). The same Fenthion - induced necrosis allowed later to point out other *P. tomato* resistant lines. Breeders are now using Fenthion in their breeding programmes. Molecular studies have shown that *P. tomato* resistance and Fenthion susceptibility are conferred not by a single gene, but two tightly linked and functionally similar genes (*Pto* and *Fen*). The separation of these genes has not been reported, but an allel of *Pto,* Pto^h from *L. hirsutum,* controls the bacterium without Fenthion sensitivity (Tanksley et al., 1996)

Various other sources of *P. tomato* resistance have been found. Pilowsky & Zutra (1986) have shown resistance in a *L. pimpinellifolium* accession and their study revealed a new gene, gene *Pto-2*. Further work concluded that this resistance is a reoccurrence of *Pto* also linked with Fenthion sensitivity (Laterrot & Miretti, 1991). Stockinger & Walling (1994) published that *Pto-1* (new symbolism for *Pto*) and *Pto-2* are loci separated by a maximum of 9 cM, on the chromosome 5.

The appearance of a pathotype adapted to the gene *Pto-1* in Canada (Lawton & Mac Neil, 1986), in Bulgaria (Bogatsevska et al.,1989) and in Italy (Buonaurio et al., 1990) involves the research of new sources of resistance. In Bulgaria, material resistant to the pathotype adapted to *Pto* was found in the progenies having different wild species in there pedigree (Stamova et al., 1990). In *L. hirsutum* Stockinger & Walling (1994) found two genes for resistance: *Pto-3* which does not confer resistance to the new pathotype and *Pto-4*, independent of *Pto-3*, which confers resistance to the race 1 adapted to the gene *Pto-1*.

In practice race 1 is rarely encountered and the gene *Pto-1* is now introduced in a increasing number of varieties, mostly F_1 hybrids for processing tomato.

1.2 *Ralstonia solanacearum* responsable of bacterial wilt

This bacterium was previously called *Pseudomonas solanacearum* and then during a short time reclassified as *Burkholderia solanacearum.* This bacterium affects many different species mainly in tropical and subtropical climates. The problems are very important for potato, tomato, eggplant, pepper, ground-nut and banana. According the host range three, then five races, are classified. In parallely a classification defined through biochemical tests distinguishes five biovars. Races and biovars are poorly correlated.

The control of this bacterium, which resides in the soil, is difficult. The grafting on resistant rootstock is a practical solution in particular conditions for tomato (and also eggplant) (Paily, 1964; Digat & Bulit, 1967; Peregrine & Kassim Bin Ahmad, 1982). Various resistant accessions of tomato and *Solanum* spp. were, and are, used in different countries but at a relatively small scale. It was observed that *Erwinia* sp. can increase wilting of the plants attacked by *Ralstonia.*

Several tomato breeding programs in various countries have developed resistant material from differing resistance sources found in *Lycopersicon pimpinellifolium* or accessions of *L. esculentum* var. *cerasiforme*. The precise origins of the various resistant lines resulting from the selection are difficult to identify (Scott, 1997)

At least two type of resistance seem to exist in tomato. One is a polygenic resistance, originating from *L. esculentum* var. *cerasiforme*. CRA66 belongs to this type and has often been used as a resistant parent in the French West Indies. The second type is a dominant resistance present in the Hawaiian material which is of simpler heredity. This resistance originated from *L. pimpinellifolium* (review of Prior et al., 1993). Molecular and clonal analysis permitted localization of quantitative trait loci important for *Ralstonia* resistance of Hawaii 7996 (Thoquet et al., 1996a). Other studies realized in parallel in field in Guadeloupe and in growth chambers permitted the detection of other QTLs (Thoquet et al., 1996b).

The multiplication of bacteria and their spatial spread in stems show that resistant material tolerates large populations of *Ralstonia*. It was demonstrated that the spread in stems is less frequent in resistant cultivars as they are able to limit the upward movement. In addition, when stems are invaded, resistant cultivars tend to tolerate the invasion without wilting. The frequency of stem infections measured at midstem was found to be positively correlated with the frequency of wilted plants in the field. This was particularly clear with CRA66 and Hawaii 7996 (Prior et al., 1993; Grimault et al., 1994).

The resistance to *Ralstonia* being not stable over locations, it was decided during the Bacteria Wilt Conference in Taiwan in 1992 to collect various sources of resistance from around the world for multilocation testing. A total of 35 bacterial wilt-resistant lines and accessions were contributed by nine researchers. Seeds of entries were multiplied in Australia and sets distributed to collaborators.

Bacterial wilt reactions of these entries were determined in naturally infested fields. The first results showed that Hawaii 7996 had the highest survival percentage over trials, but 8 other entries were not significantly different from Hawaii 7996. These lines are Hawaii 7997, Hawaii 7998, CRA66, the line BF Okitsu 101 (from Japan with resistance source from North Carolina) and 3 lines from the Philippines (Wang et al., 1996). A diallel analysis of *Ralstonia* resistance from various sources concluded that intercrossing lines that show large positive general combining ability effects, followed by selfing and selection for resistance in segregating populations might yield progeny with superior bacterial wilt resistance (Hanson et al., 1998).

The severity of bacterial wilt is increased by nematodes infection. Hence several breeding programs have been realised to introduce the gene *Mi* controlling the *Meloidogyne* spp. resistance in different *Ralstonia* resistant varieties. It was difficult to combine *Mi* with a high level of *Ralstonia* resistance (Acosta et al., 1964; Messiaen et al., 1978; Prior et al., 1993). This observation suggests that one important locus concerning the bacterium resistance may reside on chromosome 6 where *Mi* is localised. By mapping with molecular markers this hypothesis was confirmed (Aarons et al., 1993; Thoquet et al., 1993, 1996a-1996b; Danesh et al., 1994). A study realised with two pairs of *Ralstonia* resistant lines, near isogenic except for the gene *Mi*, confirms the difficulty to select a high level of bacterial resistance together with the gene *Mi* (Deberdt & Prior, 1997). It is interesting to note that these studies were realised with *Ralstonia* resistant lines from different origins.

So an important gene for resistance is situated on the chromosome 6 for the different material, notably Hawaii 7996 and CRA66.

Vitrovariations have been probed to produce *Ralstonia* resistant line. Neoformed tomato plants were regenerated after vitroculture of calli inoculated with the bacterium. Surviving resistant calli gave variant plants exhibiting tolerance to *Ralstonia.* It was no segregation for the resistance among the progeny (Nsika Mikoko, 1996).

In practice, there are few *Ralstonia* resistant varieties cultivated in the world, their number is increasing slowly. The breeders tend to cumulate in F_1 hybrids good level of resistance with large fruit size and resistance to other pathogens. Hopes are in the use of molecular markers linked to QTLs to facilitate the development of tomato genotypes adapted to different cultural conditions where *Ralstonia* is a serious problem. An important symposium was consecrated to *Ralstonia*, particularly to molecular and ecological aspects during June 1997, in Guadeloupe (Prior et al., 1998)

1.3 Clavibacter michiganensis subsp. michiganensis causing bacterial canker

Previously named *Corynebacterium michiganense,* this seed transmissible bacterium causes a worldly occurring disease affecting all plant parts and can be devastating when its occurs. The use of bacterium free seeds and some cultural practices reduce the risks.

Several sources of partial resistance to *Clavibacter* have been reported in tomato. Among *L. esculentum* genotypes several *Ralstonia* resistant lines showed a moderate resistance to *Clavibacter* (Laterrot et al., 1978). In Canada, two *Ralstonia* resistant lines, Hawaii 7998 and IRAT-L3, have given good results in a breeding program for *Clavibacter* resistance (Poysa, 1993). Some accessions of *L. esculentum* var. *cerasiforme* show a moderate resistance. *L. pimpinellifolium* is the source of several partial resistant tomato lines. Between these, Plovdiv 8/12 selected in Bulgaria (Elenkov, 1965) have been worldly used (review Laterrot et al., 1978).

Several accessions of *L. hirsutum* easily crossed with tomato were used. In Japan, the line Okitsu Sozai n°1 was obtained (Kuriyama & Kuniyasu, 1974). Its good level of resistance was confirmed in different countries and used in France in various breeding programs, particularly to obtain the line Cordeok, which has shown a good level of resistance notably in North Carolina (Gardner et al., 1989) and Italy (Crinò et al., 1995).

L. peruvianum is a rich source of *Clavibacter* resistance. Several accessions were retained (Thyr, 1969; Sotirova & Beleva, 1976). Molecular markers linked to an high level of resistance were found in an *L. peruvianum* accession (Sandbrink et al., 1995). In Bulgaria sources of resistance were found in *L. chilense* (Yordanov & Stamova, 1977) and three-genome hybrid *L. esculentum* x *L. chilense* x *L. peruvianum* (Vulkova-Achkova & Sotirova, 1981). From this latter was obtained the line Cm180 showing a monogenic resistance (Vulkova & Sotirova, 1993).

Cumulation of *Clavibacter* resistance genes has been realized. From Plovdiv 8/12 and 72.TR.4-4 (=NC72), the latter selected in North Carolina for *Ralstonia* resistance, the population Cocabul has been created (Laterrot, 1992). From Plovdiv 8/12, Okitsu Sozai N°1 and IRAT-L3, the latter selected in Martinique for *Ralstonia* resistance, the line Comech in the canning tomato type has been selected (Laterrot & Moretti, 1997). A breeding program was realized to cumulate the resistance of Cordeok obtained from Okitsu Sozai N°1 and this of *L. peruvianum* (Crinò et al., 1995).

Some studies tend to obtain resistant material from somaclone progeny (Van Den Bulk et al., 1991; Vries & Stephens, 1997). In practice several less susceptible cultivars are known but very few resistant material is cultivated. Probably after a while some resistant varieties will appear.

1.4 Xanthomonas campestris pv. vesicatoria responsable for bacterial spot

This bacterium transmissible by seeds affects all above-ground plant organs as does *Pseudomonas tomato*. It is the cause of one of the most important diseases of tomatoes and peppers in regions with high temperature during the rainy periods. The cleaning of infected seeds with sodium or calcium hypochlorite improves the situation. A good cultural protection was demonstrated notably in Florida by treatments with coper-maneb premix (Cox, 1982).

Among cultivated tomatoes varieties, some differences of susceptibility were observed in natural conditions in various subtropical and tropical regions (Crill et al., 1972; Mathew & Patel, 1975; Scott & Jones, 1984, 1986). One of the varieties less susceptible to defoliation is Campbell 28 which in large cultivation in Cuba confirms this partial resistance. Other tomato lines present a better level of resistance to defoliation, notably Ohio 4013 and Hawaii 7998 (Mc Guire et al., 1991), Hawaii 7998, selected for *Ralstonia* resistance was particularly studied in Florida (Scott & Jones, 1986). This line reacts by hypersensitivity controlled by multiple non dominant genetic factors (Wang et al., 1994a, 1994b).Three genetic factors, two localised on chromosome 1 and one on chromosome 5, appear to act independently and to have additive effects (Yu et al., 1996). Other partially resistant lines like Ohio 4013 and Campbell 28 do not show the hypersensitivity reaction (Bazzi et al., 1989).

Resistance to bacterial spot on fruit which is not associated with foliage resistance was found in PI 270248 (=Sugar) (Scott et al., 1989). The racial situation is reviewed by J. W. Scott (1997a). Three races have been identified on tomato. A race virulent on Hawaii 7998 first reported in Brazil (Nagai & Sugimori, 1986) is extending its geographic area. This race T2 hydrolyses starch unlike race T1. Several accessions resistant to this race were founded in *L. pimpinellifolium* and in *L. esculentum* PI 114490 (Scott et al., 1997b). The latter also showed a moderate resistance to race T1 and a good level of resistance to race T3. This third race was identified in Florida in 1991. T3 differed from race T1 in its ability to hydrolyse starch. T3, like T2 does not cause a hypersensitive reaction on Hawaii 7998 (Jones et al., 1995). Hypersensitivity to race 3 was observed with Hawaii 7981 and some accessions of *L. pimpinellifolium*. Hawaii 7981 presented a good level of resistance after T3 inoculation but it is susceptible to race T1. The conclusion of the Floridian studies is that a logical breeding approach would be to use a broad spectrum resistance, like that PI 114490 in future breeding efforts, either alone or combined with T1 and/or T3 resistance (Scott et al., 1997b).

Two pathotypes were distinguished in Bulgaria, a Tomato pathotype and a Pepper-Tomato pathotype. In front of the two pathotypes an high level of resistance was observed in several *Lycopersicon* wild species. A hypersensitive reaction was shown in *L. peruvianum* and *L. chilense* with the two pathotypes (Sotirova & Bogatzevska, 1993).

In the near future, the cumulation of different resistance sources and the use of a spray inoculation technique like those described by Somodi et al. (1994) or by Mello et al. (1997) will be permit creation of varieties with a stable and high level of resistance to bacterial spot.

1.5 Pseudomonas corrugata

Pith necrosis caused by *Pseudomonas corrugata* is sometimes accompanied by adventitious root formation. This disease firstly observed in greenhouses appears also in field-grown tomatoes and is always associated with high nitrogen levels in the soil and a humid environment. Affected plants sometimes wilt and collapse but frequently the less attacked plants recover. The disease seems more frequent with the vigorous cultivars. The breeders take attention to this, particularly for plastic tunnel cultivation.

2. References

Aarons S.R., Danesh D. & Young N.D. (1993) DNA genetic marker mapping of genes for bacterial wilt resistance in tomato. In: Bacterial wilt. ACIAR Proc. 45.G.L. Hartman & A.C. Hayward, eds. Watson Ferguson Co., Brisbane p.170-175.

Acosta, J.C., Gilbert, J.C. & Quinon, V.L. (1964) Heritability of bacterial wilt resistance in tomato. Proc. Amer. Soc.Hort.Sci. 84: 455-462.

Bazzi, C., Mazzucchi, U., Roncarati, R. & Sanguineti, M.C. (1989) Reactions of different tomato genotypes to *Xanthomonas campestris* pv. vesicatoria Phytopath.medit. 28: 189-194.

Bogatsevska, N.; Sotirova, V. & Stamova, L. (1989) Races of *Pseudomonas syringae* pv.tomato in Bulgaria. Tomato Genet. Coop. Rep. 39: 7

Buonaurio, R., Stravato, V.M. & Cappelli, C. (1990) Occurrence of *Pseudomonas syringae* pv. tomato race 1 in Italy on *Pto* gene-bearing tomato plants. J. Phytopathol. 144: 437-440

Cox, R.S. (1982) Control of bacterial spot of tomato in southern Florida. Plant Disease. 66(9): 870.

Cox R.S. (1985) Battling bacterial spot on tomatoes. American Veg. Growers. Dec. 85: 6-8.

Crill, P., Jones, J.B. & Burgis, D.S. (1972) Relative susceptibility of some tomato genotypes to bacterial spot. Plant Dis. Rep. 56(6): 504-507.

Crinò, P., Veronese, P., Stamigna, C., Chiaretti, D., Lai, A., Bitti, M.E. & Saccardo, F. (1995) Breeding for resistance to bacterial canker in Italian tomatoes for fresh market. In: First Int Symp. on Solanacea for fresh market, Malaga, Spain, 28-31 March 1995. Acta Horticulturae n°412: 539-545.

Danesh, D., Aarons, S., Mc Gill, G.E. & Young, N.D. (1994) Genetic dissection of oligogenic resistance to bacterial wilt in tomato. Mol.Plant Microbe Interaction. 7: 464-471

Deberdt, P.& Prior, Ph. (1997) *Mi* introgression lines as tools for the genetic analysis of bacterial wilt resistance in tomato. In: Bacterial Wilt Disease. Report Second Int. Bacterial Wilt Symposium, Gosier, Guadeloupe, France, June 22-27, 1997. Ph Prior, C. Allen, J. Elphinstone (eds.): 255-262.

Digat, B.& Bulit, J. (1967) L'emploi du greffage pour combattre le flétrissement bactérien de la tomate aux Antilles françaises. Proc. Caribbean Food Crops Soc. 5th annual meeting. Paramaribo – Surinam. 24-31/7/67: 105-110.

Elenkov, E. (1965.) Breeding of tomatoes resistant to bacterial canker. Int. Z. Landwirt, Plovdiv: 299-318.

Gardner, R.G., Shoemaker, P.G. & Echandi, E. (1989) Reactions of selected tomato lines to the foliar blight phase of bacterial canker. Tomato Genet.Coop.Rep. 39:15-16

Grimault, V., Anais, G. & Prior, Ph. (1994) Distribution of *Pseudomonas solanacearum* in the stem tissues of tomato plants with different levels of resistance to bacterial wilt. Plant Pathology. 43: 663-668.

Hanson, P.M., Licardo, O., Hanudin, Wang, J.F.& Chen, J.T. (1998) Diallel analysis of bacterial wilt resistance in tomato derived from different sources. Plant Disease 82 (1): 74-78.

Jones, J.B., Stall, R.E., Scott, J.W., Somodi, G.C.,Bouzar, H. & Hodge, N.C. (1995) A third tomato race of *Xanthomonas campestris* pv. *vesicatoria*. Plant Disease. 79: 395-398.

Kuriyama, T. & Kuniyasu, K. (1974) Studies on the breeding of disease resistant tomato by interspecific hybridization. III. On the breeding of a new tomato line resistant to bacterial canker caused by *Corynebacterium michiganense*. Bull.Veg.Ornem.Res.Stn. Japan. Ser.A (1): 93-107.

Laterrot, H. (1985) Susceptibility of the [*Pto*] plants to Lebaycid insecticide: A tool for the breeders? Tomato Genet.Coop.Rep. 35: 6.

Laterrot, H. (1992) Creation of the population Cocabul with *Corynebacterium* resistance. Tomato Genet.Coop.Rep. 42: 25-26.

Laterrot, H. (1997) Breeding strategies for disease resistance in tomatoes with emphasis on the tropics: current status and research challenges. Proc. 1st Int.Symp.Tropical Tomato Diseases 21-22 nov.; 1996. Recife, Pernambuco, Brazil: 126-132.

Laterrot, H., Brand, R. & Daunay, M.C. (1978) La resistance à *Corynebacterium michiganense* chez la tomate. Ann.Amélior.Plantes. 28: 579-591.

Laterrot, H. & Moretti, A. (1991) Allelism of *Pto* and *Pto-2*. Tomato Genet.Coop.Rep. 41: 27.

Laterrot H. & Moretti, A. (1997) In: Station d'amélioration des plantes maraîchères. Rapport d'activité 1995-96 p.78.

Lawton, M.B. & Mac Neill B.H. (1986) Occurrence of race 1 of *Pseudomonas syringae* pv. tomato on field tomato in south western Ontario. Can.J.Plant.Pathol. 8: 85-88.

Mathew, J. & Patel, P.N. (1975) Screening for genetic host resistance against the bacterial leaf spot disease in tomato by *Xanthomonas vesicatoria*. Current Science. 44(23): 867-868

Mc Guire, R.G., Jones, J.B. & Scott, J.W. (1991) Epiphytic populations of *Xanthomonas campestris* pv. *vesicatoria* on tomato cultigens resistant and susceptible to bacterial spot. Plant Disease. 75: 606-609.

Mello, S.C.M. de., Lopes, C.A. & Takatsu, A. (1997) Criterios de avaliaçào da resisténcia em tomato plants to bacterial spot. Fitopatologia Brasileira.22(4): 502-506.

Messiaen, C.M., Laterrot, H. & Kaan, F. (1978) Cumulate resistances to *Pseudomonas solanacearum* and to *Meloidogyne incognita* with determinate growth in tomato. Vegetables for hot humid tropics. 3: 48-51.

Nagai, H. & Sugimori M.H. (1986) Suscetibilidade dos tomaterios H.1998 e C-28 a pustula bacteriana (*Xanthomonas campestris* pv. *vesicatoria*) em Sao Paulo. Hort.Bras. 4(1): 62 (Abstr.)

Nisika Mikoko, E. (1996) Tolerance au flétrissement bactérien chez la tomate induite après vitroculture. Cahiers Agricultures 5: 460-462

Paily, P.V. (1964) Control of the bacterial wilts of tomato and brinjal by grafting on *Solanum torvum*. Sci. and Cult. 30: 295-296.

Pergrine, W.T.H. & Kassim Bin Ahmad (1982) Grafting – a simple technique for overcoming bacterial wilt in tomato. Tropical Pest Management. 28(1): 71-76.

Pilowsky, M. & Zutra, D. (1986) Reaction of different genotypes to the bacterial speck pathogen (*Pseudomonas syringae* pv. tomato) Phytoparasitica 14(1): 39-42.

Pitblado, R.E. (1983) Genetics of resistance to bacterial speck caused by *Pseudomonas syringae* pv. tomato in field tomatoes. Ph.D. thesis. Guelph University. 72 p.

Poysa, V. (1993) Evaluation of tomato breeding lines resistant to bacterial canker. Can.J.Pl.ant Pathol. 15: 301-304.

Prior, Ph., Allen, C. & Elphinston, J. (1998) Bacterial Wilt Disease. Second Int.Wilt Symp., Gosier,. Guadeloupe, France, 22-27 June 1997 (Hosting Institut: INRA), Springer publisher, 447 p.

Prior, Ph., Grimault, V. & Schmit, J. (1993) Resistance to bacterial wilt (*Pseudomonas solanacearum*) in tomato:Present status and prospects. In: Bacterial wilt: the disease and its causative agent=*Pseudomonas solanacearum*. Hayward and Hartman, eds. CAB. International Wallingford (UK): 209-223.

Sandbrink, J.M., Ooijen, J.W., Purimahua, C.C., Vrielink, M., Verkerk, P., Zabel, P. & Lindhout, P. (1995) Localization of genes for bacterial canker resistance in *Lycopersicon peruvianum* using RFLPs. Theoretical and Applied Genetics. 90(3/4): 444-450.

Scott, J.W. (1997) Tomato improvement for bacterial disease resistance for the tropics: a contemporary basis and future prospects. Proc. 1st Int.Symp.Tropical Tomato Diseases, 21-22/11/96. Recife.Pernambuco, 117-125.

Scott, J.W. & Jones, J.B. (1984) A source of resistance to bacterial spot (*Xanthomonas campestris* pv. vesicatoria) Tomato. Genet.Coop.Rep. 34: 16.

Scott, J.W. & Jones. J.B. (1986) Sources of resistance to bacterial spot *Xanthomonas campestris* pv. *vesicatoria*. in tomato. HortScience. 21: 304-306.

Scott, J.W., Jones, J.B., Somodi, G.C. & Stall, R.E. (1997a) Screening tomato accessions for resistance to *Xanthomonas campestris* pv. *vesicatoria*, Race 3. HortScience 30(3): 579-581.

Scott, J.W., Miller, S.A., Stall, R.E., Jones, J.B., Somodi, G.C., Barbosa, V., Francis, D.L. & Sahin, F. (1997b) resistance to race T2 of the bacterial spot pathogen in tomato. HortScience. 32(4): 724-726.

Scott J.W., Somodi G.C. & Jones J.B. (1989) Resistance to bacterial spot fruit infection in tomato. HortSci. 24: 825-7.

Somodi G.C., Jones, J.B., Scott, J.W. & Jones, J.P. (1994) Screening tomato seedlings for resistance to bacterial spot. HortScience. 29: 680-682.

Sotirova, V. & Beleva, L. (1976) Study on the resistance to *Corynebacterium michiganense* of wild tomato spp. and varieties. Gradinarska i Lozarska Nauka. 13(7): 86-91

Sotirova, V. & Bogatzevska, N. 1(993) New sources of resistance to *Xanthomonas campestris* pv. *vesicatoria* in tomato. Phytopath.medit. 32: 145-148

Stamova, L., Bogatzevska, N. & Yordanov, M. (1990) Resistance to race 1 of *Pseudomonas syringae* pv. tomato. Tomato Genet.Coop.Rep. 40: 33

Stockinger, E.J. & Walling, L.L. (1994) *Pto-3* and *Pto-4*: novel genes from *Lycopersicon hirsutum* var. *glabratum* that confer resistance to *Pseudomonas syringae* pv. tomato. Theor.Appl.Genet. 89: 879-884.

Tanksley, S., Brommonschenkel, S. & Marting G. (1996) *Pto*^{h}, allel of *Pto* conferring resistance to *Pseudomonas syringae* pv. tomato (race 0) that is not associated with Fenthion sensitivity. Tomato Genet.Coop.Rep. 46: 28-29.

Thoquet, P., Olivier, J., Sperisen, C., Rogowsky, P., Laterrot, H. & Grimsley, N. (1996a) Quantitative trait loci determining resistance to bacterial wilt in tomato cultivar Hawaii 7996. Mol.Plant Microbe Interaction. 9: 826-36.

Thoquet, P., Olivier, J., Sperisen, C., Ragowsky, P., Prior, Ph., Anais, G., Mangin, B., Bazin B., Nazer, R. & Grimsley, N. (1996b) Polygenic resistance of tomato plants to bacterial wilt in the French West Indies. Mol.Plant Microbe Interaction. 9(9): 837-842.

Thoquet, P., Stephens, S. & Grimsley, N. (1993) Mapping of bacterial wilt resistance genes in tomato variety Hawaii 7996. In: Bacterial Wilt. ACIAR Proc. N°45. G.L. Hartman and A.C. Hayward, eds., Watson Ferguson Co., Brisbane, Australia. p.176.

Thyr, B.D. (1969) Additional sources of resistance to bacterial canker of tomato (*Corynebacterium michiganense*). Plant Dis.Rep. 53(3): 234-237.

Van Den Bulk, R.W., Jansen, J., Lindhout, W.H. & Loffler, H.J.M. (1991) Screening of tomato somaclones for resistance to bacterial canker (*Clavibacter michiganensis* subsp. *michiganensis*). Plant breeding. 107: 190-196.

Vries, R.M. de & Stephens, C.T. (1997) Response of first generation tomato somaclone progeny to *Clavibacter michiganensis* subsp. *michiganensis*.Plant Science. 126(1): 69-77.

Vulkova, Z.V. & Sotirova, V.G. (1981) Trigenomic hybrid between *Lycopersicon esculentum* Mill. *L. chilense* Dun. and *L. peruvianum* var. *humifusum* Mill. - reproductive relationships and resistance to *Corynebacterium michèganense*. (Smith) Jensen. Tomato Genet. Coop Rep. 31: 19-20.

Vulkova, Z.V. & Sotirova, V.G. (1993) Study of the three-genome hybrid *Lycopersicon esculentum – L.chilense – L.peruvianum* var. *humifusum* and its use as a source for resistance. TAG. 87: 337-342.

Wang, J.F., Hanson, P.M. & Barnes, J.A. (1996) Preliminary results of world wide evaluation of international set of resistance sources to bacterial wilt in tomatoes. ACIAR Bacterial Wilt Newsletter 13: 8-10.

Wang, J.F., Jones, J.B. Scott, J.W. & Stall, R.E. (1994a) Several genes in *Lycopersicon esculentum* control hypersensitivity to *Xanthomonas campestris* pv. *vesicatoria*. Phytopathology 84(7): 702-706.

Wang, J.F. Stall, R.E. & Vallejos, C.E. (1994b) Genetics analysis of a complex hypersensitive reaction to bacterial spot in tomato. Phytopathology, 84(2): 126-132.

Yordanov, M. & Stamova, L. (1977) A new source of resistance to *Corynebacterium michiganense*. Tomato Genet.Coop.Rep. 27: 26.

Yu, Z.H., Wang, J.F., Stall, R.E. & Vallejos, C.E. (1996) Genomic localization of tomato genes that control a hypersensitive reaction to *Xanthomonas campestris* pv. *vesicatoria*. Genetics. 141(2): 675-682.

G.T. Scarascia Mugnozza, E. Porceddu & M.A. Pagnotta (Eds.)
Genetics and Breeding for Crop Quality and Resistance, 111-118, 1999

Phytopathogen resistance improvement of horticultural crops by plant-defensin gene introduction

S.V. Dolgov[1], V.G. Lebedev[1], S.S. Anisimova[2], N. Lavrova[2], L.A. Serdobinskiy[2], G.B. Tjukavin[3], S.A. Shadenkov[2] & V.G. Lunin[2]
[1]Branch of Shemyakin and Ovchinnikov Institute of Bioorganic Chemistry, RAS, 142292, Pushchino, Moscow region, Russia; [2]All-Russian Institute of Agricultural Biotechnology, RAAS; 127550, Timiryazevskaja st. 42, Moscow, Russia; [3]All-Russian Research Institute of Vegetable Breeding and Seed Production, RAAS, 143080, p/o Lesnoy Gorodok, Moscow region, Russia

Abstract: For the horticultural plants protection against fungal and microbial attack the plant defensins genes (PD) have been transferred to pear, apple and carrot. They are recently characterized as small (near 50 amino acids long) Cyst-rich antimicrobial peptides that have a complex cysteine-stabilized three-dimensional folding pattern often involving antiparallel beta-sheets. Based on previously developed transformation methods the PD gene from *Rafanus sativus* was transferred to clonal apple rootstock N545 pear variety "Burakovka" and carrot by agrobacterial mediated transformation. Obtained plant lines - 27 apple, 12 pear and 270 carrot successfully rooted on high antibiotic media and some of them demonstrated high NPT activity in greenhouse growing plants. PD gene introduction in their genoms has been confirmed by PCR analysis. Western blot analysis showed PD gene expression in 5 apple lines from 20 tested, one pear from 12 tested and 18 carrot from 20 tested.

1. Introduction

International trade in fresh temperate tree fruits rivals that of the major crops of the world. However, conventional genetic manipulation of the crops in this economically important group is difficult for the reason of the long generation times and has relied more on chance selection and propagation of elite cultivars. Genetic transformation is an alternative approach; transformation procedures allow one to make small, specific changes in the genome of cell, i.e. the addition of one or a few genes. This stands in contrast to conventional breeding, where entire sets of chromosomes are combined and desired parental genotypes are not necessarily reconstituted. Also, with generation times of years rather than months, improving woody species by breeding is temporally demanding.

Trangenic horticultural plants have been obtained from several fruit woody plant crops to date; these include apple, pear, plum and apricot, peach and some berries (Oliveira et al., 1996). Some potentials for temperate tree crop improvement are the introduction of pathogenes (insects, bacteria and fungi), frost and herbicide resistance, improvement fruit taste, production food substances, which cannot be obtained in other organisms such as bacteria or yeast.

Only limited success has been reported till recent time in the area of resistance to fungal diseases through genetic engineering. Terras et al. (1992) reported that purified, small antibiotic-like plant proteins (defensins) from *Rafanus sativus* seeds showed antimicrobial and antifungal properties. The expression of genes encoding such antifungal proteins may increase the plant resistance to fungal and microbial attack. They were characterized as small (near 50 amino acids long) Cyst-rich antimicrobial peptides which have a complex cysteine-stabilized three-dimensional folding pattern often involving antiparallel beta-sheets (reviewed by Broekaert et al., 1995). For example the plant defensins and similar genes have been introduced in apple cultivars at Cornell (USA) and Leuven (Belgium) Universities. The first results, reported by Norelli et al. (1996) and DeBondt et al. (1996) confirm their ability to improve apple resistance for fungal and bacterial diseases.

The goal of the investigation consists heterological expression of plant defensins in bacteria and transgenic plants, modification of such genes to study mechanism of PD's action and creation of the efficient chimerical forms of defensins resulting in obtaining of transgenic horticultural plants (apple, pear, carrot) which would be resistant to fungal damage.

2. Materials and Methods

2.1 *PD genes cloning*

2.1.1 Cloning and expression of nucleotide sequence coding mature Rs-AFP2 peptide

R. sativus genomic DNA was isolated by grinding 1g of radish tissue in 38 ml tube containing 15 ml of extraction buffer (Draper et al., 1988). After centrifugation supernatant was mixed with an equal volume of isopropanol. The pellet of DNA was collected by centrifugation at 8,000 r min^{-1} and dissolved in 80 mkl of sterile water. This DNA solution was used as a template in the PCR amplification. Primers used for cloning defensin gene were designed for efficient ligation and expression with the plasmid vector. The N-terminal primer was 5'-CTAGGAATTCGTAGTGATC-ATGGCTAAGTTTGCTTCTATC-3' designed to incorporate EcoRI restriction site. The C-terminal primer was 5'-TGCTCTAGAGTTAACAAGGGAAATAACAG-ATACACTTG-3', restriction site XbaI was inserted downstream of the stop codon. The PCR scheme was as follows: 3 times each at 94°C for 1 min, 56°C for 1 min and 72°C for 2 min; 30 times each at 94°C for 1 min, 60°C for 1 min and 72°C for 1 min;

one time at 72°C for 10 min, all with 100 pg. genomic DNA template. The PCR product was ligated into the pGEM4Z (Promega) vector and cloned in DH5 *E. coli* strain. Sanger sequencing procedure determined nucleotide sequence of two cloned amplification fragments.

The pRs-AFP2 vector, including the genomic region corresponding to Rs-AFP2 peptide, was used as a template for PCR amplification and cloning of the sequence coding the mature peptide Rs-AFP2. Primers for the PCR amplification were designed for expression of peptide in pQE40 (Qiagene) vector. The N-terminal primer was 5'-GCTGCTTTCGAAGAACCAACAATGGTGGATCCACAGAAG-3' designed to incorporate *Bam*HI restriction site. The C-terminal primer was 5'-GTGAGCTGTCTAGACATACTAGCAAGACCCATG-3' with *Xba*I site. PCR scheme was as above. The PCR product was ligated into the pQE40 vector and cloned in M-15 *E. coli* strain. The positive clone coding DHFR-RsAFP-2 fusion protein was expressed and purified using protocol 2 of Qiagene.

2.1.2 Preparation of the rabbits anti-Rs-AFP2 polyclonal antibodies

Anti-Rs-AFP2 IgGs were obtained by $(NH_4)_2SO_4$ precipitation of the antiserum from immunized rabbits. The IgG fraction was dialyzed against PBS buffer. The partially purified antibodies were diluted 1:1,000 in the immunoblotting procedure.

2.2 Construction of the plant expression vector

The RS-AFP-2 gene was cloned between the CaMV 35S promoter and the 35S terminator regions of the plasmid vector. The *Hin*dIII fragment of the RS-AFP-2 derivative vector (RS-AFP-2 expression cassette) was subsequently cloned in the *Hin*dIII restriction site of the binary plant transformation vector pPCV002. Resulted plasmid pPCV002rs was introduced into disarmed supervirulent *Agrobacterium tumefaciens* strain CBE 21 kindly provided Dr. E. Revenkova et al. (1993) by direct transformation.

2.3 Plant transformation

2.3.1 Apple transformation

Semidwarf clonal apple rootstock N545 was used in our experiments. The plants were cultivated *in vitro* on the QL medium (Quorin & Lepoivre, 1977) containing 1.0-1.5 mg l^{-1} BAP and 0.3 mg l^{-1} IBA and rooted on the half-strange QL medium supplemented with 0.5-1.0 mg l^{-1} IBA. Propagation and rooting of apple rootstocks were performed under the following conditions: temperature - 26/22°C; day length - 16 hours; illumination - 2,500-3,000 lux.

The leaf pieces of about 1 cm^2 from *in vitro* rooted plants were used for transformation. Co-cultivation leaf explants with agrobacteria was performed on Murashige & Skoog (1962) medium (MS) containing 3.0 mg l^{-1} 4CPU, 1.0 mg l^{-1} NAA, 1 mg l^{-1} TIBA (MSR, Dolgov & Muratova, 1995) for 2-3 days at 26°C. After co-cultivation the explants were placed on the same medium (MSR) for regeneration and selection of transformants supplemented by 500 mg l^{-1} cefotaxime. Beginning second passage 35 mg l^{-1} kanamycin were added to regeneration media. Regeneration was performed at 23°C in the darkness.

Regenerants grown to 5-7 mm in size were separated from the callus and placed on the propagation medium (section "Plant material") with 50 mg l^{-1} kanamycin. The transformants continuously propagated on the media with selective antibiotics and successfully rooted on half QL media with addition of 0.5 mg l^{-1} IBA and 25 mg l^{-1} Km. Transgenic apple plants were cultured under cultivation conditions described above.

2.3.2 Pear transformation

Regeneration of pear was provided on Nitsch & Nitsch (NN) media supplemented by 3.0 mg l^{-1} TDZ, 0.5 mg l^{-1} NAA, 0.1 mg l^{-1} GA, 500 mg l^{-1} Cef., 25 mgl^{-1} Km were added from second passage (12-14th day after cocultivation). Leaf petioles have been also used for pear transformation. Other conditions were the same as for apple transformation procedure.

2.3.3 Carrot transformation

Embryo and leaf derived calli were used for carrot transformation. These calluses were cultivated on MS media with 0.2 mg l^{-1} 2-4D and 0.2 mg l^{-1} Kin. After three day of cocultivation with *Agrobacterium* on paper filters calluses were transferred to the same media supplemented 200 mg l^{-1} Km and 500 mg l^{-1} Cef. After 2-3 passages they were regenerated on MS media with 0.1 mg l^{-1} Kin and 100 mg l^{-1} Km. Rooting were performed on liquid hormones free 1/2 MS media with 50 mg l^{-1} Km.

2.4 *Molecular biology analysis of transgenic plants*

2.4.1 DNA isolation and PCR analysis

For DNA isolation 100 mg of fresh leaves collected from greenhouse plants were ground in liquid nitrogen. The genomic DNA was isolated from the extracted material by CTAB extraction method (Gelvin & Shilperoort, 1995).

PCR analysis was used to confirm the integration of PD sequence in the genome of transgenic plants. The primers for 35S promoter sequence (5'-CTGCCGACAGTGGTCCCAAAGATGGACCC-3') and for internal part of PD gene (5'-CGTTAACCCTTAGCCACTTCATCACTTCCAGGC-3') have been used. The amplification reaction was done in a buffer containing 67 mM Tris-HCl (pH 8.8), 15 mM $(NH_4)_2SO_4$, 2.5 mM $MgCl_2$, 0.2 mM of each dNTP, 0.2 mg ml^{-1} BSA, 0.1% Triton X-100, 1.25 units of Taq-polymerase, 1.0 μg DNA, 25 pkM of each primer in a volume of 30 μl. A 40 μl amount of mineral oil was layered onto the reaction mixture. Regime of amplification was next: preliminary denaturation (94°C) - 5 min, annealing primers (52°C) - 1 min; finishing of chain building (72°C) - 1 min; denaturation (94°C) - 1 min; number of amplification cycles - 30. The PCR products were analyzed using electrophoresis in 2% agars gel. The length of the amplified fragment was 420 bp.

2.4.2 Western blot

PCR positive plants were assessed by immunoblotting. Leaf extracts were prepared by extracting of 300 mg of leaf tissue with 2 ml of extraction buffer (Tris-HCl, pH 7.0, 20 mM NaCl). Undissolved material was removed by centrifugation and supernatant was analyzed by immunoblotting. Samples were analyzed by SDS PAGE. Protein separated on the gel were electroblotted onto nitrocellulose Hybond C (Amerscham). Immunoblott was processed as described by Burnett (1981), Towbin & Gordon (1984), Tsang (1983) and product detection was performed using the enhanced hemiluminiscence method (Amerscham).

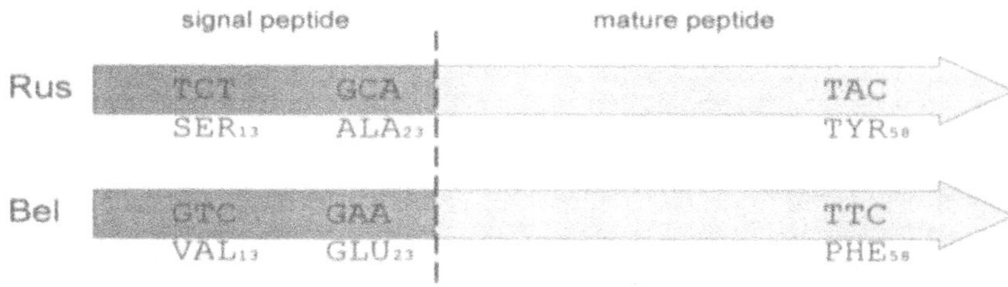

Figure 1. Nucleic acid sequence aligment of Rs-AFP2 genes.

3. Results

3.1 Cloning and analysis of the nucleotide sequence of genomic region encoding Rs-AFP2

Thirty *E. coli* clones with insertion obtained in cloning procedure were analyzed. Fifteen of these clones had the insertion of 350 bp. The nucleotide sequence of two clones was determined. It was similar to earlier published Rs-AFP sequence (Terras et al., 1995) and contains intron, approximately 100 bp, in region, coding leader peptide. Analysis of predicted amino acid sequence of cloned gene demonstrated that amino acids in positions 5 and 27 of mature form correspond to Gln and Arg, which is characteristic to Rs-AFP2 peptide. Some differences between cloned and published Rs-AFP peptide sequences were detected. There were two amino acids substitutions in the signal peptide (Val13Ser and Glu23Ala) and one substitution (Phe58Tyr) in mature peptide (Figure 1). These differences may be a result of *Rafanus sativus* subspecies diversity.

3.2 Transformation

Transformation of 309 apple leaf disks at three repetitions resulted in 28 lines formation. All obtained lines successfully proliferated on culture media with 50 mg l^{-1} Km and rooted on 25 mg l^{-1} Km. Transformation frequency varied from 1.4 to 6.0% (Table 1).

Table 1. Apple and pear transformation efficiency by plant defensin gene on pPCV002. Regeneration frequency of control apple plants - 75%, with 3.1 shoots per explant.

Variety	Explant	Leaf disks number	Regenerants formed on media 50 mg l^{-1} Km	Rooted plants on media10 mg l^{-1} Km	Western blot	transformation, efficiency, %
Apple rootstock N545	leaves	136	22	20	4	6.0
Pear Burakovka	leaves petiols	4438	57	57	1	15.817.1

Figure 2. Rooting of pear plants transformed vector construction CBE21:pPCV002rs on selective media with 25 mg l^{-1} Km, left control - nontransformed plants.

More efficient plantlets formation have been observed on pear. Cocultivation 44 leaf disks and 38 petioles resulted in formation of 5 and 7 lines respectively. These lines also successfully multiplicated on antibiotic contains media (25 mg l^{-1}) and rooted on 15 mg l^{-1} Km (Figure 2).

Agrobacterial infection of 60 carrot calluses and further their selection on selective media supplemented by 200 mg l^{-1} Km, produce more than 270 plantlets growing on 100 mg l^{-1} Km. These antibiotic concentrations completely suppressed nontransformed callus growth and regeneration. Obtained carrot plantlets were rooted on liquid half strength MS media with 50 mg l^{-1} Km and adapted to soil conditions (Table 2).

Table 2. Transformation of carrot by plant-defensin gene from *Rafanus sativus.*

Agrobacterialstrain/ plasmid	Explant (calluses)	Number of regenerants	PCRtest	Western blot
CBE21:pPCV002rs	60	270	20/20	18/20

3.3 PCR analysis

PCR analysis shown presence PD sequence in all genomes (20) of tested carrot lines (Figure 3). Usage the primer for 35S promoter sequence allowed us amplificate only transferred PD gene. No amplification products have been observed on DNA of control, nontransformed plants. This is strongly evidence of PD gene transfer and incorporation to tested plant genomes. Amplification of expected size fragment also observed in most of pear and apple transformants (data not shown).

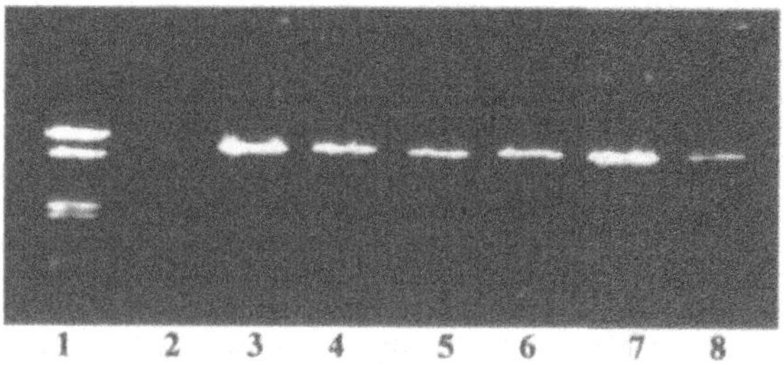

Figure 3. PCR analysis of carrot plants transformed by vector pPCV002rs. DNA was amplified using primers specific to 35S promoter and internal region of Rs AFP2 gene. Lane 1 - marker 100-700 bp; lane 2 - nontransformed control; lane 3- pPCV002rs; lanes 4-8- transgenic carrot plants.

3.4 Defensin gene expression

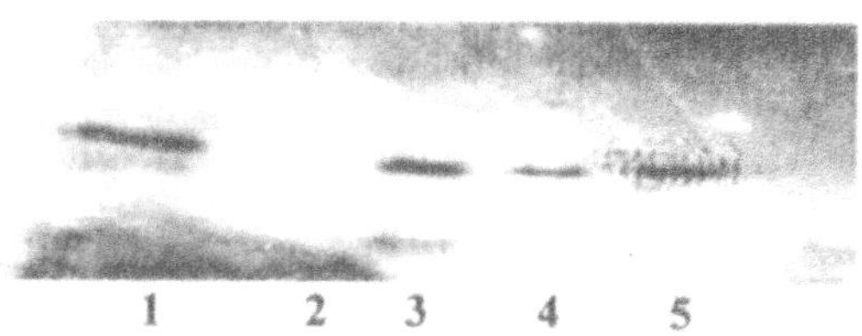

Figure 4. Western blot of total proteins of apple rootstock lines separated on 10% SDS-PAGE and probed with rabbit anti Rs-AFP2 antibody. Lane 1 contains pure Rs-AFP2 protein; lane 2 nontransformed control; lanes 3, 4, 5- transgenic apple clones.

Western blot analysis showed PD gene expression in 5 apple lines from 20 tested (Figure 4). Only in one pear line from 12 tested have been observed defensin syntesis. Contrary, most of tested carrot lines (18 from 20) demonstrated its detectable level.

4. Discussion

For the horticultural plants protection against fungal and microbial attack the plant defensins genes (PD) have been transferred to pear, apple and carrot). In contrast to antimicrobial peptides having non-plant origin (cecropin, magainin, tachyplesin and others) PDs are evidently optimal for expression in plants. Previously, we observed that fluriomertric measured activity of GUS gene included plant intron sequence were four-eight times higher in transgenic pear rootstock plants. For this reason the usage of genomic copy of PD gene with intron is more preferencial. Although most of PDs isolated to date are seed derived, evidence is now accumulating that plant defensins are also expressed in vegetative tissues. Furthermore vegetative tissues of transgenic plants with defensin gene from *Raphanus sativus* under the control of cauliflower mosaic virus 35S promoter reveal the resistance to fungal infection (Terras et al., 1995). It is essential that PDs have not activity against cells of higher eucariots.

5. References

Broekaert, W.F., Terras, F.R.G., Cammue, B.P.A. & Osborn, R.W. (1995) Plant defensins: novel antimicrobial peptides as components of the host defense system, *Plant Physiology* **108**: 1335-1358.

Burnett W.N. (1981) *Anal. Biochem.*, **112**, pp. 195-203.

De Bondt, A., Zaman, S., Broekaert, W., Cammue, B. & Keulemans, J. (1996) Genetic transformation of apple for increased fungal resistance. *EUCARPIA Symp. on Fruit Breeding and Genetics, Oxford, 1-6/9/96.* p. 5

Dolgov, S.V. & Muratova, S.A. (1995) Development of methods for genetic transformation of fruit cultures, *Abstracts of International Symposium on Engineering Plants for Commercial Products and Applications, Lexington, Kentucky, October 1-4*,1995, p. 15.

Draper, J., Scott, R. & Armitage, P. (1988) *Plant Genetic Transformatiom and Gene Expression*.Blackwell Sc. Pub. Oxford, London, Edinburgh, Boston, Melbourn,

Gelvin, S.B. & Schilperoort, R.A. (1995) *Plant molecular biology manual*, Kluwer Academic Publishers, Dordrecht, Boston, London.

Murashige, T. & Skoog, F. (1962) A revised medium for rapid growth and bioassays with tobacco tissue cultures. *Physiology Plant.* **15**, 473-497.

Norelli, J.L., Mills, J.Z., Jensen, L.A., Momol, M.T. & Aldwinckle, H.S. (1996) Genetic engineering of apple for increased resistance to fire blight. *EUCARPIA Symp. on Fruit Breeding and Genetics, Oxford, 1-6/9/96.* p. 18.

Oliveira, M.M., Miguel, C.M. & Raquel, M.H. (1996) Transformation studies in woody fruit species. *Plant tissue culture and Biothechnology*. **2**, 76- 93.

Quorin, M & Lepoivre, P. (1977) Etude de milieux adaptes aux cultures in vitro de prunus. *Acta Horticulturae* **78**, 437- 442.

Revenkova, E.V., Kraev, A.S. & Skryabin, K.G. (1993) Construction of a disarmed derivative of the supervirulent Ti plasmid pTiBo542. In: *Plant Biotech. and molecular biology*. Moscow. pp. 67-76.

Terras, F.RG., Schoofs, H.M.E., De Bolle, M.F.C., van Leuven, F., Rees, S.B., Vanderleyden, J., Cammue, B.P.A. & Broekaert, W.F. (1992) Analysis of two novel classes of antifungal proteins from radish (*Raphanus sativus* L.) seeds, *J.Biol. Chem.* **267**, 15301- 15309.

Terras, F.R.G., Eggermont, K., Kovaleva, V., Raikhel, N., Osborn, R.W., Kester, A., Rees, S.B., Torrekens, S., van Leuven, F., Vanderleyden, J., Cammue, B.P.A. & Broekaert, W.F. (1995) Small cystein-rich antifungal proteins from radish: their role in host defense, *The Plant Cell* **7**, 573- 588.

Towbin H.& Gordon J. J. (1984) *Immunol. Methods*, **72**, pp. 313- 340.

Tsang V.C.N., Peralta J.M. & Simons A.R. (1983) *Methods Emzymol*, **92**, pp. 377-391.

G.T. Scarascia Mugnozza, E. Porceddu & M.A. Pagnotta (Eds.)
Genetics and Breeding for Crop Quality and Resistance, 119-125, 1999

Plant-bacterial pathogen interaction modified in transgenic tomato plants expressing the *Gox* gene encoding glucose oxidase

R. Caccia[1], M. Delledonne[2], G.M. Balestra[3], L. Varvaro[3] & G.P. Soressi[1]
[1]Dipartimento di Agrobiologia e Agrochimica, Sez. Genetica, Università della Tuscia, Viterbo, Italy;[2] Istituto di Botanica e Genetica Vegetale, Università Cattolica del Sacro Cuore, Piacenza, Italy;[3] Dipartimento di Protezione delle Piante, Università della Tuscia, Viterbo, Italy

Abstract: Transgenic tomato (*L. esculentum* Mill.) plants containing the *Gox* gene encoding glucose oxidase from *Aspergillus niger* were obtained in two near isogenic lines (NIL) except for the *Pto* gene ("Riogrande" +/+ and "Rimone" *Pto/Pto*). T_3 transgenic plants possessing a single *Gox* gene copy were tested for resistance to *Pseudomonas syringae* pv. *tomato* to ascertain the role of H_2O_2 in plant pathogenesis response. Reactive oxygen species (ROS) such as H_2O_2 are generated following pathogen recognition, and they act both as cellular signalling molecules and as direct antimicrobial agents. In transgenic tomato plants, glucose oxidase generating H_2O_2 seems to confer to the susceptible isogenic tomato line (Riogrande) the capacity to respond actively to *P. syringae* pv. *tomato* by reducing the number of viable bacterial cells in the inoculated leaves. The effect on *Xanthomonas campestris* pv. *vesicatoria* infection and the possible synergistic action between a non-specific gene as *Gox*, and a race-specific gene as *Pto* is then discussed.

1. Introduction

Plant defence mechanisms could be assembled in two groups: (i) specific mechanisms responsible for pathogen recognition and control (specific resistance) and (ii) general mechanisms conferring resistance to a broad range of pathogens. In the first case plants respond to the attack by an avirulent bacterium with a rapid and specific process that triggers the hypersensitive disease resistance response (HR). HR is a rapid and localised death of cells within the infection site to prevent pathogen diffusion to healthy tissue (Heath, 1980). This reaction occurs in plants expressing a R-gene product when they are infected by a pathogen carrying the corresponding avirulence (*avr*) gene. Several examples of HR have been reported (Ingram, 1978; Klement, 1982; Mansfield, 1990; Baker & Orlandi, 1995; Tenhaken et al., 1995). The second mechanism occurs either in a plant that is not susceptible to a potential pathogenic bacterium or in a plant that is susceptible but able to control the pathogenic bacterium involving constitutive and inducible resistance such as pathogenesis related proteins (PR), systemic acquired resistance (SAR), etc. In both cases an early response, immediately after pathogen recognition lead to a rapid and sustained accumulation of reactive oxygen species (ROS) namely oxidative burst (Lamb & Dixon, 1997). The accumulation of hydrogen peroxide (H_2O_2) and related ROS has multiple functions in plant defence: (i) triggering local hypersensitive cell death; (ii) exerting direct antimicrobial activity against pathogens (Kim et al., 1988; Peng & Kuc, 1992); (iii) participating in the reinforcement of the plant cell wall to slow down pathogen ingress prior to the activation of the battery of defence related genes (Brisson et al., 1994; Bolwell et al., 1995). H_2O_2 acts also as a diffusible signal for the induction of cellular protectant genes in healthy tissue surrounding the infection site (Levine et al., 1994) and it mediates a systemic signal network in the establishment of plant immunity (Alvarez et al., 1998). Moreover, it has been reported that the continuous production of H_2O_2 in transgenic potato plants by ectopic expression of glucose oxidase (*Gox*) gene from *Aspergillus niger* conferred an increased level of resistance to *Erwinia carotovora* and *Phytophthora infestans* (Wu et al., 1995).

We have obtained several *Gox* transgenic lines of tomato (*L. esculentum* Mill.) expressing different levels of glucose oxidase and this material has been artificially inoculated to test its level of resistance to *Pseudomonas syringae* pv. *tomato* and *Xantomonas campestris* pv. *vesicatoria.*

2. Materials and Methods

2.1 Plant material

Two near isogenic lines (NIL) "Riogrande" (RIG,+/+), susceptible and "Rimone" (RIM, *Pto/Pto*), resistant to *P. s.* pv. *tomato* were transformed, via *Agrobacterium tumefaciens* binary vector, with a construct carrying the *Gox* gene encoding for glucose oxidase (GOD) enzyme from *Aspergillus niger* (Delledonne et al., 1993) under control of the constitutive promoter CaMV 35S.

Table 1. Tomato lines used in the experiments

Line	*Pto*gene	*Gox* gene
Riogrande (RIG)	-	-
Rimone (RIM)	+	-
RC 109	-	+
RC 115	-	+
RC 120	-	+
RC 136	+	+

Experiments were carried out on T_3 and T_4 homozygous transgenic lines possessing a single copy of the *Gox* (Table 1), and untransformed RIG and RIM were used as controls. The seedlings to be inoculated with bacteria were grown in pots under ambient temperature and light conditions in a glasshouse at Viterbo (Italy) in spring-summer.

2.2 Glucose oxidase (GOD) activity

The T_3 transgenic plants from RIG (RC 109, RC 115, RC 120) and from RIM (RC 136) together with their wild-type (RIG and RIM) were tested for the glucose oxidase activity by means of histochemical and biochemical methods. In the histochemical assay leaf discs of transgenic and control plants were placed on a medium containing starch and KI according to Olson & Varner (1993), and H_2O_2 accumulation was revealed by the formation of a dark-blue area around the leaf discs. The biochemical determination was carried out by measuring the peroxidase (POD) activity using guaiacol as hydrogen donor and H_2O_2 generated by glucose oxidase. The enzyme activity was measured as absorbance increase at 470 nm by a double-beam Perkin Elmer Lambda 3 spectrophotometer at 30°C. The assay medium contained D(+) glucose 0.46% (w/v), guaiacol 31 mM, Triton X-100 2% (w/v), POD 6 U, and 0.1 ml of plant extract.

2.3 Bacterial cultures and inoculum preparation

The *P. s.* pv. *tomato* and *X. c.* pv. *vesicatoria* strains were from the collection of the Department of Plant Protection of the University of Tuscia. Cultures of the bacteria were maintained at 4°C on nutrient agar with 2% of glycerol (NGA) slants containing. The bacteria were subcultured twice on NGA at 26 ±1°C for 24 h, collected from the second slant in sterile distilled water (SDW) and washed twice by centrifugation for 15 min at 8,500 x g at 8°C. The bacterial suspension was adjusted turbidimetrically to about 10^8 colony forming units (cfu) ml^{-1} by reference to a calibration curve relating extinction to cfu ml^{-1}. From this suspension, inoculum containing 10^8 cfu ml^{-1} and 10^6 cfu ml^{-1} respectively for *P. s.* pv. *tomato* and *X. c.* pv. *vesicatoria* strains were prepared by serial dilutions (Varvaro & Surico, 1987).

2.4 Inoculation procedure and determination of bacterial populations

The bacterial suspension was applied by means of a spray paint gun onto the upper and the lower leaf surfaces from a distance of approximately 30 cm, until visible drops were apparent (Babelegoto et al., 1988). The inoculated plants were kept at 25°C, in presence of 70% relative humidity. For *P. s.* pv. *tomato* at 3, 7, 21 days after inoculation and for *X. c.* pv. *vesicatoria* only after 7 days, samples of 2 contaminated leaflets $plant^{-1}$ were detached from 5 plants of each plot. Then, to remove as many adhering bacteria as possible from the leaf surface, leaflets were washed in SDW by an orbital shaker for 2 hours at 150 rpm. After measuring the leaflets area by means of a digital planimeter (ΔT) the leaflets were ground by an

Ultra-Turrax homogeniser. Serial dilutions were then prepared from the leaf homogenates and 0.1 ml amounts were spread onto the surface of nutrient agar plus 5% (w/v) sucrose (NSA). Colony counts were made after incubating the agar plates for 48-72 h at 26±1 °C. From these values, the number of bacteria per cm^2 of leaf were calculated. Symptom development on each tomato plant was daily observed. Leaf necrosis counts were made 7 days after inoculations.

3. Results

3.1 *Glucose oxidase (GOD) activity*

In the T_3 transgenic plants (RC 109, RC 115 RC 120, and RC 136) and the control (RIG and RIM) the GOD enzyme activity was tested. In the histochemical assay transgenic lines showed a blue colour 2-3 fold deeper than the control ones. In the same transgenic lines the GOD activity values registered following the biochemical method were 4-5 times higher than the control except for the line RC 120 in which it was only the double (Figure 1a).

3.2 *Pseudomonas syringae pv. tomato inoculation*

Two T_3 transgenic RIG lines (RC 109 and RC 115) and one T_3 RIM line (RC 136) having relatively high GOD activity (Figure 1a), and RIG (susceptible) and RIM (resistant) untransformed control plants were inoculated with *P. s.* pv. *tomato*. A remarkable difference in the growth of bacterial populations calculated as number of colony forming units (cfu) per cm^2 $leaf^{-1}$ among lines was observed (Figure 1b). In the susceptible RIG line the bacterial population showed an exponential growth with a maximum at 7 days after inoculation, whereas in transgenic RIG plants RC 109 and RC 115 the population growth was reduced like as in the resistant near isogenic control RIM.

In the transgenic RIM plants T_3 RC 136, which carry both *Gox* and *Pto*, the bacterial growth followed a trend very similar to the resistant control RIM but with a lower number of bacterial cells at 3 days after inoculation.

3.3 *Xanthomonas campestris pv. vesicatoria inoculation*

To verify a possible action of the *Gox* gene on plant defence mechanism against the pathogenic bacterium *Xanthomonas* T_4 homozygous plants derived form the T_3 lines tested against *P. s.* pv. *tomato* were inoculated with *X. c.* pv. *vesicatoria*. Since in *Gox* transgenic plants the maximum reduction of bacterial population growth was observed at 7 days after inoculation with *P. s.* pv. *tomato,* the endophytic population of *X. c.* pv. *vesicatoria* in the infected leaves was determined only 7 days after the artificial infection.

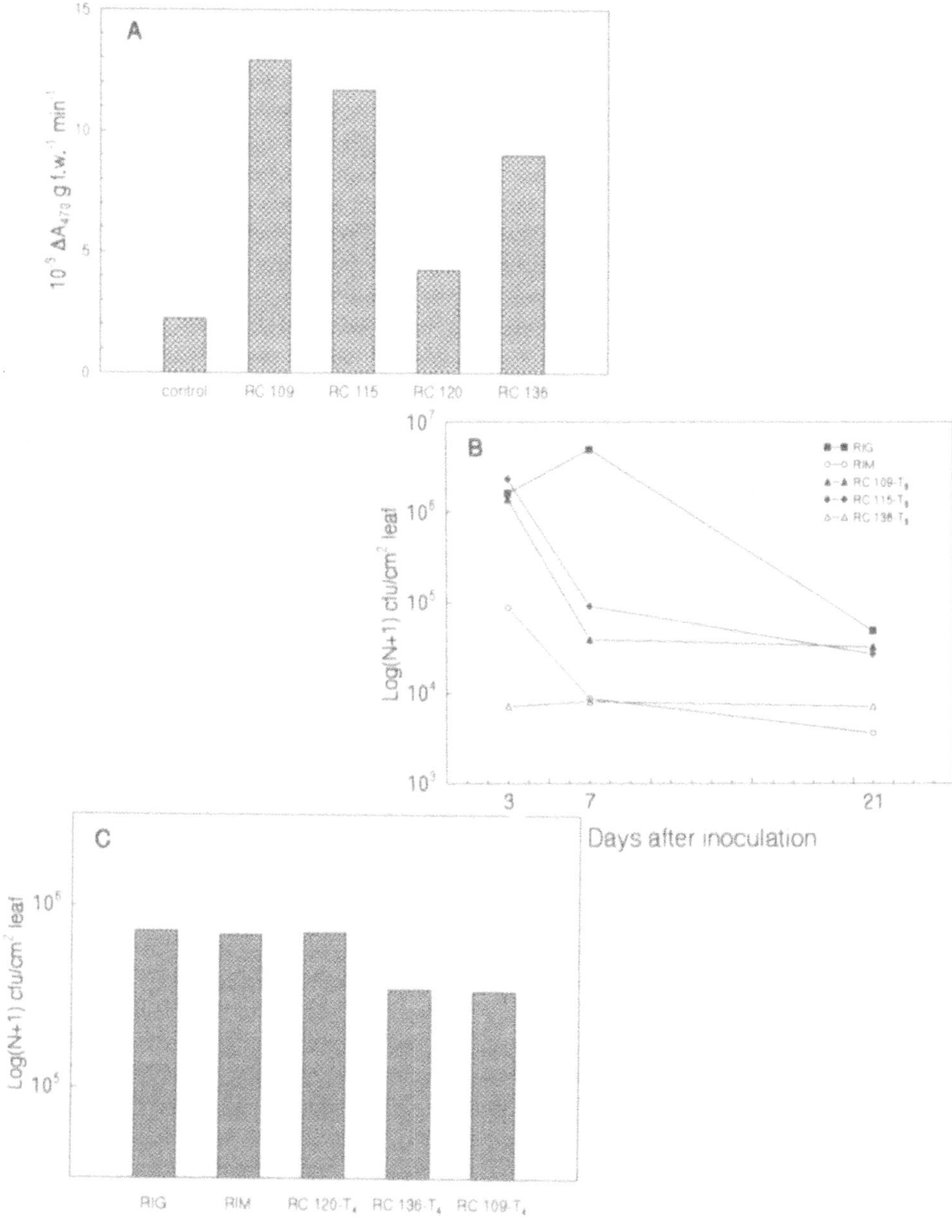

Figure 1. a) GOD activity in T_2 *Gox* transgenic plants compared with the control. b) Trend of the endophytic growth at *Pseudomonas syringae* pv. *tomato* in *Gox* transgenic lines and the susceptible (RIG) and resistant (RIM) controls at 3, 7 and 21 days after artificially inoculated (10^8 cfu ml^{-1}). c) Endophytic population of *Xanthomonas campestris* pv. *vesicatoria* in *Gox* transgenic lines and susceptible RIG and RIM at 7 days after artificial inoculation (10^6 cfu ml^{-1}).

The results obtained seems to indicate that lines (RC 136-T_4 from RIM and RC 109-T_4 from RIG) derived from plants with the highest GOD activity, are able to limit in a certain degree the growth of the endophytic bacterial population in respect of the controls and of the RC 120-T_4 line originated from a plant exhibiting the lowest GOD activity (Figure 1c).

3.4 Symptoms on leaves

In both the infection experiments with *P. s.* pv. *tomato* and *X. c.* pv. *vesicatoria,* the symptoms in the transgenic plants appeared as small necrotic spots from those characteristics of the respective controls (disease symptoms). The number of these small necroses, hereafter called hypersensitive-like necroses (HR-like), per cm^2/leaf was higher in the T_3 and T_4 transgenic line RC 109 which showed the relatively highest GOD activity and lowest bacterial growth (Table 2).

Table 2. Type and number of necrotic spots on the infected leaves 7 days after inoculation

Pseudomonas syringae pv. *tomato* (10^8 cfu ml^{-1})			*Xanthomonas campestris* pv. *vesicatoria* (10^6 cfu ml^{-1})		
Tomato line	n° necroses cm^{-2} leaf		Tomato line	n° necroses cm^{-2} leaf	
	disease symptoms	HR-like		disease symptoms	HR-like
RIG	5.6	0.0	RIG	0.4	0.0
RIM	0.0	0.0	RIM	1.3	0.0
RC 109-T_3	0.0	7.6	RC 109-T_4	0.0	1.5
RC 115-T_3	0.0	7.4	RC 120-T_4	0.7	0.0
RC 136-T_3	0.0	0.5	RC 136-T_4	0.0	0.5

4. Discussion

The *Gox* transgene from *A. niger* is regularly inherited and expressed in tomato plants. The level of GOD activity expressed in different transgenic tomato plants is one- to fivefold higher than in the control plants (Figure 1a) and confers to the plant the capacity to limit both the endophytic bacterial population growth and the disease symptoms. Infact, the T_3 transgenic RIG plants RC 109 and RC 115, even though not possessing the *Pto* gene, performed as well resistant as the RIM *Pto* line (Figure 1b).

The relatively limited growth of the endophytic bacterial population in the transgenic plants seems to prove that the *Gox* gene is inducing conditions in some way unfavourable for the multiplication of *P. s.* pv. *tomato*. This is substantiated by the resistance to *E. carotovora* observed in *Gox* transgenic potato plants (Wu et al., 1995).

The symptoms, here called hypersensitive-like necroses, on leaves inoculated with *P. s.* pv. *tomato* or *X. c.* pv. *vesicatoria* in the transgenic lines RC 109, RC 115 and RC 136 were markedly different from those observed on susceptible controls where the necrotic spots were small and scanty. Moreover, no chlorotic halo appeared on transgenic lines when inoculated with *P. s.* pv. *tomato* whereas in the control plants typical necrotic specks surrounded by an evident chlorotic area followed by a general yellowing of the leaves were observed.

The existence of a possible interaction between the non-specific *Gox* gene and the specific *Pto* resistance gene was not proved in our experiment but it cannot be excluded. Since further investigations are needed, the available *Gox* transgenic tomato NILs (RIG, +/+ and RIM, *Pto/Pto*) appear particularly useful.

5. References

Alvarez, M.E., Pennel, R.I., Mejer, P-J., Ishikawa, A., Dixon, R.A. & Lamb, C. (1998) Reactive oxygen intermediates mediate a systemic signal network in the establishment of plant immunity. *Cell* **92**:773-784.

Babelegoto, N.M., Varvaro, L. & Cirulli, M. (1988) Epiphytic and endophyitc multiplication of *Pseudomonas syringae* pv. *tomato* (Okabe) Young et al. in susceptible and resistant tomato leaves. *Phytopath. medit.* **27**: 138-144.

Baker, C.J., and Orlandi, E.W. (1995) Active oxygen in plant pathogenesis. *Ann. Rev. Phytopathol.* **33**, 299-321.

Bolwell, G.P., Butt, V.S., Davies, D.R. & Zimmerlin, A. (1995) The origin of the oxidative burst in plants. *Free Rad. Res. Comm.* **23**, 517-532.

Brisson, L.F., Tenhaken, R. & Lamb C. (1994) The function of oxidative cross-linking of cell wall structural proteins in plant disease resistance. *Plant Cell* **6**, 1703-1712.

Delledonne, M. Frigerio, L. & Fogher, C. (1993) L'enzima glucosio ossidasi: clonaggio del gene da *Aspergillus niger*. In Ricerche e innovazioni nell'industria alimentare. Chiriotti (Ed), pp. 599-608

Heath, M.C. (1980) Reaction of nonsuscepts to fungal pathogens. *Ann. Rev. Phytopathol.* **18**, 211-236.

Ingram, D.S. (1978) Cell death and resistance to biotrophs. *Ann. Appl. Biol.* **89**, 291-295.

Kim, K.K., Fravel, D.R. & Papavizas, G.C. (1988) Identification of a metabolite produced by *Talaromyces flavus* as glucose oxidase and its role in the biocontrol of *Verticillium dahlie*. *Phytopathology* **78**, 488-492.

Klement, Z. (1982) Hypersensitivity, in M.S. Mount and G.H. Lacy (eds), *Phytopathogenic Procaryotes*, Academic Press, New York, pp. 149-177.

Lamb, C. & Dixon, R.A. (1997) The oxidative burst in plant disease resistance. *Ann. Rev. Plant Physiol. Plant Mol. Biol.* **48**, 251- 275.

Levine, A., Tenhaken, R., Dixon, R. & Lamb, C. (1994). H_2O_2 from the oxidative burst orchestrates the plant hypersensitive disease resistance response. *Cell* **79**, 583-593.

Mansfield, J.W. (1990) Recognition and response in plant-fungus interactions, in R.S.S. Fraser (ed), *Recognition and Response in Plant-Virus Interactions*, Springer-Verlag, Berlin, pp. 31-52.

Peng, M. & Kùc, J. (1992) Peroxidase-generated hydrogen peroxide as source of antifungal activity *in vitro* and on tobacco leaf disks. *Phytopathology* **82**, 696-699.

Olson, P.D. & Varner, J.E. (1993) Hydrogen peroxide and lignification. *Plant Journal* **4**, 887-892.

Tenhaken, R., Levine, A., Brisson, L.F., Dixon, R. & Lamb, C. (1995) Function of the oxidative burst in hypersensitive disease resistance. *Proc. Natl. Acad. Sci. USA* **92**, 4158-4163.

Varvaro, L. & Surico, G. (1987) Multiplication of wild types of *Pseudomonas syringae* pv. *savastanoi* (Smith) Young et al. and their indoloacetic acid-deficient mutants in olive tissues, in M. Nijhoff (eds), *Plant pathogenic bacteria*, Proc Sixth Intern. Conf. Plant Path. Bact., Maryland, 2-7 June 1985, Dordrecht, pp. 556- 565.

Wu, G.S., Short, B.J., Lawrence, E.B., Levine, E.B., Fitzsimmons K.C. & Shah D.M. (1995) Disease resistance conferred by expression of a gene encoding H_2O_2-generating glucose oxidase in transgenic potato plants. *Plant Cell* **7**, 1357-1368.

G.T. Scarascia Mugnozza, E. Porceddu & M.A. Pagnotta (Eds.)
Genetics and Breeding for Crop Quality and Resistance, 127-133, 1999

Nitric oxide signalling in the plant hypersensitive disease resistance response

M. Delledonne[1], Y. Xia[2,3], R. A. Dixon[3], C. Lorenzoni[1] & C. Lamb[2]
[1]Istituto di Genetica, Università Cattolica S.C., Piacenza, 29100 Italy; [2]Plant Biology Laboratory, The Salk Institute, La Jolla, CA 92037; [3]Plant Biology Division, The Noble Foundation, Ardmore, OK 7340

Abstract: Plants have evolved several mechanisms to prevent invasion of their tissues by pathogens. A common feature of disease resistance is the hypersensitive response at and immediately surrounding infection sites, which is characterized by the formation of necrotic lesions. Following to this local resistance response, tissue distal to the infection site develops a systemic acquired resistance to a secondary infection by the same or by different pathogens. During the hypersensitive response, massive accumulation of H_2O_2 from the oxidative burst functions as a diffusible signal for the induction of cellular protectant genes, mediates a systemic signal network in the establishment of plant immunity and is required, but not sufficient, to trigger localized host cell death. Nitric oxide, a well known host-defense component in animal systems, co-operates with reactive oxygen intermediates in the induction of hypersensitive cell death, and functions independently of such intermediates in the induction of defense related genes.

1. The hypersensitive resistance response

Pathogen infection of resistant genotypes results in the hypersensitive response (HR) which is characterized by the formation of necrotic lesions at the site of infection and the restriction of pathogen growth and spread. One of the earliest events in the HR is the rapid accumulation of reactive oxygen intermediates (ROI) through the activation of an enzyme system similar to the neutrophil NADPH oxidase, and that closely resembles the oxidative burst during phagocyte activation (Groom et al., 1996; Keller et al., 1998). Activation of the oxidative burst is a central component of a highly amplified and integrated signal system which involves also salicylic acid and perturbations of cytosolic Ca^{2+}, to activate the expression of disease-resistance mechanisms (Lamb & Dixon, 1997). Different pathogens or elicitors result in different kinetics of the oxidative burst, as demonstrated in soybean cells incubated with isogenic strains of *Pseudomonas syringae* pv. *glycinea* that differ at avirulence locus *avrA*, which interacts with the soybean resistance gene *Rgm2* (Keen & Buzzel, 1991). Both avirulent and virulent strains evoke a weak oxidative burst within 1 h, although only the avirulent strain cause a second, massive oxidative burst after 3-4 h which is followed few hours later by the hypersensitive cell death (Levine et al., 1994). Host-derived oligogalacturonides and pathogen-derived nonspecific elicitors are responsible for triggering the early, weak oxidative burst, whereas avirulence genes evoke the second sustained ROI production (Baker et al., 1993). Several lines of evidence suggest that H_2O_2 accumulated during the oxidative burst drives the cross-linking of cell wall proteins to increase the resistance of the plant cell wall to digestion by wall-degrading enzymes (Bradley et al., 1992), slowing down pathogen ingress prior to the coordinate activation of a battery of defense responses in cells surrounding the infected area and restricting pathogen spread into healthy tissues (Brisson et al., 1994). The oxidative polymerization of wall proteins is also thought to contribute to pathogen killing by trapping them in the extracellular matrix of cells undergoing hypersensitive cell death. These cells also function as sites of production of secondary signaling molecules for surrounding living cells, which then secrete antimicrobial compounds for plant defense and accumulate radical scavengers counterbalancing ROI to protect themselves. H_2O_2 from the oxidative burst functions also as a diffusible signal for the induction of cellular protectant genes encoding glutathione *S*-transferase (*gst*) and glutathione peroxidase (Levine et al., 1994), and it mediates a systemic signal network in the establishment of plant immunity (Alvarez et al., 1998). However, in soybean cells ROI are not primary signals for rapid induction of defense genes encoding enzymes of phenylpropanoid biosynthesis involved in the synthesis of antibiotics, lignin and salicylic acid (Dixon & Paiva, 1995).

2. Reactive oxygen intermediates are required, but not sufficient, to trigger the hypersensitive disease resistance response

The oxidative burst required for hypersensitive cell death in soybean cell suspensions inoculated with avirulent *P. syringae* pv. *glycinea* gives a sustained accumulation of 6 to 10 μM H_2O_2 (Shirasu et al., 1997). Several inhibitors of the neutrophil NADPH oxidase, including the suicide substrate inhibitor diphenylene iodonium (DPI) (Morel et al., 1991) block the oxidative burst in plant cells and the hypersensitive cell death as well (Levine et al., 1994) indicating that the oxidative burst precedes the induction of programmed cell death and is required to orchestrate the plant hypersensitive disease resistance response. Addition of the H_2O_2-generating system glucose/glucose oxidase to the cell suspensions results in a steady state accumulation of ~6 μM H_2O_2 for >3 h, and hence closely mimics the oxidative burst induced by the avirulent pathogen. The protein phosphatase type 2a inhibitor cantharidin partially activates the signal pathway leading to the oxidative burst in the absence of pathogen avirulence factors (Levine et al., 1994); salicylic acid, which functions in signal amplification (Shirasu et al., 1997), enhances this effect, resulting in the accumulation of >30 μM H_2O_2. However, *in situ* generation of H_2O_2 in absence of the avirulence signal induces only a relatively weak cell death response (Delledonne et al., 1998). Furthermore, mutations in the *hrmA* locus of *Pseudomonas fluorescens* carrying the *hrp/hrm* region of *P. syringae* pv. *syringae* elicit production of active oxygen in inoculated tobacco suspension cells but do not cause their hypersensitive cell death, indicating that the oxidative burst does not directly lead to cell death (Glazener et al., 1996).

3. Nitric oxide

Although nitric oxide (NO) has been detected in several plant species (Wildt et al., 1997), its role has not been clearly defined yet. The few reports on the function of NO in plant suggest an important role in plant growth, development and defense (Noritake et al., 1996; Leshem & Haramaty, 1996; Cueto et al., 1996). In vertebrates NO is studied more extensively because it is involved in physiological and pathophysiological conditions, therefore its activities in animals have been explored in detail. NO is a highly reactive molecule that rapidly diffuses and permeate cell membranes. It is produced by nitric oxide synthase oxygenase (NOS) during the conversion of L-arginine to citrulline in presence of oxygen, requisite cofactors and NADPH (Crane et al., 1997). The biological half-life of NO is on the order of a few seconds because it readily forms complexes with the transition metal ions including those present in metalloproteins. Its reaction with heme containing proteins has been well characterized, especially in the case of hemoglobin and guanyl cyclase (Snyder, 1992). It has been implicated in a number of diverse physiological processes like neurotransmission, smooth muscle relaxation, penile erection, platelet inhibition and immune regulation (Stamler et al., 1992). Pathologies linked to excessive NO production include immune-type diabetes, carcinogenesis, multiple sclerosis and strokes, and pathologies linked to insufficient NO production include hypertension, arteriosclerosis, impotence and susceptibility to infections (Crane et al., 1997). NO is therefore considered a double–edged sword, beneficial as a messenger or modulator for immunologic self-defense, but potentially toxic when the antioxidant system is weak or following accumulation of reactive oxygen intermediates (Schmidt & Walter, 1994). The cytotoxic actions of activated macrophages against human cancer cell lines have been attributed to their ability to generate NO and superoxide (O_2^-) which rapidly react with NO to generate peroxynitrite (-OONO; Farias-Eisner et al., 1996). The reaction of O_2^- with NO is three times faster than the rate at which superoxide dismutase (SOD) scavenges O_2^- but it is normally minimized by high SOD activity and low O_2^- and NO concentrations. Peroxynitrite can however be generated under conditions in which concentrations of NO and O_2^- are increased and/or concentrations of SOD are decreased. ONOO- is more stable than O_2^-, and it can diffuse up to several cells diameters where it can react rapidly with lipids, proteins and DNA (Stamler et al., 1992).

4. Nitric oxide co-operates with reactive oxygen intermediates in the induction of plant hypersensitive response

It is becoming clear that the innate immune system shows several similar characteristics among vertebrates, invertebrates and plants (Medzhitov & Janeway, 1998). In the mammalian immune system NO cooperates with ROI in inducing cell death of tumor cells and macrophage killing of bacteria (Schmidth & Walter, 1994; Nathan, 1995). In plant cells challenged with avirulent pathogens, we have shown that nitric oxide plays a similar role (Delledonne et al., 1998). Treatment of soybean cell suspensions with the NO generator sodium nitroprusside (SNP) potentiates the induction of cell death by exogenous H_2O_2 in a synergistic manner, with NO promoting a ~ 10-fold increase in ROI induced cell death. NO also potentiates cell death induced by endogenous ROI. The massive oxidative burst generated by the addition of the protein phosphatase type 2a inhibitor cantharidin and salicylic acid results in the accumulation of >30 μM H_2O_2 causing only a weak induction of cell death, but the simultaneous addition of NO generators markedly enhances the response (Delledonne et al., 1998).

An enzyme similar to the neutrophil NADPH oxidase apparently contributes to the oxidative burst in plant. Now several lines of evidence indicate that another key enzyme involved in mammalian macrophage action, NO synthase, is conserved in plant where it is activated during plant-pathogen interaction (Ninneman & Maier, 1996). Tobacco mosaic virus infection of resistant, but not susceptible, tobacco results in elevated NOS activity (Durner et al., 1998), and soybean cell suspensions inoculated with *P. syringae* produce NO with a similar pattern to H_2O_2 accumulation, with a rapid stimulation by both avirulent and virulent strains followed by further strong production of NO only in cells inoculated with the avirulent strain, concomitant with the avirulence gene-dependent oxidative burst immediately prior to the onset of hypersensitive cell death (Delledonne et al., 1998). Scavenging of endogenous NO with carboxy-PTIO or blocking NO synthase with the competitive inhibitor N^{ω}-nitro-L-arginine (L-NNA), block the induction of hypersensitive cell death by avirulent *P. syringae*. Infiltration of leaves of *Arabidopsis* ecotype Col-O, which contains the R gene, with *P. syringae* carrying the corresponding *avrR* gene induces a localized HR lesion after about 24 h, accompanied by restriction of bacterial growth. Co-infiltration of L-NNA inhibits the HR and promotes the development of the spreading chlorosis observed with the isogenic virulent bacteria. Moreover, development of disease symptoms in leaves blocked for NO accumulation is accompanied by enhanced bacterial growth, although not to the levels observed with the isogenic virulent strain. Thus, the ROI/NO collaboration is necessary for the induction of the hypersensitive response (Figure 1), and plants can fine-tuning the magnitude of the signal through the regulation of levels of NO and ROI. Although nothing is known about the regulation of NO levels in plant cells, several mechanisms for scavenging reactive oxygen intermediates have been characterized and support the model of concerted action of producers/scavengers to balance the level of reactive molecules that can interact for signaling through the defense-response pathway.

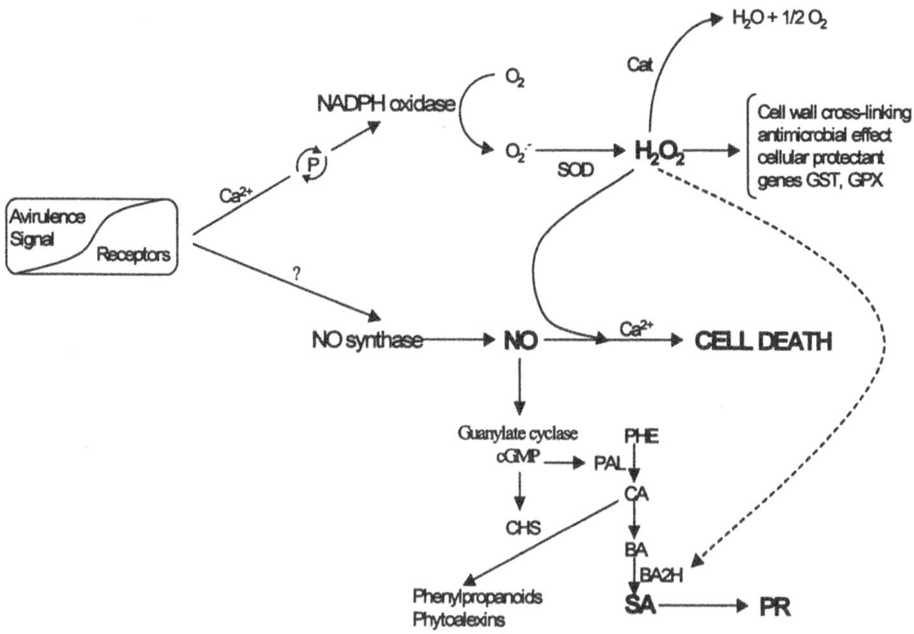

Figure 1. Model of action of NO and H_2O_2 in the hypersensitive response. (Ca^{2+}) Ca^{++} influx, (P) phosphorylation dependent step, (H_2O_2) hydrogen peroxide, (NO) nitric oxide, (GST) glutathione S-transferases, (GPX) glutathione peroxidases, (PAL) phenylalanine ammonia-lyase, (CHS) chalcone synthase, (Cat) catalases, (BA2H) benzoic acid 2-hydroxylase, (PHE) phenylalanine, (CA) cinnamic acid, (BA) benzoic acid, (SA) salicylic acid, (PR) pathogenesis-related proteins.

5. Nitric oxide is involved in transcriptional activation of defense genes

We demonstrated that NO functions along with ROI in triggering hypersensitive cell death, suggesting that NO can be involved in other defense functions complementary to, and independent of, ROI. H_2O_2 drives oxidative cross-linking of tyrosine-rich cell wall structural proteins (Bradley et al., 1992) and it is both necessary and sufficient to induce cellular protectant genes encoding glutathione *S*-transferase (*gst*) and glutathione peroxidase. However, in soybean cells ROI are not primary signals for rapid induction of defense genes encoding enzymes of phenylpropanoid biosynthesis involved in the synthesis of antibiotics, lignin and salicylic acid (Levine et al., 1994). In soybean suspension cells NOS inhibitor L-NNA has little effect on the initial induction of *gst* in response to avirulent *P. syringae*, whereas accumulation of transcripts encoding phenylalanine ammonia-lyase (*pal*), the first enzyme of phenylpropanoid biosynthesis, and chalcone synthase (*chs*), the first enzyme of the branch specific for flavonoids and isoflavonoid-derived antibiotics (Dixon & Paiva, 1995), are markedly reduced (Delledonne et al., 1998). The role of NO in the induction of defense genes is further supported by experiments in tobacco (Durner et al., 1998). Injection of tobacco leaves with mammalian NOS resulted in significant accumulation of *pal* and of pathogenesis-related 1 (*pr-1*) protein, a well-defined marker of plant disease resistance. The same pattern of accumulation has been detected in tobacco suspension cells treated with NO donors or with a membrane permeable analog of cyclic GMP. In animals cGMP is produced by guanylate cyclase, which is activated by NO binding to the heme-iron or through S-nitrosilation at critical cysteine residues (Stamler et al., 1992). NO induction of *pal* in tobacco suspension cells can be suppressed by several inhibitors of guanylate cyclase, suggesting the involvement of cGMP-dependent components in the NO-responsive defense gene activation pathway (Durner et al., 1998). Although the involvement of cGMP in several plant signal transduction pathways has been already demonstrated (Bowler et al., 1994), whether NO is the physiological activator of plant guanylate cyclase remains to be determined. It is now clear however that another important component of plant defense system has been identified. Nitric oxide plays a key role in pathogen resistance responses (Figure 1), and further studies of its regulation and mechanisms of action will offer the possibility to exploit new ways for improving plant resistance against pathogens.

6. References

Alvarez, M.E., Pennel, R.I., Meijer, P.-J., Ishikawa, A., Dixon, R.A. & Lamb, C. (1998) Reactive oxygen intermediates mediate a systemic signal network in the establishment of plant immunity. *Cell* **92**, 773-784.

Baker, C.J., Orlandi, E.W. & Mock, N.M. (1993) Harpin, an elicitor of the hypersensitive response in tobacco caused by *Erwinia amilovora*, elicits active oxigen production in suspension cells. *Plant Physiol.* **102**, 1341-1344.

Bowler, C., Neuhaus, G., Yamagata, H. & Chua, N.-H. (1994) Cyclic GMP and calcium mediate phytochrome phototransduction. *Cell* **77**, 73-81.

Bradley, D.J., Kjellbom, P. & Lamb, C.J. (1992) Elicitor- and wound-induced oxidative cross-linking of a proline-rich plant cell wall protein: A novel, rapid defense response. *Cell* **70**, 21-30.

Brisson, L.F., Tenhaken, R. & Lamb, C. (1994) Function of oxidative cross-linking of cell wall structural proteins in plant disease resistance. *Plant Cell* **6**,1703-1712.

Crane, B.R., Arvai, A.S., Gachhui, R., Wu, C., Ghosh, D.K., Getzoff, E.D., Stuher, D.J. & Tainer, J.A. (1997) The structure of nitric oxide synthase oxygenase domain and inhibitor complexes. *Science* **278**, 425-431.

Cueto, M., Hernandez-Perera, O., Martin, R., Bentura, M.L., Rofrigo, J., Lamas, S. & Golvano, M.P. (1996) Presence of nitric oxide synthase activity in roots and nodules of *Lupinus albus*. *FEBS* **398**, 159-164.

Delledonne, M., Xia, Y., Dixon, R. & Lamb, C. (1998) Nitric oxide functions as a signal in plant disease resistance. *Nature* **394**, 585-588.

Dixon, R.A. & Paiva, N. (1995) Stress-induced phenylpropanoid metabolism. *Plant Cell* **7**, 1085-1097.

Durner, J., Wendehenne, D. & Klessig D.F. (1998) Defense gene induction in tobacco by nitric oxide, cyclic GMP, and cyclic ADP-ribose. *Proc. Natl. Acad. Sci. USA* **95**, 10328-10333.

Farias-Eisner, R., Chaudhuri, G., Aeberhard, E. & Fukuto, J.M. (1996) The chemistry and tumoricidal activity of nitric oxide/hydrogen peroxide and the implications to cell resistance/susceptibility. *J. Biol. Chem.* **271**, 6144-6151.

Glazener, J.A., Orlandi, E.W. & Baker, J.C. (1996) The active oxygen response of cell suspensions is not sufficient to cause hypersensitive cell death. *Plant Physiol.* **110,** 759-763.

Groom. Q.J., Torres, M.A., Fordham-Skelton, A.P., Hammond-Kosack, K.E., Robinson, N.J. & Jones, J.D.G. (1996) *rbohA*, a rice homologue of the mammalian gp91^{phox} respiratory burst oxidase gene. *Plant J.* **10,** 515-522.

Keen, N.T. & Buzzell, R.I. (1991) New disease resistance genes in soybean against *Pseudomonas syringae* pv. *glycinea*: Evidence that one of them interacts with a bacterial elicitor. *Theor. Appl. Genet.* **81,** 133-138.

Keller, T., Damude, H.G., Werner, D., Doerner, P., Dixon, R.A. & Lamb, C. (1998) A plant homolog of the neutrophil NADPH oxidase gp91^{phox} subunit gene encodes a plasma membrane protein with Ca^{2+}-binding domains. *Plant Cell* **10,** 255-266.

Lamb, C. & Dixon, R.A. (1997) The oxidative burst in plant disease resistance. *Annu. Rev. Plant Physiol. Plant. Mol. Biol.* **48,** 251-275.

Leshem, Y.Y. & Haramaty, E. (1996) The characterization and contrasting effects of the nitric oxide free radical in vegetative stress and senescence of *Pisum sativum* Linn. foliage. *J. Plant Physiol.* **148,** 258-263.

Levine, A., Tenhaken, R., Dixon, R.A. & Lamb, C. (1994) H_2O_2 from the oxidative burst orchestrates the plant hypersensitive response. *Cell* **79,** 583-593.

Medzhitov, R & Janeway, C. A. (1998) An ancient system of host defense. *Curr. Opin. Immunol.* **10,** 12-15.

Morel, F., Doussiere, J. & Vignais, P.V. (1991) The superoxide-generating oxidase of phagocytic cells: physiological, molecular and pathological aspects. *Eur. J. Biochem.* **201,** 523-546.

Nathan, C. (1995) Natural resistance and nitric oxide. *Cell* **82,** 873-876.

Ninneman, H. & Maier, J. (1996) Indications for the occourrence of nitric oxide synthases in fungi and plants and the involvement in photoconidiation of *Neurospora crassa. Photochem. And Photobiol.* **64,** 393-398.

Noritake, T., Kawakita, K. & Doke, N. (1996) Nitric oxide induces phytoalexin accumulation in potato tuber tissues. *Plant Cell Physiol.* **37,** 113-116.

Schmidt, H.H.H.W. & Walter, U. (1994) NO at work. *Cell* **78,** 919-925.

Shirasu, K., Nakajima, H., Rajasekhar, V.K., Dixon, R.A. & Lamb, C. (1997) Salicylic acid potentiates an agonist-dependent gain control that amplifies pathogen signals in the activation of defense mechanisms. *Plant Cell* **9,** 261-270.

Snyder, S. H. (1992) Nitric oxide: first of a new class of neurotransmitters? *Science* **257,** 494-496

Stamler, J.S., Singel, D.J. & Loscalzo, J. (1992) Biochemistry of nitric oxide and its redox-activated forms. *Science* **258,** 1898-1902.

Wildt, J., Kley, D., Rockel, A., Rockel, P. & Segschneider, H.J. (1997) Emission of NO from several higher plant species. *J. Geoph. Res.* **102,** 5919-5927.

G.T. Scarascia Mugnozza, E. Porceddu & M.A. Pagnotta (Eds.)
Genetics and Breeding for Crop Quality and Resistance, 135-144, 1999

Polymorphism of inhibitors of hydrolytic enzymes present in cereal and sunflower seeds

Al.V. Konarev[1], I.N. Anisimova[2], V.A. Gavrilova[2] & P.R. Shewry[3]
[1]All-Russian Institute of Plant Protection, Podbelskogo 3, St.Petersburg, 189620, Russia. E-mail: konarev@riam.spb.su;*[2]N.I. Vavilov All-Russian Research Institute of Plant Industry, St.Petersburg, 190000, Russia. [3]IACR-Long Ashton Research Station Long Ashton, Bristol BS18 9AF, UK*

Abstract: Protein inhibitors of hydrolytic enzymes are of interest in relation to host/pathogen co-evolution, as regulators of endogenous enzymes and as markers in studies of plant evolution and diversity. We have, therefore, compared the patterns of inhibitors of proteinases and α-amylases, including endogenous and exogenous (from fungal, bacterial and insect sources) enzymes in seeds of a range of wild and cultivated species of the cereals and sunflower. Isoelectric focusing showed a high level of polymorphism in the patterns of inhibitors present in various species of the *Triticeae* with variation present both within and between different groups. For example, the spectra of inhibitors of trypsin-like proteinases of insects and fungi, inhibitors of insect amylases and cysteine proteinases are characteristic of species and genomes of di- and polyploid wheats and aegilopses. Variability of chymotrypsin/subtilisin inhibitors in wheats sometimes is higher than that of prolamins. Bifunctional inhibitors of endogenous α-amylase/bacterial subtisin showed little polymorphism in wheats but were highly variable in *Aegilops* and rye. Similarly, highly polymorphic inhibitors of insect cysteine proteinases were detected in *Secale, Elytrigia, Agropyron* and *Hordeum*. Bifunctional insect α-amylase/trypsin inhibitor was found in *H. bulbosum*. Inhibitors of trypsin and bifunctional inhibitors of trypsin and subtilisin were identified in sunflower seeds. The major trypsin inhibitor (pI ≅10, M_r 1500) also inhibited chymotrypsin and was present in all species of sunflower but absent from some genotypes of cultivated sunflower. The bifunctional inhibitors of trypsin and subtilisin had higher M_r and showed greater polymorphism between genotypes. These components also inhibited subtilisin-like proteinases from the sunflower pathogen *Sclerotinia sclerotiorum* and could, therefore, be involved in plant defence. Genetic analysis showed that the main trypsin inhibitors of sunflower were encoded by separate genes while the major bifunctional inhibitors were encoded by three linked genes with low recombination.

1. Introduction

Alpha-amylases and proteinases play key roles in assimilation of carbohydrates and proteins by insects, animals and microorganisms. The plant proteins, which inhibit these hydrolases, are considered as components of plant defence system (Shewry & Lucas, 1997). In general, inhibitors are of interest in relation to host/pathogen co-evolution, as regulators of endogenous enzymes and as markers in studies of plant diversity and evolution (Richardson, 1991; Konarev, 1996). The seeds and vegetative organs of plants contain a great number of inhibitor types, belonging to numerous evolutionary families (Kreis, et al., 1985). Some of these inhibit hydrolases of range of organisms, others are specific strictly to enzymes of insects or microorganisms while yet others inhibit only endogenous enzymes. The polymorphism and natural variability of inhibitors in many plant taxa have been studied less than their molecular structure. We have, therefore, developed simple and sensitive methods to compare the patterns of inhibitors of proteinases and α-amylases, including endogenous and exogenous (from fungal, bacterial and insect sources) enzymes, in protein extracts from seeds of a range of wild and cultivated species of the cereals and sunflower, separated by isoelectric focusing, electrophoresis and thin layer gel-filtration (Konarev, 1985, 1986; Konarev et al., 1988).

2. Materials and Methods

Seeds of about 600 wild and cultivated accessions of cereals of tribe *Triticeae* Dum. (wheats, *Aegilops* spp., rye, barley and others) were obtained from World collection of the N.I. Vavilov Institute of Plant Industry (VIR; St. Petersburg, Russia; catalogue numbers lettered by "k"), from Dr. B.L. Jonson ("g"), and from the collection of the authors ("No"). Seeds of 10 varieties of cultivated *H. annuus*, 70 inbred lines and 11 wild sunflower species were obtained from the VIR. Bovine trypsin was from "Sigma", subtilisin from "Calbiochem"; proteinases of *Aspergillus oryzae* ("P") and *A. soya* were from "Serva". Individual components of proteinase "P" of *A. oryzae*, insect gut and *Sclerotinia sclerotiorum* proteinases were purified by micropreparative isoelectric focusing (IEF) (Konarev, 1986; Konarev et al., 1988).

Proteins were extracted with water (1:2 – 1:10 w/v) from de-embryonated seeds or from endosperms after milling and then were separated by IEF in 6% PAAG with Ampholines (Pharmacia) or Servalytes (Serva) in pH ranges 5-11, 5-8 or 3-10 (Konarev, 1987a, 1993). Thin layer gel-filtration and SDS-PAGE were used for estimation of molecular weights (MW) of native inhibitors (Konarev, 1982, 1987). Proteinase inhibitors were detected by gelatin replicas method (Konarev, 1986; Konarev et al., 1988). α-amylase inhibitors were detected by starch/amylase/PAG replicas (Konarev, 1982, 1985). Single trypsin inhibitor components were purified by affinity chromatography (Konarev, 1982), IEF and reversed-phase HPLC.

3. Hydrolase inhibitors in *Triticeae* Dum. cereals

3.1 Amylase inhibitors

Diploid wheats are of interest as sources of resistance to pests and fungi (Zhukovskii, 1971) and in relation to the origin of A genome of polyploid wheats.

Water soluble endosperm proteins were separated by isoelectric focusing and thin layer gel-filtration and inhibitors of α-amylase from larvae *Tenebrio molitor* L. were detected in replicas obtained from separating gels (Figure 1). *Aegilops squarrosa* inhibitors belonging to classical groups with MW 12 and 24 kDa described by Buonocore et al. (1977) were used as a control. Most accessions of *Triticum boeoticum*, representing all main areas of its distribution (19 of 26 analyzed), did not completely inhibit insect α–amylase, as shown in Figure 1 (i). Five acessions contained inhibitors with MW 12 kDa (j) and two had inhibitors of MW 24 kDa (k). Interestingly, we have not found any published reports on presence of insect α-amylase inhibitors in *T. boeoticum*. This may be because of the different sets of material analyzed. None of the 22 accessions of cultivated *T. monococcum* (g and h) contained inhibitors of insect α-amylase that corresponds to data reported by other workers. Wild *T. urartu*, considered to be the donor of the A genome to bread wheat (Konarev, 1975; Brandolini et al., 1998), had inhibitors of MW 24 kDa that were highly variable by IEF (b-f). Accessions deficient in inhibitors may also exist in nature. Inhibitors from *T. boeoticum* of MW 24 kDa are similar in pI (isoelectric points) to some *T. urartu* inhibitors but we did not find analogues of the *T. boeoticum* k-18399 inhibitors with high pI (j) in related cereals. It may be suggested that the high intraspecific variability of amylase inhibitors in their presence/absence, isoelectric point and size reflect the diversity of diploid wheats.

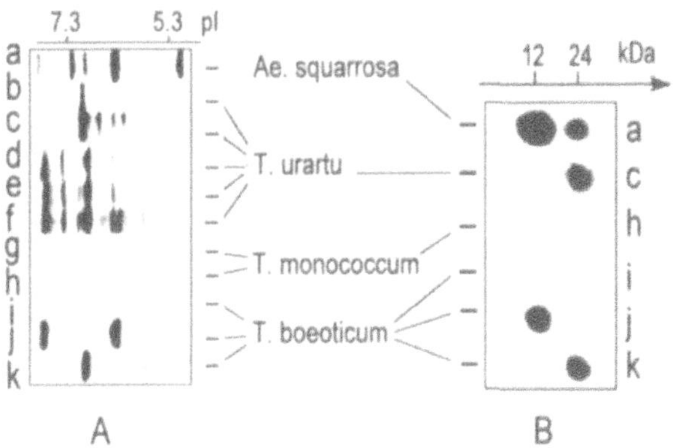

Figure 1. Polymorphism of insect α-amylase inhibitors in diploid wheats. IEF (A) and thin layer gel-filtration in Sephadex G-50 (B) of seed proteins followed by detection of inhibitors of *Tenebrio molitor* α–amylase. a, *Aegilops squarrosa* (control); b-f, *T. urartu* No 31, k-33869, k-31159, g-3221 and g-1863; g & h, *T. monococcum* k-20996 and k-48993; i-k, *T. boeoticum* k-28273, k-18399 and No 21. 12 and 24, molecular weight values in kDa.

In tetraploid wheats of the *T. timopheevii* (A^bA^bGG) and *T. turgidum* (A^uA^uBB) evolutionary groups and in hexaploid *T. aestivum* (A^uA^uBBDD) (designation as in Konarev, 1983) the variability in inhibitors of exogenous α-amylases is expressed mainly in presence or absence of several components of MW 12 kDa. In bread wheat they are controlled by chromosome 6B and similar to inhibitors of some *Aegilops* spp. of section *Sitopsis* (Konarev, 1982). Figure 2 shows variation in α-amylase inhibitors of MW 12 kDa of tetraploid wheats of the *T. turgidum* group of species (*T. durum, T. persicum, T. dicoccum,* etc.). Such components were found in

3 of 10 analysed accessions of *T. dicoccoides*, in 2 of 5 *T. dicoccum*, in 3 of 4 *T. turgidum*, and in 50 of 93 varieties of *T. durum* but were not found in any of the 12 analysed accessions of *T. persicum*. In the *T. timopheevii* evolutionary group they were present in 7 of 8 analyzed accessions of *T. timopheevii* and were absent from all 6 accessions of *T. araraticum* ssp. *araraticum* and from 2 accessions of ssp. *kurdistanicum*. 74% of the *T. durum* varieties originating from Russia and the former Soviet Union had such inhibitors but only 34% of the varieties from other parts of world contained them. Analysis of the inheritance of these inhibitors in F_2 hybrids showed monogenic control (Konarev & Mitrophanova, 1987). The tetraploid forms, containing inhibitors of MW 12 kDa are characterised by much higher activities of insect α-amylase inhibitors in endosperm compared with forms deficient in these inhibitors. Interestingly, a correlation between the activity of insect α-amylase inhibitors and pest resistance were found first in tetraploid wheats (Yetter et al., 1979). We can propose that existence of variation in inhibitor composition in tetraploid forms facilitated the demonstration of such a correlation. Bread wheat varieties do not differ to a greater extent in insect α-amylase inhibitor activity because of presence of monomorphic inhibitors of MW 12 kDa controlled by chromosome 6D (Konarev, 1982). The spectra of inhibitors of mammal and insect α-amylases of MW 24 kDa are specific for genomes G, B and D of polyploid wheats. These inhibitors are controlled by chromosomes 3B and 3D in *T. aestivum* and they are monomorphic inside species with corresponding genome composition.

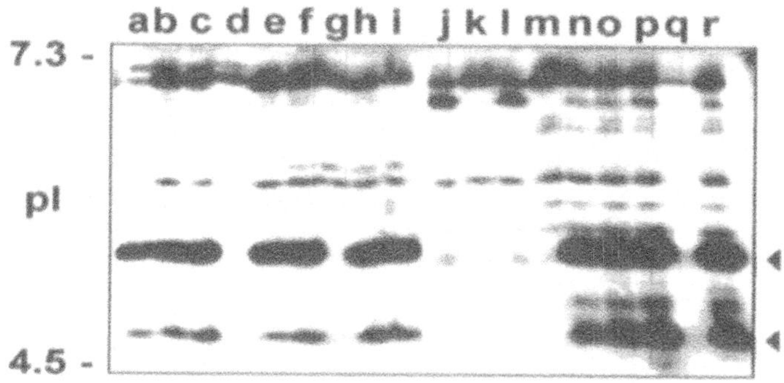

Figure 2. Variability of insect α-amylase inhibitors of MW 12 kDa (marked) in tetraploid wheats of *T. turgidum* evolutionary group. Water soluble endosperm proteins were separated by isoelectric focusing and inhibitors of *Tenebrio molitor* larvae α–amylase were detected. a, *T. turanicum* k-39319; b-d, *T. dicoccoides* k-5201, k-5196 and k-15900; g-i, *T. dicoccum* k-18971, k-81 and k-20559; e-i, *T. durum* k-43903, k-22636, k-48113, k-26489 and k-38270; i-m, *T. persicum* k-7113, k-12970, k-14940 and k-36198; n-r, *T. durum* k-44657; k-38253, k-40126, k-44670 and k-45660.

3.2 *Proteinase inhibitors*

Endosperms of wheat and related cereals contain inhibitors of serine proteinases of mammals, insects, bacteria and fungi (Figure 3). Replicas developed with well studied standard proteinases (A and C, bovine trypsin; E, subtilisin) were used as controls for the analysis of inhibitors of insect and fungal enzymes. A set of cereal accessions (differentiators) with characteristic inhibitor spectra were used for comparison of inhibitors of standard and analysed proteinases. The same endosperm proteins inhibited bovine trypsin, gut proteinase of storage pest *Rhyzopertha dominica* F. and alkaline proteinase of the fungus *A. oryzae*. Inhibitors of *A. soya* proteinase (F) and *A. oryzae* neutral proteinase (not shown) were similar to inhibitors of bacterial subtilisin (E).

Three main types of trypsin-like proteinase inhibitors which are present in the genus *Triticum* L. are represented in Figure 3A and 3B. Bands 5, 7, 8 and 9 are characteristic of endosperms of *T. aestivum* and *Ae. squarrosa* (considered as donor of genome D) and are under control of chromosome 1D (Konarev & Mitrophanova, 1987; Konarev, 1987a) (the nomenclature used for IEF bands was used as in (Konarev, 1996) with band 1 corresponding to the TI of the aleurone layer and germ not being shown). Varieties of bread wheat differ in presence of this group of components which show monogenic character of inheritance. Bands 2, 3, 4 and 6 occur in tetraploid wheats of *T. turgidum* evolutionary group and *T. aestivum*.

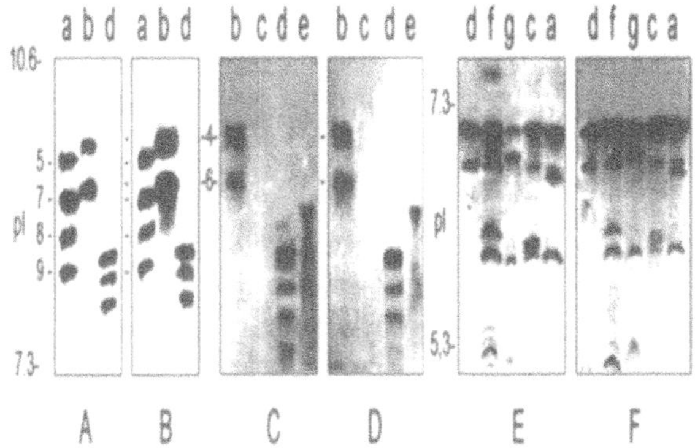

Figure 3. Comparison of inhibitors of mammal, insect, bacteria and fungal serine proteinases in cereal species. Water soluble endosperm proteins were separated by IEF in the pH range 5-11 (A-D) and 5-8 (E-F); inhibitors were detected by gelatin replicas method. Replicas were developed by bovine trypsin (A and C), wheat storage pest *Rhizopertha dominica* F. gut proteinase (B), proteinase of *Aspergillus oryzae* with pI abot 9.0 (D), subtilisin (E) and neutral proteinase of *A. soya* (F). 2-9, IEF bands of TI of *T.aestivum* endosperm. a, *Triticum aestivum* var. Bezostaya 1; b, *T. durum* var. Beloturka; c, *T. dicoccum* k-7141; d, *T. monococcum* k-39416; e, *S. cereale* var. Vyatka 2; f, *T. timopheevii* Zhuk. k-27805; g, *T. urartu* k-33869.

Components with similar pI were found in some accessions of *Aegilops* spp. of the *Sitopsis* section (Figure 4C) considered as possible donors of the genome B to wheat (Sarkar & Stebbins, 1956; Konarev, 1983; Kerby et al., 1988). Single inhibitor bands in wheats had analogues both in *Ae. speltoides* and *Ae. longissima* (Figure 4C). Figure 4A and 4B illustrate the variability of endosperm trypsin inhibitors in tetraploid wheats. The most variable were inhibitors of wild *T. dicoccoides* (b-d). The diversity of its biotypes included the majority of band combinations present in other wheats. Some forms had only bands 2 and 3, others had all four components and others did not contain any inhibitors at all. Most of *T. durum* accessions, (25 of 30) originating from various parts of the world had bands 4 and 6 (var. Beloturka, Wells, Lacota, etc.). Five varieties had the full set of components (var. Cargliano, Saba, etc.). All 9 analyzed accessions of *T. persicum* possessed only band 3. Thus, some cultivated species differing in ecologogical and morphological characters (Zhukovskii, 1971) within genetically determined species *T. turgidum* (McKey, 1975) are characterized by predominant combinations of TI bands. The variability of bands 2, 3, 4 and 6 in *T. aestivum* resembles that within tetraploids of *T. turgidum* group. Bread wheat varieties vary strongly both in the composition of TI controlled by B and D genomes and in the activity of TI in endosperms. Correlations between the activity of inhibitors and varietal resistance to the grain storage pest *R. dominica* was found. Interestingly, the gut of this pest contains only trypsin-like endoproteinases in contrast to other grain pests (Konarev, 1987a).

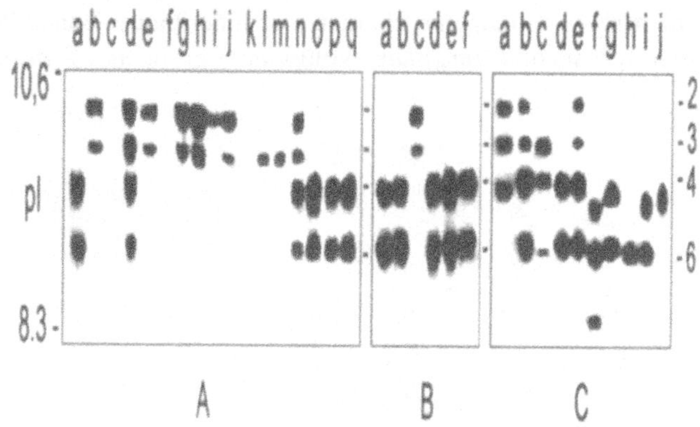

Figure 4. Variability of trypsin inhibitors in tetraploid wheats with the genome constitution A^uA^uBB and Aegilops spp of section *Sitopsis*. Water soluble endosperm proteins were separated by IEF in pH range 5-11 and inhibitors were detected by the gelatin replicas method. 2-6, positions of TI bands. **A)** a, *T. turgidum* k-42287; b-d, *T. dicoccoides* k-41966, k-20403 and k-5196; e-k, *T. dicoccum* k-20559, k-7141, k-18971, k-35926, k-7506, k-17972 and k-30093; l & m, *T. persicum* k-7113 and k-27656; n-q, *T. durum* k-45568, k-25008; k-48187 and k-11375. **B)** a, *T. durum* k-32453; b, *T.turanicum* k-39913; c, *T. ispahanicum* k-43064; d & e, *T. polonicum* k-19597 and k-9277; f, *T. aephiopicum* k-19553. **C)** a, *Ae. longissima* k-194; b & e, *T. durum k*-17769. c, *Ae. speltoides* k-12; d, *T. durum* k-32453; f-i, *Ae. longissima* k-378, k-907, k-202 and k-209; j, *Ae. speltoides* k-1593.

All diploid wheats, tetraploids of the *T. timopheevii* group and the hexaploid *T. zhukovskii* had similar TI spectra to that of *T. monococcum* (Figure 3 A-D). This spectrum of inhibitors was not found in *T. turgidum* and *T. aestivum* groups of species and, in addition, the A genomes of these polyploids have only silent genes for endosperm inhibitors of trypsin, chymotrypsin/subtilisin and exogenous α-amylases. Only some components of trypsin inhibitors from wheat leaves were found to be controlled by chromosome 3 of the A genome (Konarev, 1993). The trypsin inhibitors of *T. monococcum* differ from those of *T. turgidum* and *T.*

aestivum in being more effectively extracted by acids and, in general, being present at higher levels in endosperms, comparable with those in rye (*Secale cereale*) and barley (*Hordeum vulgare*) (Konarev, 1987a). The similarity in the size of TI of *T. monococcum*, *S. cereale* and *H. vulgare* was shown by SDS-PAGE of native TI (Figure 5) and purified and reduced TI (Konarev, 1987a, 1987b). Reduced TI of these cereals had MW about 14 kDa that corresponding to TI of rye and barley reported in (Boisen, 1983). Thus, cereals of the *Triteceae* tribe have at the least 3 types of endosperm TI which can be represented by (i) TI of *T. monococcum* or rye, (ii) TI of *T. turgidum* originating from *Aegilops* spp. of the *Sitopsis* section with MW about 19 kDa and (iii) TI of *T. aestivum* and *Ae. tauschii* with MW about 11 kDa. There is also possibly a forth TI type with MW 6.5 kDa in endosperms of *T. aestivum*. Chymotrypsin/subtilisin inhibitors are the most polymorphic and variable inhibitors in wheat (Konarev, 1988).

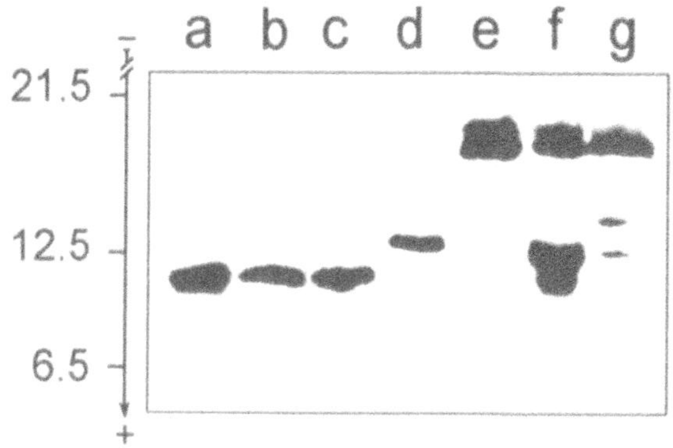

Figure 5. Trypsin inhibitors in *Triticeae* Dum. cereals. Water soluble seed proteins were heated with 1% SDS at 95°, separated by 10-30% gradient SDS-PAGE and transferred to nitrocellulose membrane. Trypsin inhibitors were detected on membrane by modified gelatin replicas method (Konarev, 1987b). a-f, seeds devoid of germ; g, germ (g). 6.5, 12.5 and 21.5 indicate the positions of MW marker proteins reduced by 2-mercaptoethanol. a, *Hordeum vulgare* var. Pirkka; b, *Secale cereale* var. Vyatka 2; c, *Triticum monococcum* k-39416; d, *Aegilops tauschii* k-28; e, *T. durum* k-17769; f & g, *T. aestivum*, var. Diamond.

Cysteine proteinase inhibitors are much less studied in cereals of the *Triticeae* than serine proteinase inhibitors although they were found as early as 1984 (Konarev, 1984). In general, cysteine proteinases of different origin (insect, wheat and rubber plant) are inhibited by the same grain protein components (Figure 6). Diploid, tetraploid and hexaploid wheats differed by inhibitor spectra. *Ae. squarrosa*, which is considered to be a donor of the D genome of hexaploid wheat, has similar inhibitor band to that of bread wheat. Analysis of nullisomic- tetrasomic lines confirmed that this band is controlled by chromosome 2 of the D genome (Konarev, 1984). In wheat each genome controls own band of cysteine proteinase inhibitor. In gel-filtration these inhibitors had mobility corresponding to MW about 12 kDa.

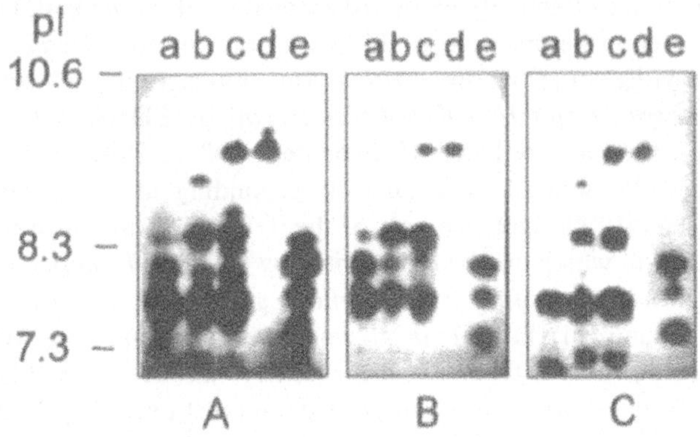

Figure 6. Inhibitors of cysteine proteinases in *Triticeae* cereals. IEF of endosperm proteins of *Triticum monococcum* (a), *T. durum* (b), *T. aestivum* (c), *Aegilops tauschii* (d) and *Hordeum vulgare* (e). Gelatin replicas were developed with *Tribolium confusum* (A), wheat grain (B) and rubber plant (ficin, C) cysteine proteinases.

In general, within the genus *Triticum* the spectra of the majority of inhibitors are clearly related to the genome compositions of polyploids and can be indicators of evolutionary relations of polyploids with diploids. The spectra of many types of inhibitors in cereals are species-specific with low or high intraspecific variability. Genetic analysis showed that the majority of bands in isoelectric point spectra of α-amylase and proteinase inhibitors have monogenic control.

4. Proteinase inhibitors in sunflower

Two main types of inhibitors were found in sunflower seeds: trypsin inhibitors (TI) with high pI and bifunctional trypsin/subtilisin inhibitors (T/SI) (Figure 7). Each type is represented by three main components (a-c). The patterns of inhibitors varied between lines and some also varied within lines. The major trypsin inhibitor (band TIa) is characteristic for all sunflower species studied but was absent from five of the 70 inbred lines, as in Figure 7 A. In such lines two other trypsin inhibitors, TIb and TIc, become visible. T/SI bands were more variable than the major trypsin inhibitor.

A series of F_1 and F_2 hybrids, generated by crossing lines VIR670xVIR648b differing in inhibitor composition, were analyzed in order to determine the inheritance and linkage relationships of the groups of inhibitors. All of the main inhibitor groups present in the two parents were expressed codominantly in F_1 seeds. Analysis of F_2 hybrids showed that the inhibitor bands TIa, T/SIa, T/SIb and T/SIc segregated in a 3:1 ratio which is consistent with monogenic inheritance. Analysis of three F_2 families indicated that the three loci encoding T/SI inhibitors were clearly linked, with the locus for the T/SIb component being located between those for the T/SIa and T/SIc inhibitors. The major TI showed no linkage to T/SIa and T/SIc and loose linkage with the T/SIb inhibitors.

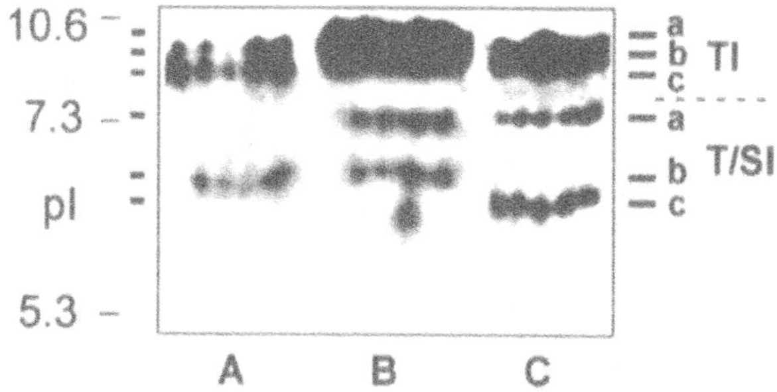

Figure 7. Polymorphism of proteinase inhibitors present in sunflower (*Helianthus annuus*) seeds. Water soluble seed proteins were separated by IEF and gelatin replica was developed by trypsin. A-C, lines VIR369, VIR340 and 648b (by 5 single seeds each). TI, trypsin inhibitors; T/SI, trypsin/subtilisin inhibitors.

The TI were found only in seeds. The main TI component was purified by affinity chromatography on trypsin Sepharose followed by reversed-phase HPLC and subjected to SDS-electrophoresis. It has a very low MW (≈1.5 kDa) which is unusual for an active plant proteinase inhibitor. Studies on structure of this inhibitor are in progress.

The T/SI present in seeds and vegetative organs were found to be active against extracellular proteinase of white rot *Sclerotinia sclerotiorum,* an important pathogen of sunflower (Konarev et al., 1998), indicating a possible protective role.

Table 1. Segregation for the pairs of loci encoding inhibitors in F_2 seeds of the cross VIR670 x VIR648b.

Pair of loci	Number of phenotypes				χ^2	P(d.f.=1)	$r \pm S_r$
	A-B-	A-bb	aaB-	aabb			
T/SIa – T/SIb	68	41	31	5	8.130	< 0.05	0.32 ± 0.039
T/SIb – T/SIc	55	33	45	3	23.124	< 0.05	0.23 ± 0.036
T/SIa – T/SIc	83	23	25	14	8.407	< 0.05	0.40 ± 0.050
TIa – T/SIa	91	20	29	6	4.020	0.025-0.05	> 0. 50
TIa – T/SIb	66	44	25	11	10.288	< 0.05	0.44 ± 0.041
TIa – T/SIc	85	26	22	12	1.971	0.10-0.25	> 0.50

5. Conclusions

Simple approaches to the analysis of inhibitors allow the range of polymorphism in the main α-amylase and proteinase inhibitors of *Triticeae* cereals and sunflower to be determined. The methods can be used to study the polymorphism of inhibitors in various plants in relation to their role in plant defence and use as markers for genetic diversity and to identify novel inhibitor types.

6. References

Boisen, S. (1983) Protease inhibitors in Cereals. (Occurence, properties, physiological role and nutritional influence), *Acta Agriculture Scandinavica* **33**, 371-381.

Brandolini, A., Empilli, S., Vaccino, P., Borghi, B. & Salamini, F. (1998):The origin of A genome in wheats as revealed by AFLP analysis of nuclear DNA, Abstr. *"Genetics and breeding for crop quality and resistance"* XV EUCARPIA General Congress, Viterbo, Italy 110.

Buonocore, V., Petrucci, T. & Silano, V. (1977) Wheat protein inhibitors of α-amylase, *Phytochemistry* **16**, 811-20.

Kerby, K, Kuspira, J & Jones, B.L. (1988) Biochemical data bearing on the relationship between the genome of Triticum urartu and the A and B genomes of the polyploid wheats, *Genome* **30**, 576-81

Konarev A.V. (1975) Differentiation of first primary genomes of polyploid wheats according to the data of immunochemical analysis of grain protein alcohol fraction, *Bulletin VIR* 47, 8-11.

Konarev, Al.V. & Mitrophanova, O.P. (1987) Analysis of inheritance of components of alpha-amylase and trypsin inhibitors in bread wheat. *Cytology and Genetics* **21**, 15-18.

Konarev, Al.V., Kochetkov, V.V., Bailey, J.A. & Shewry, P.R. (1998) The detection of inhibitors of the *Sclerotinia sclerotiorum* (Lib) de Bary extracellular proteinases in sunflower, *J. Phytopathology* **146** (in press).

Konarev, Al.V. (1986) Analysis of protease inhibitors from wheat grain by gelatine replicas method, *Biokhimiya* **51**, 195-201. (Translated in English by Plenum Publishing Corp.).

Konarev, Al.V. (1982) Component composition and genetic control of insect α-amylase inhibitors from wheat and *Aegilops* grain, *Doklady Vaskhil* 6, 42-44 (Translated in English by Allerton Press).

Konarev, Al.V. (1993) Component composition, genome and chromosome control of trypsin inhibitors from bread wheat leaves, roots and germs. Selkochozajstvennaya biologia **1**, 43-52 (In Russian).

Konarev, Al.V.: (1984) Identification of inhibitors of thiol proteinases of insects and grain in wheat and other cereals. Doklady VASKHNIL, 10, 13-15. (Translated in English by Allerton Press).

Konarev, Al.V. (1996) Interaction of insect digestive enzymes with plant protein inhibitors and host parasite co-evolution, *Euphytica* **92**, 89-94.

Konarev, Al.V. (1985) Methods for analyzing the component composition of cereal α-amylase and proteinase inhibitors, *Prikladnaya Biokhim. i Microbiol.* **21**, 92-100. (Translated in English by Plenum Publishing Corp.).

Konarev, Al.V. (1988) Polymorphism of chymotrypsin-subtilisin inhibitors and possibilities of its use in wheat variety identification, *Biochemical identification of varieties,* Materials of III International symposium ISTA., Leningrad, VIR, 176-181.

Konarev, Al.V. (1987a) Variability of trypsin-like proteinase inhibitors in wheat and related cereals in relation with resistance to grain pests, *Selkochozajstvennaya biologia* **5**, 17-24. (In Russian).

Konarev, Al.V. (1987b) Component composition of trypsin inhibitors from grain and leaves of wheat, *Doklady VASKHNIL*, N 12., 6-9.

Konarev. V.G.: (1983) The nature and origin of wheat genomes on the data of grain protein immunochemistry and electrophoresis, *Proc. 6th.Inter. Wheat Genetics Symp.*, Kyoto, Japan, 65-75.

Kreis, M., Shewry, P.R., Forde, E.G., Forde, J. & Miflin B.J. (1985) Structure and evolution of seed storage proteins and their genes with particular reference to those of wheat, barley and rye, *Oxford surveys of plant molecular and cell biology*, **2**, 253-317.

McKey, J. (1975) The boundriesand of the genus Triticum, *Proc. 12th Int. Bot. Congress.* Leningrad 509.

Richardson, M. (1991) Seed storage proteins: the enzyme inhibitors. *Methods Plant Biochem.* **5**: 259-305.

Sarkar, P. & Stebbins, G.L. (1956) Morphological evidence concerning the origin of B genome in wheat, *Amer. J.Bot.* **43**, 297-304.

Shewry, P.R. & Lucas J.A. (1997) Plant proteins that confer resistance to pests and pathogens, *Adv. Bot. Res.* **26**, 135-192.

Yetter, M.A., Saunders, R.M. & Boles H.P. (1979) α-amylase inhibitors from wheat kernels as factors in resistance to postharvest insects, *Cereal Chem.* **56**, 243-244.

Zhukovskii P.M. (1971) *Cultivated plants and their relatives,* Kolos, Moscow.

SESSION 3

RESISTANCE TO INSECTS

G.T. Scarascia Mugnozza, E. Porceddu & M.A. Pagnotta (Eds.)
Genetics and Breeding for Crop Quality and Resistance, 147-158, 1999

Cereal α-amylase/trypsin inhibitors and transgenic insect resistance

P. Carbonero, I. Díaz, J. Vicente-Carbajosa, J. Alfonso-Rubí*, K. Gaddour & P. Lara
*Laboratorio de Bioquímica y Biología Molecular, Departamento de Biotecnología-UPM, E.T.S. Ingenieros Agrónomos, 28040 Madrid, Spain.*Permanent address: Centro de Ingeniería Genética y Biotecnología C.P. 60200. Sancti Spiritus, Cuba*

Key words: α-amylase inhibitors, cereals, gene transfer, insect resistant crops, proteinase inhibitors

Abstract: Plant proteinaceous inhibitors of hydrolases from heterologous systems have been implied in plant defense and have been extensively studied at the molecular level. In wheat and barley, a substantial fraction of the total endosperm protein content is represented by toxins and inhibitors that are active against heterologous systems. More than 20 different members from a single multigene family of alpha-amylase/trypsin inhibitors, specifically expressed in endosperm, have been characterised. Their apparent molecular weigth are in the 12-16,000 range. The alpha-amylase inhibitors can be classified, according to their degree of aggregation into monomeric (as the trypsin inhibitors), dimeric and tetrameric forms. During kernel development, their synthesis precedes that of the storage proteins and they are rapidly degraded upon germination. Genes encoding these inhibitors are scattered on the Triticeae chromosomes: at least five out of the seven homeologous groups carry genes encoding different family members. Structure-Functional relationships have been pursued through site-directed mutagenesis of the wheat monomeric alpha-amylase inhibitor. In this molecule, both the carboxy- and amino-terminal ends, that are proximal in the 3D-structure due to the position of the disulphide-bridges, have been shown to be important for activity against the Tenebrio molitor amylase. NMR-structural studies in the finger-millet bifunctional inhibitor have shown that the reactive-site for trypsin is located in a loop opposite to the reactive site for alpha-amylase (in the C- and N-terminal ends of the molecule), strongly suggesting that the activity of these inhibitors can be improved by domain-swapping among different members of the family. More recently, transgenic tobacco and wheat plants have being obtained, expressing the gene encoding barley trypsin inhibitor BTI-Cme, and these where shown to be more resistant to lepidopterous *Agrotis ipsilon* and *Sitotroga cerealella* pests, respectively, than non-transformed controls. Preliminary data with other transgenically expressed members of the family confirmed the potential of these inhibitors for increasing insect resistance in genetically modified plants (GMP). This methodology is aimed at complementing an integrated pest management system, by amplifying the natural variability outside the species barrier, in a more environmental friendly Agriculture less dependant on chemical pesticides.

1. Introduction

Crop protection against insect pests is of crucial importance to modern Agriculture. With an ever increasing world population, specially in the developing countries, a concomitant increase in yield for the main crops is absolutely necessary. Close to 40% of the agricultural product world-wide is lost to pests and diseases, more than 15% being directly imputable to phytophagous insects. This damage is even greater in places were costly chemical pesticides are less available, as occurs in many countries from Central and South America, Africa and Asia. In Western Europe and North America, specially since the Second World War, abuse of chemical pesticides has led to environmental problems. For these reasons, plant breeders are focusing more and more in insect resistance genes. These can be transferred either by classical breeding methods, using as a vector the pollen from wild-species or from other compatible resistant cultivars, or by the new methodology of genetic engineering which amplifies the available variability outside the limits of pollen compatibility.

Today, not only the Solanaceae and other dicotiledoneous plants are amenable to genetic transformation, mainly utilizing the Ti-plasmid of the bacterial pathogen *Agrobacterium tumefaciens* as a vector for DNA transfer, but many other agricultural important crops, considered recalcitrant until recently, such are the cereals, can be transformed using the biolistic (gene-gun) approach (Christou, 1995).

For obtaining transgenic insect resistance, several variants of the *Cry* genes, encoding the toxic Bt crystal proteins that are found in the spores of the *Bacillus thuringiensis* bacteria, have been most widely used. Crops transformed with these *Cry* genes have already been commercialized and range from tobacco, potatoes and tomato to cotton and maize, to cite but a few (Peferoen, 1997). In response to concern about Bt resistance, other toxins produced by different bacteria have been investigated, such as that secreted by *Photorhabdus luminescens*, a parasite of entomophagous nematodes (Bowen et al., 1998).

Another approach will be to look for insect resistance genes into the plant kingdom itself. The natural defense mechanisms of plants against insect attack are physical barriers (trichomes, spines, etc.) and chemical substances, such are secondary metabolites or insecticidal proteins. In the secondary metabolite group, many different insecticidal compounds are found: pyrethroids, steroids, terpenes, phenolics, cyanogenic glycosides, etc. These are the end products of complex metabolic pathways, involving several enzymes, and in general are not easily engineered due to the high number of genes to be transferred, unless the crop to be transformed contains a close precursor of the insecticidal metabolite. Fortunately, some of the possible plant defense mechanism are based on insecticidal proteins that are the direct product of genes, and consequently are more easily engineered.

Plant proteinaceous inhibitors of heterologous hydrolases (proteinases, α-amylases, etc.) have been known for several decades and are commonly found in seeds and tubers. However, their possible role as defensive agents against herbivores was not considered until 1972, when Ryan's group demonstrated that in the Solanaceae, the synthesis of certain proteinase inhibitors were induced upon insect chewing (or mechanical damage) not only in the affected leaf but in other, systemic, leaves. The first demonstration that a transgenically expressed proteinase inhibitor increased insect resistance, came from Hilder et al. (1987) when they showed that the cowpea trypsin inhibitor (CpTI) transgenically expressed in tobacco, enhanced tolerance against *Heliothis virescens*. The CpTI protein has been proven also to confer resistance against other lepidopterous insects, depending for food digestion on trypsin-like enzymes in their mid-guts, such are *Spodoptera littoralis*, *Manduca sexta*, *Chilo suppresalis* and *Sesamia inferens*, and it has been engineered into several crops including rice and strawberries (Graham et al., 1995; Xu et al., 1996).

In wheat and barley, as in other Poaceae species, a single multigene family includes inhibitors of trypsin (a serine-protease) and of heterologous α-amylases, and this has been the subject of our interest for several time (Garcia-Olmedo, et al.,1987).

2. General characteristics

The cereal multigene family of trypsin/α-amylase inhibitors is prominently represented among the albumins and globulins from endosperm but can be extracted with several organic solvents such are 70% ethanol or chloroform-methanol mixtures. More than 20 different members from this family have been characterized (reviewed by Carbonero et al., 1993; Carbonero & Garcia-Olmedo, 1998). Their apparent molecular weights are in the 12-16 KDa range and they have nine or ten cysteine-residues. The α-amylase inhibitors can be classified according to their degree of aggregation into monomeric, dimeric, and tetrameric forms. The trypsin inhibitors are monomeric. During kernel development, their synthesis precedes that of the main storage proteins and they are rapidly degraded upon germination.

Amino acid sequences of members of this inhibitor family from wheat, barley and other cereals -directly determined or deduced from the nucleotide sequences of cDNA clones- have been aligned in Figure 1, where they have been organized into three domains (A, B, C) and grouped into ten subfamilies according to sequence similarity and *in vitro* activity. In all cases where a complete cDNA has been characterized, the major protein is preceded by a typical signal peptide of approximately 30 amino acids. This is in agreement with the observation that the synthesis of these inhibiors takes place in membrane-bound polysomes as pre-proteins that are co-translationally processed (Paz-Ares et al., 1983).

```
A
BTI-CME       --FGDSCAPGDALPH NPLRACRTYVVSQIC HQGPRLLTSD----- -----------
RTI           -SVGGQCVPGLAMPH NPLGACRTYVVSQIC HVGPRLFTWD----- -----------
RBI           -SVGTSCIPGMAIPH NPLDSCRWYVSTRTC GVGPRLATQE----- -----------
MTI           -SAGTSCVPGWAIPH NPLPSCRWYVTSRTC GIGPRLPWPEGRLE- -----------

PUP-23        -SVKDECQLGVDFPH NPLATCHTYVIKRVC GRGPSRPMLV----- -----------
PUP-88        -SVEDECQPGVAFPH NALATCHTYVIKRVC GRGPSRPMLV----- -----------
BTI-CMc       -TSIYTCYEGMGLPV NPLQGCRFYVASQTC GAVPLLPIEV----- -----------
BMAI-1        -SPGEWCWPGMGYPV YPFPRCRALVKSQ-C AG-GQVVESIQ---- -----------
RAP           -SPGEQCRPGISYPT YSLPQCRTLVRRQ-C VGRGASAADEQV--- -----------

BDAI-1        SGPWMWCDPEMGHKV SPLTRCRALVKLE-C VG-----NRVPEDVL -----------

WMAI-1        SGPWSWCNPATGYKV SALTGCRAMVKLQ-C VGSQVPEAVL----- -----------
WMAI-2        SGPWMWCDPAMGYRV SPLTGCRAMVKLQ-C VGSQVPEA------- -----------
WDAI-1        SGPWM-CYPGQAFQV PALPGCRPLLKLQ-C NGSQVPEAVL----- -----------
WDAI-2        SGPWM-CYPGQAFQV PALPACRPLLRLQ-C NGSQVPEAVL----- -----------
WDAI-3        SGPWM-CYPGYAFKV PALPGCRPVLLLQ-C NGSQVPEAVL----- -----------
WTAI-CM1      --TGPYCYAGMGLPI NPLEGCREYVASQTC GIS-ISGSAVSTEPG NT---------
WTAI-CM2      --TGPYCYPGMGLPS NPLEGCREYVAQQTC GVGIIVGSPVSTEPG NT---------
BTA1-CMa      --TGQYCYAGMGLPS NPLEGCREYVAQQTC GVT-IAGSPVSSEPG DT---------
WTAI-CM16     -IGNEDCTPWMSTLI TPLPSCRDYVEQQAC RIETPGS-------- -----------
WTAI-CM17     ---NEDCTPWTSTLI TPLPSCRNYVEEQAC RIEMPGPPYL----- -----------
BTAI-CMb      -VGSEDCTPWTATPI TPLPSCRDYVEQQAC RIETPGPPYL----- -----------
WTAI-CM3      ---SGSCVPGVAFRT NLLPHCRDYVLQQTC TFTPGSKLPEWMTSA S-IYSPGKPYL
BTAI-CMd      AAAATDCSPGVAFPT NLLGHCRDYVLQQTCAVLTPGSKLPEWMTSAELNYPGQPYL

B
BTI-CME       MKRRCCDELSAIP- AYCRCEALRIIMQGV VTWQGA--------F EGAYFK----
RTI           MKRRCCDELLAIP- AYCRCEALRILMDGV VTQQGV--------F EGGYLK----
RBI           MKARCCRQLEAIP- AYCRCEAVRILMDGV VTSSGQ--------H EGRLLQ----
MTI           LKRRCCRELADIP- AYCRCTALSILMDGA IPP-GP--------- DAQLE-----

PUP-23        -KERCCRELAAVP- DHCRCEALRILMDGV RTPEG-RWEGRLG-- ----------
PUP-88        -KERCCRELAVVP- DYCRCEALRVLMDGV RAEEGHVVEGRLG-- ----------
BTI-CMc       MKDWCCRELAGISS N-CRCEGLRVFIDRA FPPSQSQ--GAPPQL PPL-------
BMAI-1        --KDCCRQIAAIGD EWCICGALGSMRGSM YKELGVA-------- LADDKATVAE
RAP           -WQDCCRQLAAVDD GWCRCGALDHMLSGI YRELGAT-------- EAGHPMAE--
BDAI-1        --RDCCQEVANISN EWCRCGDLGSMLRSV YAALGVG-------- GGPEE-----
WMAI-1        --RDCCQQLADINN EWCRCGDS-SMLRSV YQELGVR-------- EGKE------
WDAI-1        --RDCCQQLADIS- EWPRCGALYSMLDSM YKEHGVS-------- EGQAGTG---
WDAI-2        --RDCCQQLAHIS- EWCRCGALYSMLDSM YKEHGAQ-------- EGQAGTG---
WDAI-3        --RD-CQQ------ --------------- --------------- ----------
WTAI-CM1      PRDRCCKELYDAS- QHCRCEAVRYFIGR- -RSDPN--------- SGVLK-----
WTAI-CM2      PRDRCCKELYDAS- QHCRCEAVRYFIGR- -TSDPN--------- SGVLK-----
WTA1-CMa      PKDRCCQELDEAP- QHCRCEAVRYFIGR- -RSHPD--------- WSVLK-----
WTAI-CM16     AKQQCCGELANIP- QQCRCQALRYFMGP- -KSRPD--------Q SGLM------
BTAI-CMb      AKQQCCGELANIP- QQCRCQALRFFMGR- -KSRPD--------Q SGLM------
WTAI-CM17     AKQECCEQLANIP- QQCRCQALRYFMGP- -KSRPD--------Q SGLM------
WTAI-CM3      AKLYCCQELAEIS- QQCRCEALRYFIALP VPSQPVDPRSGNVGE SGLI------
BTAI-CMd      AKLYCCQELAKIP- QQCRCEALRYFMALP VPSQPVDPSTGNVGQ SGLM------

C
BTI-CME       DSPNCPRERQTSYAA NLVTPQECNLGTIHG S-----AYCPELQPG YGVVL
RTI           DMPNCPRVTQRSYAA TLVAPQECNLPTIHG S-----PYCPTLQAG Y
RBI           DLPGCPRQVQRAFAP KLVTEVECNLATIHG G-----PFCLSLLGA GE
MTI           DLPGCPREVQRGFAA TLVTEAECNLATISG V-----AECPWILGG GTMPSK
PUP-23        DRRDCPREEQPAFAA TLVTAAECNLSSVQE P-----GVRLVLLAD G
PUP-88        DRRDCPREAQREFAA TLVTAAECNLPTVS- ------GVGSTLGAT GRWMTIELPK
BTI-CMc       AT-ECPAEVKRDFAR TLALPGQCNLPAIHG G-----AYCVFP
BMAI-1        VFPGCRTEV--MDRA VASLPAVCNQYIPNT NGT--DGVCY--WLS YYQPPRQMSSR
RAP           VFPGCRRGD--LERA AASLPAFCNVDIPNG PGG---VVCY--WLG YPRTPRTGH
BDAI-1        VFPGCQKDV--MKLL VAGVPALCNVPIPNE A-AGTRGVCY--WSA STDT
```

```
WMAI-1      VLPGCRKEV--MKLT AASVPEVCKVPIPNP SGD-RAGVCYGDWAA YPDV
WDAI-1      AFPSCRREV--VKLT AASITAVCRLPIVVD ASGDGAYVCK-DVAA YPDA
WDAI-2      AFPRCRREV--VKLT AASITAVCRLPIVVD ASGDGAYVCK-DVAA YPDA
WTAI-CM1    DLPGCPREPQRDFAK VLVTSGHCNVMTVHN A-----PYCLGLDI
WTAI-CM2    DLPGCPREPQRDFAK VLVTPGHCNVMTVHN T-----PYCLGLDI
BTA1-CMa    DLPGCPKEPQRDFAK VLVTPGQCNVLTVHN A-----PYCLGLDI
WTAI-CM16   ELPGCPREVQMDFVR ILVTPGYCNLTTVHN T-----PYCLAMEES QWS
WTAI-CM17   ELPGCPREVQMNFVP ILVTPGYCNLTTVHN T-----PYCLGMEES QWS
BTAI-CMb    ELPGCPREVQMDFVR ILVTPGFCNLTTVHN T-----PYCLAMDEW QWNRQFCSS
WTAI-CM3    DLPGCPREMQWDFVR LLVAPGQCNLATIHN V-----RYCPAVEQP LWI
BTAI-CMd    DLPGCPREMQRAFVR LLVAPGQCNLATIHN V-----RYCPAVEQP LWI
```

Figure 1. Alignment of amino acid sequences of cereal inhibitors o α-amylase/trypsin. Sequences are divided into A, B and C domains and have been grouped according to sequence similarity and *in vitro* activities. The reactive sites of trypsin inhibitors (PR▼L) at the end of the A domain are underlined. Only partial N-terminal sequences are available for WMAI-2 and WDAI-3.

The location of the disulphide bridges within these cysteine-rich molecules has been investigated to a limited extent. The four-bridge structure (9 cysteines) of the wheat dimeric amylase inhibitor WDAI-1 (syn. 0.53), represented in Figure 2, was described by Maeda et al. (1983). The presence of one additional cysteine in the sequence of the wheat monomeric amylase inhibitor WMAI-1 (syn. 0.28) was found by Poerio et al. (1991) to imply not only the formation of a fifth disulphide bridge, but a small rearrangement of the disulphide structure (Figure 2). More recently, the 3D structure of the RBI bifunctional inhibitor from ragi and that of the wheat dimeric amylase inhibitor WDAI-2 (syn. 0.19) have been reported (Strobl et al., 1995; Oda et al., 1997). These inhibitors have five disulphide bridges with the same disulphide structure as WMAI-1 (Figure 2).

The RBI inhibitor consists of a globular four-helix motif with a simple "up-and-down" topology and there is an antiparalell β-sheet motif between the 3rd and the 4th helices. The postulated location of the putative α-amylase binding site is on the face of the molecule opposite to the trypsin-binding loop, that is stabilised by two α-helices and a disulphide bridge (Strobl et al., 1995). Barley trypsin inhibitor BTI-CMe, which has ten cysteines and 55% coincident (69% similar) residues with RBI, is likely to have the same disulphide pattern and a similar 3D structure. Based on the 3D structure found for WDAI-2 (syn. 0.19; Oda et al., 1997) those of WMAI-1 and BDAI-1, with 63% and 72% similar amino-acid residues, respectively, to WDAI-2, can be predicted by computer modelling (Carbonero et al. ms in preparation).

Barley trypsin inhibitor BTI-CMe is one of the best characterized members of the family. It was first purified from barley flour as a protein of 14 KDa that specifically inhibited trypsin among many proteinases tested. Both cDNA and genomic clones for the encoding gene (*Itr1*) are also available (Mikola & Suolinna 1969; Odani et al., 1983; Rodriguez-Palenzuela et al., 1989; Royo et al., 1996). Members of the same BTI-CMe subfamily are the trypsin inhibitors from rye (RTI), maize (MTI) and ragi (RBI). The wheat homologue has not yet been isolated, although a related cDNA (pCMx) has been characterized (Sanchez de la Hoz et al., 1994). The reactive site of these inhibitors is the motif proline-arginine-leucine

(PR▼L) that is located at the right-hand border of the A domain, a region that is quite variable throughout the family.

Two homologous monomeric α-amylase inhibitors have been purified from wheat and the cDNA from one of them (WMAI-1) has been characterized (García-Maroto et al., 1991; Gomez et al., 1991; Kashlan & Richardson, 1981). It is a strong inhibitor of α-amylases of the wheat storage pest *Tenebrio molitor* (Coleoptera) and of that of *Ephestia kuehni*ella (Lepidoptera), but does not inhibit human salivary α-amylase

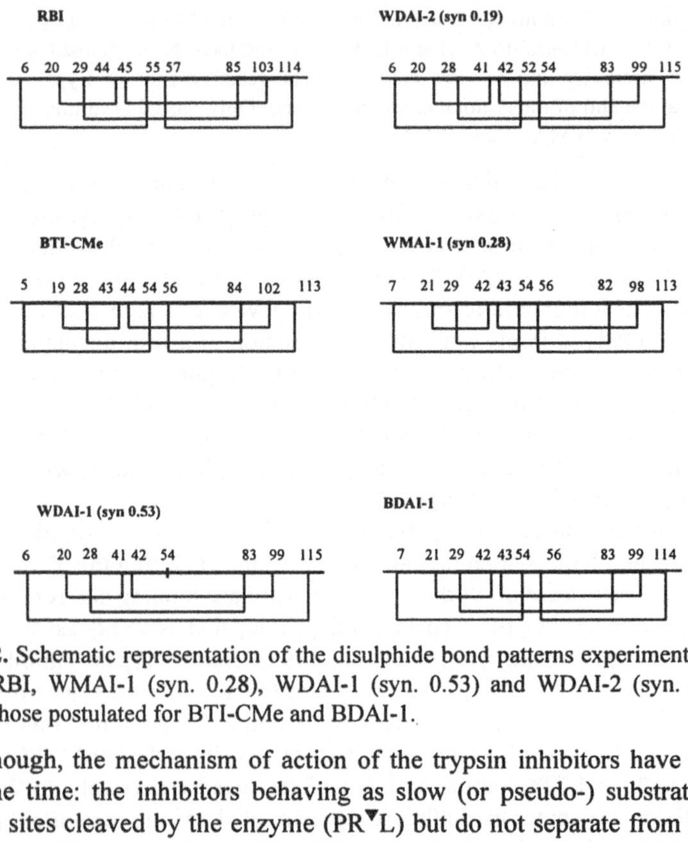

Figure 2. Schematic representation of the disulphide bond patterns experimentally found for RBI, WMAI-1 (syn. 0.28), WDAI-1 (syn. 0.53) and WDAI-2 (syn. 0.19), and of those postulated for BTI-CMe and BDAI-1.

Although, the mechanism of action of the trypsin inhibitors have been known for some time: the inhibitors behaving as slow (or pseudo-) substrates that have reactive sites cleaved by the enzyme (PR▼L) but do not separate from it, that of α-amylase inhibitors has been the subject of much speculation. This question is begining to be clarified through site-directed mutagenesis and domain-swapping (García-Maroto et al., 1991; Carbonero et al., unpublished).

3. Genetics and Evolution

The evolution of this dispersed multigene family raises a number of interesting issues that have been only investigated in a preliminary way:

i) Dispersal of the gene family over several chromosomes (five out of the seven homeology groups in the Triticeae) must have involved both intra-chromosomal duplications and inter-chromosomal translocations. Although some degree of synteny is observed in the gene locations among species, there are significant variations between closely-related species, such as wheat and barley (Figure 1; Table 1). A well-studied case of intra-chromosomal duplication is that affecting the *Itr1* locus (BTI-CMe protein) in *Hordeum vulgare* and *Hordeum spontaneum*. A single active *Itr1* gene is present in most *H. vulgare* cultivars, whereas a duplication of this locus exists in some *H. vulgare* cultivars and in most *H. spontaneum* accessions. The duplication shows divergent phenotypes leading both to more active trypsin inhibitors or to pseudogenes (Carbonero & Garcia-Olmedo, 1998). As to the possible inter-chromosomal dispersal mechanism, it is to be noted that the Itr1 gene has been shown to be located next to the long terminal repeat of the "copia-like" retro-transposon Bare-1, which suggests that transposition may have played a role in the dispersal of the members of this multigene family (Royo et al., 1996).

Table 1. Chromosomal Location of Genes Encoding Trypsin/α-Amylase Inhibitors. (*)S= short, L= long chromosome arms; W= wheat (genomes AABBDD); B= barley (genome HH); R= rye (genome RR).

Inhibitory activity against	Aggregation	Protein	Gene	Chromosome, genome, arm(*)
TRYPSIN	monomeric	BTI-Cme	*Itr1*	3HS
		RTI	*Itr-R1*	3R
		BTI-CMc	*Itr2*	7HS
α-AMYLASE	monomeric	BMAI-1	*Iam1*	2H
		WMAI-1 (syn. 0.28)	*Imha-D1*	6DS
		WMAI-2	*Imha-B1*	6BS
α-AMYLASE	homodimeric	BDAI-1	*Iad1*	6H
		WDAI-1 (syn. 0.53)	*IdhaB1.1*	3BS
		WDAI-2 (syn. 0.19)	*IdhaD1.1*	3DS
		WDAI-3	*IdhaB1.2*	3BS
α-AMYLASE	tetrameric			
1st SUBUNIT		BTAI-Cma	*Iat1*	7HS
		WTAI-CM1	*IthaD1*	7DS
		WTAI-CM2	*IthaB1*	7BS
2nd SUBUNIT		BTAI-CMb	*Iat2*	4HL
		WTAI-CM16	*IthaB2*	4BS
		WTAI-CM17	*IthaD2*	4DS
3rd SUBUNIT		BTAI-CMd	*Iat3*	4HL
(2 copies)		WTAI-CM3B	*IthaB3*	4BS
		WTAI-CM3D	*IthD3*	4D
UNKNOWN				
clone pUP23	_	_	_	6HL
clone pUP88	_	_	_	6AL, 6BL, 6DL

ii) The possible co-evolution of the plant inhibitors and the insect enzymes is suggested indirectly by the available evidence. Little intra-specific variability is observed in the inhibitor pattern, as judged by two-dimensional electrophoresis. However, there are sharp pattern differences between closely-related species. This is consistent with the idea that each inhibitor pattern represents a specific response to the main pests of the area of distribution of each species. There is also evidence of specificity changes in closely-related inhibitors. Thus, barley inhibitor BTI-CMc, which is a weaker trypsin inhibitor than BTI-CMe, is more closely-related in its N-terminal sequence (21 out of 29 coincident amino acids) to one of the subunits of the tetrameric inhibitor of α-amylase (BTAI-CMa), than to any of the cereal trypsin inhibitors described so far (Barber et al., 1986; Rodriguez-Palenzuela et al., 1989). However, the residues deduced from its cDNA at the reactive site place are proline-leucine-leucine (PL▼L), and the rest of the sequence differs sharply from that of BTI-CMe (Figure 1). Site directed mutagenesis leading to reconstitution of the canonical reactive site (PL▼L→ PR▼L) increases by 300% the trypsin inhibitory activity of BTI-CMc (our unpublished results).

iii) The evolution of the aggregative properties of inhibitor subunits also deserves attention. It has been recently reported that two very similar subunits with identical N-terminal amino acid sequences (23 residues) have strikingly different properties: one is a subunit of a tetrameric inhibitor that is active against α-amylases from the insect *Tenebrio molitor* (Coleoptera), but not towards α-amylases from other sources, such as *Ephestia kuehniella* (Lepidoptera) or human saliva, whereas the other aggregates as a homodimer and is only active against the human enzyme (García-Casado et al., 1996). A second example is that of the barley dimeric inhibitor BDAI-1, whose chromosomal location and amino acid sequence suggest that this molecule is evolutively closer to the wheat monomeric than to the wheat dimeric inhibitors (Figure 1). This would imply that BDAI-1 is a diverged form of the wheat monomeric inhibitors that has acquired the ability to self-associate.

iv) The existence of hetero-tetrameric inhibitors suggests a molecular model for intergenome heterosis in alloploids. For example, single tetrameric species were observed in *Triticum tauschii* (subunits CM1, CM3D, CM17) and in *Triticum turgidum* (CM2, CM3D, CM16), while multiple tetrameric species were observed in *Triticum aestivum*, resulting from combinations of the subunits contributed by its two parental species (Gomez et al., 1989). The three types of subunits were required for significant activity although binary mixtures involving subunit WTAI-CM1 (or the corresponding barley BTAI-CMa) also had some activity. Additional combinations of subunits were also reconstituted and their inhibitory activities ranged from 144% (CM1, CM3B, CM17) to 33% (CM2, CM3D, CM17) compared to the activity of the reconstituted inhibitor from *T. tauschii*. This, together with the established chromosomal locations of these genes, fit a model of alloploid heterosis at the molecular level (García-Maroto et al., 1990; Gomez et al., 1989).

4. In vitro inhibition of insect enzymes

Barley BTI-CMe and its maize homologue (MTI) are not only active against trypsin but also against Hageman factor XII-a of the blood-clotting cascade, and BTI-CMe is also active against Kallikrein (Chong & Reeck, 1987). BTI-CMe is inactive against chymotrypsin, papain, pepsin, bacterial and fungal proteases and the endogenous barley proteases (Mikola & Soulinna, 1969). It is also inactive against the α-amylases from such storage cereal pests as *Tenebrio molitor, Tribolium castaneum* and *Sitophilus oryzae* (Moralejo et al., 1993), which is in contrast with the bifunctional activity of its homologues from maize (MTI; Chen et al., 1992) and ragi (BTI; Shivaraj & Pattabiraman, 1981).

Differences in inhibitory activity against trypsin were found among BTI-CMe variants purified from barley cvs Bomi, Hatif de Grignon and Valticky, the two ones isolated from cv Valticky being stronger inhibitors (150%) than the BTI-CMe purified from cv Bomi (Moralejo et al., 1993).

Trypsin-like activity was detected as one of the major digestive proteases in gut extracts of *Spodoptera frugiperda* larvae, a lepidopterous pests of rice fields and other cereals in North, Central and South America. The trypsin-like activity was sensitive to the four variants of BTI-CMe tested, being similarly affected by all of them (Alfonso-Rubí et al., 1997), in contrast with data previouly obtained by Moralejo et al. (1993), towards bovine trypsin.

The monomeric, dimeric and tetrameric α-amylase inhibitors from wheat and barley differentially inhibit α-amylases from different origins (Carbonero et al., 1993). The wheat dimeric class is more active towards the α-amylase from human saliva than against the α-amylases from insect pests. The enzymes from both *Tenebrio molitor* (Coleoptera) and *Ephestia kuehniella* (Lepidoptera) are significantly more sensitive than human salivary α-amylase to the monomeric wheat inhibitor WMAI-1. A given inhibitor class may also discriminate whithin an insect group: the *Leptinotarsa decemlineata* enzyme is more affected by the homodimeric inhibitors than by the monomeric ones, while the opposite is true for the α-amylase of *Tenebrio molitor*, both coleopterous insects (Gutierrez et al., 1990). The Lepidoptera seem to be more susceptible to the tetrameric inhibitors than to the monomeric or dimeric ones (Gutierrez et al., 1993).

Insects which are able to feed on wheat endosperm have unusually high levels of α-amylase (Gutierrez et al., 1990). High inhibitor concentrations in an artifitial diet were required to affect the development of the larvae of *Tribolium confusum*, a storage pest of wheat products, while quite low concentrations were effective against *Callosobruchus maculatus*, a pest of legume seeds (Gatehouse et al., 1986).

5. Transgenic insect resistance

Plant genetic engineering makes possible to control crop pests by the insertion and expression of insect resistance transgenes into plants. Although at least 16 different inhibitors of digestive enzymes isolated from plants have been introduced into 20 crop species (Schuler et al., 1998), only 3 of them correspond to cereal genes.

The α-amylase inhibitor WMAI-1 from wheat resulted in a significant mortality of first-instar larvae of *Agrotis ipsilon* when transgenically expressed in tobacco (Carbonero et al., 1991).

The bifunctional inhibitor from maize (MTI; syn.14K-CI) showed too low levels of protein expression to enhance resistance to insects when it was engineered into tobacco (Masoud et al., 1996), although the levels of 14K-CI protein isolated from corn endosperm inhibited *in vitro* trypsin-like proteinases and/or α-amylase activities from extracts of *Tenebrio molitor*, *Hypea postica* and *Tribolium castaneum* (Masoud et al., 1996).

The barley trypsin inhibitor BTI-CMe has been transgenically expressed into tobacco (manuscript in preparation) and into wheat (Altpeter et al., 1999). BTI-CMe transformed tobacco caused significant larval mortality of the lepidopterous pests *Agrotis ipsilon* and *Spodoptera littoralis*. The genetic transformation of wheat with the Itr1 gene, encoding the BTI-CMe, under the control of the maize ubiquitine promoter has allowed to evaluate its potential for improvement of resistance against storage pests. This construct resulted in a high expression level of the inhibitor in the seeds (up to 1% of total protein) which significantly reduced the survival rate of the grain moth *Sitotroga cerealella* reared on transgenic seeds. The proteinase inhibitor had also an important effect on the weight gain of the insect, although its developmental period was not affected. However, the expression of BTI-CMe in transgenic wheat leaves was quite low (<0.3% of total protein). As expected, no effect was observed against the leaf-eating grasshopper *Melanoplus sanguinipes* (Orthoptera), an insect with predominant chymotrypsin-like enzymes in its mid-gut (Altpeter et al., 1999).

6. Conclusions

The data discussed here show that proteinaceous inhibitors of hydrolases can be successfully used both *in vitro* and transgenically to protect crops againts phytophagous insects. A number of different insecticidal proteins being tested transgenically are Bt toxins, α-amylase, serine- and cystein- protenaise inhibitors, lectins, polyphenol-oxydases, lipoxygenases, etc. The complex world of the enzymes involved in the synthesis of insecticidal secondary metabolites remain a promising field for future research and development.

Many advantages derive from transgenicaly engineered plants for insect control, such are:

i) Protection when and where desired, using appropriate plant promoters.

ii) Confinement of the insecticide within plant tissues, so that rain washes are avoided and only phytophagous insects affected.

iii) Application of chemical insecticides diminished.

iv) Widening of the avaible variability even outside the Plant Kingdom.

However, these genetic engineering practices must be integrated in pest management programmes where traditional plant breeding, crop rotations, biological pest control and a more adequate use of chemical pesticides are all to be considered, in order to increase crop yields while minimally affecting the enviroment.

Acknowledgements: This work was financed by grant Bio96-2303 from the Comision Interministerial de Ciencia y Tecnología (CICYT) Spain. The technical assistance of L. Lamoneda is acknowledged.

7. References

Alfonso-Rubi, J., Ortego, F., Sánchez-Monge, R., García-Casado, G., Pujol, M., Castañera, P. & Salcedo, G. (1997) Wheat and barley inhibitors active towards α-amylase and trypsin-like activities from *Spodoptera frugiperda, J Chem Ecol* **23**, 1729-1741.

Altpeter, F., Diaz, I., McAuslane, H., Gaddour, K., Carbonero, P. & Vasil, I.K. (1999) Increased insect resistance in transgenic wheat stably expressing trypsin inhibitor CMe, *Mol Breed* 5(1): 53-63.

Barber, D., Sanchez-Monge, R., García-Olmedo, F., Salcedo, G. & Mendez, E. (1986) Evolutionary implications of sequential homologies among members of the trypsin/α-amylase inhibitor family (CM-proteins) in wheat and barley, *Biochem Biophys* Acta **873**, 147-151.

Bowen D., Rocheleau T.A., Blackburn M., Andreev O., Golubeva E., Bhartia R. & Ffrench-Constant R.H. (1998) Insecticidal toxins from the bacterium *Photorhabdus luminescens, Science* **280**, 2129-2132

Carbonero, P., González-Hidalgo E., Royo, J., Rodriguez-Palenzuela, P., Garcia-Maroto, F., Maraña, C., Sánchez de la Hoz, P., Gutierrez, C. & Castañera P. (1991) Cereal inhibitors of insect hydrolases (α-amylases and trypsin): Genetic control, transgenic expression and insect tests. Proc. Eucarpia Symposium on Genetic manipulation in Plant Breeding, Reus/Salou, Spain, 26-30 May 1991.

Carbonero, P., Salcedo, G., Sanchez-Monge, R., García-Maroto, F., Royo, J., Gomez, L., Mena, M., Medina, J. & Diaz, I. (1993) A multigene family from cereals which encodes inhibitors of trypsin and heterologous α-amylases, in F.X. Avilés (ed.), *Innovations in Proteases and their Inhibitors.* Walter de Gruyter, Berlin and New York, pp. 333-348.

Carbonero, P. & García-Olmedo, F. (1998) A multigene family of trypsin/α-amylases inhibitors from cereals, in Casey R. and Shewry, P.R. (eds), *Seed proteins.* Chapman and Hall (in press).

Chen, M.S., Feng, G.H., Zen, K.C., Richardson, M., Valdesrodriguez, S., Reeck, G.R. & Kramer, K.J. (1992) α-Amylases from three species of stored grain Coleoptera and their inhibition by wheat and corn proteinaceous inhibitors, *Insect Biochem Mol Biol* **22**, 261-268.

Chong, G.L. & Reeck, G.R. (1987) Interaction of trypsin, β-factor XIIa and plasma kallikrein with a trypsin inhibitor isolated from barley seeds: a comparison with the corn inhibitor of activated Hageman factor, *Thrombosis Res* **48**, 211-221.

Christou, P. (1995) Srategies for variety-independent genetic transformation of important cereals, legumes and woody species utilizing particle bombardment, in A.C. Cassells and P.R. Jones (eds) *Methodology of plant genetic manipulation: Criteria for decision making,* Kluwer Academic Publishers, pp. 13-27.

García-Casado, G., Sanchez-Monge, R., Puente, X.S. & Salcedo, G. (1996) Divergence in properties of two closely related α-amylase inhibitors of barley, *Physiol Plant* **98**, 523-528.

García-Maroto, F., Carbonero, P. & García-Olmedo, F. (1991) Site-directed mutagenesis and expression in *Escherichia coli* of WMAI-1, a wheat monomeric inhibitor of insect α-amylase, *Plant Mol Biol* **17**, 1005-1011.

García-Maroto, F., Maraña, C., Mena, M., García-Olmedo, F. & Carbonero,P. (1990) Cloning of cDNA and chromosomal location of genes encoding the three types of subunits of the wheat tetrameric inhibitor of insect alpha-amylase, *Plant Mol Biol* 14,845-853.

García-Olmedo, F., Salcedo G., Sánchez-Monge, R., Gomez, L., Royo J. & Carbonero, P. (1987) Plant proteinaceous inhibitors of proteinases and α-amylases, in B. Miflin (ed), *Oxford Surveys of Plant Molecular and Cell Biology* 4. Oxford University Press-ISPMB New York, pp. 275-334.

Gatehouse, A.M.R., Fenton, K.A., Jepson, I. & Pavey, D.J. (1986) The effects of α-amylase inhibitors on insect storage pests: inhibition of α-amylase in vitro and effects on development in vivo, *J Sci Food Agric* **37**, 727-734.

Gomez, L., Sanchez-Monge, R., García-Olmedo, F. & Salcedo, G. (1989) Wheat tetrameric inhibitors of insect α-amylase: alloploid heterosis at the molecular level, *Proc Natl Acad Sci USA* **86**, 3242-3246.

Gomez, L., Sanchez-Monge, R., Lopez-Otín, C. & Salcedo, G. (1991) Wheat inhibitors of heterologous α-amylases, *Plant Physiol* **96**, 768-774.

Graham, J., McNicol, R.J. & Greig, K. (1995) Towars genetic based insect resistance in strawberry using the cow-pea trypsin inhibitor gene. *Ann Appl Biol* **127**, 163-173.

Gutierrez, C., Sanchez-Monge, R., Gomez, L., Ruiz-Tapiador, M., Castañera, P. & Salcedo, G. (1990) α-Amylase activities of agricultural insect pests are specifically affected by different inhibitor preparations from wheat and barley endosperm, *Plant Sci* **72**, 37-44.

Gutierrez, C., García-Casado, G., Sanchez-Monge, R., Gomez, L., Castañera, P. & Salcedo, G. (1993) Three inhibitor types from wheat endosperm are differentially active against α-amylases of Lepidoptera pests, *Entomol Exp Appl* **66**, 47-52.

Hilder, V.A., Gatehouse, M.A.R., Sheerman, S.E., Baker, R.F. & Boulter D. (1987) A novel mechanism of insect resistance engineered into tobacco, *Nature* **327**, 160-163.

Kashlan, N. & Richardson, M. (1981) The complete amino acid sequence of a major wheat protein inhibitor of α-amylase, *Phytochemistry* **20**, 1781-1784.

Maeda, K., Wakabayashi S. & Matsubara, H. (1983) Disulfide bridges in an α-amylase inhibitor from wheat kernel, *J. Biochem* **94**, 865-870.

Masoud, S.A., Ding, X., Johnson, L.B., White, F.F. & Reeck, G.R. (1996) Expression of a corn bifunctional inhibitor of serine proteinases and α-amylases in transgenic tobacco plants, *Plant Sci* **115**, 59-69.

Mikola, J. & Soulinna, E.M. (1969) Purification and properties of a trypsin inhibitor from barley, *Eur J Biochem* **9**, 555-560.

Moralejo, M.A., García-Casado, G., Sanchez-Monge, R., Lopez-Otin, C., Romagosa I., Molina-Cano, J.L. & Salcedo, G. (1993) Genetic variants of the trypsin inhibitor from barley endosperm show different inhibitory activities, *Plant Sci* **89**, 23-29.

Oda,Y., Matsunaga, T., Fukuyama, K., Miyazaki, T. & Morimoto, T. (1997) Tertiary and quaternary structures of 0.19 α-amylase inhibitor from wheat kernel determined by X-ray analysis at 2.06 A resolution, *Biochemistry* **36**, 13503-13511.

Odani, S., Koide, T. & Ono, T. (1983) The complete amino acid sequence of barley trypsin inhibitor, *J Biol Chem* **258**, 7998-8003.

Paz-Ares, J., Ponz, F., Aragoncillo, C., Hernandez-Lucas, C., Salcedo, G., Carbonero, P. & García-Olmedo, F. (1983) In vivo and in vitro synthesis of CM-proteins (A-hordeins) from barley (*Hordeum vulgare* L.), *Planta* **157**, 74-80.

Peferoen, M. (1997) Progress and prospect for field use of *Bt* genes in crops, *Trends Biotech.* **15**, 173-177.

Poerio, E., Caporale, C., Carrano, L., Pucci, P. & Buonocore, V. (1991) Assignment of the five disulfide bridges in an α-amylase inhibitor from wheat kernel by fast-atom bombardment mass spectrometry and Edman degradation, *Eur J Biochem* **199**, 595-600.

Rodriguez-Palenzuela, P., Royo, J., Gomez, L., Sanchez-Monge, R., Salcedo, G., Molina-Cano, J.L., García-Olmedo, F. & Carbonero, P. (1989) The gene for trypsin inhibitor CMe is regulated in trans by the Lys 3a locus in the endosperm of barley (*Hordeum vulgare* L.), *Mol Gen Genet* **219**, 474-479.

Royo, J., Diaz, I., Rodriguez-Palenzuela, P. & Carbonero, P. (1996) Isolation and promoter characterization of the barley gene *Itr1* encoding trypsin inhibitor BTI-CMe: differential activity in wild type and mutant *lys 3a* endosperm, *Plant Mol Biol* **31**, 1051-1059.

Sanchez de la Hoz, P., Castagnaro, A. & Carbonero, P. (1994) Sharp divergence between wheat and barley at loci encoding novel members of the trypsin/α-amylase inhibitors family, *Plant Mol Biol* **26**, 1231-1236.

Schuler, T.H., Poppy, G.M., Kerry, B.R. & Denholm I. (1998) Insect-resistant transgenic plants, *Trends Biotechnol* **16**, 168-175.

Shivaraj, B. & Pattabiraman, T.N. (1981) Natural plant enzyme inhibitors. Characterization of an unusual α-amylase/trypsin inhibitor from ragi (*Eleusine coracana Geartn*), *Biochem J* **193**, 29-36.

Strobl, S., Mühlhahn, P., Bernstein, R., Witscheck, R., Maskos, K., Wunderlich, M., Huber, R., Glockshuber, R. & Holak, T.A. (1995) Determination of the three-dimensional structure of the bifunctional α-amylase/trypsin inhibitor from ragi seeds by NMR spectroscopy, *Biochemistry* **34**, 8281-8293.

Xu, D., Xue, Q., McElroy D., Mawal Y., Hilder V.A. & Wu R. (1996) Constitutive expression of a cowpea trypsin inhibitor gene, CpTi, in transgenic rice plants confers resistance to two major rice insect pests, Mol Breed **2**, 167-173.

G.T. Scarascia Mugnozza, E. Porceddu & M.A. Pagnotta (Eds.)
Genetics and Breeding for Crop Quality and Resistance, 159-163, 1999

New genes for pest control

C. Tortiglione[1], C. Malva[2], F. Pennacchio [3] & R. Rao[1]
[1]Dipartimento di Scienze Agronomiche e Genetica Vegetale, Via Università 100, 80055 Portici, Italy; [2]Istituto Internazionale di Genetica e Biofisica, CNR, Napoli, Italy; [3] Dipartimento di Biologia, Potenza, Italy.

Abstract: The feasibility of using genetically modified plants for insect control has been clearly demonstrated by expressing the *Bacillus thuringensis* (*Bt*) toxins or the protease inhibitor genes. These compounds have been proved to control several insect pests in different crops. There are however many other economically important crops which are not effectively protected by known *Bt* toxins or protease inhibitors. Thus the importance to identify new gene sources for a more complete and stable control of insect pests. Neuropeptides as well as small peptidic hormones are an interesting class of molecules to be used as possible insecticides. In fact these molecules regulate several insect physiological processes and are active at very low concentration. An alteration of their title in the insect could cause severe functional modifications. Several examples exist of neuropeptides encoded by a single gene coding for multiple copies of one or more different peptides, spaced by potential endoproteolytic sites, the cleavage of which generates free bioactive molecules. A new strategy miming this natural system were developed leading to the synthesis of genes encoding for multiple copies of neuropeptides and hormones, spaced by putative endoproteolytic sites. These genes were then cloned downstream of different promoters and translational leaders, transiently expressed in tobacco protoplasts and then stably expressed in tobacco plants. The results are presented and here discussed as new approach to express in plants multiple bioactive peptides.

1. Introduction

Engineering plants to resist insects is one of the most urgent demands in agriculture. Insect pests create great costly losses for farmers and attemps to mitigate insect damage usually require applications of chemical pesticides or spray-on biological pesticides. Insecticidal applications need to be properly timed to coincide with insect infestations and do not always reach burrowing insects. The introduction of plant with built-in resistance to various insect pests will have a tremendous impact on current agricultural practices. In 1997 was introduced in the marketplace insect tolerant corn, potato and cotton plants expressing a *Bacillus thuringiensis* (*Bt*) endotoxin gene. These plants provide resistance against some of the worst crop pests.

2. Molecules suitable to confer insect resistance

Bt proteins attack directly the insect gut ephitelial cells, causing cell lysis and insect death. Other proteins such as cholesterol oxidase and lipid acyl hydrolases act in the same way, although their toxic activity, when expressed in crops has not been assayed yet (Strickland et al., 1995). Several other proteins have been identified that may be suitable for expression in transgenic plants to confer resistance to insects. Some of these, such as polyphenol oxidase and invertase (Felton et al., 1992; Purcell et al., 1994) act indirectly modifying dietary constituents, and depriving the insect of nutrients or generating toxic compounds. Lectins can bind carbohydrates in the diet and on surface in insect gut (Chrispeels & Raikhel, 1991), while other proteins, such as amylase inhibitors and proteinase inhibitors interfere with digestive enzyme of the insect gut, making digestion less efficient (Ryan, 1990). However, once an insecticidal protein is identified, it's critical to know whether or not it will be correctly expressed in a plant, in order to create an insect tolerant cultivar. The gene encoding for the protein needs to be correctly translated according to a favourable codon usage for plants and the presence of proper plant processing signals. Furthemore, expression must be high enough in order to achieve an expression level suitable to cause the desired effect on the insect pest.

3. Resistance management strategies

One concern regarding the use of transgenic crop for insect control is the possibility of resistance emerging in insect population (McGaughey & Whalon, 1992). In insect, resistance is a pre-adaptive phenomenon which develops by selection of rare individuals in a population which can survive a certain insecticide treatment. Resistence proved to be controlled by partially recessive or co-dominant (additive) genes and to involve a small number of loci. Resistance to *Bt* proteins has been studied in both laboratory and field population of insects and resistance management strategies have been developed, which, at least in theory, should retard the spread of resistance genes through a population. They aim to keep the frequency of resistance genes below the level which would require an increased insecticide usage (Gould, 1994; Mallet & Porter, 1992). The ultra high dose strategy aims to express the insecticidal protein to a level high enough to kill homozygous resistant insects and thus changing the crop to a non-host, driving the insect away from the crop. The multiple target strategy aims to target the insect pest with multiple insecticides, acting at different targets. The refuge strategy aims to express in plants high dose of insecticide which will kill heterozygous resistant insects and permit survival of homozygous resistant insects, frequency of whom could be diluted to its regular form by a regular influx of wild type genes. This could be obtained by planting mixture of transgenic and wild type seeds on which susceptible insects could survive and mate with homozygous resistante insects (Alstad & Andow, 1995).

4. A new class of molecules: the neuropeptides

Although the *Bt* proteins provide control of several insect pests in different crops, there are other serious pests that are not susceptible to the current array of *Bt* proteins. The lack of alternative genes to control these insects and provide resistance management options led us to identify a new set of molecules in insects which have the potential to become a new class of insecticides. These molecules are peptides, such as myotropin and neuropeptides, controlling several physiological processes and act as signals for important biological functions, such as inhibition of ecdysone synthesis (Growth blocking peptide), inhibition of juvenile hormon synthesis (Allatostatin), inhibition of gut motility (Proctolin), stimulation of cardiac muscle activity (β-Casomorfin) (Kelly et al., 1990). The sequence of some peptides which have been found to have a pronounced effect in physiological experiments are reported (Petrilli et al., 1994; Bavoso et al., 1995):

Proctolin:	H-Arg-Tyr-Leu-Pro-Thr-OH
Schistocerca allatostatin-5	H-Gly-Arg-LeuTyr-Ser-Phe-Gly-Leu-NH2
Locusta pyrokinin 2	H-Glu-Ser-Val-Pro-Thr-Phe-Thr-Pro-Arg-Leu-NH2
β- Casomorfin	H-Tyr-Pro-Phe-Pro-Gly-NH2

Despite their low molecular weight and apparent structural semplicity they are active at very low concentrations and different pH. Insect exposure to such peptides in unphysiological dosages or at unusual times might cause dismetabolic effects, growth inhibition and death. The use of a naturally occurring peptide to control insect pests reduce the risk of evolving resistance and the potential damage against non-target insects.

5. Digestive enzyme modulating and oostatic peptide

We focused our studies on a peptide which inhibits the synthesis of trypsin-like enzymes in the gut of the mosquito *Aedes aegipty*, and as consequence inhibits the ovary development, depriving the female of the aminoacid supply necessary for the oogenesis. This peptide, called Digestive Enzyme Modulating and Oostatic Peptide (DEMOP) is 10 aminoacid long and rich in prolines. We assayed the potential insecticidal activity of this decapeptide against the tobacco budworm (*Heliotis viriscens*) larvae by feeding on artificial diet. Bioassays conducted with the synthetic DEMOP decapeptide added at 100μg g^{-1} to the artificial diet showed that it inhibited *H. viriscens* larval growth at 20% level and caused mortality at 15% level, compared to the control. These assays proved that DEMOP produces a detrimental effect on survival and growth of neonate larvae (1st-3rd instar). The following step was to prove its effectivness when expressed in a transgenic plant, at high enough level to cause toxic effect on insects fed on it. The synthetic gene we made encodes for multiple DEMOP copies, spaced by endoproteolytic sites, in order to increase the amount of peptide produced per cell, which otherwise, given its low molecular weight, would have been very low (Figure 1).

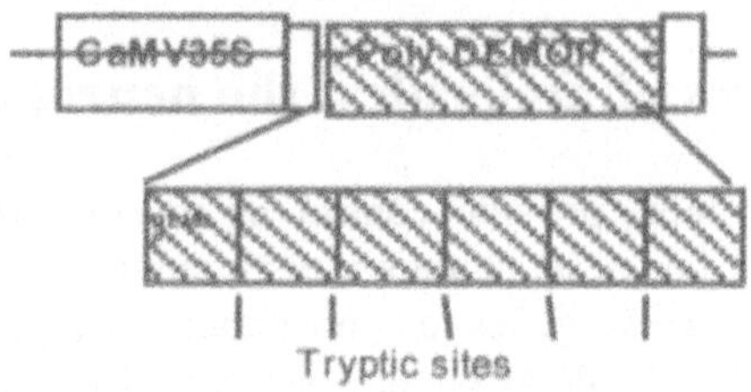

Figure 1. Scheme illustrating the construction of the synthetic gene Poly-DEMOP by single DEMOP units, and its cloning under the CaMV35S promoter.

We created a synthetic gene encoding for 6 copies of DEMOP, as described in Rao et al. (1996) generating a precursor of 13kDa, which was cloned under a plant constitutive promoter and transferred to both tobacco and potato plants by mean of *Agrobacterium tumefaciens* mediated transformation (Figure 2).

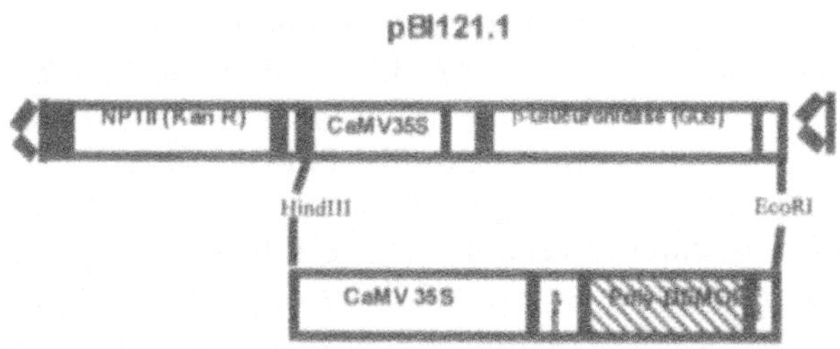

Figure 2. Bynary vector used to transform *Nicotiana tabacum*. The HindIII/EcoRI expression cassette containing the GUS gene was substitued by the HindIII/EcoRI expression cassette containing the Poly-DEMOP synthetic gene.

Tobacco plants transformed with this construct were proved by mean of molecular technologies to transcribe the transgene, but a product corresponding to the precursor size was never observed. This was due to the processing of the precoursor which released the decapeptide monomers and its multiple forms. Protein extracts from transgenic tobacco plants were assayed *in vitro* for the presence of a DEMOP specific activity. These extracts showed presence of activity in DEMOP extracts and not in control extracts made from untrasformed plants. Transgenic plants were also *in vivo* bioassayed by feeding *H. viriscens* larvae from the first instar up to the fifth instar on transgenic leaves, which produced a drammatic difference in the weight of surviving larvae. Larvae reared on DEMOP leaf disc wighed approximately 50% less than the control larvae reared on untransformed leaf. DEMOP leaf disc area consumed was reduced to 50% compared to the control leaf disc. Thus, the expression in plant of a polypeptide encoding for multiple DEMOP units showed an increased toxicity against the tobacco budworm compared to the feeding assay with the synthetic peptide, although the mortality disappeared.

This might be due to an intrinsic toxicity of the partial or total precursor proteolytic lysis. The randomly processed peptides might be more active than the single DEMOP. This finding open up the possibility to create a gene encoding for a fusion of different peptides having multiple insecticidal targets. In fact, the multiple target strategy allows a simultaneous expression into a crop of multiple gene copies. To date, the vectors available to transform plants are capable of transferring a maximum of 4 genes, but in no case they have been proved to be sincronously expressed. When using multiple insecticidal proteins, even insects homozygous for one or two resistance genes but heterozygous for another resistance gene would still be controlled by crops expressing multiple insecticidal genes, achieving in such a way one of the strategies of resistance management. The exact mechanism of action underlying the DEMOP precursor toxicity toward *Heliotis viriscens* larve is not clear, neither its effect on adults. It likely involves the inhbition of trypsin-like enzyme synthesis in the insect midgut, altering the development and the growth. Further work is also required on adults to test the putative oostatic function of DEMOP in female hatched from the surviving larvae.

6. References

Alstad, D.N. & Andow, D.A. (1995) Managing the evolution of insects resistant to transgenic plants, *Science*: **268**, 1894-1896.

Bavoso, A. Falabella, P., Giacometti, R., Halane, A.J., Ostuni, A., Pennacchio, F. & Tremblay, E. (1995) Instestinal absorpion of proctolin in *Helicoverpa armigera* (Lepidoptera, Noctuidae) larvae. *REDIA*, **78**, 173-185.

Chrispeels, M. J. & Raikhel, N. V. (1991) Lectins, lectin genes and their role in plant defense. *Plant Cell* **3**, 1-19.

Felton, G.W., Donato, K.K., Broadway, R.M & Duffey, S.S. (1992) Impact of oxidised plant phenolics on the nutritional quality of dietary protein to a noctuid herbivore, *Spodoptera exigua*, *J. Insect Physiol.* **38**, 277-285.

Gould, F. (1994) Potential problems with high-dose startegies for pesticidal engineered crops, *Biocontrol Sci.Technol.* **4**, 451-461.

Kelly, T.J., Masler, E.P. & Menn, J.J. (1990) Insect neuropeptides: new strategies for insect control. *Pesticides and Alternatives* Elseviere Science, 283-297.

McGaughey, W.H. & Whalon, M.E. (1992) Managing insect resistance to *Bacillus thuringiensis* toxin, *Science*, **258**, 1451-1455

Mallet, J. & Porter, P. (1992) Preventing insect adptation to insect resistant crops:are seed mixture or refugia the best strategy? *Proc.R.Soc.London* Series **B250**, 165-169

Petrilli, P., Caporale, C. & Caruso, C. (1994) Hydrolysis pattern of procasomorphin by gut proteases from plant parassite *Heliotis zea*, determined by sequence analyses performed on the unfractioned digestion mixtures. *Int. J. Pept. Protein Res.*, **43**, 201-204.

Purcell, J.P., Greenplate, J.T., Duck, N.B. & Sammons, R.J. (1994) Insecticidal activity of proteinases, *FASEB J.* **8**, A1372.

Rao, R., Manzi, A., Filippone, E., Manfredi, P., Spasiano, A., Colucci, G. & Monti, L. (1996) Synthesis and expression of genes encoding putative insect neuropeptide precursor in tobacco. *Gene*, **175**, 1-5

Ryan, C. (1990) Proteinase inhibitors in plants: genes improving defenses against insects and pathogens. *Annu. Rev. Phytopathol.* **28**, 425-449.

Strickland, J.A., Orr, G.L. & Walsh, T.A. (1995) Inhibition of Diabrotica larval growth by patatin, the lipid acyl hydrolase from potato tubers, *Plant Physiol.*, **109**, 667-674.

G.T. Scarascia Mugnozza, E. Porceddu & M.A. Pagnotta (Eds.)
Genetics and Breeding for Crop Quality and Resistance, 165-172, 1999

Expression of thaumatin II gene in horticultural crops

S.V. Dolgov[1], V.G. Lebedev[1], A.P. Firsov[1], S.A. Taran[1] & G.B. Tjukavin[2]
[1]Branch of Shemjakin and Ovchinnikov Institute of Bioorganic Chemistry, RAS, 142292, Pushchino, Moscow region, Russia. [2]All-Russian Research Institute of Vegetable Breeding and Seed Production, RAAS, 143080, p/o Lesnoy Gorodok, Moscow region, Russia.

Abstract: For fruit taste improvement of temperate horticultural crops - apple, pear, strawberry and carrot the gene of supersweet protein thaumatin II from *Thaumatococcus daniellii* Benth. was used in our research. Based on the coding sequence, which have been cloned in Unilever, and pBI121 we constructed the binary vector for its expression in plants. The successful transformation of apple, carrot, pear and strawberry have been obtain by usage the disarmed supervirulent agrobacterial strain CBE21 received from "Bioengineering Center" RAS. The introduction of thaumatin gene was confirmed by PCR analysis. The high thaumatin II expression was observed in callus tissues and lower one in leaves of greenhouses plants of apple, pear and carrot roots.

1. Introduction

The economic importance of horticultural crops has led to selection and breeding over thousands of years. This practice has resulted in relatively few genotypes and, therefore, in a restricted germplasm base. Recent developments in biotechnology have provided for an alternate approach to horticultural crops improvement through the introduction of genes encoding desirable traits. Most research to date has focused on genes conferring resistance to viruses, bacteria, insects and fungi. However, attention has also been given to genes that regulate such parameters as fruit taste.

Thaumatin is a product of the plant *Thaumatococcus daniellii* Benth. whose fruit produce gelatinous arils which taste extremely sweet. The sweet components of *Th. danielli* iarils are a family of proteins called thaumatins. One of most abundant, thaumatin II has an average sweetness of 2,500 times that of sucrose. They are composed of 207 amino acids in a β-pleated barrel form with eight disulphide bonds. Thaumatin II cDNA has been isolated and sequenced (Edens et al., 1982). It was introduced into potato cultivar Iwa (Witty & Harvey, 1990) and cucumber Borszczagowski (Szwacka et al., 1996). Biologically active r-thaumatin was produced in transgenic potato and cucumber plants inducing a sweet-taste phenotype. As have been marked Witty & Harvey (1990), another suitable hosts for thaumatin introduction are strawberry, melon, kiwifruit and apple. There is considerable economic interest in production of r-thaumatin in transgenic organisms. From comments made by panellists, it is believed that production of r-thaumatin in plant could improve flavour and so the insertion of the thaumatin gene into crop plants may increase their value and appeal to consumers.

2. Materials and Methods

2.1 Vector design

Plasmid pUR528 with thaumatin II gene under *lac* promoter have been kindly provided by Dr. A. Ledeboer. *Eco*RI-*Hind*III fragment containing full thaumatin II gene sequence was cloned in polylinker of plasmid pBB (Taran S.A.). The GUS gene of plasmid pBI121 was replaced on *Xba*I-*Bam*HI fragment of plasmid pBBThau carried full thaumatinII gene sequence. Resulted plasmid pBIThau with thaumatin II gene between 35S promoter and term was introduced into disarmed supervirulent agrobacterial strain CBE21 kindly provided by Dr. E. Revenkova (Revenkova et al., 1993), by direct transformation.

2.2 Plant transformation

2.2.1 Apple transformation

Apple plants of commercial variety Melba were used in our experiments. The plants were cultivated *in vitro* on the QL medium (Quorin & Lepoivre, 1977) containing 1.5-2.0 mg l^{-1} BAP and 0.3 mg l^{-1} IBA and rooted on the half-strange QL medium supplemented with 0.5-1.0 mg l^{-1} IBA. Propagation and rooting of apple plants were performed under the following conditions: temperature - 26/22°C; day length - 16 hours; illumination – 2,500-3,000 lux.

The leaves pieces of about 1 cm^2 from in vitro rooted plants were used for transformation. Co-cultivation leaf explants with agrobacteria was performed on Murashige & Skoog (1962) medium (MS) containing 3.0 mg l^{-1} 4CPU, 1.0 mg l^{-1} NAA, 1 mg l^{-1} TIBA (MSR, Dolgov & Muratova, 1995) for 2-3 days at 26°C. After co-cultivation the explants were placed on the same medium for regeneration and selection of transformants (MSR) supplemented by 500 mg l^{-1} cefotaxime. Beginning second passage 35 mg l^{-1} kanamycin were added to regeneration media. Regeneration was performed at 23°C in the darkness.

Regenerants grown to 5-7 mm in size were separated from the callus and placed on the propagation medium (section "Plant material") with 50 mg l^{-1} kanamycin. The transformants continuously propagated on the media with selective antibiotics and successfully rooted on half QL media with addition of 0.5 mg l^{-1} IBA and 25 mg l^{-1} Km. Transgenic apple plants were cultured under cultivation conditions described above.

2.2.2 Pear transformation

Regeneration of pear variety Burakovka was provided on Nitsch and Nitsch (NN) media supplemented by 3.0 mg l^{-1} TDZ, 0.5 mg l^{-1} NAA, 0.1 mg l^{-1} GA, 500 mg l^{-1} Cef., 25 mg l^{-1} Km were added from second passage (12-14th day after cocultivation). Other conditions were the same as for apple.

2.2.3 Carrot transformation.

Embryo and leaf derived calli were used for carrot transformation. These calli were cultivated on MS media with 0.2 mg l^{-1} 2-4D and 0.2 mg l^{-1} Kin. After three day cocultivation with *Agrobacterium* on paper filters callusis were transferred to the same media supplemented by 200 mg l^{-1} Km and 500 mg l^{-1} Cef. After 2-3 passages they were regenerated on MS media with 0.1 mg l^{-1} Kin and 100 mg l^{-1} Km. Rooting were performed on liquid hormones free 1/2 MS media with 50 mg l^{-1} Km.

2.2.4 Strawberry transformation.

The leaf explants of Feyerverk variety taken from *in vitro* culture plants, were incubated in bacterial suspension for 20-40 min., then transferred to co-cultivation medium consist of MS salts with addition 2.0 mg l^{-1} IAA (auxin shock). Co-cultivation was done during 2-3 days at temperature 24-26°C. After co-cultivation explants were transferred into regeneration and selection medium consist of MS salts with addition of 5.0 mg l^{-1} BAP, 0.3 mg l^{-1} IBA, 500 mg l^{-1} cefotaxime and 50 mg l^{-1} kanamycin. The explants were passaged into fresh medium with 3-weeks interval. From the fourth passage when the calluses achieved the 5-6 mm in size, the concentration of kanamycin was decreased by half. Regenerated on selective media shoots were separated from calluses and transferred into reproductive media (Boxus, 1975) supplemented by 1.0 mg l^{-1} BAP and 50 mg l^{-1} kanamycin. Transformants rooting was done on hormone free media with 50 mg l^{-1} kanamycin.

2.3 *Molecular biology analysis of transgenes*

2.3.1 DNA isolation and PCR analysis

For DNA isolation 100 mg of fresh leaves collected from greenhouse plants, were ground in liquid nitrogen. The genomic DNA was isolated from the CTAB extracted material by the method Gelvin & Shilperoort (1995).

PCR analysis was used to confirm the integration of thaumatin II sequence in the genome of transgenic plants. The primers for internal part of gene have been used. The N-terminal primer was 5'-CACCTTCGAGATCGTCAACCGGTGCTC, the C-terminal ones was 5'-CAGTAGGGCAGAAAGTGACCCTGTAGTTG. The amplification reaction was done in a buffer containing 7 mM Tris-HCl (pH 8.8), 15 mM $(NH_4)_2SO_4$, 6 mM $MgCl_2$, 0.2 mM of each dNTP, 0.2 mg ml^{-1} BSA, 0.5 Nonidet NP40, - 2 units of Taq-polymerase, 1.0 μg DNA, 25 pkM of each primer in a volume of 30 μl. A 40 μl amount of mineral oil was layered onto the reaction mixture. Regime of amplification was next: preliminary denaturation (100°C) - 2 min, annealing primers (58°C) - 1 min.; finishing of chain building (72°C) – 1.5 min.; denaturation (94.5°C) - 1 min.; number of amplification cycles - 30. The PCR products were analyzed using electrophoresis in 2% agarose gel. The length of the amplified fragment was 589 bp.

3. Results

3.1 Transformation

Transformation experiments resulted in formation of calluses, which were more sweet that controls ones. 4 apple, 4 pear, 2 strawberry and more than 100 carrot lines successfully proliferated on selective media with kanamycin have been obtained (Table 1). Transformation frequency was 2.6-5.9% for apple and 11.7-15.5% for pear. It was determined as the relationship of transgene plants regenerated at the presence of selective antibiotics to regenerants number formed on control explants. These apple, pear and strawberry plants rooted on media supplemented by 25 mg l^{-1} Km. This antibiotic concentration completely suppressed root formation of control untransformed plants (Figure 1). Carrot plants regenerated from calluses and rooted on media with 50 mg l^{-1} Km. Regenerants adapted to soil conditions and grows in greenhouse. Apple was grafted on dwarf rootstock N 62-396 and formed one-year stems. In the greenhouse all transgenic apple, pear and strawberry plants proliferated normally and appeared phenotypically similar to the control ones. Only one carrot plant has some phenotypical differences from parent variety Nantskaya. It have more dissected leaf shapes.

Table 1. Apple and pear transformation by CBE21:pBIThau35. Regeneration frequency of control plants: Burakovka-10%, with 1.0 shoot per explant. Melba-73% with 2.1 shoots per explant.

Variety	Explant number	Regenerants	Rooted on 25 mg l^{-1} Km	PCR positive lines	Transformation frequency %
Burakovka	131	2	2	2	15.3
Burakovka	168	5	2	2	11.7
Melba	33	4	3	3	5.9
Melba	25	2	1	1	2.6

Figure 1. Multiplication of trnsgenic pear plants on medium supplemented by kanamycine. 1. - control, nontransformed plants (25mg l^{-1} Km); 2,3, - transgenic clones (37, 50 mg l^{-1} Km).

3.2 PCR analysis

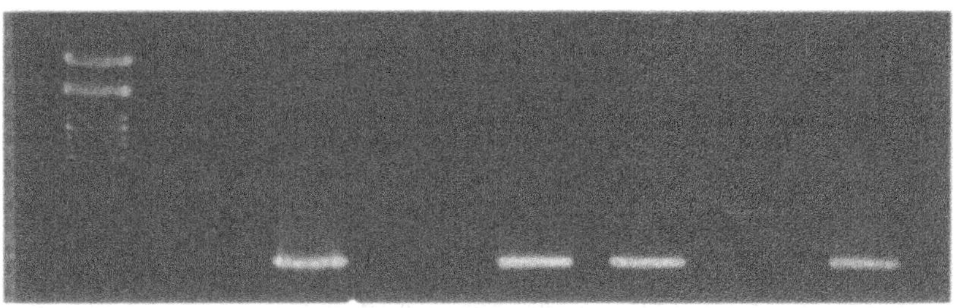

Figure 2. PCR analysis of transgenic apple plants transformed by pBIThau. DNA was amplified using primers specific to the thaumatin II gene sequence. Amplification produced an 589 bp product. Lane 1 - marker λ; lanes 2, 4 - pBIThau; lane 3 - nontransformed control; lanes 5,6,7,8- transgenic clones.

Amplification of internal 589 bp of thaumatin gene showed it presence in four apple lines (Figure 2). Line three also demonstrated amplification of thaumatin II gene sequence in repeated reaction. Including thaumatin II gene have been shown only for three pear lines (Figure 3). Presence of transferred gene observed in all tested carrot lines and both strawberry ones (Figure 4).

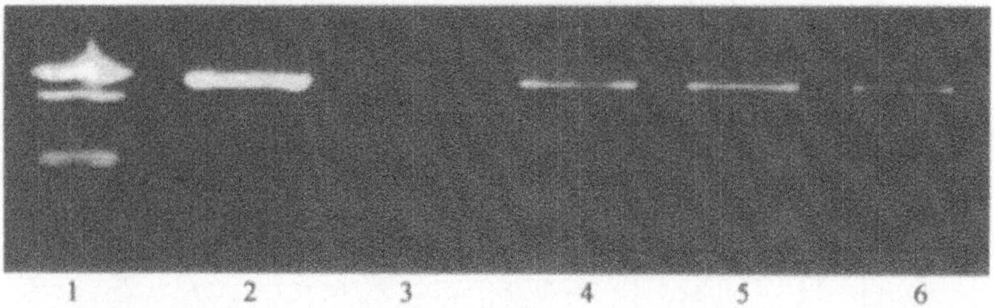

Figure 3. PCR analysis of transgenic pear plants transformed by pBIThau. DNA was amplified using primers specific to the thaumatin II gene sequence. Amplification produced an 589 bp product. Lane 1 - marker 100-700bp; lane 2 - pBIThau; lane 3 - nontransformed control; lanes 4,5,6 - transgenic clones.

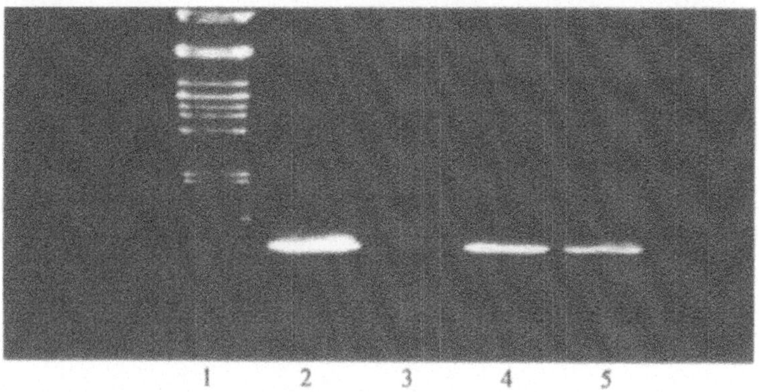

Figure 4. PCR analysis of transgenic strawberry plants transformed by pBIThau. DNA was amplified using primers specific to the thaumatin II gene sequence. Amplification produced an 589 bp product. Lane 1 - marker λ; lane 2 - pBIThau; lane 3 - nontransformed control; lanes 4,5, - transgenic clones.

3.3 Gene expression

The calluses formed on leaf surface of apple, pear and strawberry were more sweeter that on control explants. Especially this phenomena was noticeable for apple. Strawberry calluses have only slight differences from nontransformed calli. The same data have been observed for greenhouse plants. No differences can be tested for leaves and berries of transgenic strawberry plants. On the contrary, leaves of greenhouse grown apples were detectable sweeter, especially from one line and from grafted on dwarf rootstock trees. Also some of the transgenic carrot demonstrated detectable root taste changes.

4. Discussion

Thaumatin was originally found in the arils of ripe fruits from *Thaumatococcus daniellii* Benth. (van der Wel & Loeve, 1972) and, subsequently, a thaumatin-like (TL) protein was found at very low levels in persimmon (*Diospyros texana*; Vu & Huynh, 1994). A common function associated with a number of the TL proteins has been an antifungal activity, possibly mediated by membrane permeabilization (Roberts & Selitrennikoff, 1990; Vigers et al., 1991). Thaumatin-like protein isolated from *Zea mays* by D. Malehorn et al. (1994) was nearly identical with maize zeamatin and α-amylase trypsin^{-1} inhibitor. Overexpression of this protein in insect cells and two transgenic plants species showed its antifungal activity. More recently, the presence of a TL protein has also been noted in ripe cherries. The cherry TL protein accumulated in fruit, so it could be expected to have the sweet taste associated with thaumatin (Fils-Lycaon et al., 1996). Similar protein VVTL1 (*V. vinifera* TL protein 1) is a TL member of the PR-5 protein family and is encoded by single gene, which is expressed in a berry- and ripening-specific manner (Tattersall et al., 1997). It is tempting to speculate that the accumulation of PR proteins in the softening and intensely sweet grape berry serves an antifungal role. The recent results indicate a common trigger for the onset of ripening and PR protein synthesis-a phenomena that also been suggested for the cherry, another example of a nonclimacteric fruit. The concentration of soluble solids at which grape berries are resistant or susceptible to new infections correlates well with the presence or absence, respectively, of VVTL1 gene expression, suggesting that VVTL1 may play a role in the phenomena observed by Delp (1954) and Chellemi & Marois (1992). They noted that grape berries become resistant to new infections of powdery mildew above 7 0Brix (refractive index measure of total soluble solids)-the stage of high expression of TL proteins.

The sweet taste of thaumatins may be the secondary property of these PR proteins, especially for the reason of its detectability only for Old World Monkeys and Apes, including humans (Glaser et al., 1978). Thaumatin solutions taste sweet at concentration as low as 10-8 M. Below this very low sweet taste threshold thaumatin have another taste property; it enhances many kinds of other flavours (Higginbotham et al., 1981). For example, peppermint taste threshold is reduced 10-fold whereas beef extract taste threshold is only halved in the presence of 5x10-5% w/v Talin.

For these reasons the TL proteins expression in plants have at least three applications: (i) Fungal resistance improvement of agricultural crops, especially nonclimacteric fruits and berries at the ripening stage; (ii) Production forms with more sweet fruits and berries; (iii) Modification of native taste and flavour of conventional varieties.

5. References

Boxus, P., Quorin, M. & Laine, J.M. (1975) Large scale propagation of strawberry plants from tissue culture, In: *Applied and fundamental aspects of plant cell, tissue and organ culture.* Reinert, J., Bajaj, Y.P.S., eds.Springer-Verlag, Berlin. pp.130-143.

Chellemi, D.O. & Marois, J.J. (1992) Influence of leaf removal, fungicide applications, and fruit maturity on incidence and severity of grape powdery mildew, *Am. J. Enol. Vitic* **43**: 53-57.

Delp, C.P. (1954) Effect of temperature and humidity on the grape powdery mildew fungus, *Phytopathology* **44**: 615-626.

Dolgov, S.V. & Muratova, S.A. (1995) Development of methods for genetic transformation of fruit cultures, in *Abstracts of International symposium on Engineering Plants for Commercial Products and Applications, Lexington, Kentucky, October 1-4*,1995, p.15.

Edens, L., Heslinga, L., Ledeboer,A.M., Maat, J., Toonen, M.Y., Visser, Ch. & Verrips, C.T. (1982) Cloning of cDNA encoding the sweet-testing plant protein thaumatin and its expression in *Escherichia coli, Gene* **18**: 1-12.

Fils-Lycaon, B.R., Wiersma, P.A., Eastwell, K.C. & Sautiere, P. (1996) A cherry protein and its gene, abundantly exspressed in ripening fruit, have been identified as thaumathin-like, *Plant Physiology* **111**: 269-273.

Gelvin, S.B.& Schilperoort, R.A. (1995) *Plant molecular biology manual*, Kluwer Academic Publishers, Dordrecht, Boston, London.

Glaser, D., Hellekant, G., Brouwer, J.N. & van der Wel, N. (1978) The taste responses in primates to the proteins thaumatin and monellin and their phylogenetic implications, *Folia primatologica* **19**: 56-63.

Higginbotham, J.D., Lindley, M. & Stephens, P. (1981) Flavour potentiating properties of Talin sweetener (thaumatin) The quality of foods and beverages, Vol.1. Charalambous, G., Inglett G. *ed.* Academic Press., pp. 91-112.

Malehorn, D.E., Borgmeyer, J.R., Smith, K.E. & Shah, D.M. (1994) Characterization and exspression of an antifungal zeamatin-like protein (*Zlp*) gene from *Zea mays, Plant Phisiology* **106**: 1471-1481.

Murashige, T. & Skoog, F. (1962) A revised medium for rapid growth and bioassays with tobacco tissue cultures. *Physiology Plant.* **15**:473-497.

Quorin, M. & Lepoivre, P. (1977) Etude de milieux adaptes aux cultures in vitro de prunus. *Acta Horticulturae* **78**:437-442.

Revenkova, E.V., Kraev, A.S. & Skryabin, K.G. (1993) Construction of a disarmed derivative of the supervirulent Ti plasmid pTiBo542, in *Biotechnology and molecular biology.* Moscow. pp. 67-76.

Roberts, W.K. & Selitrennikoff, C.P. (1990) Zeamatin, an antifungal protein from maiz with membrane-permeabilizing activity, *J.Gen. Microbiology* **136**: 1771-1778.

Szwacka, M., Palucha, A. & Malepszy, S. (1996) *Agrobacterium tumefaciens* - mediated cucumber transformation with thaumatin II cDNA. *Genet. Pol.* **37A**: 58.

Tattersall, D.B., van Heeswijck, R. & Hoj, P.B. (1997) Identification and characterization of a fruit-specific, thaumatin-like protein that accumulates at very high levels in conjunction with the onset of sugar accumulation and berry softening in grapes, *Plant Physiology* **114**:759-769.

van der Wel, H. & Loeve, K. (1972) Isolation and characterization of thaumatin I and II, the sweet-tasting protein from *Thaumatococcus daniellii* Benth. *Eur J. Biochemistry* **31**: 221-225.

Vigers, A.J., Roberts, W.K. & Selitrennikoff, C.P. (1991) A new family of plant antifungal proteins. *Mol. Plant-Microbe Interaction* **4**: 315-323.

Vu L, Huynth, Q.K. (1994) Isolation and characterization of 27 kDa antifungal protein from the fruits of *Diospyros texana, Biochem. Biophys Res Commun.* **202**: 666-672.

Witty, M. & Harvey, W.J. (1990) Sensory evaluation of transgenic *Solanum tuberosum* producing r-thaumatin II, *New Zealand Journal of Crop and Horticulturae Science* **18**: 77-80.

G.T. Scarascia Mugnozza, E. Porceddu & M.A. Pagnotta (Eds.)
Genetics and Breeding for Crop Quality and Resistance, 173-181, 1999

Variability of the inhibitors of serine, cysteine proteinases and insect α-amylases in *Vigna* and *Phaseolus*

Al.V. Konarev[1], N.Tomooka[2], M. Ishimoto[3] & A.Vaughan[2]
[1]*All-Russian Institute of Plant Protection, Podbelskogo 3, St. Petersburg, 189620, Russia;* konarev@riam.spb.su; [2]*National Institute of Agrobiological Resources, 2-1-2, Kannondai, Tsukuba, Ibaraki 305-8602, Japan;* [3]*National Institute of Agro-Environmental Sciences, 3-1-3, Kannondai, Tsukuba, Ibaraki 305, Japan*

Abstract: Inhibitor polymorphism was studied, in the seeds of 93 accessions of 24 species in the 7 subgenera of *Vigna* and 55 accessions of 14 *Phaseolus* species, by isoelectric focusing combined with the gelatine replicas method. The majority of species examined have species specific spectra for trypsin and chymotrypsin inhibitors (TIs and CIs). In addition, considerable intraspecific variability was detected for TIs and CIs in most species, for example, *V. angularis*, *V. radiata* and *V. hirtella*. However, intraspecific variability for TIs and CIs was very low in, for example, *V. minima* and *V. mungo*. Inter- population variation for TIs and CIs were detected in *V. angularis* var. *nipponensis* from Japan. In some accessions single seeds differed by TIs (e.g. in *V. adenantha*). Differences were found in spectra of TIs and CIs from seeds and leaves of the same accession of *Vigna*. Leaf TIs and CIs also differed between *Vigna* species. Inhibitor spectra of the microbial proteinase subtilisin (SIs) from seeds were less complex and less variable than TIs. Spectra of SIs were specific for species and groups of species within *Vigna* and *Phaseolus*. Cysteine proteinase (papain-like) inhibitors (CPIs) generally show variation at the species level in both *Vigna* and *Phaseolus*. Inhibitors of diploid *V. trinerva* were closer to those of the tetraploid *V. glabrescens* than other species. Active and polymorphic insect alpha-amylase inhibitors (AIs) were found only in *Phaseolus* species not *Vigna*. The research results indicate that for *Vigna* and *Phaseolus* TIs and CIs are useful for analysis of diversity from the individual to the species level, SIs and CPIs are more useful for analysis of diversity at the species level and AIs are only useful for studying variation in *Phaseolus*. Variation revealed in this study could be useful in relation to breeding legumes for pathogen and pest resistance.

1. Introduction

Proteinase and α-amylase inhibitors are factors involved with plant resistance to insects and microorganisms (Konarev, 1987; Ryan, 1990; Ishimoto & Chrispeels,

1996; Carbonero et al., 1998). Many inhibitors in legumes, cereals and other plants are well studied biochemically and genetically (Garcia-Olmedo et al., 1987; Ryan, 1990). Our research focuses on the study of natural inhibitor variability in plants. Spectra of inhibitors provide biochemical and genetic information. In wheat, levels of amylase and proteinase inhibitory activity in seeds are related to the spectra of inhibitors (Konarev, 1984, 1987). Combination of various inhibitor systems cause synergistic effects increasing plant resistance to pests (Oppert et al., 1993) or to pathogens (Cammue et al., 1994). Proteinase inhibitors theoretically can protect amylase inhibitors from destruction by gut proteinases and preserve them as anti-nutritional factors. Inhibitors can also be useful as genetic markers in the study of plant diversity and evolution (Konarev, 1982, 1987, 1988, 1996; Kollipara & Hymowitz, 1992) in combination with conventional methods (Klozova, 1979; Jaaska, 1990; Konarev et al., 1996), in particular, for elucidation the origin of resistant plant forms. Consequently, parallel analysis of variability of all main inhibitor systems in species and genera can be highly informative.

The objectives of the present work were to study proteinase and α-amylase inhibitor polymorphism in *Vigna* and *Phaseolus* and estimate the variation in inhibitors in both genera. Although inhibitors in these genera are well studied, inhibitor variation in diverse germplasm, especially, in *Vigna* species, is still unclear. In addition, taxonomy of both genera is imperfectly understood (Marechal et al., 1978; Lumpkin & McClary, 1994; Tomooka et al., 1996). The methods used here are based on a set of methods for inhibitor identification which have been efficient in studies of inhibitors in wheat and related cereals (Konarev, 1982, 1986, 1996). Analysis of digestive enzymes of several *Coleoptera* and *Hemiptera* insect species in relation to inhibitors from cereals revealed α-amylases, trypsin-like, chymotrypsin-like and cystein proteinases (Konarev, 1984, 1986, 1996; Konarev & Fomicheva, 1991). Results from wheat studies confirm the validity of conducting most of our research on inhibitor variability using standard enzymes, such as bovine trypsin. In case of legumes we worked with insect amylases and did not work with insect proteinases, but is known that many legume pests have cysteine and serine digestive proteinases (Kitch, 1986; Ishimoto & Chrispeels, 1996).

2. Materials and Methods

The study was conducted in 93 accessions of 24 species of *Vigna* and 55 accessions of 14 *Phaseolus* species collected by authors or obtained from MAFF, Japan and USDA, USA gene banks. Species names and taxonomy were used according to Marechal et al. (1978) and Tomooka et al. (1996). Seed and leaf proteins were extracted with 20% glycerin (1:10 and 1:2 correspondingly) and were separated by isoelectric focusing in Servalyt precotes gels (Serva) pH 3-10. Proteinase inhibitors were detected by gelatin replicas method (Konarev, 1986; Konarev et al., in prep.). α-Amylase inhibitors were detected by starch/amylase/PAG replicas (Konarev, 1985; Konarev & Fomicheva, 1991).

3. Variability in hydrolase inhibitors.

3.1 Proteinase inhibitors in seeds

Trypsin-like enzymes are common in guts of many insect species and in extracellular liquids of some fungi. Inhibitors of trypsin-like enzymes are the most active and complex inhibitor system in legumes studied and they are of Bowman-Birk type (Hilder et al., 1989). Variability of seed trypsin inhibitors (TI) was studied in accessions of six subgenera of *Vigna* and in the genus *Phaseolus* (The patterns identified by number are arranged according to taxonomy in Figure 1) (Marechal et al., 1978). Not all, but majority of TI components in *Vigna* and in *Phaseolus* species were active to chymotrypsin as well (not shown). This is characteristic of Bowman-Birk type inhibitors.

Vigna subgenus *Vigna* species (Figures. 1 patterns: 2 & 3, *V. luteola*; 4, *V. marina*; 5 & 6, *V oblongifolia*; 7, *V. membranacea* and, 9-11, *V. unguiculata*) had, in general, species specific TI spectra. *V. oblongifolia* and *V. unguiculata* were polymorphic by inhibitor spectra. Some bands were common for two or three species (for example, bands with high pI in patterns 2, 6 and 10). For most *V. vexillata* accessions (12-14, subgenus Plectotropis) TI bands possessed pI below 5.3. Some results, perhaps reflected species misidentification in analysed accessions. For example, sample 13 called *V. vexillata* had TI spectra similar to *V. oblongifolia* (6) and has been now been reidentified as this species.

Main types of spectra found in *Vigna* subgenus *Ceratotropis* are shown (Figure 1). We found that the majority of *Ceratotropis* species examined had species specific spectra for TI and chymotrypsin inhibitors (CI). In addition, considerable intraspecific variability was detected for trypsin inhibitors in most species, for example, *V. angularis* (patterns 16-19), *V. hirtella* complex (24-27) and *V. radiata* (33-35). TI and CI in seeds of some *V. hirtella* accessions (g & h) were not detected (27). However, intraspecific variability for inhibitors was very low in, for example, *V. minima* and *V. mungo*.

Most species in the subgenus *Ceratotropis* are diploid except for two closely related tetraploids *V. reflexo-pilosa* subsp. *glabra* and *V. reflexo-pilosa* subsp. *reflexo-pilosa*, which are considered to be amphidiploids (Egawa et al., 1996). Both tetraploid species are resistant to bruchid beetles (Tomooka et al., 1992) and diseases (Fernandez & Shanmugasundaram, 1988) thus the origin of their genomes, which has not been clearly established, is of particular interest to plant breeders. TI of diploid *V. trinervia* (30) were closer to those of the tetraploids *V. reflexo-pilosa* subsp. *reflexo-pilosa* (31) and subsp. *glabra* (32) than other species. This result corresponds to data on izoenzymes (Egawa et al., 1996). Several TI and CI bands in tetraploid species were similar to inhibitors of accessions "a" and "b" with unclear taxonomy included to *V. hirtella* complex. It is possible that forms related to these accessions and *V. trinervia* were donors of the genomes in the allotetraploid species (Konarev, in prep.).

TI spectrum of *V. lasiocarpa* (45) differed significantly from that of *V. longifolia* (46) (subgenus *Lasiocarpa*). *V. caracalla* accessions 47 and 48 (subgenus

Sigmoidotropis) had only two similar TI bands. Three single seeds of the accession "a" of *V. adenantha* (49-51) differed strongly by TI, morphology and color. Two TI components of seed 49 were similar to those of *V. caracalla* (48). In seeds 50 and 52 (b) TI bands were very faint. High intra-accession variability for TI may be a result of outcrossing. We did not find any relationship between the TI of *V. schimperi* (53) (subgenus Haydonia) and TI of other *Vigna* representatives.

Most *Phaseolus* species possessed distinctive TI spectra, though some of them showed intraspecific variability. Some *P. vulgaris* lines (56-59) varied by TI. *P. acutifolius* accessions (60, 61, 64 and 65) differed by TI spectra as well. Some TI bands had similar pI in two or more species (for example, *P. acutifolius*, 65, and *P. angustissimus*, 66; *P. coccineus*, 67, *P. glabellus*, 69, an *P. hybrid*, 71). In the genus *Phaseolus* the same TI variants occur in several species which enables relationships between inhibitor systems within *Phaseolus* to be analysed. As to relationship between TI systems of *Vigna* and *Phaseolus*, there are some similarities in structure of TI spectra of representatives of *Vigna* subgenus *Lasiocarpa* (*V. caracalla*, acc. 47), and *P. vulgaris* (57, 79). Both *Vigna* subgenus *Lasiocarpa* and *Phaseolus* are of America (Marechal et al., 1978) which may explain their seemingly close evolutionary relationship based on TI spectra.

Subtilisin inhibitors (SI) are possible plant protective factors to fungi, because subtilisin-like extracellular proteinases are common for many plant pathogens. SI spectra are less complex and less variable than those of TI (not shown). Spectra of SI were specific for species and groups of species within subgenus *Ceratotropis* (Konarev, in prep), other subgenera of *Vigna* and for *Phaseolus* species. Some species had polymorphic SI.

Cysteine proteinase inhibitors (CPI) are poorly studied in *Vigna* compared to other inhibitors. Cysteine proteinases are very important in protein digestion of bruchid beetles which are a major pests of *Vigna* and *Phaseolus*. In this work we used the plant cysteine proteinase papain for detection of inhibitors. Papain inhibitors generally showed variation at the species level in both *Vigna* and *Phaseolus*.

Though all *Ceratotropis* species had species specific spectra of TI and other inhibitors, the majority of inhibitor components were common for two or more species. Combined data inhibitor bands in all accessions were scored (present or absent) in order to estimate evolutionary relationships between species (74 band positions were taken into account) (Konarev, in prep). It was much more difficult to find common TI components between species of *Ceratotropis* and other *Vigna* subgenera. Only in *V. vexillata* (s/g *Plectotropis*) some TI bands were similar to inhibitors of *V. hirtella* accessions "a" and "b" and both tetraploid species. The results of evaluation of evolutionary relationships between *Ceratotropis* species are shown (Figure 2) (data on all analysed inhibitor systems: TI, CI, SI and CPI for all accessions of each species, or groups of accessions with characteristic spectra were combined).

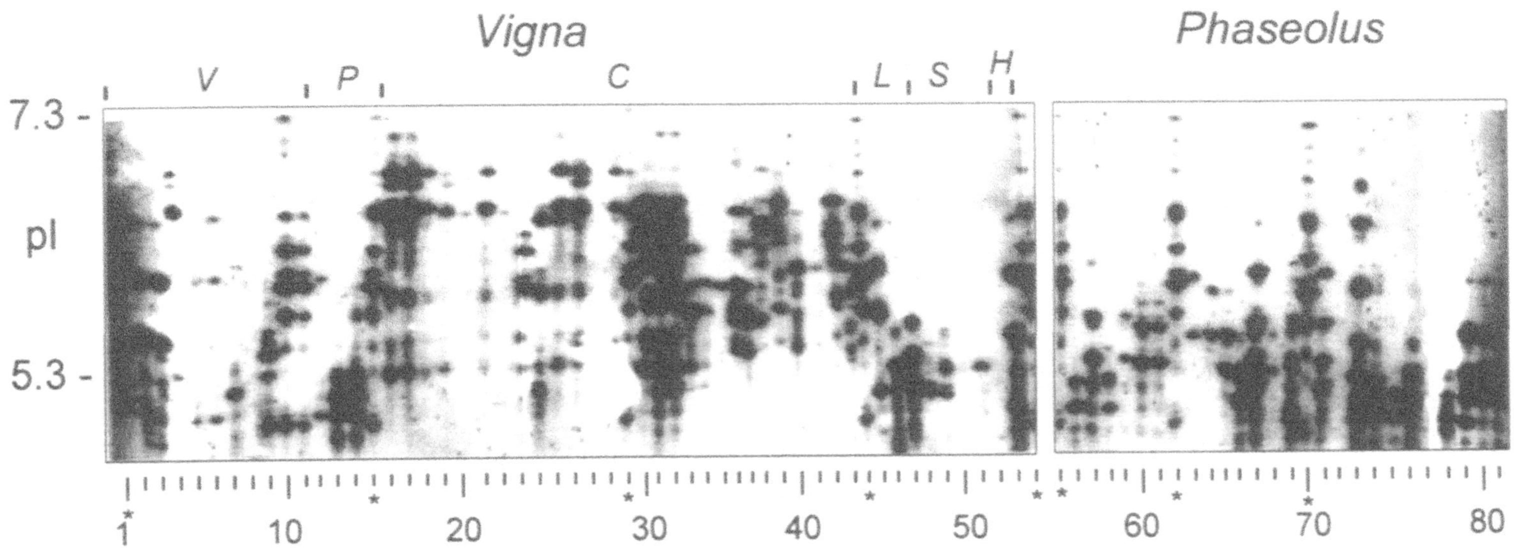

Figure 1. Variability of trypsin inhibitors in *Vigna* and *Phaseolus* species. Seed proteins extracted with water were separated in Servalyt precotes pH 3- 10 gel and trypsin inhibitors were detected by gelatin replicas method. V-M, subgenera *Vigna* a,b,. designation of accessions listed in Table 1.

V, *s/g Vigna:*
2, 3& 63, V. luteola (a&b)
4, V. marina
5&6, V. oblongifolia (a&b)
7, V. membranacea;, 8, -
9-11, V. unguiculata (a-c)
P, *s/g Plectotropis:*
12-14, V. vexillata (a, b&i)
C, *s/g Ceratotropis:*
16-19, V. angularis (a-d)
20-22, V. umbellata (a-c)
23, V. minima (c)
24-27, V. hirtella (b, c, e & h)
28, V. nepalensis (b)
30, V. trinervia
31, V. reflexo-pilosa (a)
32, V. glabrescens (a)
33-35, V. radiata (a, c&h)
36-37, V. mungo (a&d)
38-39, V. grandiflora (a&b)
40-41, V. aconitifolia (a&b)
42&43, V. stipulacea (a&b) L, *s/g Lasiocarpa:*
45, V. lasiocarpa
46, V. longifolia
S, *s/g Sigmoidotropis*
47&48, V. caracalla (a&b)
49-51 &52, V. adenantha (a&b) H, *s/g Haydonia:*
53, V. schimperi
Phaseolus:
56-59 &79-81, P. vulgaris (f-h, q, a&c)
60-61& 64-65, P. acutifolius (f, g, a&d)
66, P. angustissimus (a)
67, P. coccineus (a)
68, P. filiformis (a)
69, P. glabellus (b)
71, P. hybrid (b)
72, P. leptostachyus (a)
73, P. maculatus
74, P. microcarpus (a)
75, P. parvulus
76, P. polystachios (a)
77, P. ritensis (a)
78, P. stenolobus (a)
"*", marker mixture of samples 10 and 21

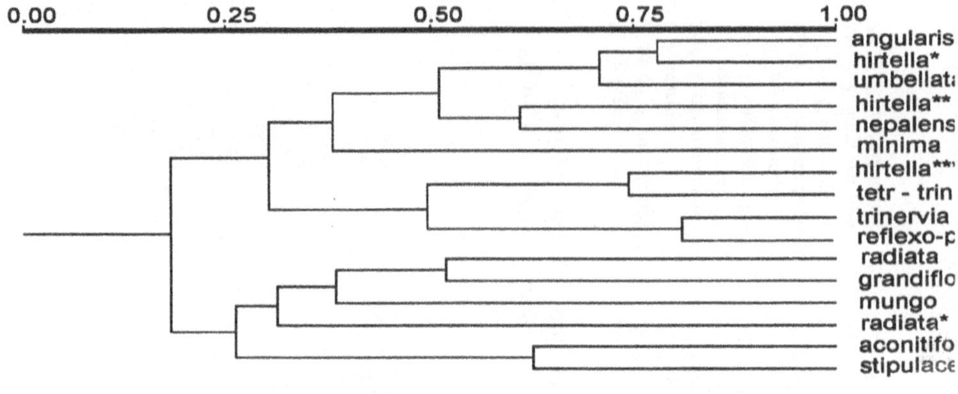

Figure 2. Relationships between *Vigna* subgenus *Ceratotropis* species on the basis of analysis of polymorphism of four inhibitor systems (TI + CI + SI +CPI). Results were estimated by UPGMA method (NTSYS 1.80). Data on all accessions of each species were combined except for clearly distinctive forms. hirtella*, acc. c-f and i-j combined; hirtella**, acc. g and h combined; hirtella***, acc. a and b combined. Tetra-trin*, hypothetical *Vigna* form constructed by subtraction of inhibitor bands characteristic of *V. trinervia* only from inhibitor set of tetraploid species. radiata*, acc. h (35 in Figure 1).

The inhibitor systems of the tetraploid and diploid forms possibly related to their putative genome donors were similar (Figure 2). This scheme coincides, in general, to schemes of evolutionary relationships between *Ceratotropis* species based on morphological characteristics and data of RAPD and isoenzyme analysis (Egawa et al., 1996; Tomooka et al., 1996), however in the case of diploid–polyploid interrelations inhibitor analysis is more informative than RAPD analysis (Tomooka et al., 1996).

3.2 *Insect α-amylase inhibitors in seeds*

Seed proteins of *Phaseolus* and *Vigna* accessions were separated by isoelectric focusing and inhibitors of mexican bean weevil, *Zabrotes subfasciatus*, and azuki and bean weevil, *Callosobruchus chinensis*, α-amylases were detected on replicas (Figure 3). Active and polymorphic insect α-amylase inhibitors were found only in some *Phaseolus* species (*P. acutifolius*, *P. coccineus*, *P. hybrid*, and *P. vulgaris*) not *Vigna*, that corresponded with data on inhibitors of *Tenebrio molitor* amylase (Puejo & Delgado-Salinas, 1997). A weak inhibitor of *C. chinensis* α-amylase was found in one accession of *P. ritensis* (Puejo & Delgado-Salinas, 1997). We detected inhibitors by their specific activity, while previous reports have identified inhibitors in gels only as low molecular glycoproteins or by immunochemical methods (Ishimoto et al., 1995; Puejo & Delgado-Salinas, 1997). Sometimes the presence of glycoproteins does not correspond to a protein with inhibitory activity. The main forms of α-amylase inhibitors characteristic of *P. vulgaris* (Ishimoto et al., 1995) (34, αAI-1; 35, αAI-2; 36, αAI-3), including some bands of specific *Z. subfasciatus* amylase inhibitors (arrowed) which are absent among azuki bean weevil inhibitors, are present in analysed accessions.

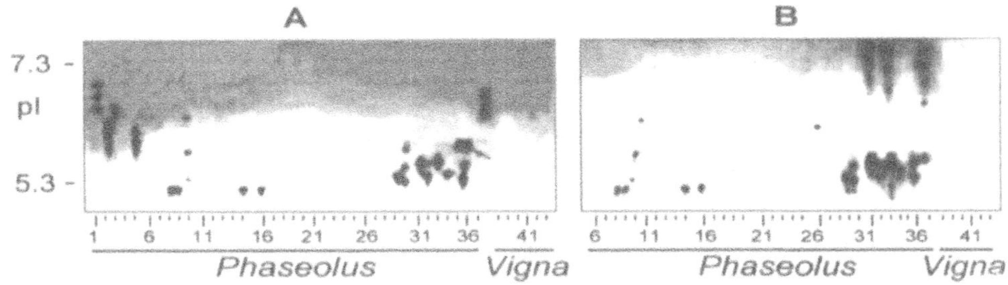

Figure 3. Insect α-amylase inhibitors in *Vigna* and *Phaseolus* species. Seed proteins extracted with water were separated by isoelectric focusing in Servalyt precotes pH 3-10. *Z. subfasciatus* and *C. chinensis* α-amylase inhibitors were detected in replicas A and B correspondingly.

1-5&37, P. acutifolius	*15&16 P. hybrid*	*23&24 P. polystachios*	*39 V. lasiocarpa*
6&7 P. angustissimus	*17-18 P. leptostachyus*	*25&26 P. ritensis*	*40 V. caracalla*
8-10 P. coccineus	*19&20 P. maculatus*	*27&28 P. stenolobus*	*41 V. schimperi*
11-12 P. filiformis	*20&21 P. microcarpus*	*29-36 P. vulgaris*	*42 V. angularis*
13-14 P. glabellus	*22 P. parvulus*	*38 V. unguiculata*	*43 V. radiata*

3.3 Proteinase inhibitors in leaves

Proteinase inhibitors in leaves of various plants are less studied than inhibitors in seeds. Wheat seeds (endosperm) and leaves contain different inhibitor systems (Konarev, 1987). In *Vigna* and *Phaseolus* species leaf TI and CI did not correspond to those in the seed and had lower pI values. In both genera species differed by spectra of TI and CI. Obtaining two gelatin replicas from the same separating gel facilitated comparison of inhibitors of both proteinases. In *V. angularis* (1-3), three groups of accessions of *V. hirtella* complex (4-9), *V. mungo* (16 & 17) and *V. stipulacea* (18) most of TI bands coincided in replicas with CI (1-9). In *V. reflexo-pilosa* (10), *V. radiata* (11-14) and *V. oblongifolia* (22-23) only a few bands of both inhibitor types coincided. In one of *V. radiata* accessions (13-14) CI bands had higher pI than TI. In *V. luteola* (20&21), *V. unguiculata* (24-26), *V. lasiocarpa* (28-29), *V. vexillata* (30-32) and *V. schimperi* (33-34) TI were detected but not CI. In normal undamaged leaves SI were not found. However, *V. angularis* seedlings damaged by mites contained active SI band with pI near 4.5 (not shown) while TI and CI spectra did not change. Thus, SI may be induced by damaging *Vigna* leaves.

4. Conclusions

Proteinase and amylase inhibitor composition is generally specific for *Vigna* and *Phaseolus* species but in some species inhibitors are highly variable. In both genera seeds and leaves contain different proteinase inhibitor systems. Variation revealed and approaches elaborated in this study can be useful in relation to breeding legumes for pest and pathogen resistance.

5. References

Cammue, B.P.A, De Bolle, M.F.C., Schoofs, H.M.E., Terras, F.R.G., Thevissen, K., Osborn, R.W., Reest, S.B. & Broekaert, W.F. (1994) Gene-encoded antimicrobial peptides from plants. Ciba Foundation Symposium 186, "Antimicrobial peptides", Wiley, Chichester, 91-106.

Carbonero, P., Diaz, I., Vicente-Carbajosa, J., Gaddour, K., Lara, P. & Fernandez-Pacios, L. (1998) Cereal amylase/trypsin inhibitors and transgenic insect resistance. Abstr. "Genetics and breeding for crop quality and resistance" XV EUCARPIA General Congress, Viterbo, Italy, 37.

Egawa, Y., Chotechuen, S., Tomooka, N., Lairungreang, C., Nakeeraks, P., Thavarasook, C. & Kitbamroong, C. (1996) Collaborative research program on mungbean germplasm (subgenus *Ceratotropis* of the genus *Vigna*) between DOA, Thailand and JIRCAS. In: Srinives, P., C. Kitbamroong & S. Miyazaki (eds.) Mungbean germplasm:Collection, evaluation and utilization for breeding program, JIRCAS, 1-8.

Fernandez, G.C.J. & Shanmugasundaram, S. (1988) The AVRDC Mungbean Improvement Program: The Past, Present and Future. In "Mungbean", Proc. 2nd Int. Symp. AVRDC. Shanhua, Taiwan, 58-70.

Garcia-Olmedo, F., Salcedo, G., Sanchez-Monge, R., Gomez, L., Royo, J. & Carbonero, P. (1987) Plant proteinaceous inhibitors of proteinases and α-amylases, Oxf.Surv.Plant. Mol.Cell.Biol. 4, 275-334.

Hilder, V.A., Barker, R.F., Samour, R.A., Gatehouse, A.M., Gatehouse, J.A. & Boulter, D. (1989) Protein and c DNA sequences of Bowman-Birk protease inhibitors from cowpea (*Vigna unguiculata* Walp.), Plant Mol. Biol. 13, 701-710.

Ishimoto, M. & Chrispeels, M.J. (1996) Protective mechanism of the mexican bean weevil against high levels of α-amylase inhibitor in the common bean, Plant Physiol. 11, 393-401.

Ishimoto, M., Suzuki, K., Iwanaga, M., Kikuchi, F. & Kitamura, K.V. (1995) Variation of seed α-amylase inhibitors in the common bean, Theor. Appl. Genet. 90, 425-429.

Jaaska, V. (1990) Isoenzyme variation in Asian beans, Botanica Acta, 103, 281-390.

Kitch, L.W., Murdock,, L.L. (1986) Partial characterisation of a major gut thiol proteinase from larvae of *Callosobruchus maculatus* F. Insect Biochem Physiol. 3, 561-575.

Klozova, E. (1979) Protein pattern of some legume taxons' seeds and its chemotaxonomic significance, in K.Muntz (ed.) Proc. Symp. "Seed Proteins of Dicotyledon Plants", Academie Verlag Berlin, 177-188.

Kollipara, K.P. & Hymowitz, T. (1992) Characterization of trypsin and chymotrypsin inhibitors in the wild perennial Glycine species, J. Agric. Food. Chem. 40, 2356-2363.

Konarev, Al.V. & Fomicheva, Yu.V. (1991) Cross analysis of the interaction of α-amylase and proteinase components of insects with protein inhibitors from wheat endosperm, Biokhimiya 56, 628-638. (Translated in English by Plenum Publishing Corp.)

Konarev, Al.V., Tomooka, N. & Vaughan. D.A. (in preparation) Proteinase inhibitor polymorphism in species of the genus *Vigna savi* subgenus *Ceratotropis*.

Konarev, Al.V. (1982) Component composition and genetic control of insect α-amylase inhibitors from wheat and Aegilops grain, Doklady Vaskhil 6, 42-44 (Translated in English by Allerton Press)

Konarev, Al.V. (1984) Identification of inhibitors of thiol proteinases of insects and grain in wheat and other cereals. Doklady VASKhNIL, 10, 13-15. (Translated in English by Allerton Press)

Konarev, Al.V. (1985) Methods for analyzing the component composition of cereal α-amylase and proteinase inhibitors, Prikladnaya Biokhim. i Microbiol. 21, 92-100. (Translated in English by Plenum Publishing Corp.)

Konarev, Al.V. (1986) Analysis of protease inhibitors from wheat grain by gelatine replicas method, Biokhimiya 51, 195-201. (Translated in English by Plenum Publishing Corp.)

Konarev, Al.V. (1987) Variability of trypsin-like proteinase inhibitors in wheat and related cereals in connection with resistance to grain pests, Selkochozajstvennaya biologia 5, 17-24. (In Russian)

Konarev, Al.V. (1988) Polymorphism of chymotrypsin/subtilisin inhibitors and possibilities of its use in wheat variety identification. In: Biochemical identification of varieties (Materials III International symposium ISTA, Leningrad, USSR, 1987) Leningrad, N.I.Vavilov Institute of Plant Industry, 176-181.

Konarev, Al.V. (1996) Interaction of insect digestive enzymes with plant protein inhibitors and host parasite co-evolution, Euphytica 92, 89-94.

Konarev, V.G., Gavrilyuk, I.P., Gubareva, N.K., Peneva, T.I., Chmeleva, Z.V., Konarev, A.V., Akhmetov, R.R., Giljazetdinov, Sh.Ja., Sidorova, V.V., Anisimova, I.N., Eggi, E.E., Vvedenskaya, I.O., Khakimova, A.G. & Kudryakova, N.V. (1996) Molecular biological aspects of applied botany, genetics and plant breeding. Series "Theoretical basis of breeding", Vol.1, St.-Petersburg, VIR.

Lumpkin, T.A. & McClary, D.C. (1994) Azuki bean: Botany, production and uses, Cambridge University Press.

Marechal, R., Mascherpa, J.M. & Stainer, F. (1978) Etude taxonomique d'un groupe complexe d'speces des genres *Phaseolus* et *Vigna* (Papilionanaceae) sur la base de donnes morphologicues et polliniques, traitees par l'analyse informatique, Boissiera, 28.

Oppert, B., Morgan, T.D., Culbertson, G. & Kramer, K.S. (1993) Dietary mixtures of cysteine and serine proteinase inhibitors exhibit synergistic toxicity toward the red flour beetle Tribolium castaneum, Comp. Biochem. Physiol. 105C, 379-385.

Pueyo, J.J. & Delgado-Salinas, A. (1997) Presence of α-amylase inhibitor in some members of the subtribe Phaseolinae (Phaseoleae: Fabaceae). American Journal of Botan 84, 79-84.

Ryan, C.A. (1990) Protease inhibitors in plants: gens for improving defenses against insects and pathogens, Annual. Rev. Phytopathol. 28, 425-449.

Tomooka, N, Egawa, Y, Lairungreang, C.& Thavarasook, C. (1992) Collection of wild Ceratotropis species on the Nansei Archipelago, Japan and evaluation of bruchid resistance, JARQ 26, 222-230.

Tomooka, N., Lairungreang, C. & Egawa, Y. (1996) Taxonomic position of wild *Vigna* species collected in Thailand based on RAPD analysis. In: Srinives, P., C. Kitbamroong & S. Miyazaki (eds.) Mungbean germplasm: Collection, ealuation and utilization for breeding program. JIRCAS, 31-4.

SESSION 4

RESISTANCE TO NEMATODE

G.T. Scarascia Mugnozza, E. Porceddu & M.A. Pagnotta (Eds.)
Genetics and Breeding for Crop Quality and Resistance, 185-193, 1999

Molecular isolation of two cyst nematode resistance genes: the *Hs1^{pro-1}* gene of beet and the *GPA2* gene of potato

W.J. Stiekema[1], A.G. van der Vossen[1], J. Rouppe van der Voort[2] J. Bakker[2] & R.M. Klein Lankhorst[1]
[1]Department of Molecular Biology, Centre for Plant Breeding and Reproduction Research (CPRO-DLO), P.O. Box 16, 6700 AA Wageningen, The Netherlands [2]aboratory of Nematology, Wageningen Agricultural University, P.O. Box 8123, 6700 ES Wageningen, The Netherlands

Abstract: Nematodes cause an estimated loss of approx. $100 billion per year in agriculture. Economically most important are sedentary nematodes of the genera *Heterodera*, *Globodera* and *Meloidogyne* that become permanently fixed in host roots after inducing specific feeding structures. The beet cyst nematode *Heterodera schachtii* has a broad host range that includes many species from different plant families such as *Chenopodiaceae* and *Brassicaceae.* The potato cyst nematodes *Globodera rostochiensis* and *G. pallida* cause large losses in potato-growing areas world-wide. Root knot nematodes (*Meloidogyne* spp.) are much more polyphagous and infect thousands of plant species and cause severe yield losses in many crops throughout the world. Environmental concerns restrict the application of nematicides. The current measures of control include crop rotations or growing catch crops. However the most promising strategy is the breeding for nematode resistant crops. Genes conferring resistance towards beet cyst and potato cyst nematodes are present in related wild species such as *Beta procumbens* and *Solanum tuberosum* spp. *andigena*. Recently, in cooperation with two other research groups, we succeeded in cloning the *Hs1^{pro1}* gene. This gene confers resistance to the beet cyst nematode in hairy roots of sugar beet In cooperation with the Wageningen University we also work on the isolation of the *GPA2* gene present in *S. tuberosum* spp. *andigena.* This gene confers resistance to a distinct population of the potato cyst nematode species *G. pallida*. Results on the analysis of the functionality of the *Hs1^{pro1}* gene in sugar beet and other plant species will be discussed. In addition our progression in cloning of the *GPA2* gene will be reported.

1. Introduction

Many plants are susceptible to infection by pathogens such as bacteria, fungi and nematodes and develop various undesirable disease symptoms upon infection, which cause retarded growth, reduced yield and consequently economical loss to farmers. Plants respond to infection with a variety of defence strategies including production of phytoalexins, deposition of lignin-like material, accumulation of cell wall hydroxyproline-rich glycoproteins, expression of pathogenesis related proteins (PR-proteins) and an increase in the activity of several lytic enzymes. Some of these responses can be induced not only directly by infection but also in some cases by exposure to exogenous chemicals such as ethylene. The full capacity of the defence mechanisms of a plant may, however, be delayed in relation to the onset of infection, and thus, plants may be severely injured before its defense mechanism reaches its maximum capacity. Also, the defence mechanism of plants may not in itself be sufficiently strong to effectively combat the infectious organism. This is in particular true for cultivated plants which have often been cultivated with the aim of increasing the yield, decreasing the climate susceptibility, decreasing the nutrient demand etc. Therefore, a normal and necessary procedure in agriculture is to treat infected plants or plants susceptible to infection with agrochemicals either as a prophylactic treatment or shortly after infection. The use of agrochemicals is neither desirable from an ecological nor from an economic point of view. Therefore other procedures have been applied to combat the infectious organism such as crop rotation. However, these strategies are not able to fully overcome the problem. A very promising third approach to solve the problem is to enhance the defence mechanism of the host plant itself by breeding for disease resistance. Introducing new and/or improved resistance genes by genetic engineering is now possible for most crop plants. The advantageous effect of this strategy would be the immediate inhibition of a phytopathogenic attack, leading to a retarded epidemic establishment of the infecting organisms in genetically engineered crop plants and thus an overall reduction in the effect of the infection.

Amongst the phytopathogenic organisms which are most widely spread are the beet cyst nematode (*Heterodera schachtii)* and the potato cyst nematodes (PCN) *Globodera pallida* and *G. rostochiensis.* These nematodes cause considerable losses in beet and potato crop growing, up to 10% of the annual yield world-wide. Because egg-containing cysts formed by these nematodes are very persistent to chemical treatment and can survive for several years in the soil, the use of nematicides and crop rotation are only moderately effective. Breeding for nematode resistance in beet and potato circumvents these drawbacks in the control of cyst nematodes.

2. Plant disease resistance loci

The majority of plant resistance (*R*-) genes are located in chromosomal bins containing other disease or insect resistance factors (reviewed in Crute & Pink, 1996). These resistance genes are dominantly inherited, are often involved in resistance processes that are characterised by a hypersensitive response (HR) and are members of multigene families hypothesised to have evolved from common ancestral genes. Most *R*-loci are characterised by the presence of DNA sequences encoding putative gene products that contain (1) a nucleotide binding site (NBS) and (2) a leucine rich repeat structure (LRR). These structural motifs are known to occur in a large number of resistance gene products. Nearly 30 resistance genes from various species have now been cloned and with the exception of two (*Hm1* and *mlo*; Johal & Briggs, 1992; Büschges et al., 1997), these genes are thought to be components of signal transduction pathways (Baker et al., 1997). On the basis of the structural similarity between the motifs of these genes, it is hypothesised that resistance genes are evolutionarily related components of a recognition system (Staskawicz et al., 1995). However, outside these structural motifs, the nucleotide sequences of disease resistance genes are unrelated and several subclasses can be distinguished (Leister et al., 1996). Genes associated with resistance to nematodes in beet and potato are likely to constitute a separate subclass of *R*-genes. However, the basic architecture hereof has not yet been uncovered. The isolation, characterisation and functional analysis of two of these nematode *R*-genes will be described here.

3. Resistance loci in sugar beet

The beet cyst nematode *Heterodera schachtii* is a severe pest in sugar beet cultivation. Typical symptoms are wilted leaves and a whiskered appearance of the roots. Genes that confer resistance to nematodes are absent in cultivated *Beta* species. The only sources of resistance are the wild species of *B. procumbens* and the related species *B. webbiana* and *B. patellaris* (Yu et al., 1984). Plants carrying the nematode resistance gene $Hs^{pro\text{-}1}$ on chromosome 1 of *B. procumbens* display an incompatible reaction between host and parasite. While the roots of these plants are invaded by J2 juveniles, most of these juveniles die before reaching maturity because of degradation of the feeding structure initiated by the nematode. In those cases that females develop they display a transparant appearance due to the absence of eggs. Thus the nematode cannot complete their life cycle.

The $Hs^{pro\text{-}1}$ resistance gene has been transferred to sugar beet by species hybridization (Savitsky et al., 1978; Heijbroek et al., 1988). However the obtained traslocation lines exhibit reduced resistance transmission rates and a poor agronomic performance, making molecular cloning of the $Hs^{pro\text{-}1}$ gene an important target.

4. Isolation and analysis of the $Hs1^{pro\text{-}1}$ nematode resistance gene

Because the absence of chromosome pairing between sugar beet and wild beet chromosomes, the translocations are inherited as a whole. No recombination between sugar beet chromosomes and the wild beet translocations has been observed in meiotic studies or through molecular marker analysis (Jung et al., 1992; Salentijn et al., 1994). Therefore the strategy for cloning this resistance gene relied on mapping chromosomal breakage points rather than on recombination mapping. Using molecular probes mapping closely to the $Hs^{pro\text{-}1}$ locus YAC (Yeast Artificial Chromosome) clones spanning the locus could be identified. Using these clones Cai et al. (1997) were able to identify a cDNA clone from a cDNA library made against root mRNA which upon transformation to hairy roots of sugar beet was able to complement these roots for nematode resistance. The phenotype of the obtained resistance was fully comparable with the resistance shown in *B. procumbens*.

DNA sequence analysis of the cDNA clone showed that it coded for a polypeptide of 282 amino acid. No sequences with homology to *Hs1^{pro-1}* were found in protein or nucleotide databases, except for three expressed sequenced tags with unknown function present in the *Arabidopsis thaliana* database. The amino acid sequence of the polypeptide can be divided in a number of regions. The two most important ones are a putative signal peptide at the NH2-terminus that may be engaged in targeting of the protein to the cytoplasmic membrane and a leucine rich region that shows poorly conserved leucine rich repeats (LRR). LRRs are suggested to be involved in protein-protein interaction and have been found in other isolated resistance genes from plants. These characteristics indicate that the *Hs1^{pro-1}* gene may function as a specific receptor for a signal from the beet cyst nematode and may be involved in a gene-for-gene relation as part of a cascade of defence reactions. However, further experiments need to be done to prove such a role for the *Hs1^{pro-1}* gene product.

Field trials with transgenic sugar beet accomodating the *Hs1^{pro-1}* gene are now performed. In addition experiments are planned to shed more light on the potential of this gene to confer resistance to *Heterodera schachtii* in oilseed rape.

5. Resistance loci in potato

Clustering of *R*-loci in potato has been reported (Leister et al., 1996). One of the large *R*-loci clusters is on the short arm of potato chromosome *5*. This cluster comprises at least five *R*-loci. Locus *R1* is associated with resistance to *Phytophthora infestans* (Leonards-Schippers et al., 1992), locus *Nb* is associated with HR type resistance to potato virus X (de Jong et al., 1997) while *Rx2* associated with an extreme type of resistance to PVX. The loci *Gpa* and *Grp1* are associated with resistance to the potato cyst nematode (PCN; Kreike et al., 1994; Rouppe van der Voort et al., 1998). The recently identified PCN R-locus *Gpa5* is also located within the *Grp1* region (Rouppe van der Voort & Van der Vossen; unpublished data). Additionally, *Gpa6* has been mapped to a region on chromosome 9 where on the homeologous region in tomato, a locus (*Sw5)*, conferring resistance to tomato spotted wilt virus, resides (Rouppe van der Voort & Van der Vossen; unpublished data).

6. The *Gpa2* nematode resistance locus

The *Gpa2* locus in potato has been found to be associated with resistance to *G. pallida* populations D383 and D372 (Arntzen et al., 1994). The presence of a single locus in potato that acts specifically to this small cluster of populations indicates that a gene-for-gene relationship underlies this plant-pathogen interaction (Rouppe van der Voort et al., 1997a; Bakker et al., 1993). Earlier work showed that the *Gpa2* locus mapped on the short arm of chromosome 12 of potato (Rouppe van der Voort et al., 1997a). However, no precise location or sequence data considering this were known.

7. Isolation of the *Gpa2* resistance gene

A map-based cloning strategy was used to isolate the *Gpa2* resistance gene. This strategy comprised of the following steps:

7.1 Genetic fine mapping of the Gpa2 locus

The *Gpa2* locus was initially mapped on chromosome 12 using information on the genomic positions of 733 known AFLP markers (Rouppe van der Voort et al., 1997a, 1997b). By use of RFLP probes, *Gpa2* was mapped more precisely between markers GP34 and CT79 on the distal end of chromosome 12 (Rouppe van der Voort et al., 1997a). This 6 cM genetic interval was previously shown to harbour the potato virus X (PVX) resistance gene *Rx*1 (Figure 1; Bendahmane et al., 1997). To confirm the assumed linkage between *Gpa2* and *Rx*1 (Rouppe van der Voort et al., 1997), segregation of both genes was followed in two different mapping populations. A tetraploid (2n = 4x = 48) mapping population was derived from a selfing of potato cv. Cara (S1-Cara), initially constructed for fine mapping of *Rx*1 (Bendahmane et al., 1997) while a diploid (2n = 2x = 24) *Gpa2* mapping population was derived from a cross between the diploid potato clones SH83-92-488 and RH89-039-16 (F1SHxRH; Rouppe van der Voort et al., 1997a, 1997b). Potato genotypes Cara and SH have the wild accession *Solanum tuberosum* ssp. *andigena* CPC 1673 in common. The S1-Cara recombinants initially chosen to confirm this linkage delimited the *Gpa2* interval between markers IPM3 and IPM5 (Bendahmane et al., 1997).

Fine mapping of the *Gpa2* locus was subsequently carried out using cleaved amplified polymorphic sequence (CAPS; Konieczny & Ausubel, 1993) markers derived from the IPM3-IPM5 interval, all of which were initially developed for the cloning of *Rx*1 (Figure 1). S1-Cara genotypes were assayed for recombination events in the IPM3-IPM5 region. In addition, F1SHxRH genotypes were subjected to a GP34/IPM5 marker screening as marker IPM3 was not informative in population F1SHxRH. Plants with recombination events between these markers were subsequently tested for all markers available in the IPM3-IPM5 region as well as for *Gpa2* resistance. This analysis showed that *Gpa2* is located between markers IPM4c and 111R (Figure 1b). Among the S1-Cara genotypes and F1SHxRH genotypes tested, only one genotype, S1-761, was identified in which a recombination event had occurred between *Gpa2* and marker IPM4c (Figure 1b).

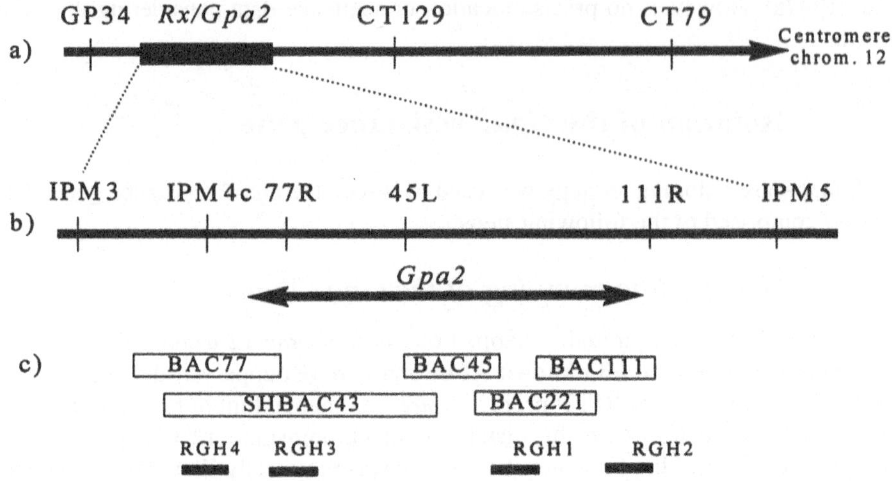

Figure 1. a). Relative position of the *Rx/Gpa2* locus on potato chromosome 12. b). Position of the *Gpa2* locus in between marker IPM4c and 111R. c). Physical map of the *Gpa2* locus. The open rectangles represent BAC clones. The closed rectangles represent the resistance gene homologues (RGH1-4) identified on the BAC clones.

7.2 Construction of a BAC contig spanning the Gpa2 locus

Four BAC clones, BAC77, BAC45, BAC221 and BAC111, which map to the 0.06 cM IPM4c-111R genetic interval harbouring the *Gpa2* locus, were isolated from a BAC library prepared from a progeny of the selfed potato cultivar Cara (Figure 1c). However these four BAC clones did not completely cover the *Gpa2* interval. Screening of the cv. Cara BAC library with CAPS markers 77R and 45L (Figure 1b) did not lead to BAC clones that spanned the region between markers 77R and 45L.

A second BAC library was constructed from the nematode resistant diploid potato genotype SH83-92-488 (SH83). Screening of the SH83 potato BAC library with CAPS markers 77R and 45L did result in the identification of a BAC clone (SHBAC43) closing the gap between these markers. In this way a contiguous physical map of the IPM4c-111R *Gpa2* interval was constructed comprising SHBAC43, BAC45, BAC221a and BAC111 (see Figure 1c). Restriction analysis of this BAC contig delimited the physical size of the *Gpa2* locus of approximately 200kb.

7.3 Identification of candidate resistance gene homologues (RGH)

As the size of the *Gpa2* locus was still too large for direct localisation of the *Gpa2* resistance gene by complementation analysis, BAC clones SHBAC43, BAC45, BAC221 and BAC111 were analysed for the presence of *R*-gene homologous sequences. Despite the general lack of DNA sequence conservation between *R*-genes, there are a few conserved amino acid sequence motifs in the NBS region present in many of these genes. Leister et al. (1996) have shown that it is possible to amplify resistance gene like sequences from potato using degenerate primers based on these homologous regions. We used the degenerate primers RG1 and RG2 (Aarts et al., 1998). The sequences of RG1 and RG2 are based on the conserved P-loop and domain 5 region of the NBS in the N, L6 and RPS2 R-genes (Whitham et al., 1994; Lawrence et al., 1995; Bent et al., 1994; Mindrinos et al., 1994). Amplification with these primers resulted in a DNA fragment of the expected size (approximately 530 bp) from BAC221. Southern analysis of *Eco*RI restricted DNA of SHBAC43, BAC45, BAC221 and BAC111 using the amplified PCR fragment from BAC221 as a probe, identified two copies of this R-gene like sequence on SHBAC43, one single copy on BAC221 and one copy on BAC111 (Figure 1c).

Subsequent sequence analysis of the complete inserts of these BAC clones showed that the previously identified *R*-gene like sequences on the BAC clones belonged to putative resistance gene homologues (RGHs; data not shown). Three of these RGH sequences were designated to be candidates for the *Gpa2* gene and selected for complementation analysis; RGH1 on BAC221a, RGH2 on BAC111 and RGH3 on SHBAC43. A fourth RGH identified on SHBAC43 contained marker IPM4c and therefore was located outside of the *Gpa2* interval (Figure 1c).

7.4 Complementation analysis

Genomic fragments of approximately 11 kb, 10.3 kb and 5.5 harbouring RGH1, RGH2 and RGH3, respectively, were subcloned from the BAC inserts into the plant transformation vector pBINPLUS (Van Engelen et al., 1995) and transferred to a nematode susceptible potato genotype using standard transformation methods. Roots of *in vitro* grown primary transformants were tested for PCN. This *in vitro* resistance assay revealed that the 10.3 kb genomic insert harbouring RGH2 was able to complement the susceptible phenotype (data not shown). RGH2 was therefore designated the *Gpa2* gene. Further experiments have to be performed to determine the spectrum of resistance of the RGH2 gene against different nematode populations.

8. Durability of PCN resistance

The durability of the resistance is determined by the extent of variation at (a)virulence loci which occur among the pathogen biotypes and the ability of the pathogen to generate novel specificity's at (a)virulence loci. For PCN, the variation at (a)virulence loci is for the majority determined by the original founders which have been introduced into Europe. PCN are endemic in the Andes region of South-America where they coevolved with their Solanaceous hosts. They are thought to have been introduced into Europe relatively recently, after 1850, together with collections of potato species which were imported for breeding purposes. Only a limited part of the genetic variation present in their centre of origin has been introduced into Europe (Folkertsma, 1977). From the moment of their introduction onwards, the genetic variation in virulence within and between European nematode populations has been determined predominantly by i) the genetic structure of the primary founders, ii) random genetic drift and iii) gene flow. Mutation and selection can be excluded as a driving force for the observed variation. Since PCN species produce only one generation in a growing season, their multiplication rate is low, the time between generations is 2 to 4 years in normal crop rotation and the active spread of the nematode is limited to several centimetres in the soil. It seems therefore highly unlikely that PCN populations have acquired other virulence characteristics than those already present at the moment of their introduction into Europe. Strategies to obtain broad-spectrum resistance against PCN are therefore based on combining a minimal number of genes with complementary or partially overlapping resistance spectra (Bakker et al., 1993). The results presented in this paper show that map-based cloning is the method of choice to isolate such a set of resistance genes able to confer durable nematode resistance to beet and potato.

9. References

Aarts, M.G.M., te Lintel Hekkert, B., Holub, E.B., Beynon, J.L., Stiekema, W.J. & Pereira, A. (1998) Identification of R-gene homologous DNA fragments genetically linked to disease loci in *Arabidopsis thaliana. Mol Plant-Microbe Interact.* **11**: 251-258.

Arntzen, F.K., Visser, J.H.M. & Hoogendoorn, J. (1994) Inheritance, level and origin of resistance to *Globodera pallida* in the potato cultivar 'Multa', derived from *S. tuberosum* ssp. *andigena* CPC1673. *Fundam. Appll. Nemat.* **16**: 155-162.

Baker, B., Zambryski, P., Staskawicz, B. & Dinesh-Kumar, S. P. (1997) Signaling in plant-microbe interactions. *Science* **276**: 726-733.

Bakker, J., Folkertsma, R.T., Rouppe van der Voort, J.N.A.M., de Boer, J.M. & Gommers, F. (1993) Changing concepts and molecular approaches in the management of virulence genes in potato cyst nematodes. *Annu. Rev. Phytopathol.* **31**: 169-190.

Bendahmane, A., Kanyuka, K. & Baulcombe, D. C. (1997) High-resolution genetical and physical mapping of the *Rx* gene for extreme resistance to potato virus X in tetraploid potato. *TAG* **95**: 153-162.

Bent, A. F. Kunkel, B. N., Dahlbeck, D., Brown, K. L., Schmidt, R., Giraudat, J., Leung, J. & Staskawicz, B. J. (1994) *RPS2* of *Arabidopsis thaliana*: A leucine-rich repeat class of plant disease resistance genes. *Science* **265**: 1856-1860.

Büschges, R., Hollricher, K., Panstruga, R., Simons, G., Wolter, M., Frijters, A., van Daelen, R., van der Lee, T., Groenendijk, J., Topsch, S., Vos, P., Salamini, F. & Schultze-Lefert, P. (1997) The barley *Mlo* gene: a novel control element of plant pathogen resistance. *Cell* **88**: 695-705.

Cai, D., Kleine, M., Kifle, S., Harloff, H-J., Sandal, N.N., Marcker K.A., Klein Lankhorst, R.M., Salentijn, E.M.J., Lange, W., Stiekema, W.J., Wyss, U., Grundler, F.M.W. & Jung, C. (1997) Positional cloning of a gene for nematode resistance in sugar beet. *Science 275:* 832-834.

Crute, I. R. & Pink, D. A. C. (1996) Genetics and utilization of pathogen resistance in plants. *Plant Cell* **8**:1747-1755.

De Jong, W., Forsyth, A., Leister, D., Gebhardt, C. & Baulcombe, D. C. (1997) A potato hypersensitive resistance gene against potato virus X maps to a resistance gene cluster on chromosome 5. *Theor. Appl. Genet.* **95**: 246-252.

Folkertsma , R.T. (1997) Genetic diversity of the potato cyst nematode in the Netherlands. PhD thesis, Wageningen Agricultural University.

Heijbroek , W., Roelands, A.J., De Jong, J.H., Van Hulst C.G., Schoone, A.H.L. & Munning, R.G. (1988) Sugar beets homozygous for resistance to the beet cyst nematode (*Heterodera schachtii*

Schm.) developed from monosomic additions of *B. procumbens* to *B. vulgaris*. Euphytica 38: 121-131.

Johal, G.S. & Briggs, S.P. (1992) Reductase activity encoded by the Hm1 disease resistance gene in maize. *Science* **258**: 985-987.

Jung, C., Koch, R., Fischer, F., Brandes, A., Wricke, G. & Herrmann, R.G. (1992) DNA markers closely linked to nematode resistance genes in sugar beet (*Beta vulgaris* L.) mapped using cheromosome additions and translocations originating from wild beets of the *Procumbentes* section. Mol. Gen. Genet. *232*: 271-278.

Konieczny, A. & Ausubel, F.M. (1993) A procedure for mapping *Arabidopsis* mutations using co-dominant ecotype-specific PCR-based markers. *Plant J.* **4**: 403-410.

Kreike, C. M., De Koning, J. R. A., Vinke, J. H., Van Ooijen, J. W. & Stiekema, W. J. (1994) Quantitatively inherited resistance to *Globodera pallida* is dominated by one major locus in *Solanum spegazzinii*. *Theor. Appl. Genet.* **88**: 764-769.

Lawrence, G. J., Finnegan, E. J., Ayliffe, M. A. & Ellis, J. G. (1995) The *L6* gene for flax rust resistance is related to the *Arabidopsis* bacterial resistance gene *RPS2* and the tobacco viral resistance gene *N*. *Plant Cell* 7:1195-1206.

Leister, D., Ballvora, A., Salamini F. & Gebhardt, C. (1996) A PCR-based approach for isolating pathogen resistance genes from potato with potential for wide application in plants. *Nature Genetics* **14**: 421-429.

Leonards-Schippers, C., Gieffers, W., Salamini, F. & Gebhardt, C. (1992) The *R1* gene conferring race-specific resistance to *Phytophthora infestans* in potato is located on potato chromosome V. *MGG* **233**: 378-383.

Mindrinos, M., Katagiri, F., Yu, G. L. & Ausubel, F. M. (1994) The A. thaliana disease resistance gene *RPS2* encodes a protein a nucleotide-binding site and leucine -rich repeats. *Cell* **78**, 1089-1099.

Rouppe van der Voort, J., Lindeman, W., Folkertsma, R., Hutten, R., Overmars, H., van der Vossen, E., Jacobsen, E. & Bakker, J. (1998) A QTL for broad-spectrum resistance to cyst nematode species (*Globodera* spp.) maps to a resistance gene cluster in potato. *Theor. Appl. Genet.* **96**: 654-661.

Rouppe van der Voort, J., Wolters, P., Folkertsma, R., Hutten, R., van Zandvoort, P., Vinke, H., Kanyuka, K., Bendahmane, A., Jacobsen, E., Janssen, R. & Bakker, J. (1997a) Mapping of the cyst nematode resistance locus *Gpa2* in potato using a strategy based on comigrating AFLP markers. *TAG* **95**: 874-880.

Rouppe van der Voort J.N.A.M., Van Zandvoort P., Eck H.J. van, Folkertsma, F.T., Hutten, R.C.B., Draaistra J., Gommers F.J., Jacobsen E., Helder J. & Bakker J. (1997b) Allele specificity of comigrating AFLP markers used to align genetic maps from different potato genotypes. *MGG* **255**: 438-447.

Salentijn, E.M.J., Sandal, N.N., Klein Lankhorst, R.M., Lange, W., De Bock, Th.S.M., Marcker, K.A. & Stiekema, W.J. (1994) Longe-range organization of a satellite DNA family flanking the beet cyst nematode resistance locus (Hs1) on chromosome-1 of B. patellaris and B. procumbens. MGG *89*: 459-466.

Salentijn, E.M.J., Sandal, N.N., Klein Lankhorst, R.M., Lange, W., De Bock, Th.S.M., Marcker, K.A., Stiekema, W.J. (1994) Long-range organization of a satellite DNA family flanking the beet cyst nematode resistance locus (*Hs1*) on chromosome-1 of *B. patellaris* and *B. procumbens*. Mol. Gen. Gen. 89: 459-466.

Savitsky H. (1978) Nematode (*Heterodera schachtii*) resistance and meiosis in diploid plants from interspecific *Beta vulgaris* x *Beta patellaris* hybrids. Can. J. Genet. Cytol. 20: 247-250.

Staskawicz, B. J., Ausubel, F. M., Baker, B. J., Ellis, J. G. & Jones, J. D. G. (1995) Molecular genetics of plant-disease resistance. *Science* **268**:661-667.

Van Engelen, F.A., Molthoff, J.W., Conner, A.J, Nap, J-P., Pereira, A. & Stiekema, W.J. (1995) pBIBPLUS: an improved plant transformation vector based on pBIN19. *Transgenic Research* 4: 288-290.

Whitham, S. Dinesh-Kumar, S.P., Choi, D., Hehl, R., Corr, C. & Baker, B. (1994) The product of the tobacco mosaic virus resistance gene N - similarity to Toll and the interleukin-1 receptor. *Cell* **78**: 1101-1115.

Yu M.H. (1984) Resistance to *Heterodera schachtii* in *Patellares* section of the genus *Beta*. Euphytica 33: 633-640.

G.T. Scarascia Mugnozza, E. Porceddu & M.A. Pagnotta (Eds.)
Genetics and Breeding for Crop Quality and Resistance, 195-202, 1999

Development and molecular characterisation of nematode-resistant rapeseed (*Brassica napus* L.)

A. Voss, W.W. Lühs, R.J. Snowdon & W. Friedt
Institute of Crop Science and Plant Breeding I, Justus-Liebig-University, Ludwigstr. 23, D-35390 Giessen, Germany

Abstract: Initiated by the lack of oilseed rape cultivars (*Brassica napus*, genome AACC) resistant against *Heterodera schachtii* (beet cyst nematode, BCN) an introgression breeding programme has been started years ago. Oilradish genotypes (*Raphanus sativus*, genome RR) possessing a high degree of nematode resistance served as gene donors. As compared to all rapeseed genotypes being susceptible to cyst nematodes some highly resistant F_1 hybrids (ACR), derived from the intergeneric cross *B. napus* x *R. sativus*, were detected. Allohexaploid plants (AACCRR) were produced by colchicine chromosome doubling and used in a backcross programme with summer type rapeseed as pollen donor. Selection of resistant individuals was performed by nematode infection tests *in vivo*. A progeny consisting of 51 BC_2 plants derived from one highly resistant BC_1 individual was tested for the degree of nematode resistance. Three resistant individuals were detected which are used to generate a BC_3 progeny. In order to overcome the laborious testing procedure *in vivo* a molecular method for detecting plants containing a gene for nematode resistance would be very useful. Marker assisted selection would accelerate the process of breeding for a highly resistant substitution or translocation line. To find a molecular marker a segregating oilradish population is produced for a bulked segregant analysis. Based on putatively homologous DNA sequences of other resistance genes, primer pairs will be developed and tested to be applied as markers to accelerate selection in backcross generations without running nematode resistance tests in every generation.

1. Introduction

Nematodes cause huge plant damages every year. Besides a number of migrating species (*Pratylenchus* spp.) under temperate climates some sedentary nematodes (*Globodera* spp., *Ditylenchus* spp., *Heterodera* spp.) are of major importance. The beet cyst nematode *Heterodera schachtii* is one of them. The broad spectrum of host plants belonging to the families *Chenopodiaceae* and *Brassicaceae* and the long persistence of the cysts lead to a sustaining problem in agriculture. In a convenient environment the cysts containing up to 500 eggs and L_2 larvae can persist over 20 years and stay infectious in the soil. Most plants can tolerate an attack and propagation of BCN. In sugar beet (*Beta vulgaris*) crop rotations the BCN (beet cyst nematode) can be a big problem. The beet develops numerous of small roots after infection with the BCN leading to substantially lowered sugar yield. To fight the BCN different attempts were made. Due to the big environmental problems caused by broad spectrum nematicides such are forbidden. A wide crop rotation including non host plants is one possibility to lower the potential as well as controlling weedy hosts. Resistant catch crops inducing the sliding of nematodes but preventing the propagation of the BCN as cysts are another very useful method to control this pest.

2. Sources of nematode resistance

Different wild allies of the sugar beet (*Beta vulgaris*) like *Beta patellaris, B. webbiana* and *B. procumbens* carry genes for nematode resistance. Such were transferred into sugar beet *via* backcross breeding and resulted in substitution and translocation lines (Savitsky, 1978). These plants were used to breed the first nematode resistant sugar beet variety 'Nematop' (BSA, 1998). Cai et al. (1997) cloned $Hs1^{pro\text{-}1}$, a nematode resistance gene against the BCN derived from *Beta procumbens*. It is well known that resistance genes are often found cluster wise in plant genomes (Rouppe van der Voort et al., 1998). According to that there are hints that $Hs1^{pro\text{-}1}$ is not the only nematode resistance gene at this locus (Sandal et al., 1998).

Within the *Brassicaceae* a number of resistant accessions exist. Baukloh (1976) screened a collection of 947 different accessions of 67 different *Brassica* species. He found resistant plants of oilradish (*Raphanus sativus*), yellow mustard (*Sinapis alba*) and *Hesperis matronalis* (common name: damask), an ancient ornamental and medical plant without of any agricultural interest. The resistant plants were used to generate resistant progenies of oilradish which led to highly resistant varieties (BSA, 1998) being cultivated as catch crops today. These varieties suffer from a relatively low yield potential and an insufficient quality as oil and fodder crops. Resistant oilseed rape would have the potential to unite a high yield potential in oil and dry matter, as well as high quality for different purposes through suitable fatty acid composition and the loss of glucosinolates. To combine the resistance of oilradish with the excellent agricultural properties of oilseed rape might lead to a very useful catch and/or oil crop.

3. Introducing nematode resistance into oilseed rape

One major source to introduce nematode resistance into oilseed rape are resistant accessions of oilradish. Both species belong to the same family, a fact that implies a simple cross to transfer the resistance. Postzygotic incompatibility reactions hinder a development of full matured seeds (Thierfelder, 1994; Paulmann & Röbbelen, 1988). The embryo is aborted in a range of 15 to 30 days after pollination (DAP).

Embryo rescue technique (ERT) in *ovulum* according to Thierfelder (1994) has been applied by preparing *ovuli* of crosses between the resistant oilradish variety 'Fortissimo' (genotype RR) and susceptible summer oilseed rape 'Lisandra' (genotype AACC) after about 20 DAP. The resulting hybrid plants were regenerated *in vitro* and were tested for nematode resistance. The selected individuals were colchicinated and crossed back with oilseed rape applying ERT. The resulting BC_1 generation suffered from a lack of fertility similar to the BC_0 plants (Pan, 1997; Tokumasu & Kato, 1988). A variation from bright yellow to white flowers and small anthers without any pollen were all in common for these individuals. After the selection of the highly resistant BC_1 individual 2062/16 with yellow flowers and a good vitality the BC_2 generation consisting of 51 individuals was produced *via* ERT again. It was to be checked in a nematode resistance test to find resistant individuals to generate the next backcross generation.

4. Testing for nematode resistance

4.1 Propagation, inoculation and assessment

To obtain a sufficient number of high infectious L_2 larvae one must establish a system for propagation and exact inoculation of the BCN. Winter oilseed rape plants were cultivated in small pots and inoculated with about 5,000 BCN larvae. After two months the brown cysts could be harvested and put on a 'Baermann Trichter' (J. Müller BBA Münster/Germany, pers. comm.) for the separation of the L_2 larvae. A 0.4 mM $ZnCl_2$ solution induced the sliding of the larvae that were caught and stored in water for a maximum of two months.

Five cuttings per individual were taken and cultivated in 96 cm^2 PVC boxes filled with sterilized sand according to Toxopeus & Lubberts (1979). Plants were fertilized with a special solution containing micro and macro nutrients twice a week. After three weeks the well established plants were inoculated with 1,000 L_2 larvae each and allowed to grow under greenhouse conditions for seven weeks. After that the plants were washed from sand and cysts. The resulting mixture of sand and cysts was separated by floating in a 1.7 M $MgSO_4$ solution. Cysts swam on the surface of the solution and were collected and quantified.

The number of cysts is an indication of the level of nematode resistance of a plant. Looking at the cysts it becomes obvious that their size differs greatly (0.8 to 2 mm). According to this the number of eggs is varying, too. Due to this fact the total number of eggs and larvae inside the cysts is a better criterion for the level of resistance of a plant because it expresses the real potential of propagation.

A corresponding suitable criterion is the P_f/P_i value, which is the quotient between the final population density (counted eggs and larvae resulting from crushed cyst suspension) and the initial population density (1,000 inoculated larvae in our case). If the quotient is >1 a propagation of the nematodes has happened, if it is <1 the population density is lowered; if it is <0.2 the plants are considered resistant (BSA, 1998).

4.2 Material and Results of Nematode Resistance Test 1997/98

The analysed material consisted of 51 BC_2 individuals and some control genotypes like susceptible and resistant oilradish and yellow mustard varieties. The major results are shown in Figure 1. There is a broad variation from an average number of 270 cysts/cutting to 0 cysts/cutting. The oilseed rape varieties were highly susceptible, the resistant mother plant 2062/16 (BC_1) was assessed at 0 cysts per cutting as well as three BC_2 individuals. Susceptible oilradish and yellow mustard (*Sinapis alba*) allowed the multiplication of cysts in a range of 10 to 27, i.e. 10 times less than oilseed rape.

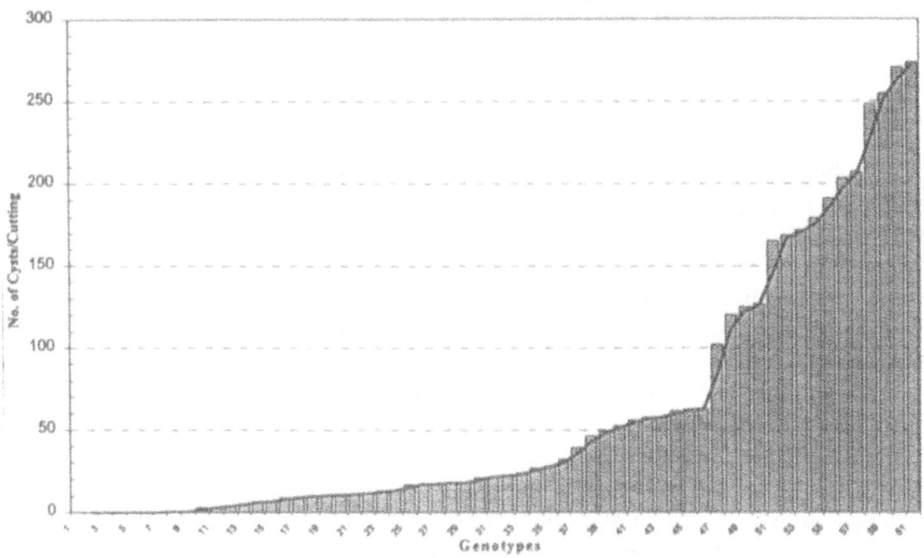

Figure 1. Results of nematode resistance test 1997/98, number of cysts/cutting.

Some individuals were excluded because of the loss of too many cuttings due to negative environmental impact during the test period. The three BC_2 individuals 2062/16/3, 2062/16/15 and 2062/16/20 showed no cysts in the root system at all; they were therefore assessed as highly resistant.

The P_f/P_i values of some selected individuals are shown in Figure 2. The ranking is similar to the "cysts per cutting" assessment. One can see that all levels of susceptibility are present within the tested material., Since they did not have cysts the three resistant BC_2 individuals had P_f/P_i values of 0, too. They will be used to produce the next backcross generation which is expected to be more fertile and subsequently more similar to oilseed rape.

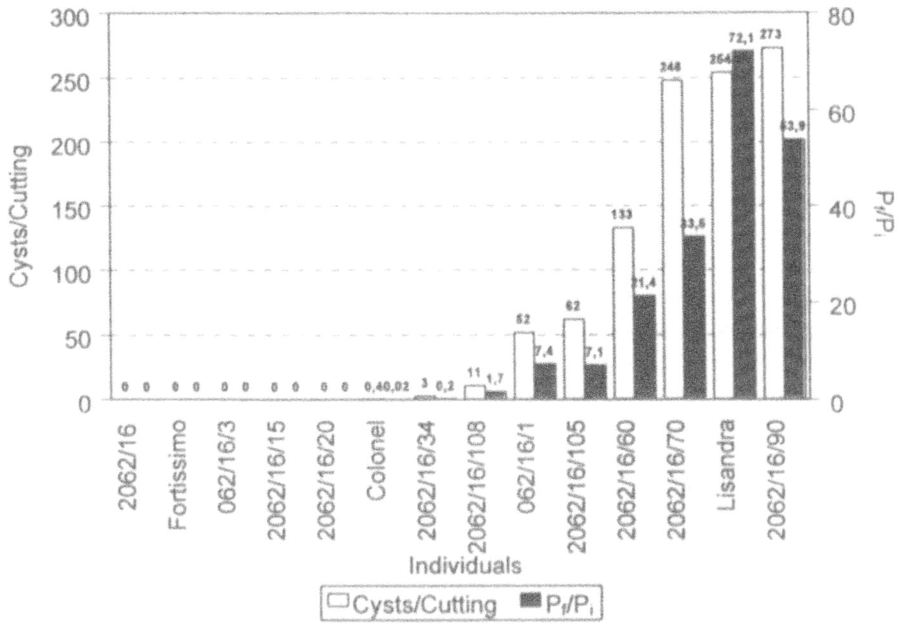

Figure 2. Mean values of cysts/cutting and P_f/P_i values of selected individuals (nematode resistance test 1997/98)

5. Cytological analysis of hybrid plants

To examine the number of remaining R chromosomes within the hybrid material of our backcross programme genetic *in situ* hybridization (GISH) was applied. One was able to distinguish between oilradish and oilseed rape related chromosomes (Snowdon et al., 1998). In the BC_1 individual 2062/16 still a complete haploid chromosome set of nine R chromosomes was existing resulting from a symmetric microspore meiosis of the BC_0 plant (AACCRR). In the BC_2 generation a deletion of four to six R chromosomes was observed in some cases without the loss of nematode resistance. Comparing the results of the cytological assay and the corresponding resistance assessments it becomes obvious that there is no correlation between the number of R chromosomes and the level of resistance. In the group of highly resistant genotypes 5-3 added R chromosomes without any influence on the nematode resistance level were found.

6. Molecular approaches

An early selection by application of molecular methods within a population of uncharacterized individuals can be very useful for a breeder in order to lower efforts and costs in a breeding programme. Especially in case of a trait being very difficult to assess because of pronounced environmental influence or a complicated breeding scheme as it is in the case of breeding for nematode resistance, molecular markers or molecular characterization in general are very helpful.

6.1 Candidate gene approach

Due to the rapidly growing number of DNA and protein sequences being analysed and sequenced it is obvious that this information must be used in breeding. In the way of a candidate gene approach (Gentzbittel et al., 1998) such is put into action. Genes with similar functions can carry some homologous DNA segments. These highly conserved regions can lead to a first hint for a molecular marker. Successful assays in this manner were carried out in oilseed rape for oil quality determining genes (Scheffler et al., 1997).

The nematode resistance gene $Hs1^{pro\text{-}1}$ was the first resistance gene against the BCN to be cloned as mentioned before (Cai et al., 1997). It was in order to get a link to the *Raphanus* related resistance gene(s) within the described backcross material., Southern analyses were carried out with the $Hs1^{pro\text{-}1}$ gene as probe. However, no signal was observed in the resistant bastard material. There was no evidence for a homology between the two examined sources of resistance. A further attempt will be the application of a number of PCR primer pairs (Botella et al., 1997) designed to cover regions of resistance genes of *Arabidopsis thaliana* being homologous with resistance genes of other species. Since *A. thaliana* belongs to the same family as oilseed rape one can suppose that there may be specific amplified fragments in common to be useful markers for the BCN resistance in future.

6.2 Bulked segregant analysis

The bulked segregant analysis (BSA; Michelmore et al., 1991) for detecting molecular markers is a method often applied (Ordon et al., 1995; Brahm, 1997). For BSA, a well characterized plant material with a homogeneous genetic background is required. Looking at the hybrid material described before one must admit that this material is not yet suited for this kind of assay because of a different number of added R chromosomes. Therefore a segregating oilradish population derived from a cross of the resistant oilradish variety 'Fortissimo' (the source of the resistance in our material) and the susceptible oilradish 'Rauola' is being developed. Afterwards molecular markers like AFLPs and RAPDs will be applied to the tested F_2 population. The identified markers will be transferred into the hybrid material for the source of the resistance is the same.

The oilseed rape like plants being highly resistant to BCN are currently backcrossed with summer oilseed rape cv. 'Lisandra' to select addition, substitution or translocation genotypes. Finally, we intent to identify BCN resistant oilseed rape lines with a high agronomic potential.

Acknowledgements: The authors like to thank Prof. J. Müller BBA Münster (Germany) for nematodes, Prof. C. Jung, Christian-Albrechts-University Kiel (Germany) for the *Hs1^{pro-1}*-probe and related plant material and the Gemeinschaft zur Förderung der privaten deutschen Pflanzenzüchtung (GFP), Bonn (Germany), for financial support.

7. References

Baukloh, H. (1976) Untersuchung zur Wirtspflanzeneignung der Kruziferen gegenüber dem Rübennematode *Heterodera schachtti* (Schmidt), unter besonderer Berücksichtigung der Resistenzzüchtung. Diss. Georg-August-Universität, Göttingen, 73p.

Botella,M.A., Coleman, M.J.,Hughes, D.E., Nishimura, M.T., Jonesm J.D.G. & Somerville S.C. (1997) Map positions of 47 *Arabidopsis* sequences with sequence similarity to disease resistance genes. The Plant Journal 12: 1197-1212.

Brahm, L. (1997) Identifizierung molekularer Marker für die Resistenz der Sonnenblume (*Helianthus annuus*) gegen den Falschen Mehltau (Plasmopara halstedii) als Basis für eine markergestützte Selektion. Diss., Justus-Liebig-Universität Gießen, 96p.

BSA (1998) Beschreibende Sortenliste 1998. Bundessortenamt, Hannover/Germany.

Cai, D., Kleine, M., Kifle, S., Harloff, H.-J., Sandal, N.N., Marcker, K.A., Klein-Lankhorst, R.M., Salentijn, E.M.J., Lange, W., Stiekema, W.J., Wyss, U., Grundler, F.M.W. & Jung, C. (1997) Positional cloning of a gene for nematode resistance in sugar beet. Science 275: 832-834.

Gentzbittel, L., Mouzeyar, S., Badaoui, S., Mestries, E., Vear, F., DeLabrouhe, D.T. & Nicolas, P. (1998) Cloning of molecular markers for disease resistance in sunflower, *Helianthus annuus* L. TAG 96: 519-525.

Michelmore, R.W, Paran, I., Kesseli, R.V. (1991). Identification of markers linked to disease-resistance genes by bulked segregant analysis: A rapid method to detect markers in specific genomic regions by using segregating populations. Proc. Natl. Acad. Sci. USA 88: 9828-9832.

Ordon, F., Bauer, E., Dehmer, K.J., Graner, A. & Friedt, W. (1995). Identification of a RAPD-marker linked to the BaMMV/BaYMV resistance gene ym4. Barley Genet. Newsl. 24: 123-126.

Pan, D. (1997) Cytologische und genetische Untersuchungen zur Entwicklung neuer Rapsadditionslinien mit Nematodenresistenz aus Kreuzungen zwischen *Brassica napus* L. und *Raphanus sativus* L. Diss., Justus-Liebig-Universität Gießen, 98p.

Paulmann, W. & Röbbelen, G. (1988). Effective transfer of cytoplasmatic male sterility from radish (*Raphanus sativus* L.) to rape (*Brassica napus* L.). Plant Breeding 100: 299-309.

Rouppe van der Voort, J., Lindemann, W., Folkertsma, R., Hutten, R., Overmars, H., van der Vossen, E., Jacobsen, E. & Bakker J. (1998) A QTL for broad-spectrum resistance to cyst nematode species (*Globodera* spp.) maps to a resistance gene cluster in potato. TAG 96: 654-661.

Sandal, N.N., Salentijn, E.M.J., Kleine, M., Cai, D., Arens-de Reuver, M., Van Druten, M., De Bock, T.S.M., Lange, W., Stehen, P., Jung, C., Marcker, K., Stiekema, W.J. & Klein-Lankhorst, R.M. (1998) Backcrossing of nematode-resistant sugar beet: a second nematode resistance gene at the locus containing Hs1^{pro-1}? Mol. Breed. 3: 471-480.

Savitsky, H. (1978) Nematode (Heterodera schachtii) resistance and meiosis in diploid plants from interspecific *Beta vulgaris* X B. *prucumbens* hybrids. Can. J. Genet. Cytol. 20: 177-186.

Scheffler, J.A., Sharpe, A.G., Schmidt, H., Sperling, P., Parkin, I.A.P., Lühs, W., Lydiate, D.J. & Heinz, E. (1997). Desaturase multigene families of *Brassica napus* arose through genome duplication. TAG 94: 583-591.

Snowdon, R., Köhler, W., Friedt, W., Lühs, W. & Köhler, A. (1998). Detecting R-genome chromatin in nematode-resistant *Brassica napus* x *Raphanus sativus* hybrids. T. LELLEY (Ed.), Current Topics in Plant Cytogenetics Related to Plant Improvement, 220-226. WUV-Universitätsverl., Wien.

Thierfelder, A. (1994). Genetische Untersuchungen für die züchterische Entwicklung neuer Rapssorten (*Brassica napus* L.) mit Resistenz gegen Nematoden (*Heterodera schachtii* Schm.). Diss., Justus-Liebig-Universität Gießen, 106p.

Tokumasu, S. & Kato, M. (1988) Chromosomal and genetic structure of *Brassicoraphanus* related to seed fertility and the presentation of an instance of improvement of its fertility. Euphytica 39: 145-151.

Toxopeus, H. & Lubberts J.H. (1979). Breeding for resistance to the sugar beet nematode (*Heterodera schachtii* Schm.) in cruciferous crops. Proc. Eucarpia Cruciferae Conf. Wageningen (Post Conf Edition), p. 151.

G.T. Scarascia Mugnozza, E. Porceddu & M.A. Pagnotta (Eds.)
Genetics and Breeding for Crop Quality and Resistance, 203-210, 1999

Integration of nematode-responsive regulatory sequences from *Arabidopsis thaliana* into nematode control strategies

N. Barthels[1], M. Karimi[1], I. Vercauteren[1], M. Van Montagu[1] & G. Gheysen[1,2]
[1] Laboratorium voor Genetica, Departement Genetica, Vlaams Interuniversitair Instituut voor Biotechnologie, Universiteit Gent, K.L. Ledeganckstraat 35, B-9000 Gent, Belgium; [2] Vakgroep Plantaardige Productie, Faculteit Landbouwkundige en Toegepaste Biologische Wetenschappen, Universiteit Gent, Coupure Links 653, B-9000 Gent, Belgium

Abstract: While for many decades fundamental research on sedentary nematode-plant parasitism encompassed mainly the ultrastructural and physiological aspects of the interaction, more recently also the area of molecular biology has started to be explored. Although the puzzle of the underlying gene regulation is far from complete, the available knowledge already enables us to evaluate the effectiveness of several envisaged genetic engineering strategies to obtain crops with nematode-resistant properties. Both antifeeding structure and antinematode approaches are considered and involve a gene complex in which a specific nematode-inducible promoter is responsible for the local production of a selected gene product aimed at inhibiting proper feeding cell and/or nematode development. Effective nematode control will greatly depend on the choice of both components in such a complex. To identify nematode-inducible plant-regulatory sequences, we have used the model system *Arabidopsis thaliana*. We report here on several transgenic lines tagged in promising regulatory sequences and discuss their potential in the scope of engineering nematode resistance into plants.

1. Introduction

The many reports on crop yield losses due to root knot and cyst nematode-plant parasitism prove its impact and have attracted the attention of many scientists working in different fields. Plant sedentary endoparasitic nematodes have developed a genius survival plan with the induction of so-called feeding cells in their host. Time lapse video-enhanced microscopy has recorded the different routes that root knot and cyst nematodes take on their infective way through the root to the site where they will start feeding and shortly after enter their sedentary stage of life (Wyss & Grundler, 1992). The ability of orientation within the root allows the root knot nematode to migrate in-between the cortex cells towards the root meristematic region after its penetration near the root elongation zone.

Once arrived, root knot nematodes turn to the opposite direction into the vascular cylinder. Although the penetration site of cyst nematodes is similar to that of the root knot nematodes, their migration path leads directly to the vascular bundle by intracellular movement through all consecutive cell layers, including the endodermis that apparently root knot nematodes cannot overcome during their intercellular movement. For both nematode species, the infection process proceeds with an exploration stage to find an appropriate initial feeding cell. High-resolution enhanced-contrast microscopy has enabled direct observation of the feeding nematode and the associated root cell alterations (Wyss, 1987; Wyss et al., 1992). In response to continuous stimulation by root knot and cyst nematodes, transfer cells with typical cell wall ingrowths are induced. Large lobate nuclei, increased metabolic activity, large amounts of endoplasmatic reticulum and mitochondria, dense cytoplasm, and a multinucleate state are common features of the feeding cells. Interestingly, structure and ontogeny of root knot and cyst nematode-induced feeding cells are different. Root knot nematodes impose repeated nuclear divisions without cytokinesis to the affected root cells that differentiate into giant cells and associated cortical cell hypertrophy and divisions cause the formation of a gall wherein the giant cells reside. The multinucleate condition of syncytia, induced by the cyst nematode, results from extensive cell wall breakdown among neighboring cells. Anatomical and cytological studies increased our knowledge on the nematode's feeding apparatus and the secretions that are held responsible for initiating the whole chain of events to establish the feeding structure. It is clear that the descriptive knowledge of this parasitic interaction is extensive. The molecular genetic foundation however remains less clear, but efforts by many researchers are more and more evolving towards a comprehension of the involved plant gene expression. Because feeding structures are indispensable to nematode survival and reproduction, one method to control this pest is to destroy the feeding cells or alter their activity by interfering with the underlying gene regulation.

2. Gene regulation in nematode feeding structures

There are several ways to study gene regulation in nematode feeding structures. The most straightforward method is using a reporter gene system to monitor the expression of cloned plant sequences with known function that are expected to play a role in the plant-nematode interaction. A widely used reporter gene is β-glucuronidase (*gus*) (Jefferson, 1987) that allows fast and easy detection of gene activity throughout the plant by the precipitation of a blue stain. Ample reports illustrate the value of this reporter gene approach. Based on the complexity of the plant-nematode interaction, one would expect the up- or down-regulation of certain plant genes or an alteration in their expression level, timing, and/or specificity. Production of structural and enzymatic proteins is expected to be enhanced to accommodate the increased cellular growth and metabolism during nematode infection. Niebel et al. (1993) found induction of the structural cell wall protein extensin in syncytia and galls upon infection of tobacco roots with *Globodera tabacum* and *Meloidogyne javanica*, respectively. Opperman et al. (1994) have proven the assumption that the difference in structure and ontogeny of syncytia and giant cells requires the onset of different gene sets. They have demonstrated high expression in giant cells induced by several *Meloidogyne* species of a tobacco gene that encodes a membrane protein functioning as a water channel. Expression in syncytia induced by *Globodera tabacum* however was not detected. Ehsanpour & Jones (1996) studied the *Parasponia andersonii* hemoglobin promoter that responds to oxygen tension that might be limited in the metabolically active giant cells; high GUS activity levels were observed in giant cells after *M. javanica* infection. Analogous work was reported on the wound-inducible gene *wun1* in potato (Hansen et al., 1996) and the *hmg2* gene coding for 3-hydroxy-3-methylglutaryl CoA reductase in tomato (Cramer, 1992). Because of its postulated involvement in cell wall thickening, we were interested in the expression pattern of the *rha1* gene that encodes a small *Arabidopsis thaliana* GTP-binding protein. *Rha1-gus* transgenic *A. thaliana* transformants were challenged with *H. schachtii* and *M. incognita*. Syncytia as well as galls showed clear GUS activity, although GUS staining was less homogenous and more transient in galls than in syncytia (Vercauteren et al., 1998). *A. thaliana* has also been chosen as the model system to study cell cycle regulation in nematode feeding sites (Gheysen et al., 1997). Because DNA synthesis occurs repeatedly in both types of feeding cells, the observed expression of *cdc2aAt* was not surprising (Niebel et al., 1996). *Cdc2aAt* encodes a catalytic subunit of a protein kinase that plays a key role in cell cycle control (Ferreira et al., 1991). *Cyc1At* encodes a mitotic cyclin and its expression is normally restricted to dividing cells (Hemerly et al., 1993). In accordance with cytological observations of repeated mitosis that is stimulated by root knot nematodes, induction of *cyc1At* in giant cells occurred rapidly. However, it was unexpected to have *cyc1At* expression during syncytium formation at very early stages of the infection process (Niebel et al., 1996), particularly because no clear mitotic figures have ever been noted in cyst nematode-induced feeding cells. The co-localization of ^{3}H-thymidine with this expression suggests that syncytial nuclei progress through the S phase and enter the G2 phase of the cell cycle, meaning that developing syncytia would progress through cycles of endoreduplication.

The analysis of genes with known function offers the opportunity to build hypotheses towards a fundamental understanding of feeding cell activity. In the quest for additional genes with a significant role in the interaction, the isolation of differentially expressed sequences was undertaken (Vercauteren et al., 1996). From displays of reverse transcribed RNA from *A. thaliana* roots infected by *M. incognita* and non-infected control roots, several RNAs, specifically present in feeding sites were identified. Many differential bands were sequenced and could be assigned a gene function by homology searches. A few potentially interesting genes among which a trypsin inhibitor and a peroxidase gene are now being studied in more detail. The differential display procedure was likewise repeated with *H. schachtii*-infected material. Several induced bands that are specific for either *M. incognita* or *H. schachtii* have been identified.

3. Nematode-inducible plant regulatory sequences

Management of sedentary endoparasitic nematodes by genetically engineered host resistance may call for the localized production of a specific gene product at the nematode feeding site. The promoter tagging strategy has proven its value in the identification of nematode-inducible plant-regulatory sequences that allow this localized expression (Barthels et al., 1997). The simple and fast detection of reporter gene expression that offers an immediate knowledge on the entire spatial pattern of promoter activity in this tagging approach surely gains an advantage over other molecular practices, despite the cumbersome preparative steps as engineering binary vectors and producing transgenic lines to be screened for the desired expression pattern.

We have made an *A. thaliana* C24 promoter tag collection by using the binary T-DNA vectors pGV1047 (Kertbundit et al., 1991) and pΔgusBin19 (Topping et al., 1991). Because of random integration into the plant genome, expression of the promoterless *gus* reporter gene is dependent of the immediate surrounding of the T-DNA insertion. A potentially interesting transgenic line has to satisfy some main criteria. T-DNA integration must have occurred so that transcription of the promoterless *gus* gene can happen, i.e. downstream from a plant-promoter sequence. Because transcriptional fusions are envisaged, in-frame insertions are not obligatory. Interestingly, due to the T-DNA integration, certain plant sequences might gain a regulatory activity without being linked with a native endogene in the wild-type context, referred to as cryptic promoters (Fobert et al., 1994). Furthermore, single-copy insertions are in favour to avoid eventual silencing events and to facilitate analysis later on. More focused to our study, the tagged promoter should be induced or at least significantly upregulated by nematode infection, with a minimum activity in plant parts besides the feeding sites. Based on their *gus* expression pattern, three *Arabidopsis* promoter tag lines, Att0001, Att0728, and Att1712 have been further analyzed. All three lines show GUS activity in galls and syncytia after inoculation of the roots with *M. incognita* and *H. schachtii*, respectively; lateral root bases show GUS as well. Minor additional staining in some other plant parts is also apparent in the Att0001 line. As mentioned above, GUS analysis allows very accurate localization of tagged promoter activity. This is exemplified by the lines Att0001 and Att1712. Although similar expression patterns were observed in whole-mount material, sections pointed to a different nature of these tags: whereas GUS is present inside the giant cells on Att0001 roots, in Att1712 galls, GUS is present in the parenchyma cells, surrounding the unstained giant cells. From Att0001, Att0728, and Att1712, left border (LB) T-DNA-flanking plant region fragments were amplified and isolated by inverse PCR. To confirm that the regulatory regions, responsible for the observed nematode-induced GUS activity, were picked up, the iPCR fragments were re-introduced into *Arabidopsis*. To have a first quick idea, a nematode-*Agrobacterium* co-inoculation system was developed (Karimi et al., 1997) that exploits the wounding by the nematode during parasitism and triggers T-DNA transfer from *A. tumefaciens* into the plant genome. This fast and mainly transient transformation system has indeed proven the promoter activity of Att0001 and Att0728, but remained unsuccessful for Att1712. However, stable *A. tumefaciens* transformation of *Arabidopsis* root explants as described by Clarke et al. (1992) with some modifications (Barthels et al., 1994; Karimi et al., 1994) resulted in a very weak GUS activity indicating that at least part of the Att1712 nematode-responsive promoter had been cloned. Efforts are now being done to obtain the original GUS activity level as in the primary Att1712 line.

The Att0001 promoter region was recently introduced into several crops (Karimi et al., 1998) to value its economical importance. Introduction of an Att0001-*gus* fusion into different plant species was performed by co-transformation of the engineered T-DNA and the *A. rhizogenes* Ri T-DNA. For the analysis of promoter activity, transformed hairy roots were incubated on callus-inducing medium (CIM) as the Att0001 promoter was shown to be active in callus tissue (Barthels et al., 1997). Analysis of the Ri-induced hairy roots on sugarbeet, mustard, and oilseed rape demonstrated clear GUS activity in the callus tissue of all three crops (Karimi et al., 1998). Expression of *gus* in hairy roots of oilseed rape on week after inoculation with *M. incognita* could also be shown.

Additionally to these tagged promoters, potentially useful promoters to direct nematode feeding cell-specific expression might also come from cDNA library screenings (Bird & Wilson, 1994; Van der Eycken et al., 1996) and differential display. Although the isolation of genes is the primary goal of these methods, corresponding promoters could be of equal value to the tagged promoters. However, information on promoter specificity is much harder to obtain and requires *in situ* hybridization or the production of *gus* fusions and transformation.

4. Integrating nematode-inducible promoters into control methods

The process from host perception to actual parasitism by the nematode involves several stages, each of which can be hampered by applying an appropriate counter-attack system. Depending on the stage one seeks to interfere with, promoters with diverse characteristics may be favoured. Promoter activities independent from nematode signals as root-specific promoters could function in the context of impeding nematode attraction to the host root system. Abolishment of the chemotaxic sensory perception by the expression of certain lectins (Marban-Mendoza et al., 1987) can interfere with the invasion process. In fact, attacking the nematode and not the feeding structure also allows the use of a constitutive promoter as the CaMV 35S promoter because antinematode proteins have no or little effect on plant cells. The production of collagenases by the host plant could destroy the integrity of the nematode's cuticle, proteinase inhibitors could interfere with the nematode's digestive system, and monoclonal antibodies against nematode saliva (Schots et al., 1992) could block migration (Smant et al., 1998) or feeding cell establishment. The constitutive expression of antinematode genes in plants however can raise biosafety or ethical concerns when plant material is destined for consumption and, in the long run, virulent pathotypes can develop, overcoming the antinematode effect. Conditional expression could be achieved by the use of wound-inducible promoters that are triggered by the necrosis that cyst nematode invasion and migration causes. The production of cytotoxic proteins however needs a strict confinement of promoter activity to the target cell. This can be exemplified by the work of Mariani et al. (1990): the specific expression of barnase in the tapetum cells of plant flowers caused the complete destruction of this cell layer, resulting in male-sterile flowers. Barnase is a very potent RNAse and any expression in non-target cells results in their destruction and should therefore be avoided.

Flexibility regarding promoter activity is tolerated to a certain extent when it comes to antisense approaches. Here, a gene is required that is naturally expressed and fulfils a key role in feeding cell establishment or functioning. The antisense RNA will bind to the complementary mRNA sequence, and prevent translation. Even when a complete protein elimination is not achieved, a weak decrease in efficiency of food supply might prevent the nematode from completing its life cycle. The most straightforward way is to combine a nematode-inducible promoter with the antisense sequence of its own coding region. The use of chimeric constructs however might be more favourable or even necessary when promoter activity occurs additionally in vital plant parts so that proper plant development and/or functioning is hindered or when a corresponding endogene is not existing (cfr. cryptic promoter). The advantage of chimeric constructs in antisense technology is that the nematode-responsive promoter does not need to be highly specific. It is however important that, besides the feeding cells, the specific RNAs, complementary to the antisense RNAs appear in plant parts different from those where the nematode-inducible promoter acts.

Opperman & Conkling (1996) were the first to report the substantial reduction of root galling on nematode-infected transgenic plants by using an antisense TobRB7 construct. This result indicates that interference with feeding site development by antisense is indeed a feasible technique to control plant-parasitic nematodes.

5. Improving promoter specificity

Certain nematode control methods call for highly specific effector gene expression. As it is very improbable that plant promoters will be tagged or isolated in any other way that act solely in nematode feeding structures, solutions have to be searched either to abolish or to circumvent promoter activity in other plant tissues. The existence of so-called nematode-responsive elements was reported by Opperman et al. (1994). Dissecting out this type of sequence can significantly re-direct the expression to the feeding structure only.

Alternatively, a dual construct could be used that results in the nematode-induced expression of a cytotoxin gene and the constitutive expression of a neutralizing gene (Sijmons et al., 1993). As opposed to the many genes that are up-regulated in feeding cells, Goddijn et al. (1993) described the down-regulation in syncytia and galls of strongly constitutive promoters from the nopaline synthetase gene, T-cyt genes, and of rooting loci of the *A. rhizogenes* T-DNA. During induction, the concentration of the cytotoxin will increase locally in the feeding cells whereas the level of the inhibitor or neutralizing gene will decrease in the same cells. Consequently, a threshold level will be reached in the feeding cell where the cytotoxin can no longer be neutralized, leading to the degeneration of the feeding cell. Plant parts outside of the feeding cell are rescued by the neutralizing gene whose expression is independent from nematode signals and thus stays up-regulated. This type of engineered resistance mimics the naturally hypersensitive response as apparent in an incompatible plant-nematode interaction. An example of a two-component system is the combination of barnase and its inhibitor barstar (Mariani et al., 1992).

6. Conclusions

The molecular information on the sedentary nematode-plant interaction is surely still too limited to allow the formulation of a comprehensive model. The availability of specific nematode-responsive regulatory plant sequences however means an important step forward, not only in the study of feeding cell functioning but also in the development of novel nematode control practices. First attempts towards nematode control are initiated and will allow us to compare the effectiveness of different chimeric genes introduced into different contexts. Reliable and durable nematode control is believed to depend on the combination of different chimeric constructs and strategies. It is clear that more engineered genes will become available in the future and also they will be evaluated for their applicability in many different plant species to finally evolve to an environmentally and socially acceptable crop protection against nematode attack.

Acknowledgements: This work is funded in part by the Ministerie van Middenstand en Landbouw (Nr. D ½ 5725A – S2) and by grants from the European Union Biotechnology Program (ERB-BIO4-CT96-0318 and FAIR3-CT96-1714). G.G. is a Postdoctoral Researcher of the Fund for Scientific Research (Flanders).

7. References

Barthels, N., Karimi, M., Van Montagu, M. & Gheysen, G. (1994) Isolation and analysis of nematode-induced genes in *Arabidopsis thaliana* through in vivo β-glucuronidase fusions, Med. Fac. Landbouww. Univ. Gent 59/2b 757-762.

Barthels, N., van der Lee, F.M., Klap, J., Goddijn, O.J.M., Karimi, M., Puzio, P., Grundler, F.M.W., Ohl, S.A., Lindsey, K., Robertson, L., Robertson, W.M., Van Montagu, M., Gheysen, G. & Sijmons, P.C. (1997) Regulatory sequences of *Arabidopsis* drive reporter gene expression in nematode feeding structures, Plant Cell 9, 2119-2134.

Bird, D. McK. & Wilson, M.A. (1994) DNA sequence and expression analysis of root-knot nematode-elicited giant cell transcripts, Mol. Plant-Microbe Interact. 7, 419-424.

Clarke, M.C., Wei, W. & Lindsey, K. (1992) High-frequency transformation of *Arabidopsis thaliana* by *Agrobacterium tumefaciens*, Plant Mol. Biol. Rep. 10, 178-189.

Cramer, C.L. (1992) Regulation of defense-related gene expression during plant-pathogen interactions, J. Nematol. 24, 586-587.

Ehsanpour, A.A. & Jones, M.G.K. (1996) Glucuronidase expression in transgenic tobacco roots with a Parasponia promoter on infection with *Meloidogyne javanica*, J. Nematol. 28, 407-413.

Ferreira, P.C.G., Hemerly, A.S., Villarroel, R., Van Montagu, M. & Inzé, D. (1991) The *Arabidopsis* functional homolog of the p34cdc2 protein kinase, Plant Cell 3, 531-540.

Fobert, P.R., Labbé, H., Cosmopoulos, J., Gottlob-McHugh, S., Ouellet, T., Hattori, J., Sunohara, G., Iyer, V.N. & Miki, B.L. (1994) T-DNA tagging of a seed coat-specific cryptic promoter in tobacco, Plant J. 6, 567-577.

Gheysen, G., de Almeida Engler, J. & Van Montagu, M. (1997) Cell cycle regulation in nematode feeding sites, in C. Fenoll, F.M.W. Grundler, and S.A. Ohl (eds.), Cellular and Molecular Aspects of Plant-Nematode Interactions (Developments in Plant Pathology, Vol. 10), Kluwer Academic Publishers, Dordrecht, pp. 120-132.

Goddijn, O.J.M., Lindsey, K., van der Lee, F.M., Klap, J.C. & Sijmons, P.C. (1993) Differential gene expression in nematode-induced feeding structures of transgenic plants harbouring promoter-gusA fusion constructs, Plant J. 4, 863-873.

Hansen, E., Harper, G., McPherson, M.J. & Atkinson, H.J. (1996) Differential expression patterns of the wound-inducible transgene wun1-uidA in potato roots following infection with either cyst or root knot nematodes, Physiol. Mol. Plant Pathol. 48, 161-170.

Hemerly, A.S., Ferreira, P., de Almeida Engler, J., Van Montagu, M., Engler, G. & Inzé, D. (1993) cdc2a expression in *Arabidopsis* is linked with competence for cell division, Plant Cell 5, 1711-1723.

Jefferson, R.A. (1987) Assaying chimeric genes in plants: the GUS gene fusion system, Plant Mol. Biol. Rep. 5, 387-405.

Karimi, M., Barthels, N., Van Montagu, M. & Gheysen, G. (1994) Identification of root knot nematode-induced genes in *Arabidopsis thaliana*, Med. Fac. Landbouww. Univ. Gent 59/2b, 751-756.

Karimi, M., Barthels, N., Van Montagu, M. & Gheysen, G. (1997) Nematode vector for introduction of *Agrobacterium* into plant roots. Abstract presented at the 5th International Congress of Plant Molecular Biology, Singapore, September 21-27, 1997.

Karimi, M., Van Poucke, K., Barthels, N., Van Montagu, M. & Gheysen, G. (1998) Analysis of nematode-inducible plant promoter-gus fusions in diffenent plant species, Med. Fac. Landbouww. Univ. Gent, in press.

Kertbundit, S., De Greve, H., Deboeck, F., Van Montagu, M. & Hernalsteens, J.-P. (1991) In vivo random ß-glucuronidase gene fusions in *Arabidopsis thaliana*, Proc. Natl. Acad. Sci. USA 88, 5212-5216.

Marban-Mendoza, N., Jeyaprakash, A., Jansson, H.-B., Damon, Jr., R.A. & Zuckerman, B.M. (1987) Control of root-knot nematodes on tomato by lectins, J. Nematol. 19, 331-335.

Mariani, C., De Beuckeleer, M., Truettner, J., Leemans, J. & Goldberg, R.B. (1990) Induction of male sterility in plants by a chimaeric ribonuclease gene, Nature 347, 737-741.

Mariani, C., Gossele, V., De Beuckeleer, M., De Block, M., Goldberg, R.B., De Greef, W. & Leemans, J. (1992) A chimaeric ribonuclease-inhibitor gene restores fertility to male sterile plants, Nature 357, 384-387.

Niebel, A., de Almeida Engler, J., Tiré, C., Engler, G., Van Montagu, M. & Gheysen, G. (1993) Induction patterns of an extensin gene in tobacco upon nematode infection, Plant Cell 5, 1697-1710.

Niebel, A., de Almeida Engler, J., Hemerly, A., Ferreira, P., Inzé, D., Van Montagu, M. & Gheysen, G. (1996) Induction of cdc2a and cyc1At expression in *Arabidopsis* during early phases of nematode-induced feeding cell formation, Plant J. 10, 1037-1043.

Opperman, C.H. & Conkling, M.A. (1996) Root-knot nematode induced TobRB7 expression and antisense transgenic resistance strategies, in G. Stacey, B. Mullin, and P.M. Gresshoff (eds.), Biology of Plant-Microbe Interactions, International Society for Molecular Plant-Microbe Interactions, St. Paul, pp. 521-526.

Opperman, C.H., Taylor, C.G. & Conkling, M.A. (1994) Root-knot nematode-directed expression of a plant root-specific gene, Science 263, 221-223.

Schots, A., Gommers, F.J. & Egberts, E. (1992) Quantitative ELISA for the detection of potato cyst nematodes in soil samples, Fundam. Appl. Nematol. 15, 55-61.

Sijmons, P.C. (1993) Plant-nematode interactions, Plant Mol. Biol. 23, 917-931.

Smant, G., Stokkermans, J.P.W.G., Yan, Y.T., de Boer, J.M., Baum, T.J., Wang, X.H., Hussey, R.S., Gommers, F.J., Henrissat, B., Davis, E.L., Helder, J., Schots, A. & Bakker, J. (1998) Endogenous cellulases in animals: isolation of ß-1,4-endoglucanase genes from two species of plant-parasitic cyst nematodes, Proc. Natl. Acad. Sci. USA 95, 4906-4911.

Topping, J.F., Wei, W. & Lindsey, K. (1991) Functional tagging of regulatory elements in the plant genome. Development 112, 1009-1019.

Van der Eycken, W., de Almeida Engler, J., Inzé, D., Van Montagu, M. & Gheysen, G. (1996) A molecular study of root-knot nematode-induced feeding sites, Plant J. 9, 45-54.

Vercauteren, I., Van der Schueren, E., Van Montagu, M. & Gheysen, G. (1996) Isolation of mRNA species expressed upon nematode infection by means of the differential display technique. Abstract presented at the 4th Annual General Meeting of the EC-AIR Concerted Action Program on "Resistance mechanisms against plant parasitic nematodes", Toledo (Spain), May 9-12, 1996, p. 15.

Vercauteren, I., Goeleven, E., Barthels, N., Van Montagu, M. & Gheysen, G. (1998) The rha1 gene, encoding a small GTP-binding protein, is induced in nematode infection sites, Arch. Physiol. Biochim., in press.

Wyss, U. (1987) Video assessment of root cell responses to dorylamid and tylenchid nematodes, in J.A. Veech, and D.W. Dickson (eds.), Vistas on Nematology. A Commemoration of the Twenty-fifth Anniversary of the Society of Nematologists, Society of Nematologists, Hyattsville, pp. 211-220.

Wyss, U. & Grundler, F.M.W. (1992) Feeding behaviour of plant parasitic nematodes, Neth. J. Phytopathol. Suppl. 2, 165-173.

Wyss, U., Grundler, F.M.W. & Münch, A. (1992) The parasitic behaviour of second-stage juveniles of *Meloidogyne incognita* in roots of *Arabidopsis thaliana*, Nematologica 38, 98-111.

G.T. Scarascia Mugnozza, E. Porceddu & M.A. Pagnotta (Eds.)
Genetics and Breeding for Crop Quality and Resistance, 211-219, 1999

Genome engineering for pest resistance in potato

H.T. Butler, A. Prevost, J. Allainguillaume & M. J. Wilkinson
Department of Agricultural Botany, School of Plant Sciences, The University of Reading, Whiteknights, PO Box 221, Reading, RG6 6AS, UK.

Abstract: The potato (*Solanum tuberosum* L.) is susceptible to many pests and diseases. There are over 200 wild relatives, many of which contain useful resistance genes. These species have been poorly utilised in potato breeding, largely because of hybridization barriers and the need for extensive backcrossing to restore agronomic performance. Two strategies to overcome these difficulties are explored. The first uses a mechanism apparently unique to the *Solanaceae*. Pollination of tetraploid *Solanum tuberosum* using pollen from diploid *Solanum* species (usually *S. phureja*) occasionally produce offspring with a diploid chromosome number (2n = 2x = 24). These plants are known as dihaploids. Dihaploids are largely maternal in origin but also contain a small amount of genetic material from the male (dihaploid inducer) parent. There are several species capable of producing dihaploid offspring. Many of these also contain desirable resistance genes. Successful exploitation of this natural introgression event partly depends on the amount of DNA transferred from the male parent. Markers generated by ISSR-PCR, RAPD and RFLP analyses are used to estimate the genetic contribution made by the male parent. Variation is observed between female (*S. tuberosum*) clones but most dihaploids were found to possess fewer than 8% of markers specific to the male parent. It is suggested that the production of resistant dihaploid inducers from several wild species could be incorporated into a dihaploid-based breeding programme. The second strategy seeks to exploit chromosome instability in callus culture. Hexaploid and aneuploid somatic hybrids between *S. tuberosum* and *S. sanctae-rosae* were compared on the basis of ISSR band profiles and morphology. It is inferred that the aneuploid had probably lost chromosomes from both parents but predominantly from *S. sanctae-rosae*. Both somatic hybrids were subjected to further regeneration through callus culture and some were found to lose ISSR bands. Principal Component Analysis of tuber characters suggested that secondary regenerants formed a continuum of variation connecting the two parents. Some individuals were found to possess tubers within the range of variation of each parent although all ISSR-PCR profiles examined differed from those of the parents. The limitations and commercial potential of each strategy are discussed.

1. Introduction

The cultivated potato (*Solanum tuberosum* L.) ranks fourth in terms of world production behind wheat, maize and rice (Hawkes, 1990). Genetic improvement for increased pest and disease resistance is made difficult by several contributing factors. The genetics of the crop is widely acknowledged to be highly complex and difficult to study. This makes it difficult to first characterize and then to integrate resistance traits into a breeding programme, particularly for resistances controlled by polygenes. Likewise, the large number of pests and diseases that affect the crop also creates difficulties. It would be impractical for a potato breeder to introduce new resistances against all major pests and at the same time, maintain selection priority for yield and quality characteristics. Breeders generally need to identify those resistance traits that are most likely to enhance the commercial success of a new cultivar. To some extent, this is governed by the needs of the target market and is sensitive to changes in legislation, climate and agricultural practice. The commercial value of a resistance trait must be weighed against the actual costs required for its inclusion in the breeding effort. The introduction of novel resistance from wild relatives can be a protracted and therefore costly process. The principal factor limiting the incorporation of any new resistance trait into commercially useful material lies in the large number of undesirable features carried by the wild relatives. These characters must be selectively removed during prebreeding to generate material that can be used as parental lines in the main breeding scheme. This process may require ten or more years to complete and is especially protracted when the material used as the source of resistance is taxonomically distant from tetraploid *S. tuberosum*. The high time costs are compounded by the fact that most *Solanum* species with resistance traits are diploid (rather than tetraploid) and that many species cannot be crossed directly with the cultivated crop. There is a need, therefore, for developing alternative strategies for the transfer of alien germplasm from wild *Solanum* species into the modern cultivated potato. In this study, the feasibility of two approaches is evaluated for limited genetic transfer: partial genome transmission associated with dihaploid induction crosses and the control of alien chromosome instability during callus-mediated plant regeneration.

2. Materials and Methods

2.1 Plant material

The following *S. tuberosum* cultivars were obtained from the potato cultivar collection housed at the Scottish Agricultural Science Agency, East Craigs, Edinburgh, UK: cv. Cara, cv. Pentland Crown, cv. Pentland Ivory. The following potato dihaploids were obtained from Dr M. De Maine at the Scottish Crop Research Institute, Dundee, UK: PDH51, PDH52, PDH55, PDH425, PDH1040, PDH1041, PDH1042, PDH1046, PDH1052, PDH1061, PDH1155, PDH1177, PDH1210, PDH1342, PDH1434, PDH1436, PDH1534.

Tubers of *S. tuberosum* cv. Brodick, *S. sanctae-rosae* (accession CPC 3779) and two somatic hybrids between cv. Brodick and *S. sanctae-rosae* (clone 83/12 and 35/9) were received from the Scottish Crop Research Institute, Dundee, UK.

2.2 Callus culture regeneration

Plants were regenerated from leaf disc explants according to the protocol described by Albani & Wilkinson (1998).

2.3 DNA extraction and ISSR-PCR analysis

DNA was extracted according to the protocol described by Doyle & Doyle (1987).

ISSR-PCR was conducted using the following primers obtained from the University of British Columbia (set#9): 807, 811, 812, 834, 841, 856, 888, 890, 891. PCR amplification was performed according to the method described by Charters et al. (1996). Electrophoresis and silver nitrate staining of the gels was conducted using the modification of the method of Charters et al. (1996) as described by Albani & Wilkinson (1998).

2.4 Principal Component Analysis

Cohorts grown from tubers of approximately the same size were grown in a random design in a heated glasshouse (minimum temperature 10°C). For the study of somatic hybrid regenerants, 29 character traits (available on request) were scored from plants eight weeks after emergence of shoots from the soil. Character scores were standardized by adjusting to zero mean and unit variance. They were then used to generate PCA plots using the Minitab statistical program package (Minitab Inc.). In the dihaploid study, characters and data analysis are described by Allainguillaume et al. (1997).

3. Results and Discussion

3.1 Genetic transfer during dihaploid induction

The cultivated potato is tetraploid (2n=4x=48) and widely acknowledged as being genetically complex. Potato dihaploids (2n=2x=24) greatly simplify genetic studies of the crop and provide alternative breeding strategies based on exerting selective pressures at the diploid rather than the tetraploid level (e.g Wenzel et al., 1979; Peloquin et al., 1990).

Dihaploid plants can be produced by anther culture or more commonly, by crossing a tetraploid potato using pollen from selected genotypes of a closely related species. Pollen parents that give rise to dihaploid offspring are known as dihaploid inducers and most frequently belong to the cultivated diploid species, *S. phureja*. Only a small proportion of the offspring from an induction cross are dihaploid and these plants bear a close morphological similarity to the female (*S. tuberosum*) parent. The discovery that dihaploids contain small quantities of DNA from the pollen parent (Clulow et al., 1991), however, raises intriguing possibilities for the rapid introgression of resistance traits from *S. phureja* and the other species known to be able to induce dihaploid potatoes.

There are two main prerequisites for the exploitation of this partial transfer event. The transferred DNA must contain expressed regions of the male genome. Secondly, the amount of DNA transferred from the male parent is critical to the usefulness of the phenomenon for breeding purposes. Sufficient DNA should be transmitted to provide a reasonable chance of transferring the target gene(s). At the same time, however, large quantities of male DNA moving into the dihaploid genome would result in the presence of excessive deleterious traits and an undesirable phenotype.

There are several lines of evidence to suggest that DNA transferred from the male parent contained expressed regions. Clulow et al. (1993) showed that several dihaploids contain isoforms specific to the male parent. Equally, morphological traits specific to the male parent such as flower colour and embryo spot seed marker have been observed in several dihaploid offspring (Caligari et al., 1988; Allainguillaume et al., 1997). PCA analysis has also revealed a quantitative influence by the male genome on the phenotype of dihaploids (Allainguillaume et al., 1997).

Table 1. Frequency of *S. phureja*-specific markers detected in dihaploids of cv. Pentland Crown.

Dihaploid	Proportion of male markers detected	Percentage of male markers detected
PDH51	5/51	10%
PDH52	3/51	6%
PDH55	2/51	4%
PDH425	1/51	2%
PDH440	10/51	20%
PDH452	1/51	2%

There is less evidence concerning the amount of DNA transferred by the male parent and the extent to which this is affected by the genotype of the female parent. Several studies have used a range of molecular, biochemical and morphological markers to demonstrate the presence of DNA from the male parent in dihaploid offspring (Clulow et al., 1991; Wilkinson et al., 1995; Clulow & Rousselle-Bourgeois, 1997; Allainguillaume, 1997; Allainguillaume et al., 1997). Here, we combine data provided by most of these works (largely from RAPD and ISSR PCR analyses) with new markers generated by ISSR PCR. Crude estimates of the genetic contribution made by the male parent can be made from the proportion of male-specific markers that appear in the dihaploid offspring. These percentages varied between 0% in two dihaploids and 20% (10/51) in PDH440. Differences between dihaploids originating from the same female parent accounted for most of the variability detected (Table 1). There were comparatively small differences in the abundance of paternal-specific markers detected in the dihaploids derived from different females. On average, dihaploids of cv. Cara contained 5.7% of the paternal-specific markers, compared with 7.2% in cv. Pentland Crown and 10% in the four dihaploids of cv. Pentland Ivory.

These values can be taken only as very rough indications of the quantity of DNA transferred and should not be viewed as definitive estimates. The fact that most of the markers generated were likely to be dominantly inherited is partly countered by the fact that the male parent used is thought to be highly homozygous. Nevertheless, the use of molecular protocols that yield mostly dominant markers would have a slight tendency to under-estimate the male contribution. Conversely, some of the works used to provide marker information aimed only to report the presence of male-specific markers in dihaploids. It is possible, therefore, that some male-specific markers that were absent from all dihaploids may have been omitted from the data set. This would have led to a slight over-estimate of the contribution by the male parent. Additional data is required to provide a more accurate appraisal of the male contribution. However, results presented here suggest that the amount of DNA contributed by the male parent could be sufficient to be of practical value for prebreeding purposes. Exploitation of the phenomenon would be dependent on several factors. First, DNA needs to be shown to occur when diploid species other than *S. phureja* are used as the paternal parent. Second, marker systems (such as the embryo spot marker) need to be transferred into an inductive species containing the desirable resistance trait.

Induction crosses typically yield low numbers of dihaploid offspring and it is unlikely that a prebreeding strategy based entirely on the transfer event could be developed in the short or medium term. A more practical approach would be to select dihaploid inducer parents from species known to contain desirable resistance genes. Embryo spot marker genes could be introgressed into these lines and effective inducer clones used to produce dihaploids as part of an established dihaploid-based breeding strategy. Dihaploids generated from resistant inducers could be routinely screened for resistance as part of the selection procedure. In this way, new resistance genes can be introduced directly into dihaploid breeding material simply by changing the inducer parent used to generate the dihaploid material. Clearly though, it is only possible to introgress DNA from species that are able to induce dihaploids in *S. tuberosum*. As such, there are resistance genes from many species that could not be introduced by this route.

3.2 Introgression by somatic hybridization and chromosome elimination

Somatic hybridization offers a far more generic procedure for the introduction of resistance genes from wild relatives of the cultivated potato. There are large numbers of works reporting the production of somatic hybrids between the cultivated potato and wild species containing useful resistance genes. The principal problem with the technology lies in the fact that the newly generated hybrids contain 72 rather than 48 chromosomes. Extensive backcrossing is therefore required as a prebreeding step to return to the tetraploid level and also to remove unwanted traits conferred by the genome of the wild species. This can be a protracted process and causes sexual reassortment of the desirable allele combinations found in the original *S. tuberosum* cultivar. The strategy being explored here is to use chromosome instability in callus culture to selectively remove the wild chromosomes carrying agronomically undesirable traits.

There have been many reports of chromosome instability in plants regenerated from *in vitro* explants (Jacobsen et al., 1983; Wheeler et al., 1985), protoplasts (Karp et al., 1982; Sree Ramulu, 1986) and also from somatic hybrids (de Vries et al., 1987; Valkonen et al., 1994; Stattman et al., 1994). There is comparatively little evidence to suggest if the loss of chromosomes is an entirely random process or whether chromosomes of one genome (or one homologue type) are lost preferentially. Pijnacker et al. (1987) used Giemsa C-banding to show that the NOR chromosomes of *S. phureja* were lost preferentially from somatic hybrids between this species and *S. tuberosum*. In contrast, De Vries et al. (1987) reported chromosome elimination in more than 80 somatic hybrids between *S. tuberosum* and *Nicotiana plumbaginifolia* but did not find that the chromosomes of either species had been lost preferentially. However, the preferential loss of unwanted chromosomes is not a prerequisite for the exploitation of chromosome instability. Random elimination of both genomes can be tolerated provided that that large numbers of regenerants can be produced and an effective selection criterion can be imposed for the identification of clones with the desired genome architecture. Nevertheless, it is doubtful whether strong preferential loss of the *S. tuberosum* chromosomes could be accommodated in this way. It is important, therefore, that some instability is demonstrated in the chromosomes from the genome being targeted for elimination. The two somatic hybrids being used in this study were generated between cv. Brodick and a PCN resistant (now extinct) clone of the diploid wild species, *S. sanctae-rosae*. One hybrid contained all chromosomes from both parents (clone 83/12, 2n=72) whereas the other had lost five chromosomes (clone 35/9, 2n=67). Band profiles generated using six ISSR-PCR primers all revealed markers specific to both parents present in each of the two hybrids.

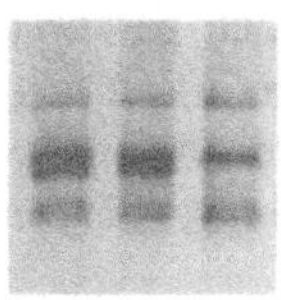

Figure 1. ISSR band profiles of cv. Brodick (lane 1), hexaploid somatic hybrid 83/12 (lane 2) and aneuploid somatic hybrid, 35/9 (lane 3) using primer 856.

The profiles of the aneuploid hybrid lacked six bands that were present in the hexaploid hybrid (e.g. Figure 1). The majority of these markers (5/6) were specific to *S. sanctae-rosae*, perhaps indicating that most of the eliminated chromosomes originated from the wild species. This inference was supported by morphological measurements taken from several cohort representatives of the parents and both hybrids. PCA plots of first principal component against the second (Figure 2a) and third (Figure 2b) principal components revealed that the aneuploid hybrid clones were distinctly closer to the *S. tuberosum* parent than the hexaploid hybrid. These data therefore suggest that the initial elimination event favoured the loss of *S. sanctae-rosae* chromosomes.

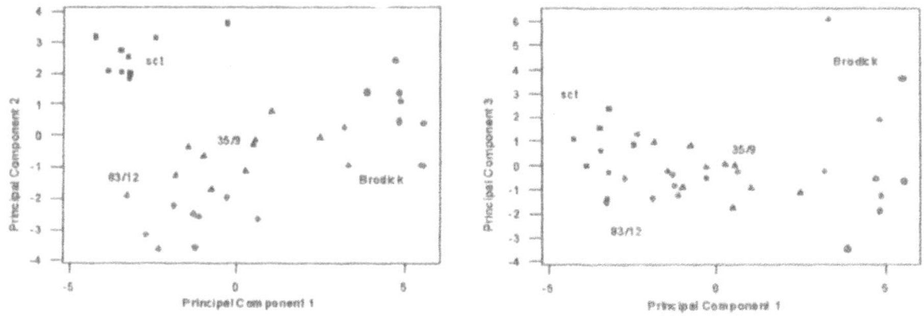

Figure 2. Principal Component Analysis comparing 9 clones of cv. Brodick (●), *Solanum sanctae-rosae* CPC3779 (□), hexaploid somatic hybrid 83/12 (♦) and aneuploid somatic hybrid 35/9 (▲). a). First Principal Component plotted against the second Principal Component. b). First Principal Component plotted, against the third Principal Component.

An attempt was then made to stimulate further elimination from the aneuploid hybrid and to lose chromosomes from the hexaploid hybrid by regenerating 200 plants from leaf discs via callus culture. Fifty regenerants were transferred to pots and grown under glasshouse conditions. ISSR-PCR analysis of 10 randomly selected regenerants identified two in which bands specific to both parents had been lost (Figure 3).

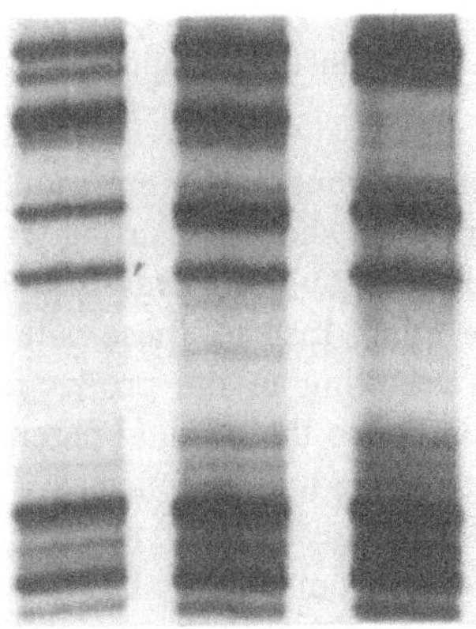

Figure 3. Band profiles generated by secondary regenerant 137 (lane 1), hexaploid somatic hybrid 83/12 (lane 2) and secondary regenerant 139 (lane 3) using ISSR primer 807. Missing bands specific to cv. Brodick are arrowed. Missing band specific to *S. sanctae-rosae* is marked by the arrow head.

None of the regenerants examined possessed a band profile identical to either parent. The moderate numbers of bands lost from these profiles is more suggestive of chromosome elimination than somaclonal variation. Albani & Wilkinson (1998) have shown that it is possible to detect somaclonal variation associated with callus regeneration using ISSR-PCR but observed very much lower rates of change to band profiles. The use of mapped markers would improve the efficiency of detecting chromosome loss but would probably be impractical to apply on a large scale for screening purposes. A more appropriate strategy would be to select regenerants on the basis of morphological similarity to the cultivated parent and by the possession of the resistance trait of interest. Selection in a breeding programme is focussed on tuber morphology. Tuber morphology of a collection of glasshouse-grown cohorts comprising 68 secondary regenerants, the aneuploid somatic hybrid from which they were derived and the original parental clones were assessed by PCA (Figure 4).

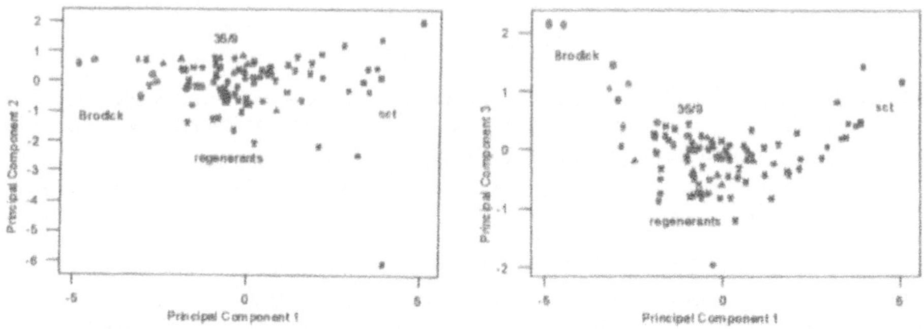

Figure. 4 Principal Component Analysis showing variation in tuber characteristics in cv. Brodick (●), *S. sanctae-rosae* (□), aneuploid somatic hybrid 35/9 (▲) and secondary regenerants of 35/9 (*). a) First Principal Component plotted against the second Principal Component. b) First Principal Component plotted against the third Principal Component.

The tubers of the secondary regenerants were more variable than those of the aneuploid somatic hybrid clones and showed a strong tendency to form a continuum of variation between the two parents. There was little evidence of random spread attributable to somaclonal variation. Instead, these data are consistent with random elimination of chromosomes resulting in the production of regenerants at either extreme that possess tubers similar to the original parents. Further work is required to develop these findings into a breeding system based on targeted elimination. In principal, however, our results suggest that it may be possible to generate large numbers of secondary regenerants from somatic hybrids and exert joint selection for parental tuber type and resistance against the targeted disease. Using this approach, it may be possible to identify clones with tuber characteristics similar to the original cultivar but also with disease resistance from the wild species.

4. References

Albani, M. & Wilkinson, M.J. (1998) Inter simple sequence repeat polymerase chain reaction for the detection of somaclonal variation. *Plant Breed.* 117, 573-575.

Allainguillaume, J. (1997) Dihaploid induction in *Solanum tuberosum* L. Ph.D. Thesis, pp 311, University of Dundee, UK.

Allainguillaume, J., Wilkinson, M.J., Clulow, S.A. & Barr, S.N.R. (1997) Evidence that genes from the male parent may influence the morphology of potato dihaploids. *Theor. Appl.Genet.* **94**, 241-248.

Caligari, P.D.S., Powell, W., Liddell, K., De Maine, M.J. & Swan, G.E.L. (1988) Methods and strategies for detecting *Solanum tuberosum* dihaploids in interspecific crosses with *Solanum phureja*. *Ann. Appl. Biol.* **112**, 323-328.

Charters, Y.M., Robsertson, A., Wilkinson, M.J. & Ramsay, G. (1996) PCR analysis of oilseed rape cultivars (*Brassica napus* L. ssp. *oleifera*) using 5' anchored Simple Sequence Repeat (SSR) primers. *Theor. Appl. Genet.* **92**, 442-447.

Clulow, S.A., Wilkinson, M.J., Waugh, R., Baird, E., De Maine, M.J. & Powell, W. (1991) Cytological and molecular observations on *Solanum phureja* induced dihaploid potatoes. *Theor. Appl. Genet.* **82**, 545-551.

Clulow, S.A., Wilkinson, M.J. & Burch, L.R. (1993) *Solanum phureja* genes are expressed in the leaves and tubers of aneusomatic potato dihaploids. *Euphytica* **69**, 1-6.

Clulow, S.A. & Rousselle-Bourgeois, F. (1997) Widespread introgression of *Solanum phureja* DNA in potato (*Solanum tuberosum*) dihaploids. *Plant Breed.* 116, 347-351.

De Vries, S.E., Ferwerda, M.A., Loonen, A.E.H.M., Pijnacker, L.P. & Feenstra, W.J. (1987) Chromosomes in somatic hybrids between *Nicotiana plumbaginifolia* and a monoploid potato. Theor. Appl. Genet. 75, 170-176.

Doyle, J.J. & Doyle J.L. (1987) A rapid DNA isolation procedure for small quantities of fresh leaf tissue. *Phytochem. Bull.* **19**, 11-15.

Hawkes, J.G. (1990) The potato: evolution, biodiversity and genetic resources. Smithsonian Institution Press, Washington DC, USA.

Jacobsen, E., Temelaar, M.J. & Bijmolt, E.W. (1983) Ploidy level in leaf callus and regenerated plants of *Solanum tuberosum* determined by cytophotometric measurements of protoplasts. *Theor. Appl. Genet.* **65**, 113-118.

Karp, A., Nelson, R.S., Thomas, E. & Bright, S.W.J. (1982) Chromosome variation in protoplast-derived potato plants. *Theor. Appl. Genet.* **63**, 265-272.

Peloquin, S.J., Werner, J.E. & Yerk, G.L. (1990) The use of potato dihaploids in genetics and breeding. In: Tsuchiya, T. and Gupta, P.K. (eds) Chromosome engineering in plants: genetics, breeding, evolution, part B. pp 79-92, Elsevier, Amsterdam, The Netherlands.

Pijnacker, L.P., Ferwerda, M.A., Puite, K.J. & Roest, S. (1987) Elimination of *Solanum phureja* chromosomes in S. tuberosum and S. phureja somatic hybrids. Theor. Appl. Genet. 73, 878-882.

Sree Ramulu, K., Dijkhuis, P., Roest, S., Bokelmann, G.S. & De Groot, B. (1986) Variation in phenotypes and chromosome number of plants regenerated from protoplasts of dihaploid and tetraploid potato. *Plant Breed.* **97**, 119-128.

Stattmann, M., Gerick, E. & Wenzel, G. (1994) Interspecific somatic hybrids between *Solanum khasianum* and *S. aculeatissimum* produced by electrofusion. Plant Cell Rep. 13, 193-196.

Valkonen, J.P.T., Xu, Y-S., Rokka, V-M., Pulli, S. & Pehu, E. (1994) Transfer of resistance to potato leafroll virus, potato virus Y and potato virus X from *Solanum brevidens* to *S. tuberosum* through symmetric and designed asymmetric somatic hybridization. Ann. Appl. Biol. 124, 351-362.

Wenzel, G., Schieder, O., Przewozny, T., Sopory. B.K. & Melchers, G. (1979) Comparison of single cell culture derived *Solanum tuberosum* L. plants and a model for their application in breeding. *Theor. Appl. Genet.* **55**, 49-55.

Wheeler, V.A., Evans, N.E., Foulger, D., Webb, K.J., Karp, A., Franklin, J. & Bright, S.W.J. (1985) Shoot formation from explant cultures of fourteen potato cultivars and studies of the cytology and morphology of regenerated plants. *Ann. Bot.* **55**, 309-320.

Wilkinson, M.J., Bennett, S.T., Clulow, S.A., Allainguillaume, J.A., Harding, K. & Bennett, M.D. (1995) Evidence for somatic translocation during dihaploid induction. *Heredity* **74**, 146-151.

SESSION 5

RESISTANCE TO VIRUS

G.T. Scarascia Mugnozza, E. Porceddu & M.A. Pagnotta (Eds.)
Genetics and Breeding for Crop Quality and Resistance, 223-232, 1999

An appraisal of pathogen-derived resistance for the control of virus diseases

G.P. Martelli, D. Gallitelli & M. Russo
Dipartimento di Protezione delle Piante, Università degli Studi and Centro di Studio del CNR sui Virus e le Virosi delle Colture Meditarranee, Via Amendola 165/A, 70126 Bari, Italy

Abstract: An account is given of the principles and use of pathogen-derived resistance (PDR) for the control of plant viruses. Strategies for obtaining PDR can be grouped in two broad categories, requiring either synthesis and accumulation of viral proteins, or of viral nucleic acid in transgenic cells. Most of the viral genes with a known function have now been vehiculated in host cells by different transforming strategies. These genes are used in the native or mutated form in an effort to find effective and biologically safe ways. In general, *Nicotiana* spp. serve as model hosts for laboratory-scale transformation experiments. Since certain genes confer resistance only to virus strains closely related to the virus of origin, alternative strategies for obtaining plants that resist a broader spectrum of viruses have been developed, using different types of non viral genes. Notwithstanding the highly satisfactory performance of transgenic plants tested in the laboratory or in small-scale field releases, very few crop plants expressing PDR to viruses are now on the market. This may not surprise, considering the necessity of a better understanding of a few key entries: (i) assessment of the biological risk; (ii) stability of the trait introduced with the transgene; (iii) cost-benefit analysis; (iv) effective marketing potentiality as dependent on the public acceptance of biotechnological applications.

1. Introduction

To date, about 800 plant viruses have been identified, described, and characterized to varying extents (Martelli, 1997). About one fourth of the known viruses elicits diseases that cause losses to agricultural crops world-wide. Losses may be relevant, originating from reduction of the quantity, quality and market value of the yield, decreased plant vigour, crop failure, increased sensitivity to adverse climatic conditions and to attacks by other pathogens and pests, and from costs for maintain crop health (Waterworth & Hadidi, 1998). Thus, the consensus is that, in terms of economic importance, virus diseases rank second only to fungal diseases. However, whereas fungal diseases can be controlled by chemicals, combating viruses is much more difficult.

Virus control is preventive and largely based on: (i) elimination of inoculum sources; (ii) production of sanitarily improved or virus-free propagative material; (iii) control of vectors; (iv) pre-immunization (cross-protection); (v) production of resistant varieties by breeding or genetic engineering.

If problems caused by plant viruses are rather severe in developed temperate countries where control measures are implemented, they are much worse in tropical areas, where virus outbreaks are often devastating, and limit the availability of basic subsistence food crops. Thus, the necessity arose to widen the range of control strategies that could assure efficient and durable results.

Breeding for resistance to viruses and/or to their vectors is a most appealing approach, which has long been used with very encouraging if not brilliant results (Khetarpal et al., 1998; Jones, 1998). However, conventional breeding methods are now being accompanied by novel procedures based on recombinat DNA techniques, whereby exogenous genes are engineered in the plant's genome. The resistance so obtained is commonly defined as "non conventional", of which two types are known: (i) pathogen-derived resistance; (ii) resistance induced by other exogenous, generally non microbial, DNA sequences.

Many comprehensive reviews on trangenic resistance are available, which readers are referred to for detailed information (Buck, 1991; Hanley-Bowdoin & Hemenway, 1992; Carr & Zaitlin, 1993; Beachy, 1993; Sturtevant & Beachy, 1993; Wilson, 1993; Baulcombe, 1994; Grumet, 1994; Kaniewski & Lawson, 1994, 1998; Jaquemond & Tepfer, 1998; Martin, 1998; Tabler et al., 1998).

2. Transgenic resistance: the principles

2.1 Pathogen-derived resistance

Pathogen-derived resistance (PDR), first advocated by Hamilton (1980), can broadly be defined as that type of resistance conferred to a plant by genes isolated from the pathogen's genome, cloned, and engineered into the plant's genome (Sanford & Johnston, 1985). This form of resistance is suggested to arise when a viral gene product is expressed in the host cells at inappropriate time, or in an inappropriate form or amount during the virus "life cycle", so as to perturb the infection process.

The great majority of the viruses against which PDR strategy has been used are single-stranded positive sense RNA viruses, whose life cycle (Figure 1) involves the following steps: (i) virus entry; (ii) decapsidation; (iii) early gene expression (replication-related proteins); (iv) genome replication; (v) late gene expression (structural proteins, movement proteins); (vi) assembly; (vii) release (spreading in the host, acquisition by vectors). PDR can act at the level of most of these steps, impairing or blocking the infection process (arrows in Figure 1).

Since its first application some twelve years ago for controlling tobacco mosaic virus (TMV) infections in *Nicotiana* (Powell-Abel et al., 1986), PDR has gone a long way, and a number of genes and strategies have been assayed to this aim. As reported in Table 1, strategies for obtaining PDR can roughly be grouped in two categories, requiring either synthesis and accumulation in transgenic cells of viral proteins (coding sequences) or of viral nucleic acid (non coding sequences).

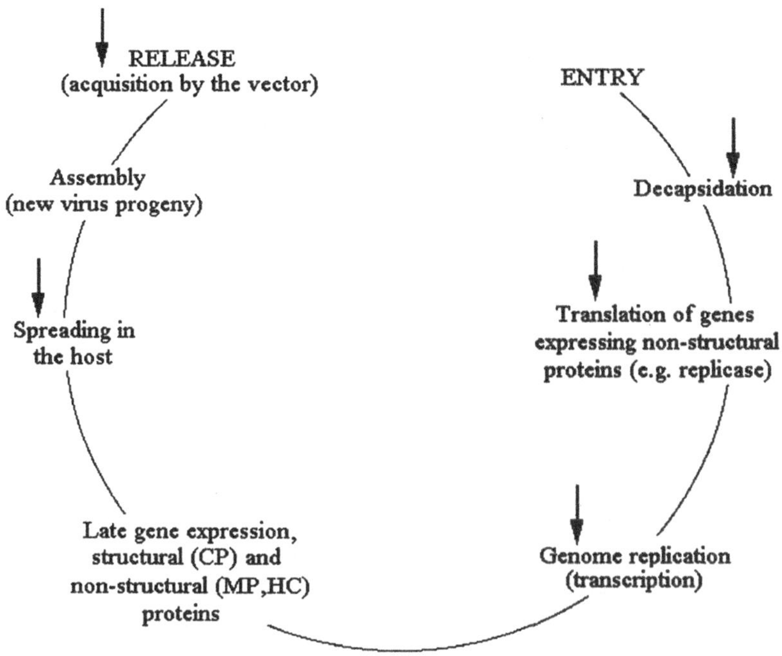

Figure 1. The "life cycle" of a plus strand RNA plant virus.

Table 1. Viral sequences used for transforming plants for resistance to viruses.

A. Coding sequences	
Capsid protein (CP)	
	Nucleocapsid (N)
	Movement protein (MP)
	Helper component (HC)
	Replicase (REP)
	Protease (PRO)
	Genome-linked protein (VPg)
B. Non coding sequences	
	Sequences made deliberately non coding
	Antisense RNAs
	Satellite RNAs (SatRNAs)
	Defective interfering RNAs (DI RNAs)

CP-mediated resistance was the first to be tested (Powell-Abel et al., 1986) and still represents a very popular approach, utilized successfully against no less than 35 viruses in 15 different taxa. With time, PDR has undergone a remarkable development, in that most of the viral genes whose function is known have been introduced into constructs to be vehiculated in the host cell by different transforming strategies, primarily via *Agrobacterium tumefaciens* and particle bombardment. These genes were used in the native or mutated form in an effort to find alternative ways that could be at the same time effective and biologically safe.

The mechanisms of action of the various gene products are diversified, and are often hypothesized rather than experimentally proven. In any case, the consensus is that multiple mechanisms are probably triggered by different pathogen-derived genes and plant-virus combinations (Kaniewski & Lawson, 1998). For protein-mediated resistance these mechanisms range from interference with virus decapsidation, long distance virus spread, or translation of viral replicase (CP); cell-to-cell transfer (CP, MP); acquisition and/or transmission by vectors (HC); genome replication (REP); polyprotein processing (PRO) (Reimann-Philipp & Beachy, 1993; Kaniewski & Lawson, 1998). Mechanisms of RNA-mediated resistance are thought to involve gene silencing, binding of antisense RNA to challenging virus genome so as to form double stranded helices, or down regulation of viral genome replication (satRNAs, DI RNAs) (Tabler et al., 1998; Jaquemond & Tepfer, 1998).

2.2 Resistance derived from non viral sequences

One of the shortcomings of certain approaches (e.g. those involving CP- or MP-based resistance) is that resistance is often conferred only to virus strains closely related to the virus from which the transgene was derived. This prompted studies for the development of alternative strategies for obtaining plants that resist a broader spectrum of viruses or that do not express genes that may recombine with other viral genomes (Martin, 1998). Table 2 lists some of the sequences that have been used to this effect at the experimental level.

Table 2. Non viral sequences used for transforming plants for resistance to viruses.

A. *Coding sequences*	
	dsRNA-specific RNase
	Mammalian 2'-5' oligoadenylate system
	Recombinant antibodies ("plantibodies")
	Toxins or suicidal genes (RIPs)
	Amber codon inhibitors
	Pseudoubiquitin
	Protein kinase
B. *Non-coding sequences*	
	Ribozymes

As with pathogen-derived resistance, the mechanisms of action of these sequences is diversified, ranging from induction of hypersensitive reaction (2'-5' oligoadenylate system, toxins, ribosome inhibiting proteins) to cleavage of viral ssRNA (ribozymes) or dsRNA molecules, to interference with early steps of virus replication (recombinant antibodies) (Conrad & Fiedler, 1994; Watanabe et al., 1995; Ogawa et al., 1996; Martin, 1998).

3. Transgenic resistance: the application

The process whereby transgenic resistant crop plants are produced involves: (i) a laboratory phase for the transformation of experimental hosts to evaluate the level of transgene efficiency, followed by transformation and regeneration of the target plant; (ii) field testing to evalute the performance and stability of the transgene; (iii) commercial release.

3.1 Laboratory phase

In general, *Nicotiana* spp. serve as model hosts for laboratory-scale transformation experiments, the outcome of which is thought to give a hint on whether or not transgenic resistance can successfully be introduced into a target crop plant. Unfortunately, the transgene efficiency expressed by *Nicotiana* does not always match the performance of the same construct in transformed crop plants. Thus, these preliminary transformation experiments are often of doubtful usefulness and, in some instances, could be skipped. Transformation of the target host is, however, necessarily carried out in the laboratory.

Table 3. Field tests of virus-resistant transgenic plants authorized by USDA since 1997 (White, 1998).

CROP	FIELD TESTS (n°)	VIRUS
A. HERBACEOUS		
Beet	1	BNYYV
Cucurbits	13	"CBI", CMV, PRSV ZYMV, WMV-2
Solanaceous	28	CMV, PLRV, PVY TRV, TYLCV, ToMV, TMV, TSWV, TEV
Leguminous	5	PEMV, BLRV, SBMV TSWV,
Cereals	4	BYDV, WSMV
B. WOODY		
Grapevine	2	"CBI"
Papaya	2	PRSV
Raspberry	1	RBDV

CBI, confidential business information (=undecleared virus and construct); BNYYV, beet necrotic yellow vein; CMV, cucumber mosaic; PRSV, papaya ringspot; ZYMV, zucchini yellow mosaic; WMV-2, watermelon mosaic 2; PLRV, potato leafroll; PVY, potato Y; TRV, tobacco rattle; TYLCV, tomato yellow leaf curl; TMV, tobacco mosaic; ToMV, tomato mosaic; TSWV, tomato spotted wilt; TEV,tobacco etch; PEMV, pea enation mosaic; BLRV, bean leafroll; SBMV, southern bean mosaic; BYDV, barley yellow dwarf; WSMV, wheat streak mosaic; RBDV, raspberry bushy dwarf.

Table 4. Field trials of some virus resistant transgenic herbaceous and woody crops since 1986 (Source: OECD, Biotrack on line, 1998).

A. HERBACEOUS CROPS	
POTATO	
Field trials:	211
viruses:	PLRV, PVY, PVX
countries:	Australia, Canada, Denmark, Finland, Germany, Netherlands, Japan, New Zealand, Russia, Spain, Switzerland, UK, USA
TOMATO	
Field trials:	68
Viruses:	CMV, PVY, TYLCV, TMV, ToMV, TSWV, BCTV
Countries:	Italy, Japan, Spain, USA, China
TOBACCO	
Field trials:	37
Viruses:	TSWV, PVY, CMV, AMV; BCTV, TEV, TMV, TGMV, TVMV
Countries:	Brasil, Bulgaria, China, Japan, USA
SUGARBEET	
Field trials:	37
Viruses:	BNYYV, BWY
Countries:	Denmark, France, Germany, UK Netherlands, Italy, Switzerland,
CUCURBITS (squash, melon, watermelon, cucumber)	
Field trials:	448
Viruses:	CMV, WMV-2, ZYMV, PRSV, SQMV
Countries:	China, France, Japan, Spain, UK,
B. WOODY CROPS	
GRAPEVINE	
Field trials:	3
Viruses:	GFLV, GLRaV-3
Countries:	France, Usa
PAPAYA	
Field trials:	3
Virus:	PRSV
Country:	USA
PLUM	
Field trials	3
Virus:	PPV
Countries:	Spain, Poland, Romania
WALNUT	
Field trials:	1
Virus:	CLRV
Country:	USA

AMV, alfalfa mosaic; BCTV, beet curly top; BWY, beet western yellows; CLRV, cherry leafroll, GFLV, grapevine fanleaf; GLRaV-3, grapevine leafroll associated 3; PPV, plum pox; PVX, potato X; SQMV, squash mosaic; TGMV, tobacco green mottle; TVMV, tobacco vein mottle. Other acronyms as in Table 3

3.2 Field testing

All the genes and sequences listed in Table 1 and 2 have been used for transforming plants of various types at the experimental level, but only some of them have been tested in the field.

From 1986, transgenic crop field trials have been conducted in 45 countries from Europe (23), America (10), Africa (3), and Australasia (9). In the US, field tests authorized by the USDA since 1997 for virus resistant plants transformed with viral (CP, MP, REP, PRO, VPg in either a native or mutated disfunctional form) or non viral genes (yeast and human dsRNase, pseudoubiquitin, and protein kinase) are as reported in Table 3 (White, 1998).

The difference in the number of trials with herbaceous (51) versus woody (5) crops is striking, but is in line with the world-wide trend (Table 4). The reasons may lie in the higher economic and nutritional relevance and wider acreage of herbaceous crops (cereals, vegetables, legumes) with respect to woody crops (fruit trees and shrubs), and to the fact that many of the herbaceous species are easier to manipulate and regenerate than woody plants.

3.3 Commercial releases

The global area of transgenic crops has increased from 1.7 million ha in 1996, to 11 million ha in 1997, to reach 27.8 million ha in 1998. Leading countries are the USA (20.5 m ha^{-1}), Argentina (4.3 m ha^{-1}), and Canada (2.8 m ha^{-1}). With 0.1 m ha^{-1} Australia accounts for 0.1% of the global area, whereas Mexico, Spain, France, and South Africa have much smaller surfaces given over to transgenic plants (James, 1997, 1998). These figures do not include China, whose acreage of transgenic crops is substantial, and refer to genetically modified crops expressing traits other than resistance to viruses. In fact, although tomato and potato varieties expressing transgenic resistance to TMV, ToMV, PVX, PVY were ready for commercial release in the USA as early as 1989, the release began only in 1998. As stated by Kaniewski (1998): “Monsanto decided not to put these products on a commercial track. The business unit did not expect to generate sufficient profit from these products to justify commercialization. Also, the mechanisms were not yet in place at this time for the introduction of genetically modified crops into commerce in the United States.”

About a decade had to elapse before the first virus-resistant transgenic plants were put to market (Table 5).

All releases in Table 5 are from US Companies or Universities. However, it should be kept in mind that for many years now tobacco and tomato expressing transgenic resistance to CMV, PVY and TMV, conferred by different genes (CP, REP, SatRNA), have been growing happily in China to the apparent satisfaction of the farmers.

The state-of-the-art of global commercialization of genetically modified plants (GMPs) for virus resistance can schematically be summarized as follows: USA and China, in progress; Japan, many field trials, deregulation in progress; European Union, some field trials, debate continues; other European countries, few or no field trials, development of framework regulations; developing countries, few field trials, gaining experience.

Table 5. Commercial releases of crop plants expressing transgenic resistance to viruses (Kaniewski, 1998; White, 1998).

CROP PLANT	VIRUS and GENE(S)
Papaya (cv.Sunup, Rainbow)	PRSV (CP)
Potato (cv. Russet Burbank)	PLRV (POL + bt/CPB)
Potato (cv. Superior)	PVY (CP + bt/CPB)
Squash	CMV + ZYMV +WMV-2 (CPs)
Squash	ZYMV + WMV-2 (CPs)

bt/CPB, resistance to Colorado potato beetle induced by the *Bacillus thuringensis* toxin

4. Transgenic resistance: benefits and risks

Transgenic resistance to viruses encompasses both benefits and risks but, as it is becoming increasingly evident, advantages greatly supersede possible disadvantages.

Benefits can be exemplified by the possibility of: (i) targeting selection of the desired trait to be introduced; (ii) cropping under high disease pressure, so as to regain areas abandoned because of recurrent destructive virus outbreaks (e.g. necrotic CMV infections in Mediterranean Europe, TYLV in the Mediterranean Basin, cocoa swollen shoots in tropical Africa, papaya ringspot in Hawaii, etc.); (iii) increasing yield and food quality under conditions of medium disease pressure; (iv) conducting effective monitoring as the introduced trait is known.

GMPs monitoring is of great relevance as it aims at evaluating critical entriers such as: (i) the stability of engineered genes after release of the GPM in the environment, and the study of their modes of expression; (ii) the effects of the modification induced in the GPM after release in the environment and its influence on the ecosystem; (iii) the risk of "gene flow" from GPMs to members of the same species and wild relatives; (iv) the risk of recombination; (v) the identification of GPMs released without notification; (vi) the analysis of food-stuff derived from GPMs.

As to risks, many are proving much less threatening that feared, as they match situations already occurring in nature. For example:

- *heterologous encapsidation*: low or no risk. Phenotypic mixing is a recurrent natural phenomenon when two viruses multiply in the same host;
- *synergism*: possible as it occurs also in nature. Worsening of the effects due to the contemporary presence of the transgene and other viruses demonstrated in a few instances;
- *recombination and reassortment*: possibile as it occurs also in nature. The ecological success of viable recombinants/reassortants is unpredictable.
- *gene flow*: already occurring with virus-resistant varieties obtained by cross-hybridization;
- *increased invasiveness of cross-pollinated wild relatives*: demonstrated only under conditions of high disease pressure;

- *genetic erosion*: already occurring with extensively cultivated varieties expressing improved agronomic and/or disease resistant traits introduced by crossing.
- *food quality*: products of virus infection (proteins and nucleic acids) are abundantly present in marketed horticultural produce.

5. Perspectives

Is there future for transgenic resistance to viruses? The remarkable increase of the global acreage planted with GMPs in the last three years indicates that the adoption of these plants is gaining momentum, and meets with farmers' acceptance. On these premises, there is little doubt that the cultivation of virus-resistant GMPs will also expand. Novel strategies will likely be pursued such as piling up multiple resistances in the same plant (e.g. to different viruses, or viruses plus insects, or viruses plus fungi, etc.) and use of promoters stronger than those currently available for expression in specific tissues or organs.

In any case, there is a number of entries that should be addressed for developing sound cultural grounds and facilitating public acceptance:

- objective evaluation of hazards with reference to the current situation and likelihood of their occurrence;
- more research aimed at identifying unpredictable risks;
- focus on traits of commercial and social importance;
- reduce risks of gene flow (e.g. through male sterility);
- improve detection and monitoring methods;
- immediate report of any problem encountered during large scale releases of virus resistant GPMs;
- international approval and harmonization of regulatory provisions;
- improve communication between regulators and scientists;
- gain public acceptance through correct information;

Presently, the last issue is one of the hottest and most difficult. A recent survey (Hoban, 1997) shows that, as compared with Europe, North American countries (USA and Canada) have a lower perception of genetic engineering as a hazard to health and are more willing to purchase food-stuff derived from GMPs. In general, EU countries maintain a very conservative, if not antagonistic, attitude towards biotechnology. If this trend does not revert promptly, Europe risks further colonization by foreign technology and invasion by foreign commodities.

6. References

Baulcombe, D. (1994) Replicase-mediated resistance. A novel type of transgenic resistance in transgenic plants. Trends in Microbiology 2, 60-63.

Beachy, R.N. (1993) Introduction: transgenic resistance to plant viruses. Seminars in Virology 4, 327-328.

Buck, K.W. (1991) Virus resistant plants, in D. Grierson (ed.) Plant Genetic Engineering, Blackie Academic and Professional, London, pp. 137-178.

Carr, J.P. & Zaitlin, M. (1993) Replicase-mediated resistance. Seminars in Virology 4, 339-347.

Conrad, U. & Fiedler, U. (1994) Expression of engineered antibodies in plant cells. Plant Molecular Biology 26, 1023-1030.

Grumet, R. (1994) Development of virus resistant plants via genetic engineering; in J. Janick (ed.) Plant Breeding Reviews, 12. Wiley & Sons, New York, pp. 47-79.

Hanley-Bowdoin, L. & Hemenway, C. (1992) Expression of plant virus genes in transgenic plants, in T. M. A. Wilson and J.W. Davies (eds.) Genetic Engineering with Plant Viruses. CRC Press, Boca Raton, pp. 251-296.

Hamilton, R.I. (1980) Defenses triggered by previous invaders: viruses, in J.G. Hosrfall and E.B Cowling (eds.) Plant Diseases: An Advanced Treatise, 5, Academic Press, New York, pp. 279-303.

Hoban,T.J. (1997) Consumer acceptance of biotechnology. Nature biotechnology 15, 232-234.

James, C. (1997) Global status of transgenic crops in 1997. ISAAA Briefs 5, 1-30

James, C. (1998) The global status of commercialized transgenic crops. Proceedings of the 5th International Symposium on biosafety results of field tests of genetically modified plants and microorganisms. Braunschweig 1998 (in press).

Jaquemond, M. & Tepfer, M. (1998) Satellite RNA-mediated resistance to plant viruses: are the ecological risks well assessed? in A. Hadidi, R.H. Khetarpal and H. Koganezawa (eds.) Plant Virus Disease Control, American Phytopathological Society Press, St. Paul, pp. 94-120.

Jones, A.T. (1998) Control of virus infection in crops through breeding plants for vector resistance, in A. Hadidi, R.H. Khetarpal and H. Koganezawa (eds.) Plant Virus Disease Control, American Phytopathological Society Press, St. Paul, pp. 41-55.

Kaniewski, W.K. (1998) From basic research to genetically improved crops. Proceedings of the Western Washington Horticultural Association Meeting, Seattle 1998, 92-95.

Kaniewski, W.K. & Lawson, E.C. (1994) Biotechnology strategies for virus resistance in plants. Journal of Phytopathology 293-296.

Kaniewski, W.K. & Lawson, E.C. (1998) Coat protein and replicase-mediated resistance to plant viruses, in A. Hadidi, R.H. Khetarpal and H. Koganezawa (eds.) Plant Virus Disease Control, American Phytopathological Society Press, St. Paul, pp. 65-78.

Khetarpal, R.K., Maisonneuve, B., Maury, Y., Chalhoub, B., Dinant, S, Lecoq, H. & Varma, A. (1998) Breeding for resistance to plant viruses, in A. Hadidi, R.H. Khetarpal and H. Koganezawa (eds.) Plant Virus Disease Control, American Phytopathological Society Press, St. Paul, pp. 1-32

Martelli, G.P. (1997) Plant virus taxa: properties and epidemiological characteristics. Journal of Plant Pathology 79, 151-171.

Martin, R.R. (1998) Alternative strategies for engineering virus resistance in plants, in A. Hadidi, R.H. Khetarpal and H Koganezawa (eds.) Plant Virus Disease Control, American Phytopathological Soc. Press, St. Paul, pp. 121-8.

Ogawa, T., Hori, T. & Ishida, I. (1996) Virus-induced cell death in plant expressing the mammalian 2'-5' oligoadenylate system. Nature Biotechnology 14, 1566-1569.

Powell-Abel, P., Nelson R.R., De, B., Hoffman, N., Rogers, S.G., Fraley, R.T. & Beachy, R.N. (1986) Delay of disease development in transgenic plants that express the tobacco mosaic virus coat protein gene. Science 232, 738-743.

Reimann-Philipp, U. & Beachy, R.N. (1993) The mechanism(s) of coat protein-mediated resistance against tobacco mosaic virus. Seminars in Virology 4, 349-356.

Sanford, J.C. & Johnston, S.A. (1985) The concept of parasite derived resistance deriving resistance genes from the parasite's own genome. Journal of Theoretical Biology 113, 395-405.

Sturtevant, AP. & Beachy, R.N. (1993) Virus resistance in transgenic plants: coat protein-mediated resistance, in A. Hiatt (ed.) Transgenic Plants. Marcel Dekker Inc. New York, pp. 93-114.

Tabler, M., Tsagris, M. & Hammond, J. (1998) Antisense RNA and ribozyme-mediated resistance to plant viruses, in A. Hadidi, R.H. Khetarpal and H. Koganezawa (eds.) Plant Virus Disease Control, American Phytopathological Society Press, St. Paul, pp. 79-93.

Watanabe, Y.,Ogawa, T., Takahashi, H., Ishida, I., Takeuchi, Y., Yamamoto, M. & Okada, Y. (1995) Resistance against multiple plant viruses in plants mediated by a double stranded-RNA specific ribonuclease. FEBS Letters 372, 165-168.

Waterworth, H.E. & Hadidi, A. (1998) Economic losses due to plant viruses, in A. Hadidi, R.H. Khetarpal and H. Koganezawa (eds.) Plant Virus Disease Control, American Phytopathological Society Press, St. Paul, pp. 1-13.

White, J.L. (1998. An overview of cultivation of virus-resistant crops in the United States. Proceedings of the 5th International Symposium on biosafety results of field tests of genetically modified plants and microorganisms. Braunschweig 1998 (in press).

Wilson, T.M.A. (1993) Strategies to protect crop plants against viruses: pathogen-derived resistance blossoms. Proceedings of the National Academy of Science USA 90, 3134-3141.

G.T. Scarascia Mugnozza, E. Porceddu & M.A. Pagnotta (Eds.)
Genetics and Breeding for Crop Quality and Resistance, 233-240, 1999

Breeding for virus resistance in barley (*Hordeum vulgare* L.)

W. Friedt, A. Schiemann, K. Scheurer, B. Pellio & F. Ordon
Institute of Crop Science and Plant Breeding I, Justus-Liebig-University, Ludwigstr. 23, D-35390 Giessen, Germany

Abstract: The soil-borne mosaic inducing viruses (BaMMV, BaYMV, BaYMV-2) and the aphid-transmitted BYDV cause severe phytopathological problems in European barley production. Therefore, the development of resistant germplasm and varieties is considered as a major task of barley breeding, today. Concerning the mosaic inducing viruses different resistance genes have been located in the barley genome and the genetic relatedness between resistant varieties was estimated. Furthermore, PCR-based markers - especially RAPDs - have been developed facilitating efficient marker based selection procedures. Corresponding activities have been initiated to study the genetics of quantitatively inherited tolerance to BYDV. Results of these studies are briefly reviewed.

1. Introduction

Virus diseases caused by the soil-borne barley mild mosaic virus (BaMMV), barley yellow mosaic virus (BaYMV, BaYMV-2; Huth, 1989; Huth & Adams, 1990) and different serotypes and strains of the aphid transmitted barley yellow dwarf virus (BYDV; Rochow, 1970), have to be considered as a serious threat to barley cultivation in Europe.

In this respect barley yellow mosaic virus disease is of special importance because the causal agents are transmitted by the soil-borne fungus *Polymyxa graminis* (Toyama & Kusaba, 1970). Therefore, chemical measures against the mosaic inducing viruses (BaMMV, BaYMV, BaYMV-2) are neither efficient nor acceptable for economical and ecological reasons, so that high yield losses can only be prevented by growing resistant cultivars. Resistant varieties have been identified among those released by the Federal Seed Board (Huth et al., 1979), soon after the first discovery of barley yellow mosaic virus disease in Europe in 1978 (Huth & Lesemann, 1978). In genetic studies it turned out that resistance of all these varieties was due to a single recessive gene, *ym4* (Friedt et al., 1990), which most probably derived from the Dalmatian landrace 'Ragusa' and is not effective against BaYMV-

2 (Huth, 1985, 1989). Even today it is presumed that resistance of commercial cultivars to BaMMV and BaYMV with the exception of cv. 'Tokyo' (Anonymus, 1998) entirely rests on this gene. Therefore, screening for and genetic analyses of resistance has been carried out followed by PCR-based molecular analysis in order to get information about the genetic relatedness of resistant germplasm and to identify molecular markers for different resistant genes facilitating efficient marker based selection procedures. Corresponding activities have been initiated to study the genetics of quantitatively inherited tolerance to a German isolate of BYDV-PAV, because the use of tolerant cultivars has to be considered as a favourable solution in integrated pest management programmes in contrast to chemical control of the vectoring aphids, only.

2. Soil-borne mosaic inducing viruses of barley

2.1 Screening for resistance and estimation of genetic relation of resistant germplasm

Due to the fact, that *ym4* is not effective against BaYMV-2 and the very narrow genetic basis of resistance, extensive screening programmes were conducted using mechanical inoculation with BaMMV in the greenhouse and tests on fields infested with BaMMV/BaYMV and BaYMV-2, respectively. In these studies it turned out, that different types of resistance to the yellow mosaic inducing viruses are present within the barley gene pool including genotypes of *H. spontaneum* derived from Turkey and Israel (Ordon et al., 1992, 1993, 1996; Götz & Friedt, 1993; Erdogan et al., 1994). In a next step some of the resistant germplasms have been evaluated for genetic relationship by RAPD-fingerprinting in order to get information about the genetic diversity present within resistant varieties (Ordon et al., 1997a). Based on the analysis of 20 RAPD-primers corresponding to 544 different fragments ranging from 2,691 to 280 bp a high genetic diversity was estimated in contrast to isozyme electrophoresis performed prior to this analysis (Le Gouis et al., 1995). Furthermore, all the varieties derived from different parts of the world, i.e. East Asia, Eastern Europe, Germany, Turkey and Israel, have been grouped according to their origin (Figure 1, Ordon et al., 1997a).

Besides information about the genetic diversity a significant correlation (r=0.87) between the genetic relatedness to high yielding adapted German cultivars and the yieldability of exotic varieties under German growing conditions based on two years' yield trials (Ordon & Friedt, 1994) was calculated. From this fact it may be concluded that varieties more closely related to German cultivars, e.g. Japanese malting barley varieties which derived from crosses to European spring barley (Muramatsu, 1976) or varieties from the Eastern part of Europe seem to be better suited to the rapid introgression of different resistance genes and conversely resistance of more distantly related varieties may be exploited by recurrent selection procedures. However, as all these varieties are of inferior agronomic performance combining resistance with good agronomic traits is a time consuming process, which may be enhanced today by marker based selection procedures.

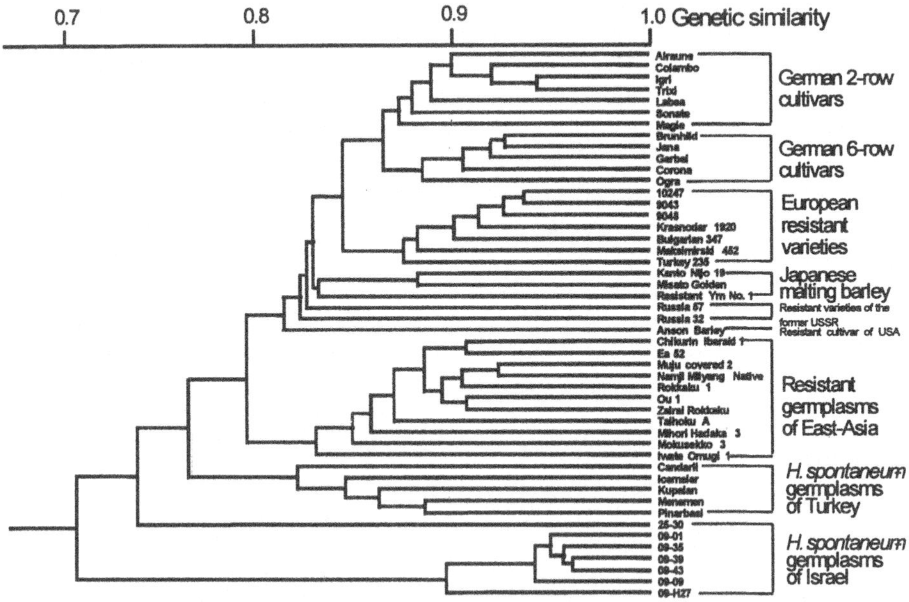

Figure 1. Relationship of barley varieties showing different reactions to BaMMV, BaYMV and BaYMV-2 revealed by cluster analysis performed on genetic similarity estimates (Nei & Li, 1979) calculated from PCR data of 20 RAPD-primers corresponding to 544 bands (Ordon et al., 1997a).

2.2 Genetics of resistance and identification of molecular markers

Besides *ym4*, which has been located to the long arm of chromosome 3 (Kaiser & Friedt, 1992; Graner & Bauer, 1993), different genes are known from Japan, i.e. *ym1* on chromosome 4, *Ym2* on chromosome 1 (Takahashi et al., 1973, Konishi et al., 1997) *ym3* on chromosome 7 (Ukai, 1984; Saeki et al., 1998) and *ym6* on chromosome 3 (Iida & Konishi, 1994). Furthermore, in genetic studies concerning BaMMV-resistance additional genes different from *ym4* and different from each other which are inherited entirely recessively have been detected (Götz & Friedt, 1993; Ordon & Friedt, 1993). In this respect the resistance gene *ym7* has been integrated into the RFLP-map of barley on chromosome 5 (Graner et al., 1995), *ym8* and *ym9* in the distal region of chromosome 4L and *ym11* in the proximal region of this chromosome (Bauer et al., 1997). Besides this, the resistance locus *ym5* derived from 'Resistant Ym No. 1' or 'Mokusekko 3,' respectively (Konishi et al., 1997), which seems to be allelic to *ym4* concerning BaMMV resistance (Götz & Friedt, 1993) but additionally confers resistance to BaYMV-2, has been mapped exactly in the same marker interval on chromosome 3 as *ym4* (Graner et al., 1995, 1998). These genes are well suited to broaden the genetic base of resistance to the soil-borne mosaic inducing viruses thereby ensuring winter barley cultivation in the growing area of infested field. As effective selection for resistance is based on reliable screening methods extensive greenhouse tests by mechanical inoculation or time consuming field tests in later generations (F_4) which depend on the climatic conditions during winter and spring time have to be conducted in order to select resistant plants. In this respect marker based selection procedures have to be considered as a very useful tool to enhance the breeding process and to combine different resistance genes in one breeding line (gene ′pyramiding′). However, as the RFLP-technique is not very well applicable in practical breeding programmes due to several reasons attempts to develop PCR-based markers - especially RAPDs - for different resistance genes have been carried out.

Concerning the resistance gene *ym4* four RAPD-markers, i.e. OP-Z04H660 and OP-L14H910 being closely linked to *ym4* as well as OP-K15H1230 and OP-N11H800 located about 25 cM proximal have been identified (Figure 2, Weyen et al., 1996).

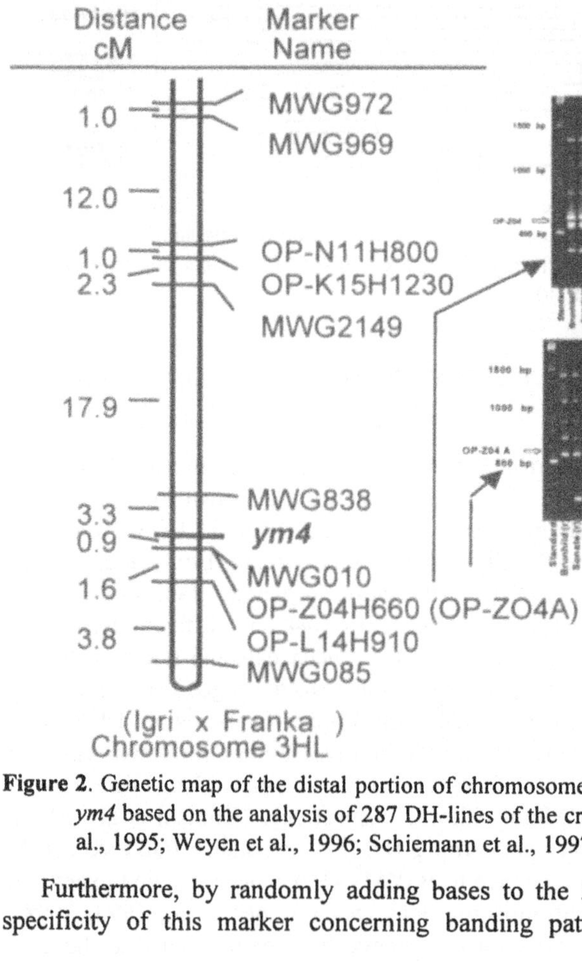

Figure 2. Genetic map of the distal portion of chromosome 3L including the resistance gene *ym4* based on the analysis of 287 DH-lines of the cross ′Igri′ x ′Franka′ (cf. Ordon et al., 1995; Weyen et al., 1996; Schiemann et al., 1997).

Furthermore, by randomly adding bases to the 3′-end of primer OP-Z04 the specificity of this marker concerning banding pattern has been enhanced and because of yielding fragments of different size on resistant and susceptible plants (Figure 2) heterozygous genotypes may be detected by using primer OP-Z04A as well (Figure 2, Schiemann et al., 1997). Like OPZO4H660 (Ordon et al., 1995) this marker perfectly discriminates between resistant (*ym4*) and susceptible cultivars and may be used as a diagnostic marker for *ym4* within commercial German winter barley cultivars, therefore.

Besides this, RAPD markers flanking the resistance locus *ym5* also located on chromosome 3 like OP-C13H590 and OP-C14H950 have been identified among others (Ordon et al., 1997b; Graner et al., 1998) enabling efficient marker based selection procedures for this locus which also confers resistant to BaYMV-2.

Contrary to *ym4* and *ym5* the resistance genes *ym9* derived from 'Bulgarian 347' and *ym11* of the two-rowed variety 'Russia 57', which is resistant to BaYMV-2 as well (Götz & Friedt, 1993) are located in the distal and proximal portion of chromosome 4L, respectively. Concerning these genes different RAPD-markers have been developed (Figure 3, Bauer et al., 1997; Schiemann et al., 1998). In this respect it has to be noted, that *ym9* is efficient against BaMMV, only and that the markers concerning *ym11* have been developed for BaMMV only also, because due to the lack of homozygous DH-lines no simultanous field tests for BaYMV and BaYMV-2 resistance respectively, could be carried out like in the case of *ym5*.

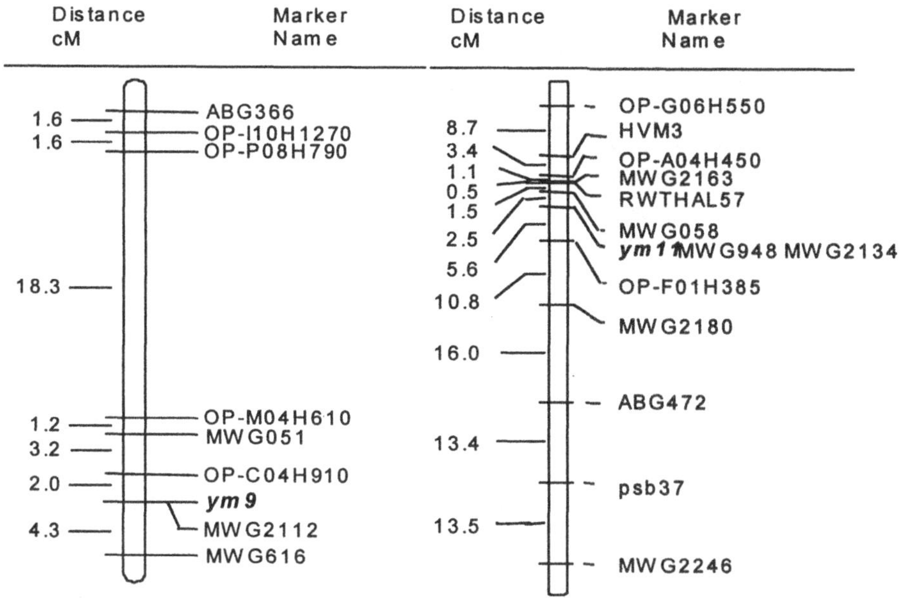

Figure 3. Genetic maps of chromosome 4L including the resistance gene *ym9* (left) and *ym11* (right, cf. Bauer et al., 1997; Schiemann et al., 1998).

Like for other pathogens of barley (cf. Graner, 1996, Ordon et al., 1998) detailed knowledge about the chromosomal location of different genes conferring resistance to the barley yellow mosaic virus complex is available as well as suitable PCR-based markers facilitating efficient marker based selection procedures thereby enhancing the process of combining resistance genes derived from exotic varieties with superior agronomic traits, e.g. in marker based backcrossing programmes. Furthermore, these marker offer the opportunity of combining different resistance genes in one cultivar ('pyramiding') may be leading to longer lasting resistances. Therefore, some PCR-based markers for different resistance genes have already found their way into practical barley breeding (cf. Förster et al., 1997; Tuvesson et al., 1998).

3. Tolerance to barley yellow dwarf virus (BYDV)

In contrast to the soil-borne mosaic inducing virus absolute resistance (immunity) to BYDV has not been found within the barley gene pool, yet (Habekuss, 1994; Huth, 1996). Most widely the *Yd2*-gene which is present e.g. in the cultivar 'Vixen' (Parry & Habgood, 1986) has been used in barley breeding. This gene for which different closely linked RFLP and PCR-based markers are present today (Collins et al., 1996; Ford et al., 1998) confers tolerance to BYDV (Skaria et al., 1985) which depends on the genetic background and the virus isolate used. Besides this, a high level of tolerance to a German isolate of BYDV-PAV has been found in other varieties, e.g. 'Post' or 'MOB 3561' (Huth, 1996). In order to get information about the genetics of tolerance of these germplasms, DH-lines derived from crosses to susceptible cultivars have been tested in field and pot experiments (Scheurer et al., 1998). For this purpose plants were artificially inoculated with virus carrying aphids in the one leaf stage and their reaction to a BYDV-infection was estimated relative to non infected plants of the same DH-line by determining the relative kernel yield, thousand kernel weight, number of ears per plant, number of kernels per ear and plant height. First preliminary results indicate a quantitative variation for these parameters in the DH-populations tested, giving hint to several genes being involved in the complex of BYDV-tolerance. As artificial inoculation procedures with virus bearing aphids - which are essential for a reliable estimation of BYDV-tolerance - are difficult to use in practical barley breeding programmes skeleton maps of the respective DH-populations using RAPDs, AFLPs and SSRs are under construction in order to carry out a QTL-analysis for this trait.

Acknowledgements: We acknowledge the excellent technical assistance of Ines Müller, Kirsten Striedelmeyer, Christiane Happel, Martin Seim and Burkhardt Lather. Thanks are due to Dr. Winfried Huth, BBA Braunschweig, for providing ELISA-antisera and infecting plants with BYDV. Furthermore, financial support of this work by the Gemeinschaft zur Förderung der privaten deutschen Pflanzenzüchtung (GFP), the Bundesministerium für Ernährung, Landwirtschaft und Forsten (G46/89HS88-HS011, G59/93-HS-92HS006, G74/97 HS), and the Deutsche Forschungsgemeinschaft (DFG Projects OR 72/1-1, OR 72/2-1) is gratefully acknowledged.

4. References

Anonymus (1998) Beschreibende Sortenliste für Getreide, Mais, Ölfrüchte, Leguminosen (großkörnig), Hackfrüchte (außer Kartoffeln). Bundessortenamt (Hrsg). Landbuch Verlagsgesellschaft, Hannover.

Bauer E, Weyen J, Schiemann A, Graner A. & Ordon F. (1997) Molecular mapping of novel resistance genes against barley mild mosaic virus (BaMMV). Theor Appl Genet 95: 1263-1269.

Collins N.C., Paltridge N.G., Ford C.M. & Symons R.H. (1996) The Yd2 gene for barley yellow dwarf virus resistance maps close to the centromere on the long arm of barley chromosome 3. Theor Appl Genet 92: 858-864.

Erdogan M., Ordon F. & Friedt W. (1994) Genetics of resistance of *Hordeum spontaneum* Koch from Turkey to the barley yellow mosaic virus complex. Barley Genet Newsletter 23: 41-43.

Ford C.M., Paltridge N.G., Rathjen J.P., Moritz R.L., Simpson R.J. & Symons R.H. (1998) Rapid and informative assays for Yd2, the barley yellow dwarf virus resistance gene, based on the nucleotide sequence of a closely linked gene. Molecular Breeding 4: 23-31.

Förster J., Knaak C. & Jäger-Gussen M. (1997) The use of molecular markers in present cereal breeding programs. In: Buerstmayer H, Ruckenbauer P (eds), Application of marker aided selection in cereal breeding programms. Book of Abstracts EUCARPIA Cereal Section, Sept 22-23 1997, Tulln Austria pp 36-37.

Friedt W., Ordon F., Götz R. & Kaiser R. (1990) Bodenbürtige Krankheiten, eine fortdauernde Herausforderung für die Pflanzenzüchtung - beleuchtet am Beispiel der Gelbmosaikvirose der Gerste. Ber Arbeitstag Saatzuchtl Gumpenstein 40: 27-38.

Götz R. & Friedt W. (1993) Resistance to the barley yellow mosaic virus complex - Differential genotypic reactions and genetics of BaMMV-resistance of barley (*Hordeum vulgare* L.). Plant Breeding 111: 125-131.

Graner A. (1996) Molecular mapping of genes conferring disease resistance: The present state and future aspects. In: Scoles G, Rossnagel B (eds.) Proc. V Intern. Oat Conf. & VII Intern. Barley Genetics Symp. Inv. Papers, Saskatoon, Canada, University Extension Press, Saskatoon, Sakatchewan, pp 157-166.

Graner A. & Bauer E. (1993) RFLP mapping of the ym4 virus resistance gene in barley. Theor Appl Genet 86: 689-693.

Graner A., Bauer E., Kellermann A., Proeseler G., Wenzel G. & Ordon F. (1995) RFLP analysis of resistance to the barley yellow mosaic virus complex. Agronomie 15: 475-479.

Graner A., Streng S., Kellermann A., Schiemann A., Bauer E., Waugh R., Pellio B. & Ordon F. (1998) Molecular mapping of the rym5 locus encoding resistance to different strains of the barley yellow mosaic virus complex. Theor Appl Genet (in press).

Habekuss A. (1994) Evaluation of winter barley for resistance to barley yellow dwarf virus. Genet Pol 35B: 199-202.

Huth W. (1985) Versuche zur Virusdiagnose und Resistenzträgererstellung in Gerste gegen Barley Yellow Mosaic Virus. Vortr Pflanzenzüchtg 9: 107-120.

Huth W. (1989) Ein weiterer Stamm des Barley yellow mosaic virus (BaYMV) gefunden. Nachrichtenbl Deut Pflanzenschutzd 41: 6-7.

Huth W. (1996) Möglichkeiten und Grenzen der Züchtung von Getreidesorten mit Resistenz gegenüber den Gelbverzwergungsviren - aus virologischer Sicht. Ber. Arbeitstag. Saatzuchtleiter, Gumpenstein, 46: 1-14.

Huth W., Lesemann DE (1978) Eine für die Bundesrepublik neue Virose an Wintergerste. Nachrichtenbl Deut Pflanzenschutzd 30: 184-185.

Huth W. & Adams M.J. (1990) Barley yellow mosaic virus (BaYMV) and BaYMV-M: two different viruses. Intervirology 31: 38-42.

Huth W., Wedler W. & Radtke W. (1979) Über die Verbreitung des Gelbmosaikvirus der Gerste in Deutschland und die Möglichkeit durch Anbau widerstansfähiger Sorten Ertragsverluste zu vermeiden. Mitt Biol Bundesanst 191: 163.

Iida Y. & Konishi T. (1994) Linkage analysis of a resistance gene to barley yellow mosaic virus strain II in two rowed barley. Breeding Science 44: 191-194.

Kaiser R. & Friedt W. (1992) Gene for resistance to barley mild mosaic virus in German winter barley located on chromosome 3L. Plant Breeding 108: 169-172.

Konishi T., Ban T., Iida Y. & Yoshimi R. (1997) Genetic analysis of disease resistance to all strains of BaYMV in a chinese barley landrace, Mokusekko 3. Theor Appl Genet 94: 871-877.

Le Gouis .J, Erdogan M., Friedt W,. & Ordon F. (1995) Potential and limitations of isozymes for chromosomal localization of resistance genes against barley mild mosaic virus (BaMMV). Euphytica 82: 25-30.

Muramatsu M.. (1976) Breeding of malting barley which is resistant to barley yellow mosaic. Proc 3rd Int Barley Genet Symp Garching, Germany, Barley Genetics III, pp 476-485.

Nei M. & Li W.H. (1979) Mathematical model for studying genetic variation in terms of restriction endonucleases. Proc Natl Acad Sci USA 76: 5269-5273.

Ordon F. & Friedt W. (1993). Mode of inheritance and genetic diversity of BaMMV-resistance of exotic barley germplasms carrying genes different from 'ym4'. Theor Appl Genet 86: 229-233.

Ordon F. & Friedt W. (1994) Agronomic traits of exotic barley germplasms resistant to soil-borne mosaic-inducing viruses. Genetic Resources and Crop Evolution 41: 43-46.

Ordon F., Bauer E. & Friedt W., Graner A (1995) Marker-based selection for the ym4 BaMMV-resistance gene in barley using RAPDs. Agronomie 15: 481-485.

Ordon F., Erdogan M. & Friedt W. (1992) Genetics of resistance of barley to soil-borne mosaic viruses. Reproductive Biology and Plant Breeding, Book of Poster Abstracts, XIIIth Eucarpia Congress Angers, France, pp 707-708.

Ordon F., Götz R. & Friedt W. (1993) Genetic stocks resistant to barley yellow mosaic viruses (BaMMV, BaYMV, BaYMV-2) in Germany. Barley Genet Newsletter 22: 44-48.

Ordon F., Weyen J., Korell M. & Friedt W. (1996) Exotic barley germplasms in breeding for resistance to soil-borne viruses. Euphytica 92: 275-280.

Ordon F., Schiemann A. & Friedt W. (1997a) Assessment of the genetic relatedness of barley accessions resistant to soil-borne mosaic inducing viruses (BaMMV, BaYMV, BaYMV-2) using RAPDs. Theor Appl Genet 94: 325-330.

Ordon F., Weyen J., Schiemann A., Pellio B., Bauer E., Graner A. & Friedt W. (1997b) RAPD-based selection in breeding for resistance against soil-borne viruses in barley. In: Buerstmayer H., Ruckenbauer P . (eds) Application of marker aided selection in cereal breeding programms. Book of Abstracts EUCARPIA Cereal Section, Sept 22-23 1997, Tulln Austria, pp 26-27.

Ordon F., Wenzel W. & Friedt W. (1998) II Recombination: Molecular markers for resistance genes in major grain crops. Progress in Botany 59: 49-79.

Parry A.L. & Habgood R.M. (1986) Field assesement of the effectiveness of a barley yellow dwarf resistance gene following ist transference from spring to winter barley. Ann Appl Biol 108: 395-401.

Rochow W.F. (1970) Barley yellow dwarf virus: Phenotypic mixing and vector specificity. Science 167: 875-78.

Saeki M., Miyazaki C., Hirota N., Saito A., Ito K. & Konishi T. (1998) RFLP mapping of BaYMV resistance gene ym3 in barley. Barley Genet Newsletter 28: 61-63.

Scheurer K., Huth W., Habekuß A., Friedt W. & Ordon F. (1998) Züchtung auf Toleranz der Gerste gegen Barley Yellow Dwarf Virus. Vortr. Pflanzenzüchtg. 42: 51-53.

Schiemann A., Graner A., Friedt W. & Ordon F. (1997). Specificity enhancement of a RAPD marker linked to the BaMMV/BaYMV resistance gene ym4 by randomly added bases. Barley Genet Newsletter 26: 63-65.

Schiemann .A, Bauer E., Graner A., Friedt W. & Ordon F. (1998) RAPD-markers linked to the BaMMV-resistance gene ym9. Barley Genet Newsletter 28: 19-22.

Skaria M., Lister R.M., Foster J.E. & Shaner G. (1985) Virus content as an index of symptomatic resistance to barley yellow dwarf virus in cereals. Phytopathology 75: 212-216.

Takahashi R., Hayashi J., Inouye T., Moriya I.& Hirao C. (1973) Studies on resistance to yellow mosaic disease in barley. I. Tests for varietal reactions and genetic analysis of resistance to the disease. Ber Ohara Inst 16: 1-17.

Toyama A. & Kusaba T. (1970) Transmission of soil-borne barley yellow mosaic virus. 2. *Polymyxa graminis* Led. as vector. Ann Phytopath Soc Japan 36: 223-229.

Tuvesson S., v Post L., Öhlund R., Hagberg P., Graner A., Svitashev S., Schehr M. & Elovsson R. (1998) Molecular breeding for the BaMMV/BaYMV resistance gene ym4 in winter barley. Plant Breeding 117: 19-22.

Ukai Y. (1984) Genetic analysis of a mutant resistant to barley yellow mosaic virus. Barley Genet Newsletter 14: 31-33.

Weyen J., Bauer E., Graner A., Friedt W. & Ordon F. (1996) RAPD-mapping of the distal portion of chromosome 3 of barley including the BaMMV/BaYMV resistance gene ym4. Plant Breeding 115: 285-287.

G.T. Scarascia Mugnozza, E. Porceddu & M.A. Pagnotta (Eds.)
Genetics and Breeding for Crop Quality and Resistance, 241-250, 1999

Inheritance of resistance to SCMV and MDMV in european maize

Th. Lübberstedt, X.C. Xia, M.L. Xu, L. Kuntze & A.E. Melchinger
Institute of Plant Breeding, Seed Science, and Population Genetics, University of Hohenheim, 70593 Stuttgart, Germany

Abstract: Sucarcane mosaic virus (SCMV) and maize dwarf mosaic virus (MDMV) cause important virus diseases of maize in Europe. In field and greenhouse experiments, 122 maize inbreds (45 flint and 77 dent lines) were evaluated for their reaction to SCMV, MDMV, and other viruses after artificial inoculation. Three dent inbreds (D21, D32, FAP1360A) displayed complete resistance against SCMV, MDMV, Johnson grass mosaic virus, and sorghum mosaic virus. All other inbreds were either partially resistant (4 lines) or completely susceptible. Resistant inbreds D21, D32, and FAP1360A were crossed with susceptible inbreds F7, KW1292, D408, and D145 to produce four F_2 and three backcrosses to the susceptible parent (BC1), and were investigated in two field environments and a greenhouse trial. RFLP and SSR markers on Chromosomes 3, 6, and 10 were used to locate resistance genes against SCMV in the maize genome. These regions are known to confer resistance to other viruses in maize such as MDMV, wheat streak mosaic virus, maize mosaic virus, and high plains virus, and also against different fungi. Segregation of SCMV resistance in F_2 and BC_1 generations fitted to different gene models depending on the environment. Marker analyses detected two genes designated *Scm1* on Chromosome 6 and *Scm2* on Chromosome 3. For cross D32xD145, 220 F_3 families were tested in two field environments and genotyped with 90 RFLP and SSR markers to detect additional SCMV resistance loci, analyse the relationship to fungal resistances, and investigate the mode of gene action for *Scm1* and *Scm2*. Both *Scm1* and *Scm2* displayed an additive mode of gene action and significant additive-dominance interactions. Presently, fine mapping of *Scm1* and *Scm2* is done employing AFLPs and BC_7 families in a "bulked segregant analysis".

1. Introduction

Sucarcane mosaic virus (SCMV) and maize dwarf mosaic virus (MDMV) cause significant losses in grain and forage yield in susceptible genotypes of maize and related crops (Fuchs & Grüntzig, 1995). Maize dwarf mosaic is the most widespread virus disease of maize in the U.S. Corn Belt (Louie et al., 1991), while in Germany SCMV is more prevalent than MDMV and of increasing importance (Fuchs et al., 1996). Both taxonomically related viruses belong to the sugarcane mosaic potyvirus subgroup (Shukla et al., 1989). SCMV and MDMV are naturally transmitted in a non-persistant manner by aphids, but artificial transmission by different mechancial inoculation methods is rather efficient.

Direct chemical control of SCMV and MDMV is impossible. The control of aphid vectors by chemical means is hampered by the non-persistant mode of virus transmission. Hence, cultivation of resistant maize varieties is the most efficient method of virus control. Resistance against MDMV/SCMV has also been successfully accomplished by genetic engineering using the coat protein approach (Murry et al., 1993). However, patent restrictions and limited acceptance of transgenic varieties by consumers might impair a widespread use of transgenic resistances. Furthermore, effective natural resistances against SCMV and MDMV have been identified in U.S. as well as in early European maize germplasm (Louie et al., 1990; Kuntze et al., 1997).

Inheritance of complete natural resistance to different strains of SCMV and MDMV as well as against wheat streak mosaic potyvirus (WSMV) has been most intensively studied for U.S. inbred line Pa405 (Louie et al., 1991; McMullen et al., 1994; Simcox et al., 1995). Depending on the susceptible parent line and environmental conditions chosen, segregation analyses indicated the presence of one to five genes conferring resistance to SCMV, MDMV, and WSMV in Pa405. Restriction fragment length polymorphism (RFLP) analyses mapped a dominant major gene (*Mdm1*) conferring resistance to MDMV to Chromosome 6S, cosegregating with the nucleolus organizer region (NOR) (Simcox et al., 1995). Three genes causing resistance to WSMV (*Wsm1*, *Wsm2*, *Wsm3*) were identified on Chromosomes 3L, 6S, and 10L (McMullen et al., 1994).

In early-maturing European maize germplasm, three dent inbreds (D21, D32, FAP1360A) displayed complete resistance to SCMV, MDMV, as well as against Johnson grass mosaic potyvirus (JGMV) and sorghum mosaic potyvirus (SrMV) (Kuntze et al., 1995). Segregation analyses suggested the presence of one to three SCMV resistance genes strongly affected by differing environmental conditions (Melchinger et al., 1998). Initial marker analyses mapped two genes involved in SCMV resistance, *Scm1* and *Scm2*, located on chromosomes 6S and 3L.

The objectives of this study were to (i) locate additional SCMV resistance genes by QTL mapping using 220 F_3 lines of cross D32 x D145 (SCMV susceptible) and 95 RFLP or SSR markers, (ii) study the mode of gene action of *Scm1* and *Scm2*, and (iii) identify closely linked markers for *Scm1* and *Scm2* by "bulked segregant analysis" (BSA) using advanced backcross generations of cross (FAP1360A x F7) x F7 (SCMV susceptible) as a first step towards genetic fine mapping of both genome regions.

2. Materials and Methods

2.1 *Plant Materials*

The materials basically involved the three European SCMV resistant inbreds D21, D32, and FAP1360A, as well as the two susceptible inbreds F7 and D145. All resistant and susceptible inbreds have been identified in a previous study by Kuntze et al. (1997). Inbreds D21 and D32 are related by pedigree with coancestry coefficient f=0.38. Both lines have Iodent background and one common dent parent (A632). Line FAP1360A has a largely different genetic background than D21 and D32 except for one ancestor in common, inbred Co125. A common origin of both highly susceptible flint lines F7 and D145 is population 'Lacaune'. The ancestor line F2 of D145 as well as F7 trace back to this population.

2.2 *Field trials*

2.2.1 Test of Allelism

The three resistant European inbreds were mutually crossed to produce F_2 populations D21 x D32, D21 x FAP1360A, and D32 x FAP1360A, tested for SCMV resistance with parent lines as checks. The experiment was conducted at Hohenheim in 1998 planting >340 seeds per F_2 in single-row plots 3 m long and 0.75 m apart.

2.2.2 QTL Analysis

Inbred lines D32 and D145 were crossed to produce 220 F_3 lines, maintained as 'Immortalized F_2' population (IF_2). Population D32 x D145 was evaluated at Eckartsweier in 1996 and at Hohenheim in 1997 for resistance to SCMV. The experiment included 230 entries in a 23 x 10 alpha design (Patterson and Williams 1976) with two replications: the 220 F_3 (1996) or IF2 (1997) families and both parent lines included five times each. Plots consisted of single rows, 3 m long and 0.75 m apart and were thinned to 9 plants m^{-2} with a total of 20 plants.

2.2.3 Bulked Segregant Analysis

Inbred FAP1360A was five times backcrossed to F7 with two generations per year. SCMV evaluation was performed during the summer seasons at Hohenheim in 1995, 1996, and 1997 in the BC_1 (backcross 1), BC_3, and BC_5 generations, respectively. BC progenies without SCMV symptoms were employed to produce BC_2 and BC_4 plants, which were randomly chosen to produce the next BC generation in the winter nursery. A total of 75 BC_5 families were planted in single-row plots of 20 plants in 1997.

2.3 *Inoculum preparation, inoculation, and SCMV scoring*

Virus inoculum for testing resistance against SCMV isolate 'Seehausen' was prepared as described by Fuchs & Grüntzig (1995). Young leaves with typical mosaic symptoms of the SCMV-infected maize variety 'Bermasil' were homogenized using 5 volumes of a 0.01 M phosphate buffer at pH 7.0. Carborund was added to the sap. During sap preparation and mechanical inoculation, the inoculum was kept at +4°C. Plants at the three- to four-leaf stage were mechanically inoculated twice at a weekly interval by an air brush technique with a tractor-mounted air compressor at a constant pressure of 800 kPa (Fuchs et al., 1996). Resistance was evaluated by scoring each plant for presence or absence of mosaic symptoms at weekly intervals, beginning 7 day after the second inoculation.

2.4 RFLP, SSR, and AFLP analyses

For QTL mapping, 95 marker loci were employed to genotype 220 F_3 lines of cross D32 x D145. Genomic DNA was extracted from ground freeze-dried leaf material according to the CTAB method (Hoisington et al., 1994). For RFLP analyses, DNA was digested with restriction enzymes *Eco*RI, *Eco*RV, *Hin*dIII, or *Bam*HI, separated by 1% agarose gel electrophoresis, and vacuum blotted onto nylon membranes. RFLP markers from the standard probe collection available at the University of Missouri, Columbia (Gardiner et al., 1993) were labeled with digoxigenin-dUTP, and DNA fragments detected by means of the chemiluminescence CSPD protocol described by Hoisington et al. (1994). Sequence of primers for SSR markers were obtained from the maize data base (http://teosinte.agron.missouri.edu/Coop/SSR_Probes/SSR1.htm). PCR amplification and MetaPhor gel electrophoresis were performed according to Lübberstedt et al. (1998). A complete list of all RFLP and SSR markers and map data employed in this study can be obtained from the authors upon request.

AFLP analyses were performed for BSA of cross FAP1360A x F7 by using the AFLP™ Analysis System I Kit (GibcoBRL, Life Technologies) with P33 labeled oligonucleotides according to the supplier's instructions and as described by Vos et al. (1995). All 54 primer combinations recommended for maize in the AFLP™ Kit were employed to investigate four DNA samples including the parents FAP1360A (Pr), F7 (Ps) as well as bulks of resistant (Br) and susceptible (Bs) BC_5 plants.

2.5 Statistical analysis (QTL mapping)

Adjusted entry means and effective error mean squares of trait data were used to compute the combined analysis of variance and covariance across environments. The resulting entry means across environments were employed for QTL mapping using the composite interval mapping approach implemented in the program PLABQTL (Utz & Melchinger, 1996). A LOD threshold of 3.0 was chosen for declaring a putative QTL. QTL positions were determined at the LOD maxima in the regions under consideration.

3. Results and Discussion

3.1 *Test of allelism*

Inbreds D21, D32, and FAP1360A showed the same performance to SCMV, MDMV, JGMV, and SrMV inoculation (Kuntze et al., 1995) as well as in tests of allelism for SCMV resistance with potyvirus resistant inbred Pa405 (Melchinger et al., 1998). This raises the question, whether the same or different genes mediate virus resistance in these three lines. Interestingly, within each of the three F_2 populations of crosses D21 x D32, D21 x FAP1360A, and D32 x FAP1360A, SCMV susceptible plants were observed (Table 1). In most cases symptoms appeared late and were frequently restricted to leaf stripes as opposed to highly susceptible genotypes with symptoms systemically spread over the whole plant.

The high number of susceptible plants in the F_2 population of cross D21 x D32 (Table 1) suggests the action of different sets of unlinked loci responsible for SCMV resistance in each of both inbreds. If only one but different dominant major genes were present in D21 and D32, the proportion of susceptible plants in the respective F_2 population would have been expected to be 1/16, whereas the observed proportion of susceptible F_2 plants was 1/5. Possible explanations are partially dominant major resistance genes or cancelled complementary epistatic interactions present between ≥ 2 resistance genes within the parent lines.

In the F_2 population of crosses D21 x FAP1360A and D32 x FAP1360A only one and eight out of 383 and 348 F_2 plants in total, respectively, were susceptible. Very likely FAP1360A contains SCMV resistance genes in the same genomic regions involved in resistance of both inbreds D21 as well as D32. Genetical explanations for the low number of susceptible F_2 plants observed are (i) different major resistance genes present in FAP1360A as compared to D21 or D32, mapping to the same genomic regions, (ii) different minor resistance genes present in the three inbreds and required for complete resistance to SCMV, if treated as threshold character. Furthermore, in some cases "resistant" genotypes might become infected as was shown by Louie (1995) after mechanically transmitting MDMV to inbred Pa405 by vascular puncture of maize kernels. Finally, the presence of false F_2 individuals due to mispollination cannot be ruled out without additional molecular investigations.

Table 1. Tests of allelism for SCMV resistance genes employing F_2 populations from crosses between the three SCMV resistant inbreds D21, D32, anf FAP1360A evaluated at Hohenheim in 1998, 21, 35, and 49 days after the second inoculation (DAI). The resistant inbreds and susceptible inbred D145 were included as checks.

Population / Inbred	Total No. of plants	No. of susceptible plants 21 DAI	35 DAI	49 DAI
F_2(D21 x D32)	341	3	25	65
F_2(D21 x FAP1360A)	383	0	0	1
F_2(D32 x FAP1360A)	348	0	0	8
D21 + D32 + FAP1360A	112	0	0	0
D145	62	57	61	62

Interestingly, for none of the three European SCMV resistant inbreds D21, D32, and FAP1360A, susceptible F_2 plants were found in tests of allelism with inbred Pa405 (Melchinger et al., 1998). A genetic explanation would be the accumulation of all SCMV resistance genes present in D21, D32, and FAP1360A within inbred Pa405. However, non-genetic reasons for the lack of susceptible F_2 plants are (i) smaller F_2 populations employed for the tests of allelism with Pa405 (200 individuals per F_2 population) and (ii) a comparatively high infection level in 1998 reducing the amount of "escapes", *i.e.*, susceptible genotypes without SCMV symptoms.

3.2 QTL analysis

Two major SCMV resistance genes, *Scm1* (Chromosome 6S) and *Scm2* (Chromosome 3L), were detected by RFLP and SSR analyses restricted to Chromosomes 3L, 6S, and 10L using a rather limited F_2 population (30 F_2 individuals) of cross D32 x D145 (Melchinger et al., 1998). In order to confirm and extend these investigations, 220 F_3 families derived from the same cross D32 x D145 were evaluated at two environments for SCMV resistance and genotyped by 94 RFLP or SSR markers covering the whole genome. Percentage of susceptible plants after mechanical SCMV inoculation was scored for seven times at weekly intervals.

The highest LOD values for each scoring date were obtained by the *Scm1* region on Chromosome 6S near SSR phi075 (Table 2). The mode of gene action of *Scm1* was partially dominant. Due to a limited resolution of QTL analyses it remains an open question, whether the *Scm1* region harbours only one major gene or ≥ 2 closely linked resistance genes as indicated for some scoring dates, comparable to the *Rp1* common rust resistance gene cluster on Chromosome 10 (Hooker, 1985). Absence of the *Scm1* allele derived from D32 was signficantly associated with a decrease in plant height and an increase in the date of flowering as well as the anthesis-silking interval. This is in agreement with the morphological changes caused by SCMV infection for susceptible genotypes such as stunting.

Table 2. QTL analysis: 220 F_3 families of cross D32 x D145 were evaluated across two environments for the percentage of infected plants 1, 3, 5, and 7 (Vir1-7) weeks after SCMV inoculation; putative QTL were displayed by their LOD values and mode of gene action (PD: partially dominant; OD: overdominant).

Chromo-	LOD values				Gene
some	Vir1	Vir3	Vir5	Vir7	Action
1	4.1	4.8	3.3	3.2	OD
3 (*Scm2*)	-	9.5	14.6	15.8	PD
5	-	-	3.0	4.2	OD
6 (*Scm1*)	38.6	32.4	34.3	29.6	PD
10	4.0	-	-	-	PD

Presence of *Scm2* close to RFLP marker umc102 was confirmed in this study showing also partially dominant gene action (Table 2). *Scm2* was not detected for the first scoring date and LOD values continuously increased until the final scoring. Although LOD values of *Scm1* generally exceeded those of *Scm2*, both loci can be considered as major loci. Both loci displayed significant digenic epistatic interactions.

LOD values of the three other genomic regions affecting SCMV resistance identified on Chromosomes 1, 5, and 10 were much smaller than those of *Scm1* and *Scm2* (Table 2). Gene action was either partially dominant (Chromosome 10) or overdominant (Chromosomes 1 and 5). The map position of the QTL located on Chromosome 10 was similar as the major resistance gene Wsm3 acting against WSMV (McMullen et al., 1994). Hence, reverting Robertson's Hypothesis (Robertson, 1985) on the relationship between major genes and QTL, the QTL for SCMV resistance identified in this study might be indicators of additional major *Scm* genes within maize germplasm.

3.3 *Bulked Segregant Analysis*

Marker-assisted selection (MAS) for improved SCMV resistance might be more efficient than conventional phenotypic selection because of (i) a strong influence of environment on the expression of SCMV resistance and (ii) the high efforts needed for resistance evaluation. An important prerequisite for reliable MAS is the availability of markers closely linked with the target gene(s). Probably the most efficient way of directed screening for tightly linked markers for major genes at present is BSA combined with AFLP markers. This approach was applied to the BC_5 generation of FAP1360A x F7, backcrossed five times to F7 with SCMV evaluation and phenotypic selection for SCMV resistant BC plants in BC_1, BC_3, and BC_5. By using a higher BC generation for AFLP-based BSA, the background noise of genomic regions not involved in SCMV resistance was minimized. Furthermore, accumulation of recombination events in BC_5 facilitates the identification of markers closely linked to the target gene(s).

The proportion of BC_5 families containing plants without SCMV symptoms was about 1/4. This is in agreement with two complementary acting genes conferring resistance to SCMV. In total, 54 AFLP primer combinations were employed for BSA resulting in 1,260 polymorphic out of 3,007 AFLP bands. Parent FAP1360A contributed 675 polymorphic bands, with 23 bands present in the resistant bulk but absent or significantly weaker in the bulk of susceptible plants (Figure 1). All these 23 AFLP marker bands mapped either to the *Scm1* region (12 markers) or to the *Scm2* region (11 markers). These findings fitted well with the phenotypic observations suggesting the segregation of two complementary acting resistance genes. In addition, the same genomic regions harbouring major SCMV resistance genes were detected for two different SCMV resistant inbreds (D32, FAP1360A) using two different approaches (QTL analysis, BSA) underlining their importance. However, recalling the test of allelism between D32 and FAP1360A, it remains to be proven or disproven that *Scm1* and *Scm2* are identical in both lines.

In this BSA, *Scm1* mapped into the RFLP marker interval umc85/bnl6.29 as did *Mdm1* (Simcox et al., 1995). One AFLP marker was absolutely linked with *Scm1* in our BC_5 materials. In contrast to the study of Simcox et al. (1995) employing a very large mapping population (>7,000 F_2 individuals), no absolute linkage was found between NOR and *Scm1* like between NOR and *Mdm1*, raising the question of identity of *Mdm1* and *Scm1*. Some BC_5 individuals without SCMV symptoms lacked any FAP1360A-derived marker band in the *Scm2* region. Probably presence of *Scm1* was sufficient for some of the BC_5 plants to be not infected by SCMV.

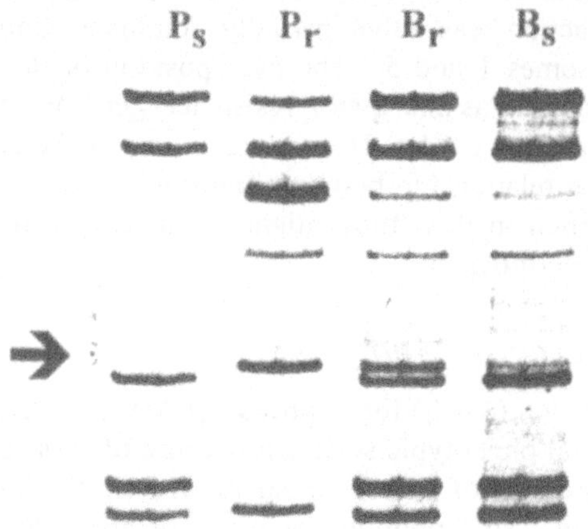

Figure 1. Example of an AFLP band (arrow) putatively linked to a SCMV resistance gene, identified by bulked segregant analysis by comparison of susceptible parent F7 (Ps), resistant parent FAP1360A (Pr), and the bulks of all SCMV resistant (Br) or susceptible (Bs) BC_5 plants.

4. Summary and Conclusions

Tests of allelism between the three European inbred lines D21, D23, and FAP1360A suggested different genetic compositions with respect to SCMV resistance. Lines D32 and FAP1360A were analysed in more detail by QTL analysis and bulked segregant analysis, respectively. Two genome regions on Chromosome 6S (*Scm1*) and Chromosome 3L (*Scm2*) were found to be most important for SCMV resistance in both inbreds. Three additional genome regions of minor importance were identified in the QTL approach on Chromosomes 1, 5, and 10. *Scm1* region displayed higher LOD values than the *Scm2* region in the QTL analysis. In addition, *Scm1* suppresses SCMV infection already in early stages in contrast to *Scm2*. Therefore, *Scm1* seems to be most valuable for application of SCMV control in maize hybrid breeding. However, a high level of SCMV resistance can be accomplished by simultaneous selection for both *Scm1* and *Scm2*. Due the partial dominance of SCMV resistance of both loci/regions, presence of *Scm1* and *Scm2* in only one of both hybrid parents should be sufficient. AFLP markers closely linked to *Scm1* and *Scm2* were produced by AFLP-based bulked segregant analysis. However, the question of presence of only one major gene in each of both genomic regions (*Scm1*, *Scm2*) or alternatively clustering of resistance genes within these regions needs further investigations.

Acknowledgements: The authors thank Dr. Klein and S. Pluskat at Eckartsweier, F. Mauch, F. Oeynhausen, and B. Devezi-Savula at Hohenheim for conducting field experiments. We are grateful to Prof. Dr. E. Fuchs and Dr. M. Grüntzig, Martin-Luther-University of Halle-Wittenberg, for serological tests as well as to Prof. Dr. H.F. Utz for biometrical support. This work was supported by the "Deutsche Forschungsgemeinschaft" (DFG), grant LU601/2.

5. References

Fuchs, E. & Grüntzig, M. (1995) Influence of sugarcane mosaic virus (SCMV) and maize dwarf mosaic virus (MDMV) on the growth and yield of two maize varieties, *J. Plant Dis. Protect.* **102**, 44-50.

Fuchs, E., Grüntzig, M., Kuntze, L., Oertel, U. & Hohmann, F. (1996) Zur Epidemiologie der Potyviren des Maises in Deutschland, *In*: Bericht über die 46. Arbeitstagung der Arbeitsgemeinschaft der Saatzuchtleiter im Rahmen der "Vereinigung österreichischer Pflanzenzüchter" **46**, 43-49.

Gardiner, J.M., Coe, E.H., Melia-Hancock, S., Hoisington, D.A. & Chao, S. (1993) Development of a core RFLP map in maize using an immortalized F2 population, *Genetics* **134**, 917-930.

Hoisington, D.A., Khairallah, M.M. & Gonzales-de-Leon, D. (1994) Laboratory protocolls, CIMMYT Applied Molecular Genetics Laboratory, CIMMYT, Mexico.

Hooker, A.L. (1985) Corn and sorghum rusts, *In*: A.P. Roelfs and W.R. Bushnell (eds.) The Cereal Rusts, Academic Press, New York.

Kuntze, L., Fuchs, E., Grüntzig, M., Schulz, B., Hennig, U., Hohmann, F. & Melchinger, A.E. (1995) Evaluation of maize inbred lines for resistance to sugarcane mosaic virus (SCMV) and maize dwarf mosaic virus (MDMV), *Agronomie* **15**, 463-467.

Kuntze, L., Fuchs, E., Grüntzig, M., Schulz, B., Klein, D. & Melchinger, A.E. (1997) Resistance of early-maturing European maize germplasm to sugarcane mosaic virus (SCMV) and maize dwarf mosaic virus (MDMV), *Plant Breed.* **116**, 499-501.

Louie, R. (1995) Vascular puncture of maize kernels for the mechanical transmission of maize white line mosaic virus and other viruses of maize, *Phytopathology* **85**, 139-143.

Louie, R., Knoke, J.K. & Findley, W.R. (1990) Elite maize germplasm: reaction to maize dwarf mosaic and maize chlorotic dwarf viruses, *Crop Sci.* **30**, 1210-1215.

Louie, R., Knoke, J.K. & Findley, W.R. (1991) Genetic basis of resistance in maize to five maize dwarf mosaic virus strains, *Crop Sci.* **31**, 14-18.

Lübberstedt, T., Dußle, C. & Melchinger, A.E. (1998) Application of microsatellites from maize to teosinte and other relatives of maize, *Plant Breed.* (in press).

McMullen, M.D., Jones, M.W., Simcox, K.D. & Louie, R. (1994) Three genes control resistance to wheat streak mosaic virus in the maize inbred Pa405, *Mol. Plant-Microbe Interact.* **7**, 708-712.

Melchinger, A.E., Kuntze, L., Gumber, R.K., Lübberstedt, T. & Fuchs, E. (1998) Genetic basis of resistance to sugarcane mosaic virus in European maize germplasm, *Theor. Appl. Genet.* **96**, 1151-1161.

Murry, L.E., Elliott, L.G., Capitant, S.A., West, J.A., Hanson, K.K., Scarafia, L., Johnston, S., Deluca-Flaherty, C., Nichols, S., Cunanan, D., Dietrich, P.S., Mettler, I.J., Dewald, S., Warnick, D.A., Rhodes, C.A., Sinibaldi, R. & Brunke, K.J. (1993) Transgenic corn plants expressing MDMV strain-B coat protein are resistant to mixed infections of maize dwarf mosaic virus and maize chlorotic mottle virus, *Bio/Technol.* **11**, 1559-1564.

Patterson, H.D. & Williams, E.R. (1976) A new class of resolvable incomplete block designs, *Biometrika* **63**, 83-92.

Robertson, D.S. (1985) A possible technique for isolating genic DNA for quantitative traits in plants, *J. Theor. Biol.* **117**, 1-10.

Shukla, D.D., Tosic, M., Jilka, J., Ford, R.E., Toler, R.W. & Langham, M.A.C. (1989) Taxonomy of potyviruses infecting maize, sorghum, and sugarcane in Australia and the United States as determined by reactivities of polyclonal antibodies directed towards virus-specific N-termini of coat protein, *Phytopathology* **59**, 699-702.

Simcox, K.D., McMullen, M.D. & Louie, R. (1995) Co-segregation of the maize dwarf mosaic virus resistance gene, Mdm1, with the nucleolus organizer region in maize, *Theor. Appl. Genet.* **90**, 341-346.

Utz, H.F. & Melchinger, A.E. (1996) PLABQTL: A program fro composite interval mapping of QTL. J. Quant. Trait Loci. http://probe.nalusda.gov:8000/otherdocs/jqtl.

Vos, P., Hogers, R., Bleeker, M., Reijans, M., van de Lee, T., Hornes, M., Frijters, A., Pot, J., Peleman, J., Kuiper, M. & Zabeau, M. (1995) AFLP: a new technique for DNA fingerprinting. *Nuc. Acids Res.* **23**, 4407-4417.

G.T. Scarascia Mugnozza, E. Porceddu & M.A. Pagnotta (Eds.)
Genetics and Breeding for Crop Quality and Resistance, 251-256, 1999

Strategies to search for new Citrus Tristeza Virus resistant genotypes in a germplasm bank

M.J. Asins, P.F. Mestre, L. Navarro & E.A. Carbonell
I.V.I.A. Apdo Oficial; 46113 Moncada (Valencia), Spain

Abstract: Virulent isolates of citrus tristeza virus (CTV) are continuously arising and their spread threatens world citrus industry. Methods for effective utilisation of material deposited in germplasm banks are needed. Two objectives are pursued: the search of new CTV resistant genotypes within a citrus germplasm bank and to test two sampling strategies for this search. One is based on a study of genetic relationships among species of the orange subfamily and the other based on scoring genotypes at molecular marker loci known to be linked to the CTV resistant locus *Ctr*. Sampled plants were graft inoculated with a mild and two virulent CTV isolates. Susceptibility, i. e. multiplication of CTV in plants, was checked by DAS-ELISA and DTBIA at four, six and ten months after inoculation. All cultivars of *Poncirus trifoliata* tested, *Severinia buxifolia*, *Atalantia ceylanica* and *Fortunella crassifolia* behave as CTV resistant. Given that *F. crassifolia* yields edible fruits, the finding of CTV resistance in this species makes the improvement for CTV resistance of scion varieties, now feasible. It is shown that the sampling strategy based on phylogenetic data can be very useful to search new genotypes expressing an agronomically important trait. The marker-assisted criterion may only be worthwhile when the search is restricted to accessions of the species where linkage analysis was carried out. Therefore, a good documentation system that allows quick and easy sampling of accessions to form core collections in order to carry out the search of new and useful genes, is suggested to enhance genetic resources.

1. Introduction

Citrus Tristeza Virus (CTV) is the causal agent of one of the most important diseases of citrus (Bar-Joseph et al., 1989). This floem limited closterovirus exists in a great variety of isolates differing in biological properties such as symptoms in the field (Aubert & Bové, 1984; Da Graça et al., 1989; Roistacher & Moreno, 1991), reaction on indicator plants (McClean, 1974) and aphid transmissibility (Roistacher & Bar-Joseph, 1984; Hermoso de Mendoza et al., 1988). Since its outbreak in the early 1930s, tristeza has caused the dead of millions of trees grafted on sour orange all around the world (Bar-Joseph et al., 1989).

The genus *Citrus* belongs to the subtribe Citrinae, tribe Citreae, subfamily Aurantioideae of the family Rutaceae. All Aurantioideae species are trees or shrubs with persistent leaves except for the three monotypic genera *Poncirus, Aegle* and *Feronia,* three species of *Clausena* and one of *Murraya.* Citrus are one of the major fruit crops in world. Hybridization, apomixis and many centuries of cultivation have complicated *Citrus* taxonomy with the result that very different number of species have been proposed. Genetic diversity and relationships have been reviewed in the orange subfamily Aurantioidea (Herrero et al., 1996). Main cultivated species are: sweet oranges (*Citrus sinensis* (L.) Osb), tangerines (*C. clementina* Hort. ex Tan. and *C. unshiu* (Mak.) Marc., mainly), grapefruits (*C. paradisi* Macf.) and lemons *(C. limon* L. Burm. f.*).* The cultivars of these species are always propagated vegetatively by bud-grafting onto a seedling rootstock in order to obtain a more uniform and early yielding tree.

CTV infects all citrus species and varieties, most hybrids and some citrus relatives. There are only three *Citrus* relatives which have been reported to be resistant to CTV (Yoshida et al., 1983; Garnsey et al., 1987; Bar-Joseph et al., 1989): *Severinia buxifolia* (Poir.) Tenore, *Swinglea glutinosa* (Blanco) Merr. and *Poncirus trifoliata* (L.) Raf. The resistance to CTV in *P. trifoliata* is conferred by two dominant genes and molecular markers linked to this gene have been reported (Mestre et al., 1997a, 1997b). However this reported CTV resistant species (the only sexually compatible species of the three with *Citrus* spp.) is distantly related to citrus (Herrero et al., 1996). This has negative repercussions on the viability of some genetic combinations in progenies of crosses involving *Citrus* and *Poncirus,* and on the ratio genetic/physical distance between the CTV resistance gene and marker loci. Therefore, the search for new, CTV-resistant genotypes more closely related to citrus would provide genetically diverse sources for durable resistance and allow citrus breeding programs and map-based cloning experiments to be more efficient.

Germplasm collections of major crop plants continue to grow in number and size around the world. Today, better access to and use of the genetic resources in collections have become important issues. However, the very large size and heterogeneous structure of collections have hindered efforts to increase the use of germplasm-bank material in plant improvement. Identification of different resistance genes requires allelism tests which involve extensive crossing and progeny evaluation after inoculation. Strategies to efficiently find resistant genotypes would enhance the use of genetic resources.

Two objectives were pursued in the present paper: to find new CTV-resistant genotypes and to evaluate two searching strategies. One is a sampling strategy based on choosing only those species related to previously known CTV-resistant species following a study of genetic relationships among *Citrus* and *Citrus*-related species (Herrero et al., 1996); the other is a marker-assisted screening using molecular markers known to be linked to the CTV-resistance locus of *Poncirus trifoliata* (Mestre et al., 1997a). Most results included is this paper are exhaustively detailed in Mestre et al. (1997c).

2. Materials and Methods

Species and cultivars sampled belong to the Citrus Germplasm Bank at IVIA (Valencia, Spain):

Atalantia ceylanica (Arn.) Oliv.
Severinia buxifolia (Poir.) Tenore
Microcitrus australis (Planch) Swing.
Fortunella crassifolia Swing.
Fortunella hindsii (Planch) Swing.
Poncirus trifoliata (L.) Raf. cv Benecke
" cv Hiryu
" cv Pomeroy
" cv Rich 7-5
Citrus grandis (L.) Osb. cv Cuban Shadock

They are mature, virus (and virus-like) free plants grown in containers kept in a screenhouse (Navarro et al., 1988). The choice of this material followed the relationships deduced from a minimum spanning tree based on the chord distance of Cavalli-Sforza and Edwards that included the main species of the orange subfamily Aurantioideae (Herrero et al., 1996). Hence, the nearest species to previously described resistant ones (*Severinia buxifolia*, *Swinglea glutinosa* and *Poncirus trifoliata*) were sampled to build up a putative CTV-resistant core collection. Thus, *Microcitrus australis* (Planch) Swing., the closest species to both *Swinglea glutinosa* and *Severinia buxifolia*, was sampled; similarly, *Atalantia ceylanica* (Arn.) Oliv. that is the closest to *Severinia buxifolia* and *Fortunella crassifolia* Swing. and *F. hindsii* (Planch) Swing., the most closely related species, in this order, to *P. trifoliata* were also sampled. Although *Murraya paniculata* (L.) Jack is also close to *S. glutinosa,* it was not included in this survey because it is so distantly related to *Citrus* that certain clones, like the one we have, are not graft-compatible with citrus. *Citrus grandis* (L.) Osb cv "Cuban Shaddock" was included because it had been previously reported as CTV resistant and it is closely related to the cultivated *Citrus.*

Craft propagations of all plants were done on healthy "Rough Lemon" (*Citrus jambhiri* Lush) as rootstock and then cultured in containers in a greenhouse at 18-26°C. Each plant was graft inoculated with two patches of infected tissue and pruned to force new growth. Three CTV isolates were used: T-346, T-388, and T-305. T-346 is a mild isolate, while both, T-388 and T-305 are virulent ones.

Multiplication of CTV in plants was checked by DAS-ELISA (Double Antibody Sandwich Enzyme-Linked Immunosorbent Assay) using the mix of monoclonal antibodies 3CA5 and 3DF1 as described in Sánchez-Vizcaíno and Cambra (1987). Additionally, detection of CTV by direct tissue blot immunoassay (DTBIA) was also performed for all the samples following the procedure described in Garnsey et al. (1993). These analysis of virus multiplication were carried out at four different times: four, six, ten and twelve months after inoculation.

The clones cE20, cK16, cW18 and cG18, which reveal RFLP (Restriction Fragment Length Polymorphism) loci linked to the CTV resistance gene in *P.*

trifoliata were used as non-radioactive labelled probes in the RFLP analysis as described by Mestre et al. (1997a).

3. Results and Discussion

All rootstocks ("Rough lemon"), where species and cultivars were grafted to be challenged with CTV, gave a clear positive reaction, demonstrating the effectiveness of the inoculation method. All cultivars of *Poncirus trifoliata* tested, as well as *Severinia buxifolia* and *Atalantia ceylanica* were resistant to the three CTV isolates; *Fortunella crassifolia* (Meiwa kumquat) resists two of them. Differences in response were found only in *F. crassifolia*, where T-346 was not detected while the T-388 CTV isolate was, although at a low titer. Here, it has been proved that the strategy based on phylogenetic analysis by Herrero et al. (1996) is an efficient searching way. Two accessions sampled using this strategy were found to be CTV resistant. This, additionally confirms the relationships between *A. ceylanica* and *S. buxifolia* and among *P. trifoliata, F. crassifolia* and *F. hindsii* and suggests that resistance genes, from *A. ceylanica, S. glutinosa,* and *P. trifoliata,* were lost from *M. australis* and *F. hindsii.* Confirming again the efficiency of this searching strategy, Yoshida (1996) has just found no evidence of CTV infection in *Murraya paniculata* which agrees with its phylogenic grouping with *Swinglea glutinosa.*

The marker-assisted screening strategy must be taken with caution. All *P. trifoliata* cultivars tested behave as resistant ones and present RFLP and microsatellite alleles linked, in coupling phase, to the resistance allele *R* in all the linked markers assayed. The only difference among these accesions concerned the RFLP allele putatively linked to the susceptibility allele *r*. Among other cultivars tested, only *A. ceylanica* shows a RFLP allele (with cE20 only) identical to that linked, in coupling phase, to the resistance gene in *P. trifoliata.* This coincides with the resistance of this accession to all CTV isolates used. Therefore, this strategy seems to be very useful when the search is carried out within the species where the linkage analysis was developed, although it may fail due to recombination or sequence changes.

CTV resistance has been reported in *F. crassifolia* (Mestre, 1997c). The plant-pathogen interaction between *F. crassifolia* and CTV seems to be complex and very variable; thus, Yoshida et al. (1983) also reported an accession of *F. crassifolia* as susceptible to a severe CTV-SY (seedling yellows) strain. Although the resistance found in the accession we have used seems not to be effective against all severe CTV isolates, there may be other accessions that resist a wider spectrum of CTV isolates like the variability reported for CTV resistance among accessions of *S. buxifolia.* Up to now there is no citrus cultivar with edible fruits resistant to CTV. Although *P. trifolialata* is commonly used in breeding programs for citrus rootstock as CTV resistance donor, it can not be used for variety improvement because it is a wild species with no edible fruits and fruit characters would be lost by recombination and hybridization. In contrast to *P. trifoliata, Fortunella crassifolia* (Meiwa kumquat) yields edible fruits. The genus *Fortunella* is more closely related to *Citrus* than *Poncirus* and it is the closest genus to most important scion cultivars, sweet orange and mandarins (Herrero et al., 1996) then, a new possibility is open for their CTV resistance improvement by means of sexual hybridization. Besides, large segregant families where genetic and physical distances are likely more coincident should be possible, given the relatedness of *Fortunella* to *Citrus*. This would represent a very favourable situation for fine mapping the CTV resistance gene in order to attempt its map-based cloning. Efforts should be focused on the evaluation of more accessions of *F. crassifolia* for CTV resistance and the study of the resistance: its inheritance, genetic location, functioning and efficiency against different CTV isolates.

Finding new, resistant genotypes, very different from each other as revealed by isozyme, RFLP and microsatellite analysis, will help in the fight against CTV. New, resistant genotypes may differ in resistance alleles and could be used to control the continuously growing genetic variability of the virus. This strategy is specially important in the case of an RNA virus because of the high mutation rate. Cases of resistance-breaking isolates have been reported for several virus-host systems (Fraser, 1990), and some authors have suggested the combined use of different resistance sources as an efficient way to fight RNA viruses. Finding new, resistant genotypes will also help in designing more experiments to comparatively study the virus-plant interaction and unveil how the resistance gene of the plant makes the virus unable to multiply or to move through the plant.

Information on phylogeny has been shown to be very useful to establish sampling strategies to quickly find new genetic sources for plant breeding and achieve an efficient use of germplasm banks. Our results suggest that phylogenetic analysis should be included in the documentation system of germplasm banks and that they should be updated when accessions are added.

4. References

Aubert, B. & Bové, C. (1984) Mild and severe strains of citrus tristeza virus in Reunion island. In: Garnsey SM, Timmer LW, Dodds JA, eds. Proceedings of the 9th Conference of the International Organization of Citrus Virologists, IOCV, Riverside, 57-61.

Bar-Joseph, M., Marcus, R. & Lee, R.F. (1989) The continuous challenge of Citrus Tristeza Virus control. Annu Rev Phytopatol 27:291-316.

Da Graça, J.V., Marais, L.J. & Von Broemsen, L.A. (1984) Severe tristeza stem pitting decline of young grapefruit in South Africa. In: Garnsey SM, Timmer LW, Dodds JA, eds. Proceedings of the 9th Conference of the International Organization of Citrus Virologists, IOCV, Riverside, 62-65.

Fraser, R.S.S. (1990) The genetics of resistance to plant viruses. *Annu.Rev.Phytopatol.* 28, 179-200.

Garnsey, S.M., Barret, H.C. & Hutchison, D.J. (1987) Identification of Citrus Tristeza Virus resistance in *Citrus* relatives and its potential applications. Phytophylactica 19: 187-191.

Garnsey, S.M., Su, H.J. & Tsai, M.C. (1997) In: *Proceedings of the 13th Conference of the International Organization of Citrus Virologists*, eds. Da Graça, J.V., Moreno, P. & Yokomi, R.K. (IOCV, Riverside).

Herrero, R., Asins, M.J., Pina, J.A., Carbonell, E.A. & Navarro, L. (1996) Genetic diversity in the orange subfamily Aurantioideae. II. Genetic relationships among genera and species. Theor Appl Genet 93: 1327-1334.

Hermoso de Mendoza, A., Ballester-Olmos, J.F. & Pina, J.A. (1988) Comparative aphid transmission of a common citrus tristeza virus isolate and a seedling yellows isolate recently introduced in Spain. In: Timmer LW, Garnsey SM, Navarro L, eds. Proceedings of the 10th Conference of the International Organization of Citrus Virologists, IOCV, Riverside, 68-70.

Mestre, P.F., Asins, M.J., Pina, J.A., Carbonell, E.A. & Navarro, L. (1997a) Molecular markers flanking citrus tristeza virus resistance gene from *Poncirus trifoliata* (L.) Raf. Theor Appl Genet 94: 458-464.

Mestre, P.F., Asins, M.J., Carbonell, E.A. & Navarro, L. (1997b) New gene(s) involved in the resistance of *Poncirus trifoliata* (L.) Raf. to citrus tristeza virus. Theor. Appl. Genet. 95:691-695.

Mestre, P.F., Asins, M.J., Pina, J.A. & Navarro, L. (1997c) Efficient search for new resistant genotypes to the citrus tristeza closterovirus in the orange subfamily Aurantioideae. Theor. Appl. Genet. 95: 1282-1288.

Navarro, L., Juarez, J., Pina, J.A., Ballester, L.F. & Arregui, J.M. (1988) The citrus variety improvement program in Spain after eleven years. Proc tenth Conf Intern Organization Citrus Virol IOCV, 1988. pp 400-406.

Roistacher, C.N. & Bar-Joseph, M. (1984) Transmission of tristeza and seedling yellows tristeza virus by Aphis gossypii from sweet orange, grapefruit and lemon to Mexican lime, grapefruit and lemon. In: Garnsey SM, Timmer LW, Dodds JA, eds. Proceedings of the 9th Conference of the International Organization of Citrus Virologists, IOCV, Riverside, 9-18.

Roistacher, C.N. & Moreno, P. (1991) The worldwide threat from destructive isolates of citrus tristeza virus-a review. In: Brlansky RH, Lee RF, Timmer LW, eds. Proceedings of the 11th Conference of the International Organization of Citrus Virologists, IOCV, Riverside 7-19.

Sanchez-Vizcaino, J.M. & Cambra, M. (1987) *Enzyme immunoassay techniques, ELISA, in animal and plant diseases*. Eds: Office International des Epizooites and Instituto Nacional de Investigaciones Agrarias. Technical Series N° 7. 72 pp.

Yoshida, T. (1996) Graft compatibility of Citrus with plants in the Aurantioideae and their susceptibility to citrus tristeza virus. Plant Disease 80: 414-417.

Yoshida, T., Shichijo, T., Ueno, I., Kihara, T., Yamada, Y., Hirai, M., Yamada, S., Ieki, H. & Kuramoto, T. (1983) Survey for resistance of citrus cultivars and hybrid seedlings to citrus tristeza virus (CTV). Bull Fruit Tree Res Stn B (Okitsu) 10: 51-68.

SESSION 6

QUALITY

G.T. Scarascia Mugnozza, E. Porceddu & M.A. Pagnotta (Eds.)
Genetics and Breeding for Crop Quality and Resistance, 259-269, 1999

Genetic manipulation of lipoxygenases for the agrifood industry

R. Casey
John Innes Centre, Norwich Research Park, Norwich NR4 7UH, U.K.

1. Introduction

Lipoxygenases (LOX) are a group of non-haem iron-containing enzymes that are widely distributed within Nature and can be found in algae, fungi, animals and plants. In animals they are recognized as catalyzing the first step on the pathway from arachidonic acid to leukotrienes, important compounds that mediate inflammatory and allergic responses. In plants, they were for a long time considered to be an enigma, but now have clearly-defined roles in defence reactions, may be involved in apoptosis, have a role in metabolic homeostasis and also influence the quality of foods in several ways. This article describes: the multiplicity of LOX in plants; the significance of such multiplicity in terms of generation of different fatty acid hydroperoxides; the various ways in which these hydroperoxides are metabolized; the ways in which we have attempted to manipulate the amounts of LOX in plants and what this has told us; and the consequent "balancing act" that needs to be performed, between our perceived requirements in terms of food quality and the requirements of plants in terms of growth, development and health.

2. The distribution and multiplicity of plant lipoxygenases

LOX are found in a wide range of plant organs, sometimes in high concentrations, as in potato tubers and grain legume seeds. It has been suggested (Axelrod, 1974) that LOX is ubiquitous in plants and that the inability to demonstrate LOX activity in particular species, or organs, is a consequence of the sensitivity of the detection method, rather than the absence of activity. Enzyme activity has been purified from a very wide range of plant sources and more than 30 sequences of LOX derived from cloned DNA have been reported, from species as diverse as soybean, lentil, pea, French bean, potato, tomato, rice, maize, barley and *Arabidopsis*. Pairwise comparisons of these sequences show a range of identities from about 45% to about 90%. Analysis of the expression of LOX genes shows them to be active in stems, roots, tubers, leaves, flowers, fruits, seeds, nodules, etc., depending on developmental stage and/or biotic or abiotic stimulation. Although all plant LOX catalyze similar chemical reactions, there are subtle differences in substrate preference, products, and subsequent metabolism of these products, that lead to LOX being involved in a number of biological phenomena.

3. The biochemistry of LOX

LOX catalyze the formation of (*Z,E*)-conjugated hydroperoxides from any compound containing a (*Z,Z*)-1,4-pentadiene, or methylene-interrupted double bond, system. The common plant molecules to contain such a moiety are the polyunsaturated fatty acids linoleic (18:2) and linolenic (18:3); arachidonic acid (20:4) also is a substrate and the LOX from potato tubers has been reported to convert it, *via* hydroperoxide, to leukotrienes (Shimizu et al., 1984).

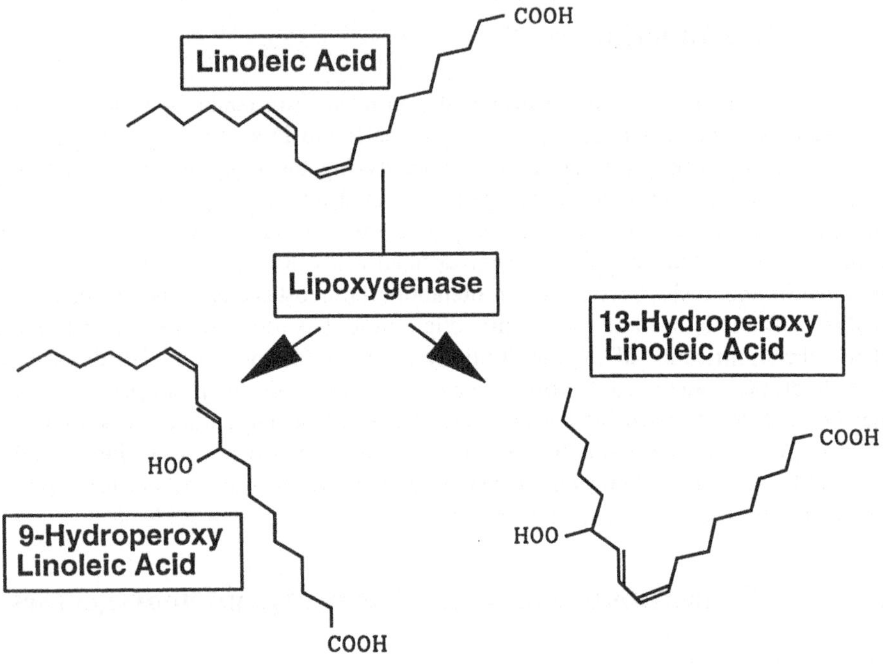

Figure 1. Formation of 9- or 13-hydroperoxides from linoleic acid catalyzed by lipoxygenase.

The combination of different substrates, different enzymes and different reaction conditions can give rise to variation in the position of insertion of the hydroperoxide, the geometry (*Z,E*) of the product and its absolute (R,S) stereochemistry, all of which will be important in determining the range and properties of the compounds derived from them by secondary reactions. The specificity of positional attack is a function of the source and type of enzyme; for example, soybean seed LOX-1 at alkaline pH generates almost exclusively 13-

hydroperoxides (Axelrod et al., 1981), potato tuber LOX at pH 6 produces 9-hydroperoxides (Royo et al., 1996) and pea seed LOX-3 at pH 6.5 produces a 2:1 mixture of 9- to 13-hydroperoxides (Hughes et al., 1998), from linoleic acid (Figure 1).

4. Fatty acid hydroperoxides in plant defence

One obvious consequence of the above reaction is the destruction of polyunsaturated fatty acids, essential components of human diet. The other is the formation of highly reactive, cytotoxic hydroperoxides. Such compounds may play a role in programmed cell death during hypersensitive response to pathogen attack. Rancé et al. (1998) have shown that the incompatible interaction between *Phytophthora parasitica* var. *nicotianae* race 0 and tobacco is suppressed in transgenic plants expressing antisense LOX sequences. The implication from this work is that a specific tobacco LOX has a role in resistance to fungal challenge, possibly mediated through fatty acid hydroperoxides.

5. Secondary metabolites of fatty acid hydroperoxides

5.1 Jasmonate

One well-established route by which a specific hydroperoxide - the 13-hydroperoxide of linolenic acid - is detoxified is by conversion to jasmonate (Figure 2), an important plant growth regulator that activates defence and other genes. The activity immediately following LOX on the pathway is allene oxide synthase, a cytochrome P450 homologue (Song et al., 1993), that produces allene oxide. Subsequent cyclization and reduction, thought to take place in the chloroplast, to produce 12-oxophytodienoic acid, are followed by β-oxidation to produce jasmonate. The pathway is activated by wounding and has been described as the octadecanoid, or oxylipid, signalling pathway (Farmer & Ryan, 1992). The jasmonate produced on wounding activates the genes for chalcone synthase, phenylalanine-ammonia lyase, so-called vegetative storage proteins (acid phosphatases and lipoxygenases) and proteinase inhibitors (pin2 and cdi); these last are thought to be the basis of a defence reaction against insect attack that occurs both in wounded and non-wounded ('systemic') leaves.

The formal involvement of LOX in wound-induced jasmonate synthesis in *Arabidopsis* has been demonstrated by down-regulation of a plastidial LOX through co-suppression (Bell et al., 1995), which eliminated wound-induced jasmonate production and jasmonate-responsive gene expression. The role of allene oxide synthase has been shown through overexpression of a flax sequence in transgenic potato (Harms et al., 1995); although such overexpression appreciably increased the amounts of leaf jasmonate, induction of pin2 gene expression required an additional increase by wounding, suggesting that there may be more than one jasmonate pool in leaves.

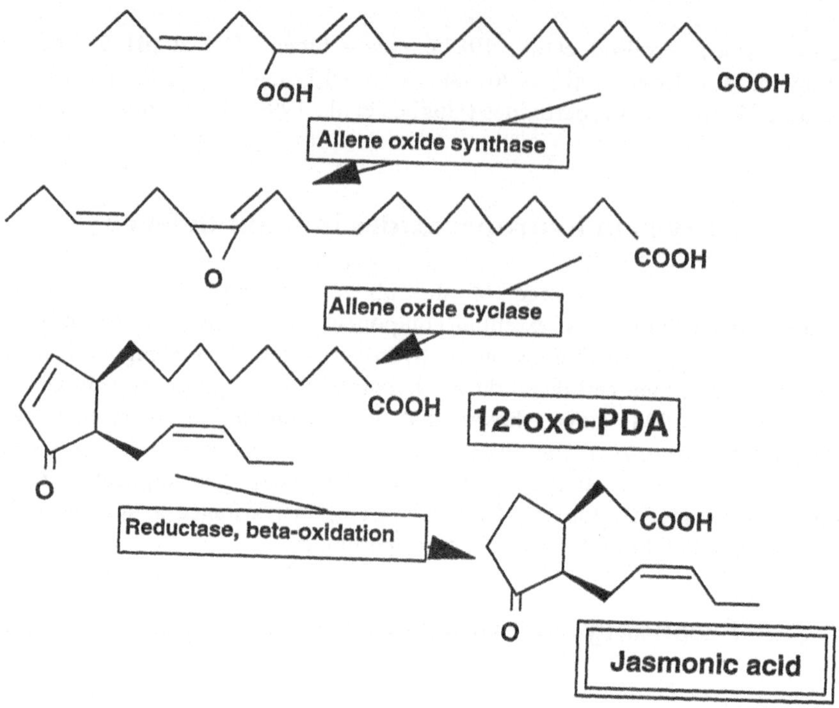

Figure 2. The pathway to jasmonate from the 13-hydroperoxide of linolenic acid.

5.2 Aldehydes

A second route by which hydroperoxides may be detoxified is by lyase-mediated cleavage (Figures 3 and 4). Hydroperoxide lyases also are cytochrome P450 homologues (Matsui et al., 1996; Bate et al., 1998), which casts an interesting evolutionary slant on the two detoxification pathways, both beginning with LOX activity followed by cytochrome P450-like catalysis, both producing plant growth regulators and both having defence implications. In the case of the lyase pathway, the growth regulator is a 12-carbon compound known as traumatic acid (Zimmerman & Caudron, 1979), which plays a role in wound sealing through accelerated cell division. The remainder of the parent hydroperoxide molecule is a six-carbon moiety, released as aldehyde, which can then be converted either to isomeric forms or reduced, through alcohol dehydrogenase, to alcohols. The aldehydes act as plant defence molecules. In an incompatible reaction between *Phaseolus vulgaris* and *Pseudomonas syringae*, bacteriocidal amounts of volatile aldehydes, in particular *Z*-3-hexanal and *E*-2-hexenal, are produced to kill the invading bacteria in a remarkable, rapid defence response (Croft et al., 1993). In this case the involvement of LOX has not been formally proven by transgenic knockout or mutagenesis.

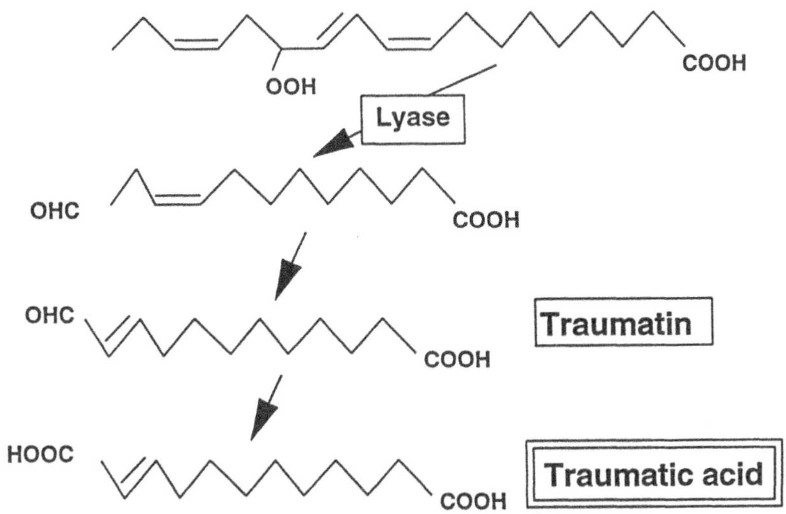

Figure 3. Production of traumatin from the action of hydroperoxide lyase on the 13-hydroperoxide of linoleic acid.

6. Aroma compounds and genetic manipulation of LOX

Such volatile aldehydes are much more than defence agents; they form the basis of the keynote aromas and flavours of a range of fruits and vegetables (Robinson et al., 1995), including mushrooms, tomatoes, potatoes, melons, cucumbers, bananas and avocados. *E*-2-hexenal, for instance, mentioned above as a defence compound, is one of the major flavour volatiles of fresh tomatoes (Buttery, et al., 1987). Their production is extraordinarily rapid, as is obvious from the almost instant appearance of 'green' aroma on damaging a tomato leaf, or mowing a lawn. Some of these aldehydes, such as *E*-2-hexenal, are perceived as pleasant whereas others, such as *n*-hexanal, are not; the chemical difference between unpleasant and pleasant is minimal.

6.1 Soybean

n-hexanal is the 'grassy-beany' off-flavour associated with soybean products. Soybean seed has three major LOX isoforms and genetic removal of LOX-2 reduces hexanal production and improves the consumer acceptability of soymilk and soy flour (Davies et al., 1987). LOX-2 is a major source of *n*-hexanal because it is a 13-LOX (see Figure 4). Removal of LOX-1 or LOX-3 does not improve the quality of soybean products and there is evidence to suggest that LOX-3 removal is counterproductive in terms of hexanal production (Hildebrand et al., 1990). On the other hand, removal of both LOX-2 and LOX-3 has the greatest reported effect on reduction of off-flavours (Nielsen, 1993, cited in Evans et al., 1997).

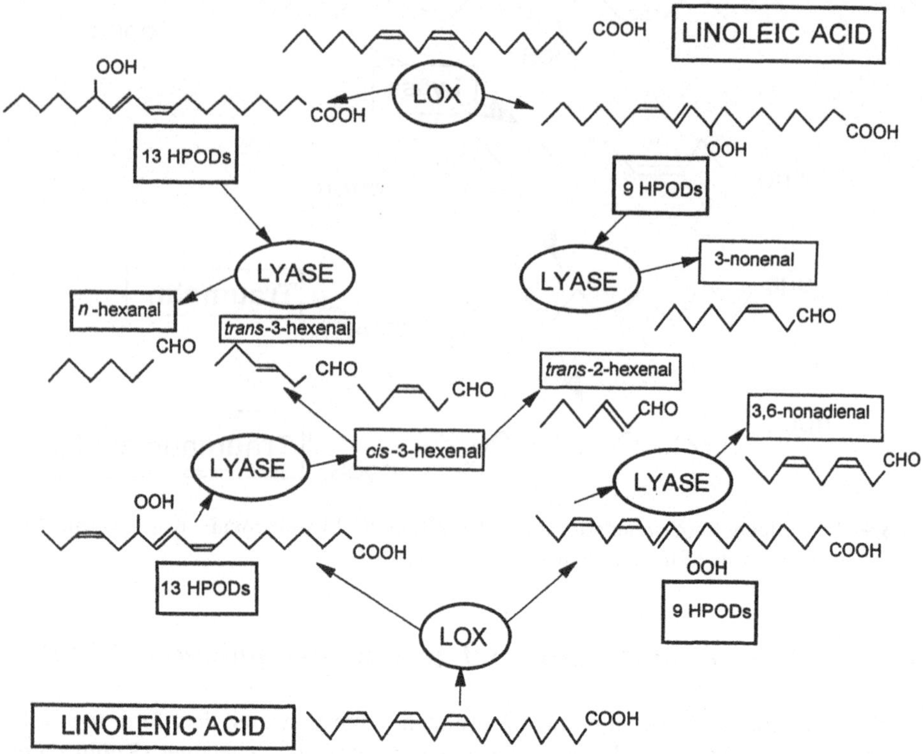

Figure 4. General scheme for the production of 6- and 9-carbon taste and aroma compounds through the action of lipoxygenases and lyases on polyunsaturated fatty acids.

6.2 *Peas*

Peas contain two major seed LOX polypeptides that are similar in sequence to soybean LOX-2 and LOX-3 (Casey et al., 1985; Ealing & Casey, 1988, 1989). Each is encoded by two or three genes (Domoney et al., 1990, 1991) clustered at a single genetic locus (North et al., 1989). A genetic variant of *Pisum fulvum* has been identified that lacks LOX-2 polypeptides (North, 1990) and the null mutation has been introduced into a standard *Pisum sativum* cultivar by repeated back-crossing to produce near-isogenic lines that either have, or completely lack the LOX-2 polypeptides. Analyses of reaction products show that crude LOX preparations from these lines behave very differently, reflecting the differences between LOX-2 and -3 from peas (Hughes et al., 1998) In particular, LOX-3 appears to readily release a fatty acid peroxyl radical before it has been reduced to hydroperoxide, whereas LOX-2 does not show this behaviour (Hughes et al., 1998) This means that LOX-3 is an effective co-oxidizing enzyme, whereas LOX-2 is not; this in turn means that all the LOX activity in the LOX-2-null mutant has co-oxidizing potential that is significant to food quality because it can compromize antioxidant status, bleach pigments and modify proteins.

6.3 Potatoes

Potato tubers have very large amounts of 9-LOX activity that prefers linoleic to linolenic and arachidonic acids as substrates at pH 6 (Royo et al., 1996). The role of 9-lipoxygenases in plants is poorly understood, but the action of lyases on 9-hydroperoxides gives rise to 9-carbon aldehydes and oxoacids that are part of the characteristic flavours and odours of fruits and vegetables. Down-regulation of the amounts of tuber LOX to about 10% of normal levels in the cultivar 'Desiree' had no significant effect on the profile of volatile aroma compounds from cooked potatoes (Ames, Casey, Duckham and Edwards, unpublished), suggesting that potato tuber 9-LOX may not play an important role in the generation of flavour volatiles; it should, however, be borne in mind that the residual 10% of activity might still be significant in the context of aldehyde generation because potatoes contain such high starting levels of activity.

6.4 Tomatoes

The LOX pathway has long been cited as a likely route to volatile flavour compounds, but this has not been formally proven. The major aroma volatiles in tomato fruit are 6-carbon aldehydes and alcohols that would be expected to arise from 13-hydroperoxides, but by far the majority of the LOX activity in tomato fruits is 9-LOX. This means that either the important activity in the context of 6-carbon volatile formation is a minor 13-LOX activity; or that there is 13-, but no 9-lyase, activity in tomato fruits. It is also possible that the aldehydes and alcohols are derived through a completely different pathway. By overexpressing and down-regulating the activity of alcohol dehydrogenase 2 (ADH) in ripening tomato fruits, Speirs et al. (1998) could demonstrate alterations in aldehyde to alcohol ratio that related to ADH levels. In a preliminary taste trial, fruit with elevated ADH activity and higher levels of alcohols were identified as having a more intense "ripe fruit" flavour (Speirs et al., 1998), demonstrating a possible role for aldehyde:alcohol ratio in consumer perception of tomato flavour. It remains to be seen whether or not the amounts of LOX and lyase can influence this ratio by altering the amounts of aldehydes.

7. Co-oxidation and food quality

The ability of some LOX to catalyze co-oxidation reactions is significant to food quality for several reasons. Fatty acid hydroperoxide radicals will oxidize antioxidants and thereby compromise their amounts in foods; these will include pigments, which will therefore become bleached, leading to loss of food colour. This may or may not be a problem, depending on circumstances; some consumer populations prefer white, and others yellow, pasta, for instance. Loss of colour in vegetables and fruits is usually considered to be undesirable, but bleaching of pigments in breadmaking is a positive issue, leading to a 'brighter' loaf. Pigment breakdown by oxidation can give rise to unpleasant aroma compounds.

Probably the most important application of LOX co-oxidation in the food industry is its use in breadmaking. Excessive energy input during the mixing process can lead to overmixing, loss of dough quality and a reduction in loaf volume, problems that can be overcome by the addition of enzyme-active soybean flour in the form of so-called bread 'improver'. One of the main active components of this improver with respect to tolerance to overmixing is LOX; mixing in nitrogen, or in the presence of LOX inhibitors, negates the benefits of adding the improver (Frazier, 1979). It is difficult to determine the underlying molecular basis of such improvement in rheology, but it is likely to relate to the ability of LOX to co-oxidise the thiol groups in high-molecular-weight (HMW) glutenins, some of the major proteins of wheat flour responsible for the visco-elastic behaviour of wheat gluten (see Shewry, 1995). The structure of the HMW glutenins includes internal repeats that can form β-spiral structures which are thought to have spring-like properties (Shewry, 1995) and it is possible that interchain disulphide bond formation catalyzed by LOX leads to larger, stronger spring-like networks. The necessity for addition of soybean, or other, LOX relates to the fact that although wheat contains a number of LOX isoforms (Shiba et al., 1991), they are not effective co-oxidizers.

There are several biotechnological approaches to bread improvement through the use of LOX (see Casey, 1998). One possibility is to express a co-oxidizing LOX in baker's yeast in such a way that it is secreted into the dough. Such expression was achieved by using a phosphoglycerate kinase promoter driving pea LOX sequences, with the intention of achieving secretion through the use of an invertase signal peptide. Although pea LOX was produced in baker's yeast containing such constructs, it was targeted to the vacuole instead of being secreted. Pea seed LOX is cytosolic and does not contain a signal peptide; by adding such a sequence, the LOX was sequestered into the secretory pathway, whereupon it was recognized as a vacuolar sequence as a consequence of cryptic vacuolar targeting motifs within its coding sequence (Knust & von Wettstein, 1992).

A second approach is to produce recombinant co-oxidizing LOX by fermentation in a micro-organism such as *E. coli.* Pea seed LOX-3 is a good candidate because it is an effective co-oxidizer, is a 9-LOX and therefore less likely to produce 6-carbon taste compounds, and is relatively stable (Busto et al., 1998); it has been produced as native enzyme with identical properties to that from pea seed (Hughes et al., 1998).

The third possibility is to transform bread-making wheat to produce a co-oxidizing LOX within its endosperm during seed development, under the control of an endosperm-specific promoter such as that for HMW glutenin polypeptides. Transformation efficiencies are adequate (Barro et al., 1997) to produce reasonable numbers of transgenic wheat expressing pea or soybean LOX for evaluation.

8. Conclusions

LOX in plants are the products of complex gene families that are active in several parts of the plant under a variety of circumstances. They play clear roles in defence through the production of hydroperoxides and their metabolites such as jasmonate and aldehydes. They also are produced in circumstances that suggest that they have other, as yet undefined, roles in growth and development. For example, several different LOX are detected in developing pea root nodules, apparently in response to the presence of *Rhizobium* and the consequent process of nitrogen fixation; they can be detected at the nodule apex, in the vicinity of nodule vascular tissue and in the lumen of *Rhizobium*-induced infection threads (Gardner et al., 1996) and each may play a different role in the symbiosis. Some members of the soybean LOX gene family have been proposed to function in nitrogen partitioning and storage, accumulating to high levels in the vacuoles of paraveinal mesophyll, bundle-sheath and adjacent cells under conditions of sink limitation (Stephenson et al., 1998). It is also clear that particular LOX genes are activated in response to water deficit (Bell & Mullett, 1991).

Some LOX, such as those from pea and soybean seeds, may be redundant in terms of plant health and development, since they apparently can be removed without any agronomic consequences (Pfeiffer et al., 1992). On the other hand, some of the flavour compounds generated by plants are identical to defence chemicals generated under other circumstances; in a sense, flavour volatile generation can be considered as a defence response to the damage of being eaten. It is therefore important that we are circumspect in our attempts to alter the amounts of LOX in order to understand the process of flavour generation in fruits and vegetables; we must bear in mind their potential as components of defence pathways. They may be involved in plant protection in ways that we do not yet understand; there are many uncertainties over the role of LOX in the production of volatile aroma compounds and at present the only clear evidence that a particular LOX is involved in 6-carbon aldehyde generation *in vivo* comes from the soybean LOX-2 mutant described above. In other instances it may be that minor LOX species which have not yet been identified play a major role in flavour generation.

We also need to know more about other components of the pathway, such as lipases, lyases, isomerases and dehydrogenases. Our current understanding of lyases and isomerases in particular is slim and would benefit from an investigation of their role in aldehyde generation.

9. References

Axelrod, B. (1974) Lipoxygenases, American Chemical Society Advanced Chemistry Series **136**, 324-348.

Azelrol, B., Cheesbrough, T.M. & Lasko, S. (1981) Lipoxygenase from soybeans, Methods in Enzymology 71, 441-451.

Barro, F., Rooke, L., Bekes, F., Gras, P., Tatham, A.S., Fido, R., Lazzeri, P.A., Shewry, P.R. & Barcelo, P. (1997) Transformation of wheat with high molecular weight subunit genes results in improved functional properties, Nature Biotechnology **15**, 1295-1299.

Bate, N.J., Sivasankar, S., Moxon, C., Riley, J.M.C., Thompson, J.E. & Rothstein, S.J. (1998) Molecular characterization of an *Arabidopsis* gene encoding hydroperoxide lyase, a cytochrome P-450 that is wound inducible, Plant Physiol. **117,** 1393-1400.

Bell, E., Creelman, R.A. & Mullet, J.E. (1995) A chloroplast lipoxygenase is required for wound-induced jasmonic acid accumulation in *Arabidopsis*, Proceedings of the National Academy of Sciences USA **92,** 8675-8679.

Bell, E. & Mullet, J.E. (1991) Lipoxygenase gene expression is modulated in plants by water deficit, wounding, and methyl jasmonate Molecular and General Genetics **230,** 456-462.

Busto, M.D., Owusu-Apenten, R.K., Robinson, D.S., Wu, Z., Casey, R. & Hughes, R.K. (1998) Food Chemistry, in press.

Buttery, R.G., Teranishi, R. & Ling, L.C. (1987) Fresh tomato aroma volatiles: a quantitative study, Journal of Agricultural and Food Chemistry **35,** 540-544.

Casey, R., Domoney, C. & Nielsen, N.C. (1985) Isolation of a cDNA clone for pea (Pisum sativum) seed lipoxygenase Biochemical Journal **232,** 79-85.

Casey, R. (1998) Lipoxygenases in breadmaking, Proceedings of the 1st European Symposium on Enzymes and Grain Processing TNO, The Netherlands pp. 188-194.

Croft, K.P.C., Jüttner, F. & Slusarenko,. A.J. (1993) Volatile products of the lipoxygenase pathway evolved from *Phaseolus vulgaris* (L.) leaves inoculated with Pseudomonas syringae pv phaseolicola Plant Physiology **101,** 13-24.

Davies, C.S., Nielsen, S.S. & Nielsen, N.C. (1987) Flavor improvement of soybean preparations by genetic removal of lipoxygenase-2, Journal of the American Oil Chemists' Society **64,** 1428-1433.

Domoney, C., Firmin, J.L., Sidebottom, C., Ealing, P.M., Slabas, A. & Casey, R. (1990) Lipoxygenase heterogeneity in Pisum sativum, Planta **181,** 35-43.

Domoney, C., Casey, R., Turner, L. & Ellis, N. (1991) Pisum lipoxygenase genes, Theoretical and Applied Genetics **81,** 800-805.

Ealing, P.M. & Casey, R. (1988) The complete amino acid sequence of a pea (Pisum sativum) seed lipoxygenase predicted from a near full-length cDNA, Biochemical Journal **253,** 915-918.

Ealing, P.M. & Casey, R. (1989) The cDNA cloning of a pea (Pisum sativum) seed lipoxygenase. Sequence comparisons of the two major pea seed lipoxygenase isoforms, Biochemical Journal **264,** 929- 932.

Evans, D.E., Tsukamoto, C. & Nielsen, N.C. (1997) A small scale method for the production of soymilk and silken tofu, Crop Science, **37** 1463-1471.

Farmer, E.E. & Ryan, C.A. (1992) Octadecanoid precursors of jasmonic acid activate the synthesis of wound-inducible proteinase inhibitors, The Plant Cell **4,** 129-134.

Frazier, P.J. (1979) Lipoxygenase action and lipid binding in breadmaking, Bakers Digest **53,** 8-29.

Gardner, C.D., Sherrier, D.J., Kardailsky, I.V. & Brewin, N.J. (1996) Localization of lipoxygenase proteins and mRNA in pea nodules: identification of lipoxygenase in the lumen of infection threads, Molecular Plant-Microbe Interactions **9,** 282-289.

Harms, K., Atzorn, R., Brash, A., Kuhn, H., Wasternack, C., Willmitzer, L. & Pena-Cortes, H. (1995) Expression of a flax allene oxide synthase cDNA leads to increased endogenous jasmonic acid (JA) levels in transgenic potato plants but not to a corresponding activation of JA-responding genes, The Plant Cell **7,** 1645-1654.

Hildebrand, D.F., Hamilton-Kemp, T.R., Longhrin, J.H., Ali, K. & Anderson, R.A. (1990) Lipoxygenase 3 reduces hexanal production from soybean seed homogenates, Journal of Agricultural and Food Chemistry **38,** 1934-1936.

Hughes, R.K., Wu, Z., Robinson, D.S., Hardy, D., West, S.I., Fairhurst, S.A. & Casey, R. (1998) Characterization of authentic recombinant pea-seed lipoxygenases with distinct properties and reaction mechanisms, Biochemical Journal **333,** 33-43.

Knust, B. & von Wettstein, D. (1992) Expression and secretion of pea-seed lipoxygenase isoenzymes in Saccharomyces cerevisiae, Applied Microbiology and Biotechnology **37,** 342-351.

Matsui, K., Shibutani, M., Hase, T. & Kajiwara, T. (1996) Bell pepper fruit fatty acid hydroperoxide lyase is a cytochrome P450 (CYP74B), FEBS Letters **394,** 21-24.

North, H. (1990) Pea seed lipoxygenase variants, Ph.D. thesis, University of East Anglia, U.K..

North, H., Casey, R. & Domoney, C. (1989) Inheritance and mapping of seed lipoxygenase polypeptides in Pisum, Theoretical and Applied Genetics **77,** 805-808.

Pfeiffer, T.W., Hildebrand, D.R. & TeKrony, D.M. (1992) Agronomic performance of soybean lipoxygenase isolines, Crop Science **32,** 357-362.

Rancé, I., Fournier, J. & Esquerré-Tugayé, M.-T. (1998) The incompatible interaction between Phytophthora parasitica var. nicotianae race 0 and tobacco is suppressed in transgenic plants expressing antisense lipoxygenase sequences, Proceedings of the National Academy of Sciences USA **95**, 6554- 6559.

Robinson, D.S., Wu, Z., Domoney, C. & Casey, R. (1995) Lipoxygenases and the quality of foods, Food Chemistry **54**, 33-43.

Royo, J., Vancanneyt, G., Pérez, A.G., Sanz, C., Störmann, K., Rosahl, S. & Sánchez-Serrano, J.J. (1996) Characterization of three potato lipoxygenases with distinct enzymatic activities and different organ- specific and wound-regulated expression patterns, The Journal of Biological Chemistry **271**, 21012-21019.

Shewry, P.R. (1995) Plant storage proteins, Biological Reviews **70**, 375-426.

Shiba, K., Negishi, Y., Okada, K. & Nagao, S. (1991) Purification and characterization of lipoxygenase isozymes from wheat germ, Cereal Chemistry **68**, 115-122.

Shimizu, T., RΔdmark, O. & Samuelsson, B. (1984) Enzyme with dual lipoxygenase activities catalyzes leukotriene A_4 synthesis from arachidonic acid, Proceedings of the National Academy of Sciences USA **81**, 689-693.

Song, W.-C., Funk, C.D. & Brash, A.R. (1993) Molecular cloning of an allene oxide synthase: A cytochrome P450 specialized for the metabolism of fatty acid hydroperoxides, Proceedings of the National Academy of Sciences USA **90**, 8519-8523.

Speirs, J., Lee, E., Holt, K., Yong-Duk, K., Scott, N.S., Loveys, B. & Schuch, W. (1998) Genetic manipulation of alcohol dehydrogenase levels in ripening fruit affects the balance of some flavor aldehydes and alcohols, Plant Physiology **117**, 1047-1058.

Stephenson, L.C., Bunker, T.W., Dubbs, W.E. & Grimes, H.D. (1998) Specific soybean lipoxygenases localize to discrete subcellular compartments and their mRNAs are differentially regulated by sink-source status, Plant Physiology **116**, 923-933.

Zimmerman, D.C. & Coudron, C.A. (1979) Identification of traumatin, a wound hormone, as 12-oxo-trans- 10-dodecanoic acid, Plant Physiology **63**, 536-541.

G.T. Scarascia Mugnozza, E. Porceddu & M.A. Pagnotta (Eds.)
Genetics and Breeding for Crop Quality and Resistance, 271-282, 1999

Molecular marker and genetic engineering strategies to improve wood quality in poplar[1]

W. Boerjan & M. Van Montagu
Laboratorium voor Genetica, Departement Genetica, Vlaams Interuniversitair Instituut voor Biotechnologie, Universiteit Gent, K.L. Ledeganckstraat 35, B-9000 Gent, Belgium

Abstract: The genetic improvement of trees is a slow process in comparison to that of annual crops. Tree breeding though is important, given the ever increasing demand for wood and wood products. It is the aim of this paper to show that the classical genetic improvement of trees by breeding and selection can be assisted and accelerated by the application of molecular biology tools that have been developed over the last decade. First, it is now possible to develop a set of diagnostic markers that predict the characteristics of new hybrids soon after they have germinated, thus long before the traits are displayed. Second, genetic engineering allows the modification or addition of a given trait that would be difficult or impossible to obtain by conventional breeding. Case studies in both fields, with respect to disease resistance and wood quality, are presented.

1. Introduction

In comparison to annual crop breeding, breeding of forest trees is still in its infancy. The first selections and crosses of annual crops have been made at the beginning of agriculture, approximately 10,000 years ago. In contrast, forest tree-breeding programs have only started approximately 50-100 years ago. The genetic improvement of trees by breeding is a tedious work because of the long reproductive cycles typical for trees and the fact that many traits are only displayed at maturity. For instance, the most advanced programs for conifer improvement, such as for Radiate pine, are into their fourth generation. For deciduous trees such as poplar, tree breeding is only in the third generation. The slow genetic improvement of forest trees species is worrisome, given the ever increasing consumption of wood and wood products, and the declining forest cover. According to the State of the World's Forests report (FAO, 1997), the consumption of wood has increased by

[1] Slightly modified version of the article published in Agro-Food-Hi-Tech 9 (1), 8-11 (1998) (with permission).

30% between 1970 and 1994. According to the same report, the forest cover decreases annually with 12 million ha or 0.3% of the global forest area. The continuous population growth and the higher rate of economic growth make that it will be difficult to slow down this process. It has been estimated that in the next 15 years, an additional 90 million ha of land will be put into agriculture, mainly in developing countries, and hence, for a large part at the expense of forest area (FAO, 1997). For the above mentioned reasons, it is necessary to develop methods that allow a faster improvement of the quality and the yield of wood for tree species that can be grown in agro-forestry systems.

The aim of this paper is to show that molecular biology offers numerous possibilities for the improvement of trees; first, by the development of diagnostic tests to assist conventional breeding programmes and, second, by the application of genetic engineering strategies. Examples of our progress in both areas are presented, with respect to the genetic improvement of poplar.

2. Poplar as a model system

Progress in biology is achieved when the research community agrees to focus on a given species that has major advantages at the experimental level, without being too far related from a commercial crop. Poplars (*Populus* spp.) have a number of characteristics that makes them attractive both as a model tree and as a commercial crop; they are fast-growing trees yielding yearly up to more than 20 m^3 ha^{-1} of wood (Steenackers et al., 1993); they can be propagated vegetatively by cuttings, enabling the establishment of large monoclonal plantations where all trees have the same characteristics. Furthermore, approximately 30 species of the genus *Populus* are known, most of which can be sexually crossed to generate new superior hybrids (Stanton & Villar, 1996; Stettler et al., 1996). In addition, poplar has the advantage to be susceptible to genetic transformation by *Agrobacterium tumefaciens* and to have a small genome size, which is only approximately 4-fold that of *Arabidopsis*, the model of choice for experimental research on annual plants. Poplars have become one of the most planted tree species. In 1995, the area of poplar plantation has been estimated to cover more that 1 million ha in Europe and 1.3 million ha in China (FAO, 1997). Poplar wood has a range of end uses, including pulp and paper, sawn timber, plywood, pallets, soft board, and hard board. There is also an increasing interest for cultivation of poplar as a biomass crop (Pearce, 1995).

3. Towards marker-assisted selection in poplar breeding

At the Institute of Forestry and Game Management (IBW; Geraardsbergen, Belgium), breeding of poplar was initiated in 1948. The aim was to generate new hybrids that produce suitable wood for the production of veneer. Since its start, the IBW has generated numerous poplar hybrids, several of which have superb growth characteristics and are planted throughout Europe. Most of the breeding effort has been focused on disease resistance, fast growth, rooting capacity, shape, frost tolerance, and photoperiodic response.

It takes the breeder 10 to 15 years of evaluation and selection before new elite clones can be released on the market for commercial plantations. The obvious reasons for this delay are that several characteristics, such as diameter increment, branching, and height, have to be followed for a decade, but also that certain traits can only be scored at maturity. If methods were available to predict these characteristics at the seedling stage, the selection of superior hybrids could be accelerated significantly. These molecular techniques are now becoming available.

In the past 10 years, molecular biologists have created tools to undoubtedly identify and discriminate individual organisms, based on their DNA sequence. These tools are called DNA fingerprinting methods. One of the most promising methods is called amplified fragment length polymorphism (AFLP™) (Vos et al., 1995). This technique generates highly informative fingerprints and is very reproducible. In addition, these DNA fingerprinting methods allow the identification of DNA fragments that are co-inherited with a particular trait in a pedigree. Importantly, the presence or absence of such DNA fragments, called molecular markers, can be scored at very early stages of development of the plant, e.g., a few days or weeks after germination. DNA fingerprinting methods thus allow the development of diagnostic tools for early selection of superior hybrids. In particular, this will be useful in selecting for traits that are difficult to evaluate or that are only displayed at maturity.

4. Identification of diagnostic markers for leaf rust resistance - a case study

Populus nigra, which is native to Europe, is susceptible to one of the most damaging pathogens of poplar in Europe, the leaf rust *Melampsora larici-populina* (Pinon, 1992). Infections cause early leaf fall with a reduction in biomass increment, a decreased frost tolerance, and a reduced wood quality as a consequence. Repeated infections over successive years can result in a complete loss of the plantation. Therefore, rust resistance has always been one of the major selection criteria for poplar breeding in Europe. To generate poplars resistant to rust, *P. nigra* has been crossed with *P. deltoides*, native to Eastern North America. From the *P. deltoides* provenances, clones have been identified that were resistant to this fungus. In a number of *P. deltoides x P. nigra* crosses, this resistance trait was transmitted to 50% of the F_1 progeny trees, whereas the other 50% remained susceptible, suggesting that the *P. deltoides* parent was heterozygous for a single dominant locus (a gene or a cluster of genes) that confers this resistance. By bulked segregant analysis (BSA) (Michelmore et al., 1991), DNA markers can be identified that are genetically linked with the resistance trait and that are, therefore, present in the resistant individuals and absent in the susceptible individuals. The BSA technique holds that DNA fingerprints are made from the resistant parent and a pooled set of resistant individuals and compared to DNA fingerprints from the susceptible parent and a pooled set of susceptible individuals (Figure 1). DNA bands present in the resistant parent and the 'resistant bulk' and absent in the susceptible parent and the 'susceptible' bulk are candidate diagnostic markers. These markers have to be evaluated in all individual plants that constitute both bulks (Figure 1). Such markers, which were generated by the AFLP™ DNA fingerprinting technique (Cervera et al., 1996; Villar et al., 1996), can be used to select progeny trees that are resistant to infection by *M. larici-populina* without going through artificial inoculations. The presence of the marker in a given DNA sample of an individual is then indicative for resistance of this individual, whereas the absence is indicative for susceptibility for leaf rust. It is evident that an early selection of seedlings via DNA fingerprinting becomes more efficient and more cost effective, the more traits can be scored by diagnostic markers.

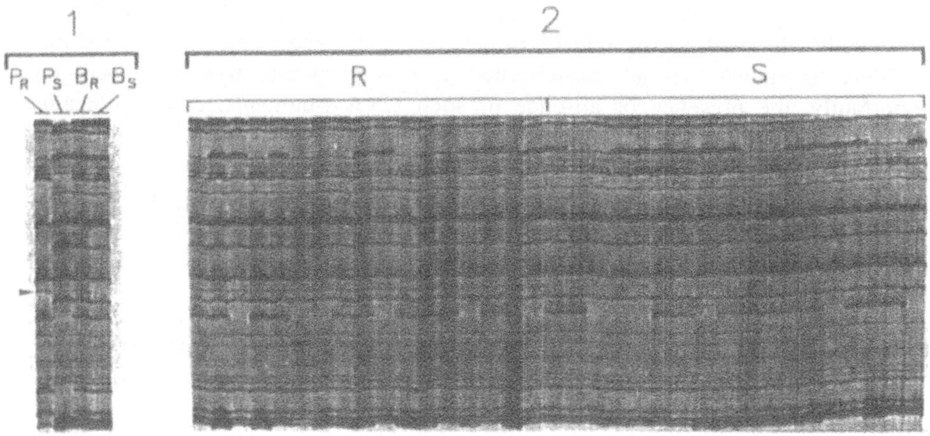

Figure 1. AFLP™ markers linked to leaf rust resistance. 1, the bulked segregant analysis (BSA) is presented as a set of four DNA fingerprints: resistant parent *P. deltoides* (P_R), susceptible parent *P. nigra* (P_S), resistant bulk (B_R), and susceptible bulk (B_S). 2, the AFLP marker indicated by an arrow is present in the resistant (R) but absent in the susceptible (S) F_1 progeny (reproduced with permission from Cervera et al.,1996).

5. The importance of genetic linkage maps

Many traits are however not determined by single dominant genes. Most traits, such as height growth, wood specific gravity, and stem diameter, are polygenic, i.e., more than one gene is involved in determining the final expression of the trait. This is also the case, for example, for resistance of poplar towards bacterial canker, caused by the bacterium *Xanthomonas populi*, another important pathogen for poplar in Europe. To dissect polygenic traits into their Mendelian components and to develop diagnostic markers for polygenic traits, genetic linkage maps have been created (Bradshaw et al., 1994; Grattapaglia & Sederoff, 1994; Bradshaw, 1996;

Cervera et al., 1997). A genetic linkage map is a representation of all individual chromosomes of the plant where each chromosome itself is represented by a series of genetically linked DNA markers. Computer-assisted analyses of the trait values of all individuals from a particular cross will subsequently show which chromosomal regions on the genome map contain (a) gene(s) that play(s) an important role in the expression of the trait. Such a region is called a quantitative trait locus (QTL). In addition, such analyses give information on correlations between traits, e.g. a QTL involved in determining leaf area in poplar coincides with a QTL for stem diameter (Bradshaw & Stettler, 1995; Bradshaw, 1996). All this information should result in a more judicious choice of parents for new crosses and in the ability to make a pre-selection of new superior hybrids at very early stages of development. A pre-selection allows an increased intensity of selection and, therefore, an increased probability of identifying rare clones of superior quality that possess most of the desired traits.

Because the genomics area in plants is rapidly evolving, it can be expected that it will be possible in the near future to identify the genes behind the QTLs. This will significantly improve our understanding on the fundamental processes that influence the different traits. It will then be very interesting to modify the expression of these genes by genetic engineering.

6. Fundamental and applied research by genetic engineering

In contrast to conventional breeding where thousands of genes are combined, genetic engineering provides a way to significantly improve an organism by the introduction of only a limited number of genes. By genetic engineering, single plant genes can be expressed at a higher or a reduced level, hence resulting in the production of more or less of a particular protein, respectively. In addition, genes foreign to plants can be introduced into the plant genome. The importance of genetic engineering lies not only in the fact that traits can be generated that cannot be obtained by conventional breeding, but also, and especially for trees, that improvements can be made in a fraction of the time it would take to reach the same goal by conventional breeding.

At the Laboratory of Genetics (Gent, Belgium), in collaboration with research groups throughout Europe and with the financial aid of the European Union, research is being conducted to improve wood for the pulp and paper industry via genetic engineering. One of the main obstacles in the production of high-quality paper is the removal of lignin from the pulp. Lignin is an aromatic polymer that is present mainly in the cells that constitute the wood. The cellulose fibres, from which paper is made, are embedded in this matrix. During mechanical pulping, the wood chips are ground and subsequently bleached and lignin is not separated from cellulose. Paper that is derived from this pulping method is used as e.g. newsprint, weekly magazines, wrapping paper, or other low-quality paper. Because it contains substantial amounts of lignin however, this kind of paper is sensitive to sunlight upon which it turns yellow. During Kraft pulping, which is the mostly used chemical pulping method, the bulk of lignin is separated from cellulose resulting in high-quality white paper. Such chemical pulping methods are environmentally unfriendly, consume vast amounts of energy, and are expensive. In addition, chemical pulping results in a low biomass utilization, given the fact that lignin can constitute up to 35% of the dry weight of wood (Higuchi, 1985; Biermann, 1993). Therefore, more environmentally clean and efficient pulping methods are necessary. One way to improve the extractability of lignin from the pulp is via genetic engineering of the trees. A modification of the lignin content or composition is expected to result in altered pulping characteristics.

To obtain this goal, a basic insight into the biosynthesis of lignin is needed. The overall biochemical route towards lignin has been studied extensively (Higuchi, 1997; Baucher et al., 1998) (Figure 2), although novel alternative or bypass pathways are still discovered (Ye et al., 1994; Lee et al., 1997; Ralph et al., 1997). Lignin is made by the polymerization of mainly three cinnamyl alcohols, *p*-coumaryl alcohol, coniferyl alcohol, and sinapyl alcohol. These units give rise to *p*-hydroxyphenyl (H), guaiacyl (G), and syringyl (S) units in the lignin polymer, respectively. Woody angiosperm lignin consists of mainly G and S units (G-S lignin), whereas gymnosperm lignin consists of mainly or exclusively G units (G lignin); grass lignin consists of all three units (H-G-S lignin). From studies of spontaneous mutants of maize and sorghum, it was known that reductions in the activity of the enzymes called bispecific caffeic acid/5-hydroxyferulic acid-*O*-methyltransferase (COMT) and cinnamyl alcohol dehydrogenase (CAD), two important enzymes in the pathway (Figure 2), resulted in a reduced synthesis of lignin in the plant (Grand et al., 1985; Pillonel et al., 1991; Vignols et al., 1995). Therefore, reducing the activities of COMT and CAD in trees by genetic engineering seemed a good strategy to reduce the amount of lignin in wood.

The antisense strategy is a technology that is commonly used in plant as well as animal molecular biology to diminish the production of a particular protein. In this approach, the coding sequence of the target gene is cloned in reverse orientation behind a strong promoter. Subsequently, the target organism is transformed with this construct and those individuals that show less target protein activity compared to the wild-type plants are studied in more detail. This approach was used for COMT and for CAD in poplar.

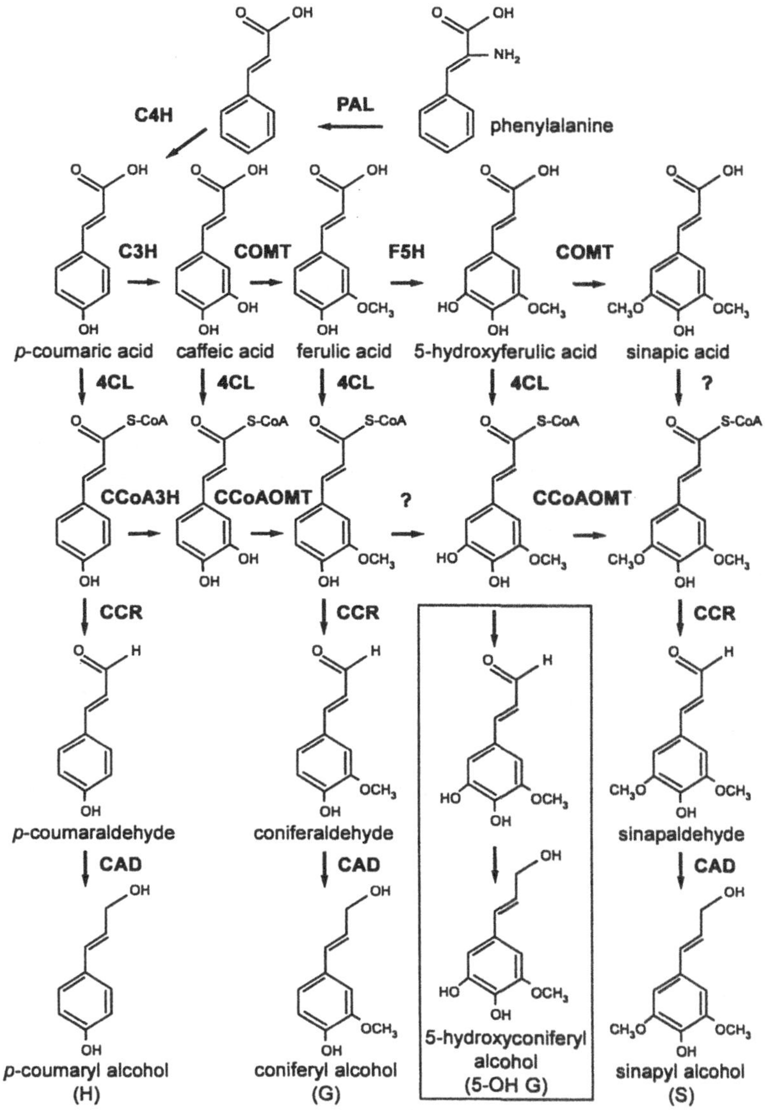

Figure 2. Lignin biosynthesis pathway. PAL, phenylalanine ammonia-lyase; C4H, cinnamate-4-hydroxylase; C3H, 4-coumarate-3-hydroxylase; COMT, bi-specific caffeic acid/5-hydroxyferulic acid-*O*-methyltransferase; F5H, ferulate-5-hydroxylase; CCoA3H, coumaroyl-CoA-3-hydroxylase; CCoAOMT, caffeoyl-CoA-*O*-methyltransferase; 4CL, 4-hydroxycinnamate-CoA ligase; CCR, cinnamoyl-CoA reductase; CAD, cinnamylalcohol dehydrogenase. 5-OH G, 5-hydroxyconiferyl alcohol. The boxed part is present in transgenic poplars downregulated for COMT and is undetectable in wild-type plants.

An antisense construct for COMT was generated and introduced into poplar via an *Agrobacterium*-mediated gene transfer system (Leplé et al., 1992; Gheysen et al., 1998). All transgenic plants obtained were subsequently analysed for their COMT activity in stems and two plants were found to have a 95% reduction in the activity of COMT (Van Doorsselaere et al., 1995). In contrast to mutants of maize, defective in the gene encoding COMT (Vignols et al., 1996), the transgenic poplars had no reduced amount of lignin, as measured by the Klason method (Effland, 1977). Instead, a considerable change in the lignin composition was detected, characterized by a reduction in the number of S units and an increase in the number of G units, as measured by thioacidolysis (Lapierre et al., 1986; Rolando et al., 1992). Thioacidolysis measures only those units that are linked by β-O-4 ether linkages (Figure 3). Because β-O-4 ether linkages represent only 50% of the linkages in the polymer, the obtained results can not be generalized to the whole lignin polymer. In addition, 5-hydroxyguaiacyl, corresponding to the substrate of COMT, was incorporated into the lignin polymer (Figure 2). This unit was not present in lignin of non-transformed control poplars. Simulated Kraft pulping experiments of the transgenic wood indicated a more difficult delignification process, i.e., less lignin could be extracted using a given amount of chemicals. The relatively higher number of G units incorporated into the lignin might be the direct reason for the more difficult pulping: the carbon-5 position of the aromatic ring can generate carbon-carbon linkages that are difficult to break during chemical pulping, while S units are substituted at this position preventing the formation of this kind of carbon—carbon linkages (Figures 2 and 3). Consistent with this hypothesis is that the lignin of gymnosperm wood, which is almost exclusively of the G type, is more difficult to extract than that of angiosperm wood. Therefore, to achieve an improved pulping, it seems necessary to obtain the opposite, i.e. a reduced incorporation of G units and an increased incorporation of S units. Down-regulation of 4-hydroxycinnamate-CoA-ligase (4-CL) and overproduction of ferulate-5-hydroxylase (F5H) do result in a higher S/G ratio in annual model plants (Figure 2) (Lee et al., 1997; Meyer et al., 1997), but their effect in trees on lignin extractability has still to be evaluated. An enzyme that is possibly involved in controlling the number of G units is caffeoyl-CoA-*O*-methyltransferase (CCoAOMT; Figure 2). A similar antisense approach to reduce CCoAOMT activity might result in an improved pulping, even when the total amount of lignin remains the same.

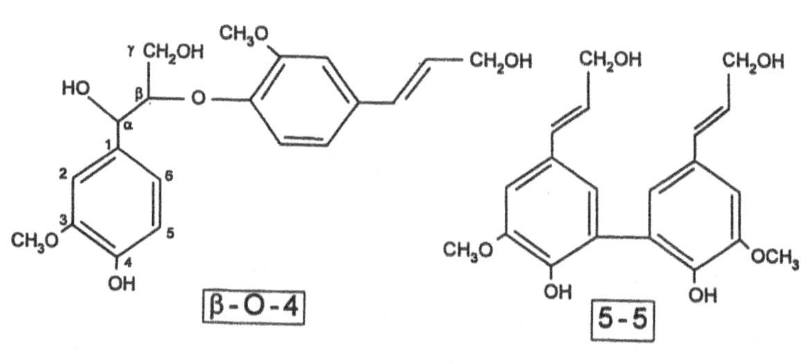

Figure 3. Schematic representation of two frequent bonds in the lignin polymer: the β-*O*-4 ether linkage and the 5-5 carbon-carbon linkage. Ether linkages are easily broken during chemical pulping whereas carbon-carbon linkages are not. For convenience, the structures are represented as dimers. For an overview of the different linkages, see Higuchi (1985; and references therein).

Transgenic poplars with a 70% reduced CAD activity had no large reductions in lignin content either. Simulated Kraft pulping experiments, however, indicated an improved lignin extractability; less chemicals are necessary to extract a given amount of lignin from the pulp (Baucher et al., 1996; Lapierre et al., 1999). The xylem of these transgenic poplars is red, instead of the whitish yellow colour of wild-type poplar wood. It has been shown by Higuchi et al. (1994), that *in vitro* synthesized lignin polymers of coniferyl alcohol and coniferyl aldehyde, the substrate for CAD, are red, whereas those made with coniferyl alcohol alone are brown. Although difficult to prove, the incorporation of aldehydes into the lignin polymer was suggested to result in alterations in the chemical linkages between the lignin monomers, resulting in a different susceptibility of the polymer to chemical degradation (Baucher et al., 1996). A similar reddish colour was also observed in the maize and sorghum mutants with a reduced COMT and CAD activity, which were therefore called 'brown midrib' mutants.

From these experiments, it can be deduced that COMT in poplar plays a major role in determining the amount of S units, but not the amount of G units, and that the S/G composition of lignin probably plays an important role in determining the extractability of lignin. Furthermore, the results with CAD transgenic poplars show that the extractability of lignin during chemical pulping can be improved without a large reduction in the total amount of lignin. The positive results with antisense CAD poplars have convinced Shell Forestry to transfer similar constructs into eucalyptus, another tree species grown on large scale for pulp and paper (John Purse, personal communication). It can be expected that within a short time transgenic trees will have been generated that are particularly suited for pulp and paper production. As we know more about the function of the lignin biosynthesis enzymes, it will be possible to make transgenic trees that have a reduced or an enhanced activity of more than one lignin biosynthesis enzyme. It can be anticipated that such trees would have less lignin and, in addition, a higher lignin extractability because of alterations in the lignin composition. There is no doubt that such engineered trees will result in a significantly reduced consumption of chemicals and energy. It would take numerous generations of conventional tree breeding, if ever, to obtain the same improvements in lignin extractability. Furthermore, when positive results are obtained in one species, the technology can be transferred to any other species for which transformation procedures are available.

Manipulation of lignin content might become increasingly more important because of the growing interest in short rotation biomass plantations as a renewable source of energy. Lignin has a calorific value that is 50% to 60% higher than that of cellulose and hemicellulose (Brown, 1985). Higher amounts of lignin should thus result in high-quality fuel wood and charcoal. Coppice plantations form an alternative for fossil energy sources. It has been estimated that burning 1 ton of dried coppice would generate as much electricity as 650 kg of coal, but would save the release of up to 500 kg of carbon into the atmosphere (Pearce, 1995). The annual consumption of fuel wood and charcoal has increased by 60% between 1970 and 1994 and demands more than 40% of the world's total wood production (FAO, 1997).

From fundamental studies in the model plant *Arabidopsis* and from large genome sequencing efforts, the function of many new genes is now being unravelled. These genes offer a seemingly unlimited number of applications, many of which can result in large improvements in the biomass and quality of crops. It is often underestimated that genetic engineering offers many opportunities for environmentally safe production methods. For example, many complex molecules that are now synthesized by often polluting chemical synthesis are made by plants. Plants can be engineered into factories that make large amounts of these complex molecules in an environmentally safe way.

In conclusion, both molecular marker technologies as well as genetic engineering strategies hold promise for the genetic improvement of trees and must be considered for application and integration into current breeding programmes.

Acknowledgements: This work has been supported by grants from the Commission of the European Communities (ECLAIR-OPLIGE AGRE-0021-C(EDB), AIR-IRPI CT92-0349, the Human Capital Mobility program 41AS8694), and the Flemish Government (BNO//BB/6/1994, 1995, 1996).

7. References

Baucher, M., Chabbert, B., Pilate, G., Van Doorsselaere, J., Tollier, M.-T., Petit-Conil, M., Cornu, D., Monties, B., Van Montagu, M., Inzé, D., Jouanin, L. & Boerjan, W. (1996) Red xylem and higher lignin extractability by down-regulating a cinnamyl alcohol dehydrogenase in poplar (*Populus tremula x P. alba*), *Plant Physiol.* **112**, 1479-1490.

Baucher, M., Monties, B., Van Montagu, M. & Boerjan, W. (1998) Biosynthesis and genetic engineering of lignin, *Crit. Rev. Plant Sci.* **17**, 125-197.

Biermann, C.J. (1993) *Essentials of Pulping and Papermaking.* San Diego, Academic Press.

Bradshaw, H.D. Jr. (1996) Molecular genetics of *Populus*, in R.F. Stettler, H.D. Bradshaw Jr., P.E. Heilman and T.M. Hinckley (eds.), *Biology of Populus and its Implications for Management and Conservation*, NRC Research Press, Ottawa, pp. 183-199.

Bradshaw, H.D. Jr & Stettler, R.F. (1995) Molecular genetics of growth and development in *Populus*. IV. Mapping QTLs with large effects on growth, form, and phenology traits in a forest tree, *Genetics* **139**, 963-973.

Bradshaw, H.D. Jr, Villar, M., Watson, B.D., Otto, K.G., Stewart, S. & Stettler, R.F. (1994) Molecular genetics of growth and development in *Populus*. III. A genetic linkage map of a hybrid poplar composed of RFLP, STS, and RAPD markers, *Theor. Appl. Genet.* **89**, 167-178.

Brown, A. (1985) Review of lignin in biomass, *J. Appl. Biochem.* **7**, 371-387.

Cervera, M.-T., Gusmão, J., Steenackers, M., Peleman, J., Storme, V., Vanden Broeck, A., Van Montagu, M. & Boerjan, W. (1996) Identification of AFLP molecular markers for resistance against *Melampsora larici-populina* in *Populus*, *Theor. Appl. Genet.* **93**, 733-737.

Cervera, M.-T., Storme, V., Liu, B., Gusmão, J., Steenackers, M., Ivens, B., Michiels, B., Van Montagu, M. & Boerjan, W. (1997) AFLP™ genome mapping of poplar, *Med. Fac. Landbouww. Univ. Gent* **62/4a**, 1435-1441.

Effland, M.J. (1977) Modified procedure to determine acid-insoluble lignin in wood and pulp, *Tappi* **60**, 143-144.

FAO (1997) *State of the World's Forests 1997*. Rome, FAO.

Gheysen, G., Angenon, G. & Van Montagu, M. (1998) *Agrobacterium*-mediated plant transformation: a scientifically intriguing story with significant applications, in K. Lindsey (ed.), *Transgenic Plant Research*, Harwood Academic Publishers, Amsterdam, pp. 1-33.

Grand, C., Parmentier, P., Boudet, A. & Boudet, A.M. (1985) Comparison of lignins and of enzymes involved in lignification in normal and brown midrib (bm_3) mutant corn seedlings, *Physiol. Vég.* **23**, 905-911.

Grattapaglia, D.& Sederoff, R. (1994) Genetic linkage maps of *Eucalyptus grandis* and *Eucalyptus urophylla* using a pseudo-testcross: mapping strategy and RAPD markers, *Genetics* **137**, 1121-1137.

Higuchi, T. (1985) Biosynthesis of lignin, in T. Higuchi (ed.), *Biosynthesis and Biodegradation of Wood Components*, Academic Press, Orlando, pp. 141-160.

Higuchi, T. (1997) *Biochemistry and Molecular Biology of Wood* (Springer Series in Wood Science), Springer, Berlin.

Higuchi, T., Ito, T., Umezawa, T., Hibino, T. & Shibata, D. (1994) Red-brown color of lignified tissues of transgenic plants with antisense CAD gene: wine-red lignin from coniferyl aldehyde, *J. Biotechnol.* **37**, 151-158.

Lapierre, C., Rolando, C. & Monties, B. (1986) Thioacidolysis of poplar lignins: identification of monomeric syringyl products and characterization of guaiacyl syringyl-lignin fractions, *Holzforschung* **40**, 113-118.

Lapierre, C., Pollet, B., Petit-Conil, M., Toval, G., Romero, J., Pilate, G., Leplé, J.-C., Boerjan, W., Ferret, V., De Nadaï, V. & Jouanin, L. (1999) Structural alterations of lignins in transgenic poplars with depressed cinnamyl alcohol dehydrogenase or caffeic acid *O*-methyltransferase activity have opposite impact on the efficiency of industrial Kraft pulping, *Plant Physiol.* 119: 153-164.

Lee, D., Meyer, K., Chapple, C. & Douglas, C.J. (1997) Antisense suppression of 4-coumarate:coenzyme A ligase activity in *Arabidopsis* leads to altered lignin subunit composition, *Plant Cell* **9**, 1985-1998.

Leplé, J.C., Brasileiro, A.C.M., Michel, M.F., Delmotte, F. & Jouanin, L. (1992) Transgenic poplars: expression of chimeric genes using four different constructs, *Plant Cell Rep.* **11**, 137-141.

Meyer, K., Cusumano, J.C., Ruegger, M.O., Bell-Lelong, D.A. & Chapple, C.C.S. (1997) Regulation and manipulation of lignin monomer composition by expression and over expression of ferulate 5-hydroxylase, a cytochrome P450-dependent monooxygenase required for syringyl lignin biosynthesis. Abstract presented at the Second International Wood Biotechnology Symposium, Canberra (Australia), March 10-12, 1997, paper 9.

Michelmore, R.W., Paran, I. & Kesseli, R.V. (1991) Identification of markers linked to disease-resistance genes by bulked segregant analysis: a rapid method to detect markers in specific genomic regions by using segregating populations, *Proc. Natl. Acad. Sci. USA* **88**, 9828-9832.

Pearce, F. (1995) Seeing the wood for the trees, *New Scientist* **145 (1960)**, 12-13.

Pillonel, C., Mulder, M.M., Boon, J.J., Forster, B. & Binder, A. (1991) Involvement of cinnamyl-alcohol dehydrogenase in the control of lignin formation in *Sorghum bicolor* L. Moench, *Planta* **185**, 538-544.

Pinon, J. (1992) Variability in the genus *Populus* and sensibility to *Melampsora* rusts, *Silvae Genetica* **41**, 25-33.

Ralph, J., MacKay, J.J., Hatfield, R.D., O'Malley, D.M., Whetten, R.W. & Sederoff, R.R. (1997) Abnormal lignin in a loblolly pine mutant, *Science* **277**, 235-239.

Rolando, C., Monties, B. & Lapierre, C. (1992) Thioacidolysis, in S.Y. Lin and C.W. Dence (eds.), *Methods in Lignin Chemistry*, Springer-Verlag, Berlin, pp. 334-349.

Stanton, B.J. & Villar, M. (1996) Controlled reproduction of *Populus*, in R.F. Stettler, H.D. Bradshaw Jr., P.E. Heilman and T.M. Hinckley (eds.), *Biology of Populus and its Implications for Management and Conservation*, NRC Research Press, Ottawa, pp. 113-138.

Steenackers, J., Steenackers, V., Van Acker, J. & Stevens, M. (1993) Stem form, volume and dry matter production in a twelve-year-old circular Nelder plantation of *Populus trichocarpa x deltoides* 'Beaupré', *For. Chron.* **69**, 730-735.

Stettler, R.F., Zsuffa, L. & Wu, R. (1996) The role of hybridization in the genetic manipulation of *Populus*, in R.F. Stettler, H.D. Bradshaw Jr., P.E. Heilman and T.M. Hinckley (eds.), *Biology of Populus and its Implications for Management and Conservation*, NRC Research Press, Ottawa, pp. 87-112.

Van Doorsselaere, J., Baucher, M., Chognot, E., Chabbert, B., Tollier, M.-T., Petit-Conil, M., Leplé, J.-C., Pilate, G., Cornu, D., Monties, B., Van Montagu, M., Inzé, D., Boerjan, W. & Jouanin, L. (1995) A novel lignin in poplar trees with a reduced caffeic acid/5-hydroxyferulic acid *O*-methyltransferase activity, *Plant J.* **8**, 855-864.

Vignols, F., Rigau, J., Torres, M.A., Capellades, M. & Puigdomènech, P. (1995) The *brown midrib3 (bm3)* mutation in maize occurs in the gene encoding caffeic acid *O*-methyltransferase, *Plant Cell* **7**, 407-416.

Villar, M., Lefèvre, F., Bradshaw, H.D. Jr. & Teissier du Cros, E. (1996) Molecular genetics of rust resistance in poplars (*Melampsora larici-populina* Kleb/*Populus sp.*) by bulked segregant analysis in a 2x2 factorial mating design, *Genetics* **143**, 531-536.

Vos, P., Hogers, R., Bleeker, M., Reijans, M., van de Lee, T., Hornes, M., Frijters, A., Pot, J., Peleman, J., Kuiper, M. & Zabeau, M. (1995) AFLP: a new technique for DNA fingerprinting, *Nucleic Acids Res.* **23**, 4407-4414.

Ye, Z.-H., Kneusel, R.E., Matern, U. & Varner, J.E. (1994) An alternative methylation pathway in lignin biosynthesis in *Zinnia*, *Plant Cell* **6**, 1427-1439.

G.T. Scarascia Mugnozza, E. Porceddu & M.A. Pagnotta (Eds.)
Genetics and Breeding for Crop Quality and Resistance, 283-290, 1999

Advanced backcross QTL analysis: a method for the systematic use of exotic germplasm in the improvement of crop quality

S. Grandillo[1,3], D. Bernacchi[4], T. M. Fulton[1], D. Zamir[2] & S.D. Tanksley[1]
[1]Department of Plant Breeding, Cornell University, Ithaca, new York 14853, USA;[2] Department of Field and Vegetable Crops, Faculty of Agriculture, The Hebrew University of Jerusalem Rehovot 76100, Israel;[3] Research Institute for Vegetable and Ornamental Plant Breeding, 80055 Portici, Italy;[4] Dekalb Argentina, Estacion Camet, Ruta Nacional 226 km7, Mar del Plata, Argentina

Abstract: Most traits of agronomic interest, including those related to quality, are polygenically inherited. Exotic germplasm represents a valuable and unique source of genetic variation, but it has rarely been used for the genetic improvement of quantitative traits. We have efficiently screened for new and beneficial quantitative trait loci (QTL) able to improve the agronomic performance of elite varieties. We refer to this method as Advanced Backcross QTL (AB-QTL) analysis. The strategy has been tested in tomato over the past 5 years (using 5 different wild species donors) and in rice for 2 years (one wild species donor). In each case, QTL from the wild species have been identified which are able of substantially increasing quality or productivity of the cultivated species. We have created lines, in the genetic background of an elite processing tomato cultivar, which contain wild QTL alleles for several quality attributes important for the tomato processing industry including soluble solids content, fruit colour, viscosity and firmness. These lines have been subjected to replicated field trials in five locations world-wide. A number of these lines significantly outperformed the elite variety by 22%, 33% and 48% for soluble solids, fruit colour and yield, respectively.

1. Introduction

For most tomato breeding programs yield and adaptability are the primary objectives to be pursued; however, in many cases considerable effort has been invested to develop cultivars with improved fruit quality. For processing tomatoes fruit quality parameters include soluble solids, colour, pH, total acidity, viscosity and firmness. The relative importance of each of these traits depends upon the processed product for which the cultivar is to be used. For example, tomato paste standards are based upon final soluble-solids content of the finished product, and therefore the main objective is to develop high soluble-solids cultivars which yield a greater amount of product from a fixed quantity of freshly harvested fruit, and require less energy in concentration than do low solids cultivars. On the other hand, for a product such as catsup, high viscosity (or consistency), which is determined by insoluble solids, may be the primary quality attribute. For tomato products that are sold on the basis of solid contents, the higher the solids of the raw product, the greater the value of crop yield. For example, an increase in solids of 1%, as from 5% to 6%, represents a 20% increase in yield for products whose processed product yield is directly influenced by solids constituent (Kalloo, 1991).

The importance of fruit soluble solids content has justified extensive research to improve this quality parameter (Rick, 1974). Despite the efforts invested in the improvement of this important quality trait, there is a clear difficulty in combining increased level of soluble solids with high yield in processing cultivars, because of the polygenic nature of the trait, the large effect of the environment, and the negative relationship between soluble solids content and fresh yield (Stevens & Rudich, 1978).

2. Genetic gains for processing tomatoes

In order to get an estimate of the genetic gains achieved so far in yield and quality traits of processing tomatoes by conventional breeding programs, we analysed the data of replicated field trials that have been conducted over the past 20 years by the California Tomato Research Institute and the Israeli Ministry of Agriculture (Grandillo et al., 1998). The trials have been designed with the objective of identifying superior varieties that would benefit farmers and the processing industry; however, since identical check varieties were tested in consecutive years, it was possible to estimate the role of genetics by regressing the data as percent of the check against years.

In Israel, the data indicate a significant ($P<0.001$) annual genetic gain of 0.40% yr^{-1} for yield, and of 0.53% yr^{-1} for brix and of 0.9% yr^{-1} for the derived trait brix x yield. On the other hand, in California, no significant genetic gain was found for brix during the entire period (1977-94), while an overall significant ($P<0.001$) annual genetic gain of 1.5% was reported for yield and for brix x yield. The data also suggest that approximately 30% and 38% of the total genetic gains achieved in yield during the past 20 years in California and Israel, respectively, can be attributed to the replacement of inbred varieties with hybrid varieties, which was especially rapid starting in, 1983-85 both in California and Israel.

3. Molecular maps and the exploitation of exotic germplasm

Up to date, breeders have mostly relied on repeated intercrossing of adapted elite genotypes for the improvement of quantitative traits, like yield and quality. This has led to the narrow genetic basis characteristic of many crops, particularly of self-pollinated crops including tomato and rice (Ladizinsky, 1985; Miller & Tanksley, 1990; Wang et al., 1992). However, broad-based genetic materials are essential to meet a number of breeding objectives. For the genetic improvement to succeed we need to be able to exploit the wealth of genetic variation provided by nature and currently warehoused in our seed repositories. These wild and unadapted germplasm resources can potentially fuel crop plant improvement efforts for many years into the future.

Unadapted germplasm has been mostly used as a source for major genes for disease and insect resistances, which can be incorporated via backcross breeding, but very limited has been its use for the improvement of quantitative traits. Undesirable effects, in fact, are often associated with introgressed segments, perhaps in some cases due to pleiotropy but probably more often due to linkage; and much of the wild germplasm is phenotypically inferior to modern cultivars for many of the quantitative traits (e.g. yield, vigour and quality) that breeders would like to improve.

However, recent research, focusing on the tomato as a model system, has shown that despite their inferior phenotypes, wild species are likely to contain QTLs that can substantially increase the yield and quality of elite cultivars (de Vicente & Tanksley, 1993; Eshed & Zamir, 1995). Using genetic markers, one can determine the chromosomal location and phenotypic effect of a gene and the mode of inheritance of the trait(s) influenced by the gene. Pleiotropy can be distinguished from close linkage, and recombinants can be identified in which close linkages are broken.

Despite the large amount of QTL mapping work reported, the impact of QTL mapping on the development of improved varieties has been minimal. Based on these considerations, Tanksley & Nelson (1996) proposed a new molecular breeding strategy based on QTL mapping that integrates the processes of QTL analysis and variety development. This new procedure is referred to as Advanced Backcross QTL analysis (AB-QTL). This strategy uses molecular markers to identify beneficial alleles from unadapted germplasm that have the potential to improve the agronomic performance of elite cultivated lines. These QTL alleles are simultaneously transferred into near isogenic lines (NILs) which are then field tested in replicated trials. Thus, a cycle of AB-QTL breeding (i.e. QTL discovery and NIL development and testing) represents a direct test of the underlying assumption of QTL breeding: that beneficial QTL alleles identified in segregating populations (BC_2, BC_3 in AB-QTL analysis) will continue to exert their positive effect when transferred to elite lines.

So far the AB-QTL analysis projects have produced mapping results for a battery of agronomic traits of interest in processing tomatoes. Populations involving crosses with five *Lycopersicon* species (*L. pennellii*, *L. hirsutum*, *L. pimpinellifolium*, *L. parviflorum*, *L. peruvianum*) have ben field and laboratory tested in a number of locations around the world. Here we will report the results obtained for quality and yield traits from the first three AB-QTL studies which used as donor wild parents *L. pimpinellifolium* (accession LA1589) (denoted PM)

(Grandillo & Tanksley, 1996; Tanksley et al., 1996), *L. peruvianum* (accession LA1706) (denoted PV) (Fulton et al., 1997) and *L. hirsutum* (accession LA1777) (denoted H) (Bernacchi et al., 1998a).

4. Wild *Lycopersicon* species as a source of valuable quality QTLs

Exotic germplasm accessions for the use with AB-QTL method have been selected on the basis of genetic uniqueness, in order to sample accessions representing the broadest possible spectrum of wild species and races present in the seed banks. This should increase the probability of identifying in each separate study a high proportion of new and useful QTLs. The red-fruited species *L. pimpinellifolium* represents the most closely related wild species of the cultivated tomato (*L. esculentum*), whereas the two green-fruited species, *L. hirsutum* and *L. peruvianum*, are more distantly related. All these three species have been the sources of many major resistance genes, however no effort has been made to take full advantage of the high level of genetic variation available for the improvement of quantitative traits (Kaloo, 1991).

For the PM and H studies BC_2 populations were used, whereas in the PV study a BC_3 population was tested (Tanksley et al., 1996; Fulton et al., 1997; Bernacchi et al., 1998a). The number of phenotyped families were 170 BC_2/BC_2F_1 and BC_2/BC_3 in the PM study, 315 BC2/BC3 in the H study, and 200 BC_3/BC_4 for the PV study, with marker genome coverages higher than 90% in all cases. Between 20 and 35 traits of agronomic interest were evaluated in 3-4 locations worldwide. In all cases total yield (YDT), red yield (YDR) and main fruit quality characteristics, including soluble solids content (SSC), fruit colour (FC), viscosity (VIS), firmness (FIR) and fruit pH (pH) were measured. Due to the frequent negative relationship characterising soluble solids content and yield the derived parameter brix x yield (BY) has been considered as a more comprehensive biological and agricultural estimate for the productivity of processing tomatoes (Eshed & Zamir, 1995; Tanksley et al., 1996).

For all traits it was possible to detect favourable wild QTL alleles (Figure 1), and in all three wild species tested, the highest number of favourable wild QTLs were detected for SSC and to a lesser extent for FC. Overall a lower number of favourable wild QTLs were identified for yield.

Specific segments of the wild genome were chosen for QTL-NIL development. Targeted regions were selected to contain QTL for which the wild allele was expected to have a favourable effect on one or more traits and no major negative effect on other traits of agronomic interest (Figure 2) (Bernacchi et al., 1998b). For example, the top of chromosome 3 of *L. pimpinellifolium* was selected for carrying potential favourable QTL alleles for fruit colour and firmness, and the centromeric region of chromosome 9 for carrying a QTL allele predicted to increase fruit weight. For *L. hirsutum* one of the regions chosen for NIL development was the bottom of chromosome 4 for the positive effects on both fruit colour and brix x red yield. For

L. peruvianum, several regions of potential breeding value have been identified for which QTL-NIL are now being developed (Fulton et al., unpublished data).

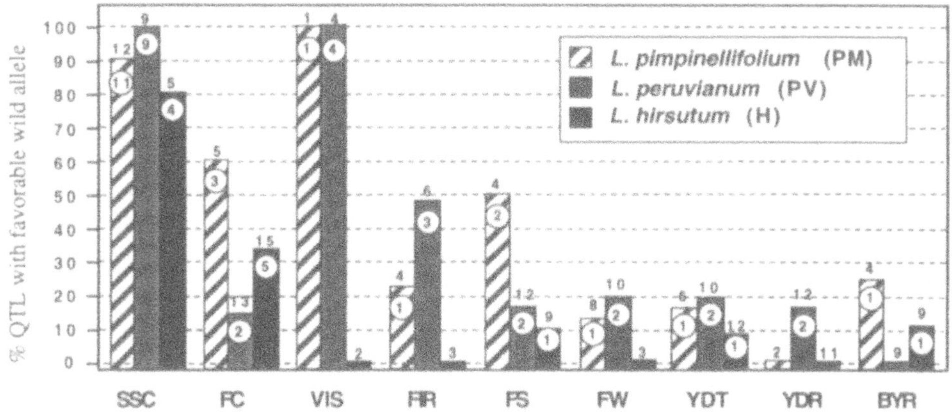

Figure 1. Percentage of QTLs with favourable allele (from an horticultural perspective) detected for *L. pimpinellifolium* (striped bars) (Tanksley et al., 1996), *L. peruvianum* (gray bars) (Fulton et al., 1997) and for *L. hirsutum* (black bars) (Bernacchi et al., 1998a). The total number of QTLs detected for the trait are indicated above the bars; the total number of QTLs with favourable alleles are circled. SSC = soluble solids content, FC = fruit colour, VIS = viscosity, FIR = firmness, FS = fruit shape, FW = fruit weight, YDT = total yield, YDR = red yield, BYR = brix x yield.

The performance of the lines developed for PM and H has been confirmed under different growing conditions around the world (Tanksley et al., 1996; Bernacchi et al., 1998b). Per location gains over the elite control (E6203) ranged from 6% to 22% for SSC; 14% to 33% for FC; 7% to 22% for VIS; 17% to 34% for FIR; 20% to 28% for YDT; 15% to 48% for YDR and 9% to 59% for BYR.

The magnitude of these improvements is substantial; and this is especially clear if we compare the potential genetic gains per unit time that can be achieved by means of the AB methodology with the historical genetic gains achieved for soluble solids and yield in processing tomatoes with conventional breeding. For example, after four years required for the first cycle of AB-QTL breeding using *L. hirsutum* as donor parent, the gains obtained for brix x yield ranged from 14% to 33%, resulting in yearly gains between 3.5% to 8%. For the same trait the estimates of annual genetic gains obtained with conventional breeding are of 0.9% and 1.5% for the Israeli and California data, respectively (Grandillo et al., 1998).

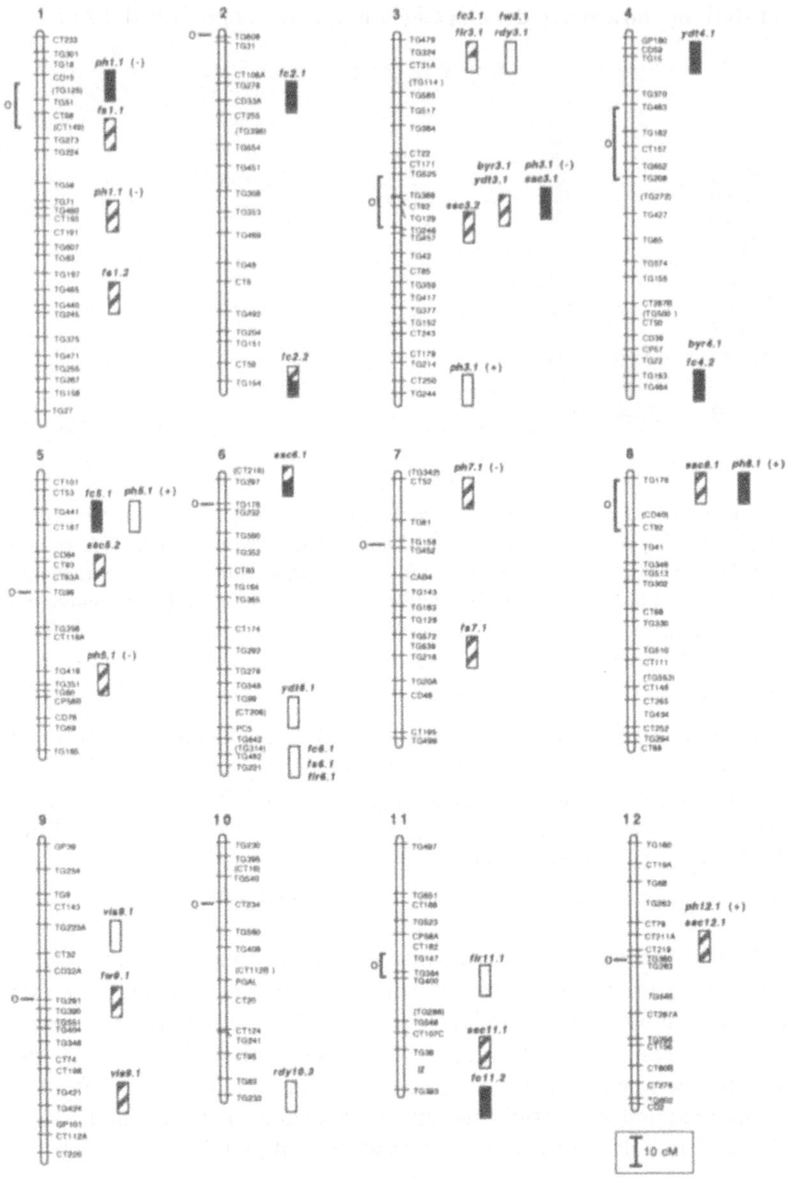

Figure 2. Map location of putative QTLs with favourable wild alleles and with no major negative effect on other traits of agronomic interest (Tanksley et al., 1996; Fulton et al., 1997; Bernacchi et al., 1998a). CentiMorgan distances are based on the molecular linkage map of tomato (Tanksley et al., 1992); the centimorgan scale is below chromosome 12. The most likely position of the QTLs is indicated by the boxes to the right of the chromosomes. PM QTLs are indicated by striped bars, PV QTLs by gray bars and H QTLs by black bars. QTLs are designated according to trait abbreviations (see Figure 1) and followed by chromosome number; a second number is used to

distinguish QTLs mapping to the same chromosome and affecting the same trait. For pH, +/- sign indicate an increase or decrease, respectively.

Of particular interest is the improvement in red fruit (attributable to the pigment lycopene) considering that the wild tomato *L. hirsutum* is green-fruited and lacks an active enzyme for the last step in the pathway and cannot synthesise lycopene. This results can be explained hypotisizing that the wild tomato contains genes (alleles) that can enhance earlier steps in the biosynthetic pathway leading to lycopene which, when combined with an active form of the gene for lycopene synthesis from cultivated tomato, results to even higher levels of pigment production in the interspecific cross. Similarly, fruit size of fruit size has been increased in cultivated tomato lines by the introduction of genes identified through molecular mapping from the small-fruited ancestor *L. pimpinellifolium* (Tanksley et al., 1996).

Currently, AB-QTL studies are also under way in rice (Xiao et al., 1997) and maize (McCouch, personal communication). Since these QTLs are linked to molecular markers, marker assisted selection can be readily applied and the different NILs produced by AB QTL breeding can be easily intercrossed to pyramid different QTL alleles for the same traits or different traits to obtain even greater potential improvement. This method not only results in improved elite varieties, but also is a general enrichment of cultivated germplasm.

5. References

Bernacchi, D., Beck-Bunn, T., Eshed, J., Lopez, J., Petiard, V., Uhlig, J., Zamir, D. & Tanksley, S.D. (1998a) Advanced backcross QTL analysis of tomato. I: Identification of QTL for traits of agronomic importance from *Lycopersicon hirsutum, Theor. Appl. Genet.* 97(3):381-397.

Bernacchi, D., Beck-Bunn, T., Emmatty, D., Eshed, J., Inai, S., Lopez, J., Petiard, V., Sayama, H., Uhlig, J., Zamir, D. & Tanksley, S.D. (1998b) Advanced backcross QTL Analysis of tomato. II: Evaluation of near-isogenic lines carrying single donor introgression desirable wild QTL-alleles derived from *Lycopersicon hirsutum* and *L. pimpinellifolium, Theor. Appl. Genet.* 97:1191-1196

de Vicente, M.C. & Tanksley, S.D. (1993) QTL analysis of transgressive segregation in an interspecific tomato cross, *Genetics* **134**, 585–596.

Eshed, Y. & Zamir, D. (1995) An introgression line population of *Lycopersicon pennellii* in the cultivated tomato enables the identification and fine mapping of yield associated QTLs, *Genetics* **141**, 1147-1162.

Fulton, T.M., Beck-Bunn, T., Emmatty, D., Eshed, Y., Lopez, J., Petiard, V., Uhlig, J., Zamir, D. & Tanksley, S. D. (1997) QTL analysis of an advanced backcross of *Lycopersicon peruvianum* to the cultivated tomato and comparisons with QTLs found in other wild species, *Theor. Appl. Genet.* **95**, 881-894.

Grandillo, S. & Tanksley, S.D. (1996) QTL Analysis of horticultural traits differentiating the cultivated tomato from the closely related species *Lycopersicon pimpinellifolium, Theor. Appl. Genet.* **92**, 935-951.

Grandillo, S., Zamir, D. & Tanksley, S.D. (1998) Genetic Improvement of Processing Tomatoes: a 20 Years Perspective, *Euphitica* (in press).

Kalloo, G. (1991) *Genetic Improvement of Tomato*, Springer-Verlag Publishers Berlin Heidelberg.

Ladizinsky, G. (1985) Founder effect in crop–plant evolution, *Econ. Bot.* **39**, 191–199.

Miller, J.C. & Tanksley, S.D. (1990) RFLP analysis of phylogenetic relationships and genetic variation in the genus *Lycopersicon, Theor. Appl. Genet.* **80**, 437–448.

Rick, C.M. (1974) High soluble-solids content in large-fruited tomato lines derived from a wild green-fruited species, *Hilgardia* **42**, 493–510.

Stevens, M.A. & Rudich, J. (1978) Genetic potential for overcoming physiological limitations on adaptability, yield, and quality in the tomato, *Hort. Science* **13**, 673–678.

Tanksley, S.D., Ganal, M.W., Prince, J.P., de Vicente, M.C., Bonierbale, M.W., Broun, P., Fulton, T.M., Giovannoni, J.J., Grandillo, S., Martin, G.B., Messeguer, R., Miller, J.C., Miller, L., Paterson, A.H., Pineda, O., Röder, M.S., Wing, R.A., Wu, W. & Young, N.D. (1992) High density molecular linkage maps of the tomato and potato genomes, *Genetics* **132**, 1141–1160.

Tanksley, S.D. & Nelson, J.C. (1996) Advanced backcross QTL analysis: a method for the simultaneous discovery and transfer of valuable QTLs from unadapted germplasm into elite breeding lines, *Theor. Appl. Genet.* **92**, 191-203.

Tanksley, S.D., Grandillo, S., Fulton, T.M., Zamir, D., Eshed, Y.,Petiard, V., Lopez, J. & Beck-Bunn, T. (1996) Advanced backcross QTL analysis in a cross between an elite processing line of tomato and its wild relative *L. pimpinellifolium*, *Theor. Appl. Genet.* **92**, 213-224.

Wang, Z.Y., Second, G. & Tanksley, S.D. (1992) Polymorphism and phylogenetic relationships among species in the genus *Oryzae* as determined by analysis of nuclear RFLPs, *Theor. Appl. Genet.* **83**, 565-581.

Xiao, J., Li, J., Grandillo, S., Ahn, S., McCouch, S.R., Tanksley, S.D. & Yuan, L. (1996) Genes from wild rice improve yield, *Nature* **384**: 223-224.

G.T. Scarascia Mugnozza, E. Porceddu & M.A. Pagnotta (Eds.)
Genetics and Breeding for Crop Quality and Resistance, 291-299, 1999

QTLs for organoleptic quality in fresh market tomato

V. Saliba-Colombani[1], M. Causse[1], J. Philouze[1], M. Buret[2], S. Issanchou[3] & I. Lesschaeve[3]

[1]INRA–Unité de Génétique et d'Amélioration des Fruits et Légumes BP, 94 – 84143 Montfavet, Cedex, France; [2]INRA–Station de Technologie des produits végétaux, BP 94 – 84143 Montfavet, Cedex, France; [3]INRA, Laboratoire de Recherche sur les Arômes, 17, rue Sully, 21034 Dijon, France.

Abstract: The organoleptic quality of tomato fruit is determined by many characters. Therefore, plant breeders often find difficulties to improve such a characteristic. A program of QTL detection for physical, chemical and sensorial traits has been achieved, in order to understand the genetic determinism of tomato organoleptic quality. One hundred and forty-four recombinant inbred lines (RILs), derived from an intraspecific cross, were analyzed with segregating molecular markers. An almost saturated map was constructed with RFLP, AFLP and RAPD marker. The RILs were also evaluated for fruit chemical (sugar, pigment and acid contents) and physical traits (color, firmness and fruit size). These analyses were combined with fruit evaluations by a panel of trained judges. A sensorial profile was thus defined for each RIL. Strong correlations have been detected between some chemical and sensorial traits. As an example, the fruit sweetness was positively correlated to the sugar content. Such a correlation is very useful since it allows tomato breeders to improve organoleptic quality indirectly. Significant quantitative trait locus (QTL) associations with marker loci were identified for each trait. Phenotypic variation explained by each QTL ranged from 7.9% to 45.6%. A few chromosomal regions appeared to explain a major part of the variation for many traits, one of the most important being on the long arm of chromosome 2. Some of these QTLs, with large identified effects, will be useful for marker aided selection.

1. Introduction

Tomato is one of the major vegetable crop for many countries and fruit quality is becoming of paramount importance for consumer acceptance. In recent years tomato breeding has focused on increased yield, improved plant adaptation, better disease resistance, improved external fruit quality and firmness. This selection has favoured varieties with attractive appearance but which were often poor in taste. Nowadays, the increase of consumer complaints makes fruit organoleptic quality one of the main objective for fresh tomato breeding. However, organoleptic quality is a complex characteristic, with several components, only few of them have been studied. We began a research project in order to study chemical, physical and sensorial components of organoleptic quality and to use the molecular markers to map QTLs affecting these components. The use of molecular markers allowed us to assess the number of loci involved in quality components and to explore how these components interact together.

Hereafter, we present the construction of an intraspecific molecular map and the major QTLs controlling variation of tomato quality traits.

2. Materials and Methods

2.1 *Plant material*

One hundred and forty four recombinant inbred F_7 lines (RILs) were developed from a cross between two inbred lines, "C" (*Lycopersicon esculentum*, var. *cerasiforme*), a cherry tomato (8g) chosen for its good taste and aromatic intensity and "L" (*Lycopersicon esculentum* Mill.), a line with bigger fruits (130g) and common taste.

2.2 *Molecular markers*

Polymorphism between the two parents has been evaluated with RFLP, RAPD and AFLP markers.

RFLP: Survey blots with six restriction enzymes (*Eco*RI, *Hin*dIII, *Eco*RV, *Sst*I, *Xba*I, *Dra*I) were hybridized with 600 probes (kindly supplied by Dr. S. D. Tanksley), chosen from the high-density tomato molecular map (Tanksley et al., 1992). Among the 164 probes (30%) showing polymorphism between C and L, 136 were mapped on the segregating population.

RAPD: One hundred and twenty six primers were surveyed on parental lines, 48 revealed at least one polymorphic band. Only primers showing reproducible and clearly scorable polymorphism were used to amplify RIL DNA. We firstly mapped 28 RAPD loci on a subset of the RILs population. Then loci providing a complementary information were mapped on the whole population.

AFLP: AFLP analyses were carried out as described by Vos et al., 1995. Digestions were performed with the restriction enzymes, *Mse*I (frequent cutter) and *Hin*dIII (rare cutter). A preamplification was performed with *Mse*I primer + 1 selective nucleotide, M02: 5'-GAT GAG TCC TGA GTA A C- 3' and *Hin*dIII primer +1 selective nucleotide, H01: 5' – GAC TGc GTA CCA GCT T A- 3'.

Amplifications involving primers with three selective nucleotides were then conducted. A total of 46 primer combinations were tested on the parents. Primer combinations which were easy to score and which showed a number of polymorphic bands higher than 5, were chosen for the segregation analysis (Table 1).

A single morphological locus, *hairs absent* (*h*), was scored on the RIL population.

Table 1. Primer combinations scored on the RILs. Each three selective nucleotide combination is labeled with a code.

Code		M47 CAA	M48 CAC	M49 CAG	M50 CAT	M51 CCA	M52 CCC	M53 CCG	M54 CCT	M55 CGA	M56 CGC	M57 CGG	M58 CGT	M59 CTA	M60 CTC	M61 CTG	M62 CTT
H33	AAG	X		X	X	X	X		X	X				X		X	X
H35	ACA	X	X			X										X	X
H38	ACT	X												X			X
H42	AGT	X															X

Chemical components: Titratable acidity (TA), pH, sugar content (SC), soluble solids content (SSC), dry matter weight (DMW), carotene content (CAR) and lycopene content (LYC).

Sensorial components: Sensory analysis with a panel of 90 trained judges was used to evaluate consumer perception of fruits. Each line was evaluated by about 18-23 judges in an incomplete randomized bloc design trial. Twelve descriptors were chosen, among which only six will be presented here:

Flavor: sweetness (SWE) and sourness (SOU).

Aroma: overall aroma intensity (ARO).

Texture: Firm texture (FIT), starchiness (STA) and juiciness (JUI).

3. Results

3.1 Genetic map

The map presented on figure 1, includes one morphological, 87 RFLP, 3 RAPD and 137 AFLP markers. Deviation from the expected segregation ratio concerned 8% and 18% of RFLP and AFLP markers, respectively. The overall map length was 1260 cM and the average distance between markers was 9.7 cM. Fourteen linkage groups were obtained as chromosomes 1 and 4 are both represented by two linkage groups. Due to the low level of polymorphism, 4 intervals in telomeric regions (in the range of 34 –58 cM) were still not covered on chromosomes 5, 7, 9 and 11. Twenty AFLP primer combinations have been assessed on this population and produced 174 non randomly distributed markers. These markers formed 47 linkage groups of 5 cM each (with 1-34 markers per group), or 82 linkage groups of 1 cM each (with 1-27 markers per group).

The largest clusters mapped on chromosomes 4 and 9, with 22 and 34 markers, respectively (Figure 1). Most of the clusters were localized around putative centromeric regions. The RAPD markers also clustered around putative centromers. The largest RAPD cluster contained 11 loci within 10 cM (Figure 1). At last, only 3 RAPD and 30 AFLP markers were useful to complete the initial RFLP map.

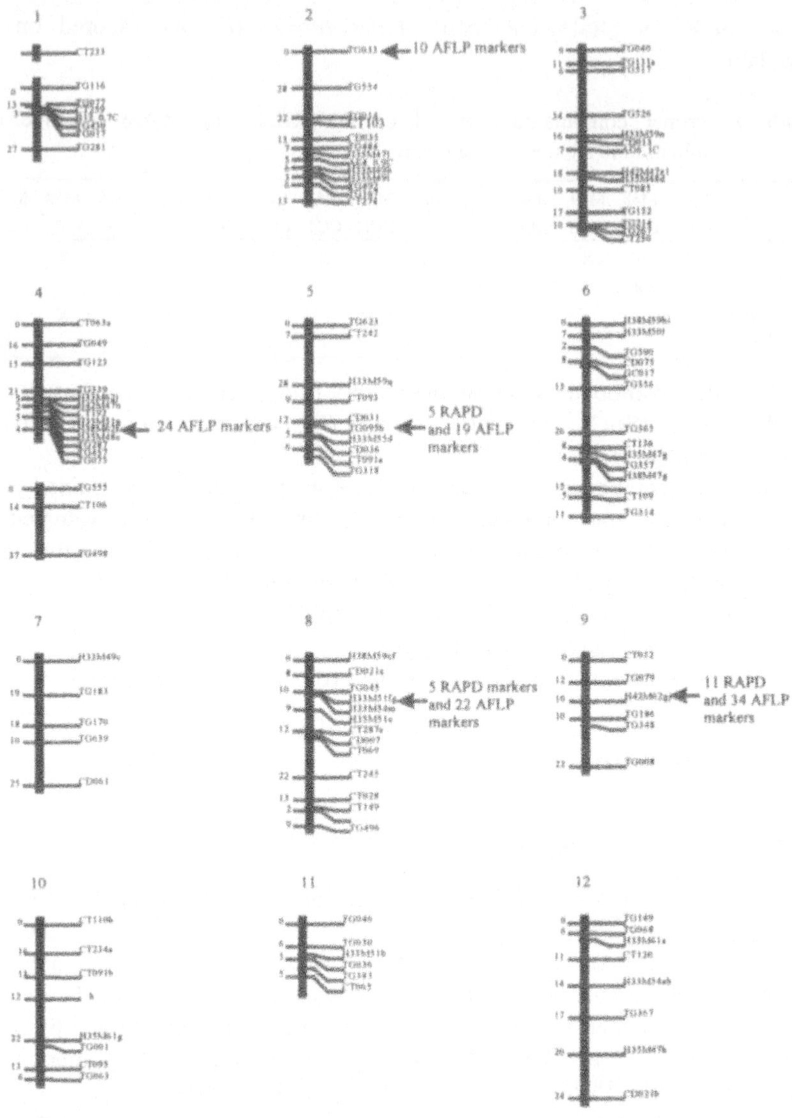

Figure 1. Genetic molecular map. Chromosome numbers are shown above the bars, locus name on the right and distances (Kosambi cM) on the left. AFLP markers are designated by the 3 selective nucleotide codes of *Mse*I and *Hin*dIII (Table 1) followed by a letter from a to z depending on molecular weight. RAPD markers are designated by the operon name, the molecular weight of the band and the parent exhibiting the band (C or L). Clusters of AFLP and RAPD markers are indicated by arrows.

3.2 *Distributions of quantitative traits*

For each trait, the average value over the six weeks was estimated. Some distributions deviated from normality. As an example, fruit weight showed a significant skewed distribution consistent with other observations (Grandillo & Tanksley, 1996). In order to approach normality for QTL detection, we used $\log_{10}$ or square root transformation. For all traits, except fruit weight and fruit size, extreme individuals (relatively to parents) were observed suggesting transgressive segregations.

3.3 *Correlations between traits*

Significant phenotypic correlations were observed between traits. Table 2 displays the correlation matrix among major quality traits. Among the chemical components, the strongest positive correlations were observed between dry matter weight and sugar content (r=0.79) and between dry matter weight and titratable acidity (r=0.67). For the physical traits, highly positive correlations were found between fruit size, fruit weight, elasticity and firmness (FIR1). A negative correlation was detected between fruit weight and dry matter weight. The sweet taste rated by the panel of judges was negatively correlated to sourness. The overall aroma intensity was positively correlated to sweetness and sourness with r=0.40 and 0.46, respectively. Some chemical traits were positively correlated to sensorial traits like sweetness which was correlated to sugar content (r=0.65) and sourness which was correlated to titratable acidity (r=0.82). Instrumentally measured firmness was not correlated to the firmness evaluated by judges.

Table 2. Significant correlations among major quality traits ($p<0.01$).

	SWE	SOU	ARO	FIT	JUI	STA	SUG	pH	DMW	LYC	CAR	TA	FW	FS	FIR1	FIR2	ELA
SOU	-0.43																
ARO	0.40	0.46															
FIT		0.29															
JUI																	
STA	-0.30		-0.32	-0.29	-0.52												
SUG	0.65		0.67			-0.24											
PH						-0.25											
DMW	0.39	0.46	0.77	0.23			0.79										
LYC							0.27										
CAR			0.24			-0.24	0.26		0.34	0.48							
TA		0.82	0.55				0.23	-0.26	0.67								
FW		-0.44	-0.52				-0.45	0.27	-0.67	-0.32	-0.40	-0.57					
FS		-0.42	-0.55			0.22	-0.48	0.26	-0.70	-0.31	-0.44	-0.59	0.98				
FIR1		-0.31	-0.40				-0.41		-0.63		-0.29	-0.55	0.53	0.57			
FIR2				0.29	-0.20					0.23		-0.22	-0.22	-0.22	0.26		
ELA		-0.41	-0.40		-0.25		-0.34		-0.58		-0.35	-0.57	0.50	0.54	0.56	0.58	
L		-0.35	-0.38				-0.33		-0.56			-0.46	0.23	0.26	0.55	0.39	0.44

3.4 *QTL analyses*

QTLs were detected for all traits. Fifty five significant QTLs were detected both by single-point ($p<0.005$) and interval mapping (LOD>2.5) analyses (Table 3). The number of significant QTLs per trait ranged from 1 to 5. The percentage of phenotypic variation explained by each QTL ranged from 7.9% to 45.6% for colour parameter (a*) and fruit size, respectively. Allelic effects opposite to that predicted by the parental phenotype value were detected for 13 QTLs (18% of the QTLs). QTLs are briefly described hereafter.

Table 3. QTLs detected by interval mapping at LOD>2,5. R^2=percent of phenotypic variation explained by the QTL. Last column represents the parent (C or L) showing higher allelic effects.

Trait	Chr	Marker	R^2 (%)	LOD	Parent
Dry matter weight	2	AE4.0.9C	40.60	12.8	C
	9	CT032	9.30	2.7	C
Soluble solids content	2	CD035	26.00	8.4	C
	2	AE4.0.9C	30.70	10.29	C
Sugar content	2	CD035	25.00	7.85	C
	2	AE4.0.9C	29.50	9.42	C
	10	TG063	10.00	3	C
	11	TG383	12.10	3.65	C
Titratable Acidity	1	TG077	12.40	4.13	C
	2	TG492	14.30	4.52	C
	3	CT085	15.60	4.6	C
	9	CT032	13.40	3.77	C
	12	CT120	9.60	2.86	C
PH	11	TG383	15.10	4.44	L
	12	CD021b	16.40	3.29	L
Carotene content	2	TG167	14.50	4.21	C
Lycopene content	4	TG457	11.20	3.37	C
	11	CT065	20.30	6.42	C
Fruit size	2	H33M49h	45.60	16.21	L
	3	CT085	17.30	4.56	L
	11	CT065	15.20	4.82	L
Fruit weight	2	H33M49h	43.20	14.82	L
	3	CT085	20.20	5.4	L
	11	CT065	15.60	4.96	L
Fruit elasticity	2	TG167	19.70	6.25	L
	4	TG457	13.70	4.45	C
Firmness 1	2	TG167	28.10	9.58	L
	4	TG457	19.70	5.89	C
Firmness 2	4a	TG498	17.40	3.15	C
	4b	TG457	32.00	11.93	C
L	2	H35M47l	11.60	3.55	L
	4	TG457	23.50	7.04	C
	9	CT032	15.30	4.59	L
a*	4	TG457	24.70	8.3	C
	9	CT032	7.90	2.5	L
b*	2	H35M47l	13.50	4.11	L
	4	TG457	19.90	6.17	C
	9	CT032	21.20	6.64	L
Sweetness	2a	CD035	9.30	2.8	C
	2b	AE4.0.9C	12.30	4.03	C
	3	CT085	9.40	2.83	L
	6	TG356	14.50	2.67	L
	11	TG383	11.40	3.41	C
Sourness	1	TG077	11.00	3.6	C
	2	TG554	17.60	3.18	C
	9	CT032	25.10	6.32	C
Overall aroma intensity	2	AE4.0.9C	22.70	7.58	C
Firm texture	5	CD031	10.40	2.83	C
	9	H35M51f	39.60	11.35	C
Starchiness	2	TG167	17.30	4.85	L
	3	CT085	9.50	2.53	C
	4	TG457	15.40	4.82	L
Juiciness	2	TG167	11.00	3.09	C
	5	CD031	11.50	3.15	L
	12	TG367	13.50	2.63	C

3.4.1 Chemical components

The dry matter weight had the largest QTL effect on chromosome 2 with R^2=40.6%. In the same chromosomal region, 2 highly significant QTLs were identified for sugar content with R^2=29.5 and 25. Titratable acidity showed several QTLs on chromosomes 1, 2, 3, 9 and 12, which explained 12.4%, 14.3%, 15.6%, 13.4% and 9.6% of phenotypic variation. Carotene content was controlled by one

QTL on chromosome 2 explaining 14.5% of phenotypic variation and lycopene content was controlled by 2 QTLs on chromosomes 4 and 11 with R^2=11.2 and 20.3, respectively.

3.4.2 Physical components

Three QTLs were detected for fruit weight on chromosomes 2, 3 and 11 which explained 43.2%, 20.2% and 15.6% of the phenotypic variation, respectively.

QTLs of fruit size were localized in the same regions. Fruit firmness was affected by 4 QTLs, one on chromosome 2 and three on chromosome 4. The most significant was on chromosome 4 and explained 32% of the phenotypic variation. Three loci were implicated in the variation of colour parameters on chromosome 2, 4 and 9.

3.4.3 Sensorial components

Six QTLs for sweetness were detected. The 2 QTLs on chromosome 2 affected the phenotypic variation with 9.3% and 12.3%. For overall aroma intensity, one QTL was identified on chromosome 2 explaining 22.7% of phenotypic variation.

Sourness was controlled by 3 QTLs on chromosomes 1, 2 and 9. The major QTL was located on chromosome 9 and explained 25.2% of sourness variation.

Two QTLs were found to influence firm texture, distributed on chromosomes 5 and 9 and affected the trait with 10.4% and 39.6%, respectively.

3.5 Colocalizations of QTLS

QTLs were mainly distributed on chromosomes 2, 3, 4, 9, 11 and 12. The region characterized by the largest cluster of QTLs affecting fruit quality traits was found at the lower end of chromosome 2 (a region of 40 cM). The major QTLs for dry matter weight, sugar content, sweetness, juiciness, sourness and aroma intensity were in this cluster. In the same region, we found QTLs which exert the strongest effect on fruit weight and size. QTLs of titratable acidity were localized in the same region as sourness QTLs on chromosomes 1 and 9. On chromosome 4, QTLs of instrumentally measured firmness are colocalized with QTLs of color parameters. On chromosome 11 a QTL of fruit weight was colocalized with one of sugar content.

4. Discussion and Conclusions

Deviation from the expected segregation ratio was detected for 8% and 18% of RFLP and AFLP markers, respectively. This level of distortion, was lower than those observed in interspecific crosses with *L. pimpinellifolium and L. cheesmanii*, the closest species to the cultivated tomato (Paran et al., 1995; Grandillo & Tanksley, 1996). The intraspecific molecular map covered 85% of the genome. Order of RFLP markers was in agreement with the tomato high-density molecular map (Tanksley et al., 1992). However, two large gaps, on chromosomes 1 and 4, showed distances higher than expected.

The RAPD markers were non randomly distributed and clustered around putative centromers, as observed in another population by Grandillo & Tanksley (1996). The AFLP method revealed a high number of polymorphic bands per primer combination. However, their non-random distribution restricts their efficiency in tomato mapping (Saliba et al., 1998)

We found QTLs for all the traits. These QTLs were mainly localized on chromosomes 2, 3, 4, 9, 11 and 12. Fruit weight, soluble solids, sugar content and titratable acidity have been studied in other tomato crosses. Depending on the population under study, QTLs of fruit weight and SSC have been localized on chromosomes 2, 3, 4, 6, 9 and 12 (Tanksley et al., 1996; Goldman et al., 1995). For these two traits, we found the same QTL locations on chromosomes 2 and 3. However, on chromosome 6 we did not find any of the QTLs expected in the fore-mentioned studies. The fruit weight QTL is localized on the same region of chromosome 2 as *fw2.2* (Alpert et al., 1995). This QTL maps to the same position in the green-fruited wild tomato species *Lycopersicon pennellii* and in the red-fruited wild tomato species *L. pimpinellifolium.*

Eighteen percent of QTLs showed allelic effects opposite to that predicted by the parental phenotype value. This can explain the occurrence of transgressive phenotypes as discussed in de Vincente & Tanksley (1993).

QTLs of correlated traits were frequently localized in the same regions, as in the case of sweetness and sugar content, and titratable acidity and sourness. This confirms that in spite of all possible sources of variation, QTLs of chemical traits are concordant with QTLs of equivalent traits evaluated by sensorial analyses. However, sensory firmness had not any related trait among the instrumentally measured firmness measures. Thus, the sensorial analysis could not always be replaced by physical or chemical traits.

Overall aroma intensity, sweetness and sourness are important traits related to organoleptic quality. However, QTLs of these traits were often colocalized with QTLs of fruit weight, but with opposite effects. Several studies reported this negative relationship between fruit weight and dry matter weight (Tanksley et al., 1996; Fulton et al., 1997). Fine mapping experiments will be necessary to determine whether these QTLs exhibit pleiotropic effects or are tightly linked. On chromosome 10, a QTL of sugar content was detected in a region where no QTL of fruit weight was detected. Marker assisted selection will be helpful in transferring QTLs of fruit organoleptic quality without loosing other quality traits especially fruit weight.

Acknowledgements: This investigation was a collaborative work between INRA and GIE Clause-Limagrain, Gautier associated partner and was funded by the French Ministry of Agriculture. We would like to acknowledge P. Duffé and M. Milesi for their excellent technical assistance.

5. References

Alpert, K.B., Grandillo, S. & Tanksley, S.D. (1995) *fw2.2*:a major QTL controlling fruit weight is common to both red- and green-fruited tomatoes species, *Theor Appl Genet* **91,** 994-1000.

Fulton, T.M., Beck-Bunn, T., Emmatty, D., Eshed, Y., Lopez, J., Petiard, V., Uhlig, J., Zamir, D. & Tanksley, S.D. (1997) QTL analysis of an advanced backcross of *Lycopersicon peruvianum* to the cultivated tomato and comparisons with QTLs found in other wild species, *Theor Appl Genet* **95,** 881-894.

Goldman, I.L., Paran, I. & Zamir, D. (1995) Quantitative trait locus analysis of a recombinant inbred line population derived from a *Lycopersicon esculentum* x *Lycopersicon cheesmani* cross, *Theor Appl Genet* **90,** 925-932.

Grandillo, S. & Tanksley, S.D. (1996) Genetic analysis of RFLPs, GATA microsatellites and RAPDs in a cross between *L. esculentum* and *L. pimpinellifolium*, *Theor Appl Genet* **92,** 957-965.

Paran, I., Goldman, I., Tanksley, S.D. & Zamir, D. (1995) Recombinant inbred lines for genetic mapping in tomato, *Theor Appl Genet* **90,** 542-548.

Saliba, V., Duffé, P., Gervais, L. & Causse, M. (1998) Efficiency of AFLP markers to saturate a tomato intraspecific map, *TGC Report*.

Tanksey, S.D. (1997) QTL analysis of an advanced backcross of Lycopersicon peruvianum to the cultivated tomato and comparisons with QTLs found in other wild species, *Theor Appl genet* **95,** 881-894.

Tanksley, S.D., Ganal, M.W., Prince, J.P., de Vicente, M.C., Bonierbale, M.W., Broun, P., Fulton, T.M., Giovannoni, J.J., Grandillo, S., Martin, G.B., Messeguer, R., Miller, J.C., Miller, L., Paterson, A.H., Pineda, O., Röder, M.S., Wing, R.A., Wu, W. & Young, N.D. (1992) High density molecular linkage maps of the tomato and potato genomes, *Genetics* **132,** 1141-1160.

Tanksley, S.D., Grandillo, S., Fulton, T.M., Zamir, D., Eshed, Y., Petiard, V., Lopez, J. & Beck-Bunn, T. (1996) Advanced backcross QTL analysis in a cross between an elite processing line of tomato and its wild relative *L. pimpinellifolium*, *Theor Appl Genet* **92,** 213-224.

de Vincente, M.C. & Tanksley, S.D. (1993) QTL analysis of transgressive segregation in an interspecific tomato cross, *Genetics* **134,** 585-596.

Vos, P., Hogers, R., Bleeker, M., Reijans, M., Van Delee, T., Hornes, M., Frijters, A., Pot, J., Peleman, J., Kuiper, M. & Zabeau, M. (1995) AFLP: a new technique for DNA fingerprinting, *Nucleic Acids Research* **23,** 4407-4414.

G.T. Scarascia Mugnozza, E. Porceddu & M.A. Pagnotta (Eds.)
Genetics and Breeding for Crop Quality and Resistance, 301-306, 1999

Genetic engineering of parthenocarpic vegetable crops

G.L. Rotino, G. Donzella[1], M. Zottini[2], H. Sommer[3], N. Ficcadenti[4], C. Cirillo, S. Sestili[4], E. Perri, T. Pandolfini[2] & A. Spena[2]

Research Institute for Vegetable Crops, 26836 Montanaso Lombardo (LO), Italy;[1]Ente Sviluppo Agricolo Sicilia SOPAT N° 36, 97019 Vittoria (RG), Italy;[2]Faculty of Science, University of Verona, 37134 Verona, Italy;[3]MPI für Züchtungsforschung, 50829 Cologne, Germany;[4]Research Institute for Vegetable Crops, 63030 Monsampolo del Tronto (AP), Italy

Key words: Auxin, *Lycopersicon esculentum*, *Nicotiana tabacum*, parthenocarpy, *Solanum melongena*

Abstract: Parthenocarpy, referred to as the ability to develop fruits in absence of fertilization, is a desirable genetic trait in vegetable crops grown for the commercial value of their fruits. The advantages conferred by parthenocarpy are fruit production under environmental condition prohibitive for fertilization and production of seedless fruit. Methods for achieving parthenocarpic development use either synthetic growth factors, genetic mutants, or plant altered in their ploidy level. Nevertheless, parthenocarpic cultivars have so far been available only to a limited degree. Therefore, flower buds treatment with synthetic phytohormones is commonly used to induce parthenocarpic vegetable fruit development for early or late production under protected cultivation. Transgenic parthenocarpic tobacco, eggplant and tomato plants are described which contain in their genome the coding region of the *iaaM* gene from *Pseudomonas syringae* pv. *savastanoi* under the control of the placental- ovule-specific *defh9* gene regulator sequences from *Anthirrhinum majus*. Expression of the chimeric *defH9-iaaM* transgene starts during early flower development and causes production of marketable eggplant and tomato fruits from both emasculated and pollinated flowers under environmental conditions prohibitive for fruit setting in the untransformed controls, which did not set fruit at all. Under normal environmental conditions, production of marketable fruits took place from pollinated and unplanted transgenic flowers, while untransformed plants produced fruits of marketable size only from fertilized flowers. Field tests, performed in a cold plastic-greenhouse in the Mediterranean area, showed that three experimental transgenic parthenocarpic eggplant hybrids were able to give satisfactory winter production without need of phytohormones treatment.

1. Introduction

Parthenocarpy, referred to as the ability to set and develop fruits in absence of fertilization, is a desirable trait for many horticultural crops cultivated for the commercial value of their fruits. Parthenocarpic marketable fruits can be obtained under environmental condition which, normally, are limiting for fruit setting and growth. Moreover, the parthenocarpic fruits are, often, seedless. This feature, at least for some species, improve fruit quality and acceptance by consumers.

Many enviromental factors hamper processes such as formation, dispersal and germination of pollen, fertilization and seeds maturation which are crucial for fruit set and growth. In particular, excessive or low temperature and humidity, low light intensity, heavy rains and strong winds negatively affect the reproductive process impairing fruit production (Osborne & Went, 1953; Romano & Leonardi, 1994).

To overcome these problems, classical methods based on agronomical and genetic tools have been used. Agronomical techniques have been focused on the modification of environmental conditions by cultivation under greenhouse (heated or unheated), and by using other devices to protect cultivation from adverse climatic effects (Bailey & Hunter, 1988). Moreover, pollinator insects or exogenous phytohormones sprays are quite popular technique to improve fruit set in many horticultural crops (Sarma & Barman, 1977; Saavedra, 1979; Khishnamoorthy, 1981). All these approaches are costly and/or labour-intensive adding extra costs to the production process.

The development of parthenocarpic and/or seedless varieties represents the most cost effectively solution to improve the capacity of fruit set and development under adverse climatic condition. Mutants have been identified in several plant species (De Ponti, 1976; Philouze, 1985; Hennart, 1996) and several breeding programs to release parthenocarpic varieties are currently in progress (i.e. tomato, cucumber). Alteration of ploidy level is another breeding strategy to obtain seedless fruit (i.e. triploid watermelon, *citrus* spp.).

However, a large use of parthenocarpic and/or seedless varieties has been restricted by several problems. The most important are the reduction of fruit size and, for certain species, of fruit set percentage (Mapelli et al., 1978). The parthenocarpic trait is, often, polygenic and, therefore, more difficult to deal with in a breeding program. Phenotypic expression of natural genetic parthenocarpy is, sometimes, associated with a loss of fruit quality (i.e. fruit firmness in tomato).

The ideotype of the parthenocarpic trait, to improve the productivity of vegetable crops, has to satisfy the following features: i) production of marketable fruit without pollination; ii) percentage of fruit setting under adverse environmental conditions should be similar to that obtained under favourable growth conditions; iii) phenotypic expression of the trait should not display any negative effect on both intrinsic and extrinsic fruit quality.

Fruit setting and development is associated with an increase in the level and/or activity of endogenous growth factors. It is well known that the developing ovules and the embryos are a source of auxinic phytohormones (Archbold & Dennis, 1985) which lead to fruit setting and growth, and external application of phytohormones can replace the developing ovules to sustain fruit growth (Nitsch, 1950).

To genetic engineer parthenocarpic fruit development the expression of the foreign gene coding for a plant phytohormone has to take place during early floral and fruit development. Several phytopathogenic bacteria possess in their genome information able to increase auxin content and activity in plant cell (Spena et al., 1992). We have used genetic information derived from the bacteria *Pseudomonas syringae* pv. *savastanoi* and from the plant *Anthirrhinum majus* to construct a chimeric gene able to induce parthenocarpic fruit development in transgenic plants. The chimeric gene, named *defh9-iaaM*, was able to induce parthenocarpy in transgenic tobacco, eggplant, tomato and muskmelon. *Defh9-iaaM* gene mimicks the hormonal effects of pollination and embryo development by increasing the content and/or the activity of auxin specifically in the ovule.

2. Chimeric gene d*efH9-iaaM* as genetic tool to induce parthenocarpy

The *defH9*, *deficiens* homologue 9, gene is a MADS-box containing gene from *Antirrhinum majus*. It is specifically expressed in the developing placenta and ovules of the female reproductive organ. Its level of expression in vegetative tissues is below the detection limit of northern blot analysis (Rotino et al., 1997). The promoter and the regulatory sequences of the *defh9* gene was linked to the coding region of the *iaaM* gene from *Pseudomonas syringae* pv. *savastanoi* to obtain an increase in auxin content and/or activity in cells of female reproductive organ. *iaaM* gene codes for the enzyme tryptophanmonoxigenase which convert tryptophan to indolacetamide a precursor of IAA (indolacetic acid), the major form of auxin in plants (Yamada et al., 1985). Indolacetamide is slowly converted to IAA, in the plant cell, either chemically or enzimatically by hydrolases (Kawaguchi et al., 1991).

The chimeric gene *defh9-iaaM* was inserted in the plasmid pPCV002, which contains also the selective marker NPTII conferring resistance to the antibiotic kanamycin (Koncz & Schell, 1986). The recombinant plasmid pPCV002-*defh9-iaaM* was introduced in the *Agrobacterium tumefaciens* strain GV3101 for plant genetic transformation.

3. Phenotypic and molecular analysis of the parthenocarpic trait

Transgenic plants were obtained in tobacco, eggplant and tomato with an efficiency similar to that obtained using constructs containing only marker genes (i.e. *nptII* and GUS). Parthenocarpic transgenic plants showed normal vegetative growth and a phenotype indistinguishable from that one of seed-derived plants. They produced normal fruit with seeds when pollinated, while emasculated flowers developed fruit containing only aborted seeds. This confirmed that the chimeric gene *defh9-iaaM* does not interfer on the *in vitro* regeneration process and has no effect on phenotype of transgenic plants. The highly specific expression of the *defh9-iaaM*, which has a no or very low basal level of constitutive expression, is a crucial aspect. The genetic information introduced into the plant cells controls the level and activity of the growth factor IAA. This could cause detrimental modifications of other developmental and physiological processes if expression should have taken place in other parts of the plant. Therefore, the promoter *defh9* is particularly suited to engineer parthenocarpic fruit development.

In tobacco, the *defh9-iaaM* gene causes production of parthenocarpic capsules, containing aborted seeds, from emasculated flowers. The expressivity of the parthenocapic trait showed a phenotypic variation, estimated by monitoring the percentage of fruit setting and the capsule size after emasculation with respect to the capsules derived from selfed flowers, which correlates with the steady state level of the *defh9-iaaM* mRNA present in young flower buds (between 3 and 5 x 10^{-8} of the total population of mRNA).

In all the transgenic parthenocarpic eggplant analysed, fruit setting and normal fruit development was achieved from both emasculated and selfed flowers under very limiting environmental condition (low temperature, low light intensity and duration). In the same condition, the untrasformed control plants did not set fruit even from hand-pollinated flowers (Rotino et al., 1997).

The different behaviour between the two *Solaneceous* species tobacco and eggplant could be due to the different type of fruit, capsules and berries, which may require a different level of *defh9-iaaM* gene expression to reach complete phenotypic expression of the parthenocarpic trait (size of non-pollinated fruits similar to that one of selfed fruits).

In eggplant and tomato, experiments performed under normal environmental conditions (spring-summer) allowed to compare fruit setting and fruit characteristics in transgenic and untransformed plants. The weight and the size of parthenocarpic eggplant fruits obtained from not pollinated flowers was about fourfold higher than that one of fruits derived from the emasculated flowers of the corresponding untransformed controls. These results showed that engineered parthenocarpic eggplant were superior to eggplants with natural tendency to parthenocarpy.

In tomato, flowers from all transgenic plants developed seedless parthenocarpic berries, whilst control plants did not set fruits at all. All parthenocarpic tomato plants showed similar percentages of fruit setting in pollinated and unpollinated flowers. Moreover, no gross variation in qualitative characteristics of fruit was noted (Ficcadenti, unpublished data).

Expression of the *defh9-iaaM* gene, analysed by competitive RT-PCR on polyA$^+$RNA, in transgenic young flower buds, opened flowers, and fruit at various growth stages confirmed the presence of an amplicon corresponding to the 5' end of the spliced *defh9-iaaM* mRNA in all the analysed organs of the transgenic plants while no amplicon was detected in the untransformed control plants (Rotino et al., 1997; Spena, unpublished data). The steady state level of expression in flower buds from independent transgenic plants of the three species has been estimated to be in the range of $1x10^{-8}$-$2x10^{-9}$ of the total mRNA population. Thus, expression of the *defh9-iaaM* gene starts during early flower development and is protracted to later stages of fruit development.

4. Outseason winter eggplant production

Field test, under cold plastic-greenhouse, was carried out to evaluate the winter production of transgenic eggplant hybrids (Not. N° B/IT/97/29). The trial was done at Vittoria (Sicily), a typical zone for outseason eggplant production, where yearly more than 1,000 hectares of eggplant are cultivated under unheated plastic greenhouse. To obtain marketable eggplant winter production, growers regularly treat open flower buds with commercial formulation of plant growth regulators.

Three transgenic parthenocarpic eggplant hybrids were compared to the corresponding control hybrids and the commercial parthenocarpic cultivar Talina. All the genotypes were subjected or not to sprays with phytohormones. The normal cultural practises were followed. The results showed that: i) transgenic hybrids gave a significantly higher production than untrasformed ones; ii) the treatment of floral buds with exogenous plant growth regulators did not affected the production of transgenics, on the contrary the production of all the untrasformed hybrids was positively influenced by phytohormone sprays. On average the parthenocarpic transgenic hybrids outyielded non-transgenic phytohormone-sprayed control hybrids of 30 tons ha^{-1}.

In conclusion, the hybrids transgenic for the gene *defh9-iaaM* allowed to increase the winter production and, at the same time, to reduce the total cultivation cost of about 10% due to the labour and materials saved for carrying out phytohormone treatments.

5. Perspectives

The innovative methods for inducing parthenocarpic fruit development can be efficiently utilized in other vegetable crops (watermelon, pepper, cucumber, zucchini, etc.) and fruit trees (cherry, grape, lemon, orange, mandarin, kiwi, plum, prickly pear, etc.) both for freshmarket and postharvest processing use. The expected benefits are: higher early production because of the improved fruit setting under adverse climatic conditions; an improved quality of the product due to the absence or reduction of number of seeds; yield stability over the years, especially for species which suffer of pollination problems (self-incompatibility); a better management of harvesting time because of elongation of *in planta* commercial life of fruit.

Experiments are in progress to verify the effect of the *defh9-iaaM* gene in several of the aforementioned species.

Acknowledgments: Partially financed by the program "Biotecnologie Vegetali" of the MiPA (Ministero Politiche Agricole) and by the program "Biotecnologie" of the C.N.R..

The authors thank the G.IN.E.ST.R.A (Genetic INitiative European STrategical Research in Agriculture) for encouragement and support.

6. References

Archbold, D.D.& Dennis, F.G. (1985) Strawberry receptacle growth and endogenous IAA content as affected by growth regulator application and achene removal, *J. Amer. Soc. Hort. Sci.* **110**, 816-820.

Bailey, B.J. & Hunter, A. (1988) Plant response and energy use in five high thermal resistance greenhouse, *Acta Hort.* **229**, 165-171.

De Ponti, O.M.B. (1976) Breeding parthenocarpic pickling cucumbers (*Cucumis sativus* L.): necessity, genetical possibilities, environmental influences and selection criteria, *Euphityca* **25**, 29-40.

Hennart, J.W. (1996) Sélection de l'aubergine, *PHM Revue Horticole* **374**, 37-40.

Kawaguchi, M., Kobayashi, M., Sakurai, A. & Syono, K. (1991) The presence of an enzyme that converts indole-3-acetamide into IAA in wild and cultivated rice, *Plant Cell Physiol.* **32(2)**, 143-149.

Koncz, C. & Schell, J. (1986) The promoter of TL-DNA gene 5 controls the tissue-specific expression of chimaeric genes carried by a novel type of *Agrobacterium* binary vector, *Mol Gen Genet* **204**, 383-396.

Khishnamoorthy, H.N. (1981) *Plant growth substance-including application in agriculture*, Tata McGray Publ. Comp. Lmt, New Delhi.

Mapelli, S., Frova, C., Torti, G. & Soressi, G.P. (1978) Relationship between set, development and activities of growth regulators in tomato fruits, *Plant Cell Physiol.* **19(7)**, 1281-1288.

Nitsch, J. (1950) Growth and morphogenesis of the strawberry as related to auxin, *Amer. J. Bot.* **37**, 211-215.

Osborne, D.L. & Went, F.W. (1953) Climatic factors influencing parthenocarpy and normal fruit set in tomatoes, *Bot. Gaz.* **114**, 312-322.

Philouze, J. (1985) Parthénocarpie naturelle chez tomate. II. Etude d'une collection variétale, *Agronomie* **5(1)**, 47-54.

Romano, D. & Leonardi, C. (1994) The responses of tomato and eggplant to different minimum air temperature, *Acta Hort.* **366**, 57-66.

Rotino, G.L., Perri, E., Zottini, M., Sommer, H. & Spena, A. (1997) Genetic engineering of parthenocarpic plants, *Nature Biotech.* **15**, 1398-1401.

Saavedra, E. (1979) Set and growth of *Annona cherimolia* Mill. Fruit obtained by hand-pollination and chemical treatments, *J. Am. Hort Sci.* **104 (5)**, 668-673.

Sarma, C.M. & Barman, T.S. (1977) Production of parthenocarpic fruits in brinjal (*Solanum melongena* Linn.) by the application of β-naphthoxyacetic acid (β-NOA), *Ind. J. Hort.* **34 (4)**, 422-425.

Spena A., Estruch J.J. & Schell J.J. (1992) On microbes and plants: New insights in phytohormonal research, *Curr. Op. Biotech.* **3**, 159-163.

Yamada, T., Palm, C.J., Brooks, B. & Kosuge, T. (1985) Nucleotide sequence of the Pseudomonas savastanoi indoleacetic acid genes show homology with *Agrobacterium tumefaciens* T-DNA, *Proc. Natl. Acad. Sci. USA* **82**, 6522-6526.

G.T. Scarascia Mugnozza, E. Porceddu & M.A. Pagnotta (Eds.)
Genetics and Breeding for Crop Quality and Resistance, 307-312, 1999

Control of melon ripening by genetic engineering

M. Guis, M. Ben Amor, R. Botondi, R. Ayub, A. Latché, M. Bouzayen & J.C. Pech
ENSAT, Avenue de l'Agrobiopole BP 107, Auzeville Tolosan, 31326 Castanet Tolosan Cedex, France

1. Introduction

The melon, *Cucumis melo* L. is an ancient plant, likely originating from the tropical Africa/Middle-East regions. It then spread out to the Mediterranean, India and Asia (Withaker & Davis, 1962) from where a large diversity of fruit types have now emerged (Kirkbride, 1993). Within the *Cucumis melo* species, a wide range of ripening behaviour exists. For example, some varieties tend to have a sharp climacteric respiration associated with a fast fruit ripening and abscission rate (Lyons et al., 1962). In contrast, other varieties do not abscise at maturity, and the climacteric may be absent (Nukaya et al., 1986) or extended over several days (Pratt et al., 1977).

The Cantaloupe Charentais melon, which belongs to the cantaloupensis group of *Cucumis melo* (Naudin, 1859), is the most cultivated melon type in France. It is a round shaped, medium size fruit, with smooth skin, orange flesh and good organoleptic traits. It represents a good source of carbohydrates, β-carotene, vitamin C and minerals. The edible quality of these melon fruits depends on the colour and texture of the flesh along with the typical aromatic flavour and sweetness (Aulenbach & Worthgton, 1974; Yamaguchi et al., 1977). The Charentais melon fruit should be left on the vine till it is fully mature as the quality attributes of the fruit develop mainly in the later stages of ripening. Premature harvesting results in a low sugar content and poor consumer acceptability. On the other hand, harvest at full maturity results in poor storage capacity, overripening, and loss of quality attributes (Seymour & McGlasson, 1993). The shelf life of the typical Charentais type melon is, in general, less than four days. However, in recent years, breeders have generated new genotypes that exhibit longer shelf-life (Gry, 1996) but these have lost some of their original quality attributes. The extension of the shelf life of Charentais type melons while maintaining its eating quality remains a challenge for geneticists.

The rapid ripening, within a few days, of the typical Charentais melon is associated with a sharp increase in the production of the plant hormone, ethylene, synthesised by the ripening fruit (Seymour & McGlasson, 1993). Ethylene stimulates the expression of many genes involved in several biochemically important pathways leading to changes in colour, flavour, texture and aroma (for review see Lelièvre et al., 1997). But ethylene also accelerates the senescence process resulting in poor keeping quality. Due to its key role in the ripening process, ethylene is an attractive target for controlling fruit ripening. Most of the enzymes involved in ethylene biosynthesis have been characterised and the corresponding genes cloned in melon (ACC oxidase, Balagué et al., 1993; ACC synthase, Yamamoto et al., 1995). As melon genetic transformation can be easily performed by *Agrobacterium* (for review see Guis et al., 1998a), genetic engineering strategies are applicable for the manipulation of ethylene production in this crop.

Strong inhibition of ethylene synthesis has been achieved in Cantaloupe Charentais melons by down regulating the gene encoding for the last enzyme of the ethylene biosynthetic pathway, ACC oxidase (Ayub et al., 1996). An engineered antisense ACC oxidase line with a 99% inhibition of ethylene production exhibited a strong but reversible inhibition of fruit ripening. In the present paper, we review the current state of our work on the molecular, biochemical and physiological characterisation of this line both at pre- and post-harvest levels.

2. Results

2.1 *Effect of ethylene inhibition in melon ripening*

The wild type Charentais melon undergoes yellowing of the rind and develops a peduncular abscission zone during ripening. In contrast, the rind of the transgenic line remains green throughout development on the vine (Ayub et al., 1966; Guis et al., 1997a). Degreening of the rind is due to chlorophyll degradation. As chlorophyllase is known to be positively regulated by ethylene (Trebitsh et al., 1993), it is likely that chlorophyllase activity is not stimulated in ethylene-inhibited fruit, resulting in the persistence of the green colour. The peduncular abscission zone also fails to develop so that transgenic fruit never detach from the vine while wild type fruit abscises 38 days after pollination.

The Cantaloupe Charentais melon displays extensive softening during ripening in association with the degradation of cell wall polysaccharides (Rose et al., 1998). Changes in cell wall polymers occur at level of both pectic and hemicellulosic matrices. Modification of tightly bound hemicellulose seems to be one of the early changes occurring to the cell wall at the onset of fruit ripening, while pectin degradation occurs later in the process (Rose et al., 1998). Suppression of ethylene production in transgenic fruits resulted in a strong reduction in fruit softening. No changes in the molecular weight of polysaccharides could be observed, suggesting that pectins and hemicelluloses were not substantially degraded (Guis et al., 1998b). On the contrary, in wild type fruit all polysaccharide fractions exhibited a shift from high to lower molecular weight compounds. A number of cell wall degrading enzymes such as α-L-arabinosidase, β-D-galactosidase and endo-polygalacturonase showed much higher activity in the wild type than in transgenic fruit (Guis et al., 1997b), indicating that they may play an important role in cell wall disassociation. Other enzymes had similar activity in both control and transgenic fruits (pectin methyl esterase and exo-polygalacturonase), implying that they may not be critical in the cell wall degradation process. A similar picture has been observed for the expression of genes encoding cell wall modifying enzymes. For example, polygalacturonases (*CmPG1* and *CmPG3*) or xyloglucan endotransglycosylases (*CmXET1* and *CmXET3*) are expressed at very low levels in the transgenic fruits, whereas they are abundantly transcribed in wild type fruits at the peak of ethylene production (Guis et al., 1998b). The expression of *CmPG2* was not affected by ethylene suppression and was observed in both type of fruits at the early stages of ripening, before the onset of ethylene production (Guis et al., 1998b). These data clearly demonstrate that melon softening involves both ethylene-independent and ethylene-dependent gene expression.

In melon flesh, the intensity of the orange colour is an important quality attribute, which primarily results from the biosynthesis of carotenoids (Reid et al., 1970). The importance of these pigments in fruit quality is not only related to their natural colorant properties, but also to their role as antioxidants and as precursors of vitamin A (Edge et al., 1997; Bartley & Scolnik, 1995). We have observed that the colour of the flesh and the carotenoid content of Charentais melons is not affected by ethylene-suppression. This is consistent with the demonstration by Karvouni et al., 1995, that the expression of the phytoene synthase gene, which catalyses one of the first steps of carotenoid biosynthesis occurs at an early stage of melon fruit development, well before the onset of the ripening climacteric.

The quality of Charentais melon is also closely correlated with its sugar content (Alavoine et al., 1988). Sugars, mainly sucrose, fructose and glucose accumulate primarily during the final stages of fruit development and during fruit ripening (Bianco et al., 1977). For this reason, fruits that are kept on the vine for longer accumulate more sugars, although there is a risk of overripening. We have observed that the rate of sugar accumulation is similar in wild type and antisense fruit. However, as ethylene-inhibited fruits fail to develop a peduncular abscission zone, they remain attached to the plant and thus can accumulate higher amounts of soluble sugars and attain better sensory qualities. Interestingly, the loss of titrable acidity, which also plays part in sensory quality, is greatly delayed in transgenic fruit compared to the wild type. Together, these parameters result in an improved taste in the transgenic fruits.

2.2 Post-harvest behaviour of antisense ACC oxidase melons

We have also compared the behaviour of wild type and transgenic fruits after harvest and during storage at 25°C. It appears that, in both fruits, the sugar and carotenoid contents remain unchanged after several days of storage. However, in contrast to the wild type, the transgenic fruits stay green and free of fungal attacks, even after 12 days of storage (Guis et al., 1997c). Wild type fruit harvested at the commercial picking date and stored under the same conditions ripen very rapidly and is completely rotten after only 8 days of storage. A significant softening is observed in the transgenic melon after harvest, although fruit remain about twice as firm as control fruit. The slight increase in internal ethylene concentration that has been observed upon detachment (Guis et al., 1997a could trigger some aspects of cell wall degradation. However, at the same time, no yellowing of the rind could be observed, indicating that different threshold values of ethylene production are required for the induction of each type of metabolic process (chlorophyll degradation and cell wall disruption).

The perishability of harvested horticultural products, including fruits, is generally proportional to the respiration rate. Melons exhibit a large increase in the rate of respiration during ripening (Hadfield et al., 1995) However, harvested antisense ACC oxidase fruit failed to develop a respiratory climacteric, consistent with the previous demonstration that ethylene plays an important role in stimulating respiration in plants (Biale & Young, 1981).

As with most tropical and subtropical fruits, melons are sensitive to chilling injury (Dunlap et al., 1990). Storage at low, non freezing temperatures, induces water-soaking of the flesh and browning and pitting of the rind. Interestingly, antisense ACC oxidase melons display high resistance to chilling injury during low temperature storage at 2°C. They also fail to accumulate ethanol and acetaldehyde as a result of metabolic dysfunctioning. This suggests that antisense ACC oxidase melon fruits can be stored at low temperatures, therefore allowing a longer storage capability. It has been demonstrated that complete restoration of ripening could be achieved after low temperature storage by treating with ethylene.

2.3 Restoration of ripening by exogenous ethylene treatment

Exposure to exogenous ethylene enables the restoration of the wild type phenotype by inducing the ethylene-dependent pathways (Guis et al., 1997a). However, because autocatalytic ethylene production is impossible, the response is proportional to the dose and duration of applied ethylene. During a 4-day period, low concentrations of applied ethylene (between 0 and 1 ppm) fail to stimulate most of the ethylene-dependent ripening parameters. At concentrations between 1 and 5 ppm, intermediate ripening can be achieved, while at 50 ppm or more, full restoration is reached after 4 days. In addition, the rate of change at 50 ppm is identical in both wild type and ethylene-treated transgenic fruits. This demonstrates that the ethylene perception and transduction pathway has not been altered in transgenic fruits. These data show that it is possible to "titrate" the various ripening parameters to the desired level.

3. Conclusion

Reducing ethylene production in the melon by antisense inhibition of ACC oxidase greatly delays most of the ripening processes. However, some pathways (coloration of the flesh, sugar accumulation) or enzyme activities (pectin methyl esterase, exo-polygalacturonase) remain unaffected and can be considered as ethylene-independent events. In other words, the ripening of climacteric fruit comprises non-climacteric events. Because transgenic fruits can stay on the vine longer without risk of overripening they offer the advantage of a more flexible picking date and a reduced number of harvests. In addition, as the rind remains firm and is therefore less susceptible to damage during handling, mechanical harvest becomes conceivable. Long term storage at low temperature and resistance to pathogens are other potential benefits of transgenic antisense ACC oxidase melons. However, postharvest handling must also comprise a treatment with ethylene at a duration and concentration that allows the fruit to reach the optimum sensory quality. Such a strategy needs to be tested to determine if this is practical at an applied level. We are currently evaluating the possibility of introducing the ethylene-suppressed characteristic in a commercial breeding programme.

Acknowledgements: The authors would like to acknowledge the French Italian Galileo programme between INP-ENSAT, France and University of Viterbo, Italy, INRA for a fellowship to Ben Amor M., and EU for a FAIR-CT 96-1138 grant.

4. References

Alavoine, F., Crochon, M., Fady, C., Fallot, J. & Pech, J.C. (1988) La qualité gustative des fruits. Méthodes pratiques d'analyse, CEMAGREF, Antiny.

Aulenbach, B.B& Worthgton, J.T. (1974) Sensory evaluation of muskmelon: Is soluble solids content a good quality index?, *HortScience* **9**, 136-137.

Ayub, R., Guis, M., Ben Amor, M., Gillot, L., Roustan, J.P., Latché, A., Bouzayen, M. & Pech, J.C. (1996) Expression of an ACC oxidase antisense gene inhibits ripening of cantaloupe melons fruits, *Nature Biotechnology* **14**, 862-864.

Balagué, C., Watson, C.F., Turner, A.J., Rougé, P., Picton, S., Pech, J.C. & Grierson, D. (1993) Isolation of a ripening and wound induced cDNA from *Cucumis melo* L., with homology to the ethylene-forming enzyme, *Eur. J. Biochem.* **212**, 27-34.

Bartley, G.E. & Scolnik, P.A. 1995. Plant carotenoids: Pigments for photoprotection, visual attraction, and human health. *Plant Cell* 7, 1027-1038

Biale, J.B. & Young, R.E. (1981) Respiration and ripening in fruits, in J. Friend and M.J.C. Rhodes. (eds.), *Recent advances in the biochemistry of fruits and vegetables*, Academic Press, London, pp. 1-39.

Bianco, V.V. & Pratt, H.K. (1977) Compositional changes in muskmelon during development and in response to ethylene treatment, *J. Amer. Soc. Hort. Sci.* **102**, 127-133.

Dunlap J.R., Lingle S.E. & Lester G.E. (1990) Ethylene production in netted muskmelon subjected to postharvest heating and refrigerated storage, *HortScience* **25**, 207-209.

Edge, R., McGarvey, D.J. & Truscott, T.G. (1997) The carotenoids as anti-oxidants: A review, *Journal of Photochemistry and Photobiology* **41**, 189-200.

Gry, L. (1996) Vers un melon de qualité, *Semences et Progrès* **86**, 14-23.

Guis, M., Botondi, R., Ben Amor, M., Ayub, R., Bouzayen, M., Pech, J.C. & Latché, A. (1997a) Ripening-associated biochemical traits of Cantaloupe Charentais melons expressing an antisense ACC oxidase transgene, *J. Amer. Soc. Hort. Sci.* **122**, 748-751.

Guis, M., Bouquin, T., Zegzouti, H., Ayub, R., Ben Amor, M., Lasserre, E., Botondi, R., Raynal, J.,Latché, A., Bouzayen, M., Balagué, C. & Pech, J.C. (1997b) ACC Oxidase genes in melon, in Kanellis A.K., Chang C., Kende H., and Grierson D. (eds), *Biology and Biotechnology of the Plant Hormone Ethylene*, Kluwer Academic Publishers, Dordrecht, pp. 327-337.

Guis, M., Ben Amor, B., Botondi, R., Ayub, R., Latché, A., Bouzayen, M. & Pech, J.C. (1997c) Ethylene and biotechnology of fruit ripening. Pre- and Post-harvest behaviour of transgenic melons with inhibited ethylene production, *Acta Horticulturae* **463**, 31-37.

Guis, M., Roustan, J.P., Dogimont, C., Pitrat, M. & Pech, J.C. (1998a) Melon Biotechnology, in M.P. Tombs (ed.), *Biotechnology and Genetic Engineering Reviews*, Intercept Limited, Hants, pp. 289-311.

Guis, M., Latché, A., Bouzayen, M., Pech, J.C. & Rose, J.K.C., Hadfield K.A., Bennett A.B. (1998b) Understanding the role of ethylene in fruit softening using antisense ACC oxidase melons, in Kanellis A.K., Chang C., Grierson D., Klee H. and Pech J.C (eds), *Biology and Biotechnology of the Plant Hormone Ethylene II*, Kluwer Academic Publishers, Dordrecht, in Press.

Hadfield, K.A., Rose, J.K.C & Bennett, A.B. (1995) The respiratory climacteric is present in Charentais (*Cucumis melo* cv. reticulatus F1 Alpha) melons ripened on or off the plant, *J. Exp. Bot.* **46**, 1923-1925.

Karvouni, Z., John, I., Taylor, C.F., Watson, A.J., Turner, A.J. & Grierson, D. (1995) Isolation and characterization of a melon cDNA clone encoding phytoene synthase, *Plant Molec. Biol.* **27**, 1153-1162.

Kirkbride, J.H. (1993) *Biosystematic monograph of the genus Cucumis (Cucurbitaceae)*, Parkway Publishers, Bonne.

Lelièvre, J.M., Latché, A., Jones, B., Bouzayen, M. & Pech, J.C. (1997) Ethylene and fruit ripening, *Physiol. Plant.* **101**, 727-739.

Lyons, J.M., McGlasson, W.B. & Pratt, H.K. (1962) Ethylene production, respiration, and internal gas concentration in cantaloupe fruits at various stages of maturity, *Plant Physiol.* **37**, 31-36.

Naudin C. (1859) Essai d'une monographie des especès et des variétés du genre *Cucumis*, *Ann. Sci. Nat. Ser. Bot.* **11**, 5-87.

Nukaya, A., Ishida, A., Shigeoka, H. & Ichikawa, K. (1986) Varietal difference in respiration and ethylene production in muskmelon fruits, *HortScience* **21**, 853.

Pratt, H.K., Goeschl, J.D. & Martin, F.W. (1977) Fruit growth and development, ripening, and the role of ethylene in the Honey Dew muskmelon. *J. Amer. Soc. Hort. Sci.* **102**, 203-210.

Reid, M.S., Lee, T.H., Pratt, H.K. & Chichester, C.O. 1970. Chlorophyll and carotenoid changes in developing muskmelons, *J. Amer. Soc. Hort. Sci.* **95**, 814-815.

Rose, J.K.C., Hadfield, K.A., Labavitch, J.M. & Bennett, A.B. (1998) Temporal sequence of cell wall disassembly in rapid ripening melon fruit, *Plant Physiology* **117**, 345-361.

Seymour, G.B. & McGlasson, W.B. (1993) Melons, in. G. Seymour, M.A.Taylor and G.A. Tucker (eds.), *Biochemistry of fruit ripening*, Chapman & Hall, London, pp. 273-290.

Trebitsh, T., Goldschmidt, E.E. & Riov, J. (1993) Ethylene induces *de novo* synthesis of chlorophyllase, a chlorophyll degrading enzyme, in Citrus fruit peel, *Proc. Natl. Acad. Sci.* **90**, 9441-9445.

Withaker, T.W. & Davis, G.N. (1962) *Cucurbits. Botany, cultivation and utilization*, Interscience Publishers, New York.

Yamaguchi, M., Hughes, D.L., Yabmoto, K. & Jennings, W.G. (1977) Quality of cantaloupe muskmelons: Variability and attributes, *Scientia Hort.* **6**, 59-70.

Yamamoto, M., Miki, T., Ishiki, Y., Fujinami, K., Yanagisawa, Y., Nakagawa, H., Ogura, N., Hirabayashi, T. & Sato, T. (1995) The synthesis of ethylene in melon fruit during the early stage of ripening, *Plant and Cell Physiol.* **36**, 591-596.

G.T. Scarascia Mugnozza, E. Porceddu & M.A. Pagnotta (Eds.)
Genetics and Breeding for Crop Quality and Resistance, 313-322, 1999

Breeding of vegetables with minimum pollutant accumulation

A. Kilchevsky[1], L. Khotylyova[2], V. Peshich[3], L. Kogotko[1], A. Schoor[1],. A. Gavrilov[4] & A. Kruk[4]

[1]Belarussian Agricultural Academy, Gorky, Mogilev region, 213410 Belarus; [2]Belarussian Institute of Genetics and Cytology, Minsk, 220072 Belarus; [3]University of Belgrad, 11080, Yugoslavia; [4]University of Gomel, 246029, Belarus

Abstract: Problem of agricultural production quality now has a new aspect: pollution of radionuclids, heavy metals, nitrates ets, especially in Belarus after Chernobyl accident. We investigated the genetical basis of cadmium, lead and nitrate accumulation in tomato in open ground using the diallel analysis method. Differences on 2 - 5 times were found between the varieties according to their nitrate and heavy metals accumulation. It was established that the main type of pollutant accumulation inheritance is superdominance aimed at decreasing nitrate, cadmium and lead accumulation in tomato fruits. It leads to conclusion that heterosis breeding is a method of lowering the pollutant accumulation in tomato fruits. Study of uptake and distribution of ^{137}Cs and ^{90}Sr in vegetable varieties (tomato, onion, carrot, cabbage) in Gomel region (Chernobyl zone) revealed the intra species genetic variation and possibility to select the genotypes with low content of radionuclides. Creation of varieties with minimum accumulation of pollutants is the most radical and cheap way of obtaining ecologically safe plant produce.

1. Introduction

At present the quality of farm produce is determined not only by the content of useful substances (proteins, fats, carbohydrates, vitamins, etc.), but by the pollutant accumulation (nitrates, heavy metals, radionuclids, pesticides, etc.). All high nutrition value of any farm produce can become useless if it containts some toxic substances exceeding hygienic norms.

The problem of pollution is connected with anthropogenic contamination of agrolandscape (industry, transport, agriculture) and with Chernobyl catastrophe. In Belarus about 20% of the territory is contaminated with radionuclids. Considerable regions of the Ukraine and Russia are contaminated as well. Vegetables can actively accumulate some pollutants in the productive part. Problems of pollutants in vegetable growing in Belarus are complecated by the fact that more than 80% of vegetables is grown on households including those situated in pollutants contaminated areas.

The process of pollutant accumulation in farm produce depends on three main factors (Kilchevsky & Khotylyova, 1997): (i) genetical (crop and variety characters, determing the input, transport, accumulation and detoxication of the pollutant); (ii) environmental (proximity to the source of pollution and the intensity of pollution, abiotic and biotic factors of environment, landscape etc); (iii) agrotechnical (doses and terms of fertilizer and pesticide application, control of pollutant intake by plants from the soil using agrotechnical methods etc).

A number of authors established interspecies and intervariety variability in pollutant accumulation in productive plant parts: nitrates – Zhuchenko & Andryushchenko (1980), Andryushchenko (1981), Agapov et al. (1990), Mineev (1990), Pyshnaya (1990), Klimashevsky (1991), Kilchevsky et al. (1993), Kilchevsky & Kogotko (1995), Kilchevsky & Khotylyova (1997); heavy metals – Baker (1981), Hunesly et al. (1982), Saric (1983), Popp (1983), Schat et al. (1993), Gamzikova et al. (1993), Li et al. (1995), Kilchevsky et al. (1998); radionuclids – Annenkov & Yudinceva (1991), Grodzinsky (1989), Prister et al. (1991), Aleksahin et al. (1992), Bogdevich (1997). Andryushchenko (1981), Gamzikova (1992), Kilchevsky & Khotylyova (1997) think that the most radical and the cheapest way of lowering pollutant accumulation in produce is plant breeding. This way is possible in determining interspecies and intervariety genotype variability in pollutant accumulation, finding out their genetic determination, donors with minimum pollutant accumulation, the development of breeding strategy in these traits.

So the object of our research is to study interspecies and intervariety variability of pollutant accumulation in productive parts of vegetable crops, as well as the character of inheritance of nitrate and heavy metals accumulation in tomato fruits in the open ground.

2. Materials and Methods

2.1 Nitrates

To study nitrate content inheritance in tomato fruits by diallel analysis method we tested in 1991 - 1992 in open ground in Gorky Mogilev region 28 hybrid tomato combinations between 8 initial forms using two agrophones of mineral food (normal and higher than normal). Parent material was Talalihin(1), Dohodny(2), P - 7(3), Beta(4), Sub - arctic mini(5), Line -7(6), *L. pimpinellifolium*(7), Torosa(8).The nitrate content in fruits (mg kg^{-1}) was analysed by the potenciometric method. The combining ability of samples was evaluated in method 2 of Griffing (1956), genetic inheritance parameters – in Hayman's method (1954).

2.2 *Heavy metals*

In experiment were studied parents Talalihin(1), Dohodny(2), Sprint(3), Line - 7(4), Opus(5), Povarek(6), Radek(7) and the hybrids between them in the open ground in 1995. During the vegetative period plants were sprayed with the salt solution of cadmium and lead in doses 0.25 of MPC (maximum permissible concentration) in the soil. Heavy metals content was evaluated in mg kg^{-1} of dry matter using the method of atomic absorbtion. The combining ability of samples was evaluated in method 2 of Griffing (1956), genetic inheritance parameters – in Hayman's method (1954).

2.3 *Radionuclids*

Studies were made of the 5 varieties of 4 vegetable crops (tomato, cabbage, carrot, onion) in the open ground in Bragin district Gomel region in 1997-1998 at the density of the soil pollution of 10 Ku km^{-2} (^{137}Cs) and 1 Ku km^{-2} (^{90}Sr). ^{137}Cs and ^{90}Sr content was determined in productive organs of vegetables in spektrometrical and radiochemical methods respectively.

3. Results and Discussion

3.1 *Nitrates*

Nitrate content in tomato varieties and hybrids on the experimental plot (control phone) ranges from 14.2 to 55.0 mg kg^{-1} in 1991, from 10.0 to 37.5 mg kg^{-1} in 1992 (Table 1). On the higher phone nitrate content in 1991 changed from 13.5 to113.3 mg kg^{-1}, in 1992 - from 10 to 55.3 mg kg^{-1} (Table 2). The lowest effect GCA had *L. pimpinellifolium* and Torosa, the highest - Beta, Line - 7 and P-7. The MPC of nitrate content in tomato fruits is 100 mg kg^{-1}.

Heterosis effect can be evaluated according to the level of dominance Hp (Table 3). In the majority of cases (39.3 - 46.4%) the effect of negative superdominance is manifested, that is heterosis toward decreasing nitrate content in fruits. There has been frequent cases of intermediate inheritance (32.1-46.4). Positive superdomince manifests itself seldom (10.7-21.4 %).

Analysis of Hayman diagram shows that inheritance in most cases follows the line superdominance. Most varieties are bearers of dominant genes, and Beta variety - of recessive. The averagy level of dominance H/D ranged from 0.95 to 2.74 (Table 4) taking into account testing medium, it also confirms the manifestation of dominance and superdominance. On the higher agrophones the frequance of dominant genes increases because the parameters $[\sqrt{(4DH_2)}+F]/[\sqrt{(4DH_2)}-F]$ overreach 2.17-5.24.

Table 1. Contents of nitrates in tomato fruits at the control phone of mineral nutrition.

Back-ground	N° Genotypes	1	2	3	4	5	6	7	8	Effect GCA	Variance SCA
	1	15.7	20.4	22.3	14.2	28.0	28.0	19.2	12.8	-2.4	0
	2		21.1	24.1	26.1	19.1	22.2	21.8	16.0	0.7	0
	3			19.2	21.9	27.9	55.0	18.5	15.8	2.4	75.9
1991*	4				34.6	25.1	16.0	15.7	27.0	1.5	33.2
	5					22.3	22.1	15.4	19.8	1.0	0
	6						37.5	18.9	20.4	5.8	105.1
	7							18.6	16.4	-3.7	0
	8								16.1	-3.9	0
	1	18.5	16.2	14.5	15.7	13.5	37.5	18.5	27.2	2.4	42.0
	2		23.2	15.0	12.3	15.8	33.5	16.3	18.3	1.8	17.9
	3			17.7	15.2	21.5	14.2	18.5	17.3	-0.5	12.5
1992**	4				17.2	13.8	13.5	13.8	19.3	-1.8	5.8
	5					14.7	15.7	11.8	12.7	-2.2	3.1
	6						24.8	16.2	14.2	3.8	53.9
	7							12.7	10.0	-2.6	0
	8								13.7	-1.0	13.2

*,** Significantly at $P<0.05$, 0.01 respectively

Correlation between trait meaning and the sum variance and covarience is more often positive which testifies the prevailing of dominance of minimum nitrate content in fruits. Meanings of fully dominant D_{max} and fully recessive R_{max} parent prove it. The studied varieties differ in one group of genes.

Table 2. Contents of nitrates in tomato fruits at the high phone of mineral nutrition.

Back-ground	N° Genotypes	1	2	3	4	5	6	7	8	Effect GCA	Variance SCA
	1	22.9	27.4	35.6	17.4	19.7	23.3	15.5	17.7	-4.2	21.2
	2		24.6	19.3	16.4	31.1	39.0	19.5	14.7	-2.8	50.4
	3			24.9	29.5	25.5	69.4	19.5	33.9	3.8	173.4
1991*	4				113.3	22.0	31.8	16.1	16.3	13.2	751.9
	5					27.4	25.5	18.5	23.3	-2.4	6.6
	6						40.6	17.0	23.2	6.6	155.7
	7							24.2	18.6	-7.1	25.0
	8								13.5	-7.0	11.8
	1	32.7	30.3	35.7	20.3	32.3	26.8	15.3	10.0	0.3	39.9
	2		25.3	18.5	21.0	18.5	55.3	14.5	24.3	0	89.7
	3			29.2	27.8	25.8	50.7	17.7	22.8	2.4	40.5
1992*	4				49.2	27.7	25.0	18.8	20.0	2.6	75.9
	5					31.0	34.7	16.7	26.8	1.1	11.2
	6						37.8	19.8	29.5	8.4	116.1
	7							14.8	13.7	-8.7	2.3
	8								11.3	-6.3	13.0

* Significantly at $P<0.05$

Thus nitrate accumulation is more often inherited by heterosis type toward their minimum content in fruits. In connection with it we may consider heterosis breeding for receiving hybrid F_1 as a method of decreasing nitrate accumulation in the produce.

Table 3. Degree of dominance at the character "nitrate content in tomato fruits"

Year	Background	Hybrids	Hp<-1	-1<Hp<1	Hp>1
1991	Control	Quantity	11	10	7
		%	39.3	35.7	25.0
	High	Quantity	13	12	3
		%	46.4	42.9	10.7
1992	Control	Quantity	13	9	6
		%	46.4	32.1	21.4
	High	Quantity	12	13	3
		%	42.9	46.4	10.7

Table 4. Parameters of Hayman at character "nitrate content in tomato fruits".

Background	Year	$\sqrt{H_1/D}$	$\frac{H_2}{4H_1}$	$\frac{\sqrt{4DH_2}+F}{\sqrt{4DH_2}-F}$	r	D_{max}	R_{max}	$\frac{h^2}{H_2}$
Control	1991	0.95	0.27	1.30	0.65	14.96	34.31	0.21
	1992	2.74	0.22	1.45	-0.37	18.53	13.60	0.06
High	1991	1.10	0.12	5.24	0.98	16.87	123.13	0.69
	1992	1.08	0.19	2.17	0.89	12.96	61.02	0.90

3.2 *Heavy metals*

In experiment at air pollution with cadmium and lead the combining ability of tomato genotypes on lead content in fruits was evaluated (Table 5). Between initial forms maximum content was marked in Line -7, Sprint, Talalihin. The same forms had high effects of general combining ability (GCA) and variencies of specific combining ability (SCA). Minimum content of lead was noted in the varieties Opus and Radek. We determined differences between tomato varieties in lead accumulation which equal 3,3 times (Table 5).

Table 5. Content of the heavy metals in tomato fruits at the air pollution.

Background	N° Genotypes	1	2	3	4	5	6	7	Effect GCA	Variance SCA
Lead*	1	2.03	0.37	0.39	1.14	0.48	1.13	1.08	0.09	0.28
	2		1.25	0.43	1.24	0.88	0.53	0.63	-0.14	0.19
	3			2.58	0.43	0.53	0.31	1.16	0.06	0.72
	4				2.97	0.21	0.41	0.93	0.27	0.71
	5					0.89	0.34	0.46	-0.35	0.20
	6						1.74	1.88	0.02	0.45
	7							1.08	0.05	0.15
Cadmium**	1	0.41	0.65	0.24	0.42	0.64	0.89	0.44	0.05	0.04
	2		0.13	0.36	0.13	0.52	0.60	0.32	-0.09	0.04
	3			0.54	0.24	0.21	0.22	0.19	-0.12	0.04
	4				0.51	1.67	0.28	0.61	0.09	0.21
	5					0.69	0.26	0.13	0.14	0.22
	6						0.57	0.33	0.01	0.06
	7							0.42	-0.08	0.03

*,** Significantly at P<0.05, 0.01 respectively

Lead content in all initual varieties exceeded the permissible concentration (0.5 mg kg^{-1}) while hybrids had increased permissible concentration only in 11 cases out of 21.

In cadmium accumulation in fruits a tomato varieties differed 5.4 times. Maximum accumulation was noted in the varieties Opus, Povarek, Sprint, Line -7; minimum – in the variety Dohodny. Cadmium content in fruites of all varieties and hybrids exceeded permissible concentration (0.03 mg kg^{-1}) 4 - 30 times and even more, which testifies the weakness of detoxication of this pollutant in tomato plants and high potential danger when cadmium penetrates into the agrocenosis by air. The least ranks of GCA were marked by the genotypes Sprint, Dohodny and Radek, the same forms had low variances of SCA. Noticeable, that the ranks of varieties and ranks of this GCA effects in character "cadmium content in tomato fruits" did not coincide. So variety Sprint marks the higher content of cadmium in fruits but low effect of GCA in this trait. All hybrids with this variety had accumulation less cadmium then the initial forms.

Analysis of dominance direction in studied traits (Table 6) shows that both in lead accumulation (90.4 % of cases) and in cadmium accumulation (42.8 % of cases) the main type of inheritance is superdominance (negative heterosis) towards decrease of heavy metals accumulation in fruits. It permits to consider heterosis breeding using the method of lowering of heavy metals in tomato fruits in polluted territories. Similar results were found by Li et al. (1995) at sunflower. Twentyseven of the 36 crosses registered negative heterosis, indicating breeding for low kernel Cd in sunflower hybrids should be feasible.

Hayman's parameters (Table 7) give additional information on inheritance of studied traits. Judging the parameter $\sqrt{(H_1/D)}$ superdominance manifests in loci. In studied populations dominant alleles are prevail because the parameter is $[\sqrt{(4DH_2)}+F]/[\sqrt{(4DH_2)}-F]>1$

Table 6. Degree of dominance at the characters "lead and cadmium content in tomato fruits" at the artificial pollution.

Hybrids	Hp<-1	-1<Hp<1	Hp>1
Content of lead			
Quantity	19	1	1
%	90.4	4.8	4.8
Content of cadmium			
Quantity	9	6	6
%	42.8	28.6	28.6

Table 7. Parameters of Hayman at the character "lead and cadmium content in tomato fruits" at the artificial pollution.

Character	$\sqrt{H_1/D}$	$\frac{H_2}{4H_1}$	$\frac{\sqrt{4DH_2}+F}{\sqrt{4DH_2}-F}$	r	D_{max}	R_{max}	$\frac{h^2}{H_2}$
Content of lead	1.34	0.22	1.96	0.98	0.90	3.10	1.85
Content of cadmium	1.82	0.21	1.72	0.82	0.24	0.98	0.08

Correlation coefficient r gives very important information on the direction of dominance. It testifies that dominant alleles decrease the meaning of both traits. This fact is also confirmed by the comparison of maximum dominant D_{max} and maximum recessive R_{max} genotypes. Considering parameters h^2 H_2^{-1}, variability in lead accumulation in the population is controlled by 2 loci. The same parameter doesn't give the objective information on the inheritance of cadmium accumulation.

Thus heterosis breeding can be a means of getting hybrids with a lower heavy metals accumulation in tomato fruits in contaminated areas.

Early we have made similar conclusions about the possibility of the use of heterosis breeding for lowering of nitrate accumulation in tomato fruits. One can suppose that there are general biochemical and physiological mechanisms leading to the decrease of pollutant accumulation in the produce. We suppose this common mechanism can be the increase of general resistance of a hybrid organism to stresses as a result of adaptive heterosis. Besides there can be "biological delution" of pollutants in hybrids in comparison with parents which is a consequense of heterosis on yielding capasity.

3.3 *Radionuclids*

Table 8. Content of ^{137}Cs at the varieties of vegetables (Bq kg^{-1}).

Crops	Varieties	Content of ^{137}Cs		
		1997	1998	Averagy
	Sprint	7.00	8.33	7.66
	Peramoga 165	10.33	11.00	10.66
Tomato	Kalinka	16.67	12.00	14.34
	Dohodny	23.67	22.00	22.84
	Ruzha	12.00	14.67	13.34
	Significantly at P	<0.05	<0.05	
	'1 Gribovsky 147	40.67	16.33	28.50
	Belorusskaya 85	7.00	9.33	8.16
Cabbage	Rusinovka	23.00	19.67	21.34
	Amager 611	19.00	12.00	15.50
	Turkiz	36.33	23:33	29.83
	Significantly at P	<0.05	<0.01	
	Nantskaya	7.33	10.00	8.66
	Losinoostrovskaya	20.67	18.00	19.34
Carrot	Vitaminnaya	15.67	17.67	16.67
	NIIOH 336	29.33	23.33	26.33
	Shantene	27.67	22.00	24.84
	Significantly at P	<0.05	<0.01	
	Vetraz	33.00	26.33	29.66
	Shutgarten Rizen	30.00	26.67	28.34
Onion	Iantarny	36.33	33.67	35.00
	Krivitcky ruzhovy	31.00	24.33	27.66
	Skvirsky	26.33	29.67	28.00
	Significantly at P	<0.01	n.s.	

n.s. = non significantly

In experiment using 4 vegetable crops we established differences between varieties in ^{137}Cs content in productive organs of all crops in all years of fruits

(Table 8). Maximum differences between varieties were: tomato - 2.6-3.4, cabbage - 2.5-5.8, carrot - 2.3-4.0, onion -1.4 times. Tomato varieties which accumulated least of all of ^{137}Cs were Sprint, Peramoga 165; carrot - Nantskaya and Vitaminnaya 6, cabbage - Belarusskaya 85, Amager 611 and onion-Krivitcky ruzhovy and Skvirsky.

Marked differences between varieties in ^{90}Sr accumulation in productive organs in 1997 (table 9) were found. Maximum difference between varieties was: tomato-2,3; cabbage-2,4; carrot-1,6; onion-2,7 times.

In their ability to accumulate ^{137}Cs vegetables can be placed in the order of priority in the following way: onion, cabbage, carrot, tomato. The same line can be made for vegetables in their ability to accumulate ^{90}Sr: onion, carrot, cabbage, tomato. The PMC for vegetables in ^{137}Cs at Belarus is 100 Bq kg^{-1}, the content of ^{90}Sr in vegetables don't normalize.

Table 9. Content of ^{90}Sr at the varieties of vegetables (Bq kg^{-1}).

Crops	Varieties	Content of ^{90}Sr
	Sprint	3.0
	Peramoga 165	5.9
Tomato**	Kalinka	3.6
	Dohodny	2.5
	Ruzha	3.1
	N°1 Gribovsky 147	14.8
	Belorusskaya 85	32.0
Cabbage**	Rusinovka	30.1
	Amager 611	28.2
	Turkiz	34.9
	Nantskaya	25.6
	Losinoostrovskaya	31.6
Carrot**	Vitaminnaya	27.9
	NIIOH 336	36.6
	Shantene	41.0
	Vetraz	65.1
	Shutgarten Rizen	177.9
	Iantarny	84.9
Onion**	Krivitcky ruzhovy	87.4
	Skvirsky	87.5

** significantly at $P< 0.01$

Revealed differences in radionuclids accumulation in varieties allow selecting varieties to lower radionuclid intake together with vegetables 2-5 times. The presence of genetic variability makes it possible to carry out breeding aimed at lowering radionuclid accumulation in vegetables.

4. Conclusion

Analysis of literature and results of carried out experiments show that intervariety variability in pollutant accumulation in vegetables are quite enough to single out genotypes lowering their intake with food 2-5 times. Analysis of inheritance of nitrate and heavy metals accumulation in tomato fruits shows that the main type of inheritance in conditions of produce contamination is superdominance towards decrease of pollutant content. It allows to admit that heterosis breeding is the method of lowering pollutant accumulation in tomatoes.

In our opinion (Kilchevsky & Khotylyova, 1997) general strategy of breeding towards lowering pollutant accumulation in agricultural produce should include three main stages:

1) Evaluation of the initual material corresponding to a number of useful traits and pollutant accumulation in contaminated areas, selection of initual forms for hybridization which correspond to tasks of breeding.
2) Selection in early generations (F_2-F_5) according to the traits of usefulness as well as to traits of correlation connected with accumulation of pollutants in contaminated areas.
3) Carrying out of competitive and ecological tests on the polluted and clean territory to evaluate selection results.

High price of carrying out analysis of pollutant content will hardly permit to control their accumulation in many samples in F_2-F_5. That's why it is very important to study correlation ties of these traits with other morphobiological and physiological traits for indirect selection. Breeding aimed at lowering pollutant accumulation in agricultural produce is the most radical, cheap and economically based means of lowering pollutant intake with food.

5. References

Agapov A.S., Shmanaeva T.N. & Pyshnaya O.N. (1990) Metodical directions of tomato breeding for greenhouses on low nitrate content in fruits, VNIISSOK, Moscow (in Russian).

Aleksahin R.M., Vasiljev A.V., Dikarev V.G. & Egorova V.A. (1992) Agricultural radioecology, Agropromizdat, Moscow (in Russian).

Andryushchenko V.K. (1981) Methods of optimisation of biochemical breeding of vegetables, Shtiinca, Kishinev (in Russian).

Annenkov B.N. & Yudinceva E.V. (1991) Elements of agricultural radiology, Agropromizdat, Moscow (in Russian).

Baker A.G.M. (1981) Accumulators and exluders – strategies in the response of plants to heavy metals, J. Plant Nutrition, 3, 643-654.

Bogdevich I.M. (1997) Direction on agricultural production in conditions of soil radioactive pollution at Belarus in 1997-2000. Ministry of Agriculture, Minsk (in Russian).

Gamzikova O.I., Barsucova V.S. & Koval S.F. (1993) Possibilities of the use of the near –isogenic lines for studing common wheat resistance to calmium and nichel, in: Genetical collections of plants, Inst. Cyt. Genet., Rus. Acad. Sci., Novosibirsk, pp. 116-131(in Russian).

Gamzikova O.I. (1992) Genetical aspects of wheat responsibility to conditions of mineral nutrition, Dis. Dr. Sci., Novosibirsk (in Russian).

Griffing B. (1956) Concept of general and specific combining ability in relation to diallel crossing systems, Austr. J. Biol. Sci. 9, 463-493.

Grodzinsky D.M. (1989) Plant radiobiology, Naukova dumka, Kiev (in Russian).

Hayman B.I. (1954) The theory and analysis of diallel crosses, Genetics 39, 789-809.

Hunesly T.D., Alexander D.E., Redborg K.E. & Ziegler E.R. (1982) Differential accumulation of cadmium and zink by corn hybrids grown on soil amended with sewage sludge, Agronomy J., 74, 469-474.

Kilchevsky A.V. & Khotylyova L.V. (1997) Ecological plant breeding, Technalogia, Minsk (in Russian).

Kilchevsky A.V. & Kogotko L.G. (1995) Genetical control of nitrate accumulation in tomato fruits, Vesti AN of Belarus, 4, 40-44 (in Belarussian).

Kilchevsky A.V., Pinchuk I.I., Kogotko L.G. & Bashirova L.N. (1993) Ecological approach to tomato breeding, in L. Stamova (ed), Proc. XII EUCARPIA Meeting on Tomato Genetics and Breeding, Maritsa Vegetable Crops Research Institute, Plovdiv, pp. 125-129.

Kilchevsky A.V., Schoor A.V. & Martiniak - Przybuszewska (1998) Accumulation of heavy metals in tomato fruits. Proc. Conf. "Best life", Olshtyn, pp. 22-25 (in Polish).

Klimashevsky E.L. (1991) Genetical aspect of mineral plant nutrition, Agropromizdat, Moscow (in Russian).

Li Y.-M.,Chaney R.L., Schneiter A.A. & Miller J.F. (1995) Combining ability and heterosis estimates for kernel cadmium level in sunflower, Crop Sci., 35, 1015-1019.

Mineev V.G. (1990) Chemistrization of agriculture and natural environment, Agropromizdat, Moscow (in Russian).

Popp M. (1983) Genotypic differences in the mineral metabolism of peans adapted to extrem habitals, Plant and Soil, 72, 261-273.

Prister B.S., Loshchilov N.A., Nemec O.F. & Pojarkov V.A. (1991) Elements of agricultural radiology, Urozhay, Kiev (in Russian).

Pushnaya O.N. (1990) Elaboration of evaluation method of tomato breeding material for greenhouses in nitrate content in fruits, Dis. Ph. D., VNIISSOK, Moscow (in Russian).

Saric M.R. (1983) Theoretical and practical approaches to the genetical specifity of mineral nutrition of plants, Plant and Soil, 72, 137 -150.

Schat H., Kuiper E., Ten Bookum W.M. & Vooijs R. (1993) A general model for the genetic control of copper tolerance in Silene vulgaris: evidence from crosses between plants from different tolerant populations, Heredity, 70, 142-147.

Tivo P.F. & Saskevich L.A. (1990) Nitrates: talks and reality, Uradzhay, Minsk (in Russian).

Zhuchenko & Andryushchenko (1980) Possibilities of decrease of nitrate contents in vegetables by the methods of plant breeding, Vestnik of Agr. Sci., 12, pp. 62-71 (in Russian).

G.T. Scarascia Mugnozza, E. Porceddu & M.A. Pagnotta (Eds.)
Genetics and Breeding for Crop Quality and Resistance, 323-330, 1999

Genetic modification of erucic acid biosynthesis in *Brassica napus*

W.W. Lühs[1], A. Voss[1], J. Han[2], A. Gräfin zu Münster[2], D. Weier[3], F.P. Wolter[2], M. Frentzen[3] & W. Friedt[1]

[1]Institut für Pflanzenbau und Pflanzenzüchtung I, Justus-Liebig-Universität, Ludwigstr. 23, D-35390 Giessen, Germany; [2]Institut für Allgemeine Botanik und Botanischer Garten, Universität Hamburg, Ohnhorststr. 18, D-22609 Hamburg, Germany; [3]Institut für Biologie I, Spezielle Botanik, Rheinisch-Westfälische Technische Hochschule, Worringer Weg 1, D-52056 Aachen, Germany

Abstract: Erucic acid (C22:1), the major component of traditional rapeseed (*Brassica napus* L.) triacylglycerols (TAG), is detrimental to the food quality of the oil. Progress in rapeseed breeding has led to the development of double-low rapeseed (Canola) cultivars. Conversely, for industrial purpose the C22:1 content of rapeseed oil should be as high as possible. Molecular biologists have demonstrated that the development of high-erucic acid rapeseed (HEAR) genotypes containing considerable amounts of erucic acid in the *sn*-2 position of the TAG, is an elementary step in order to increase the C22:1 content of rapeseed oil. Furthermore, transgenic HEAR bearing trierucin in its seed oil needs improvement by stimulating the biosynthetic capacity for fatty acid elongation. With regard to the molecular basis of C22:1 inheritance it is supposed that true erucic acid alleles are expressing different isoforms of the elongase condensing enzyme (β-ketoacyl-CoA synthase). In this concern, further work aims in the identification, molecular characterization and accumulation of more effective alleles for C22:1 synthesis in *B. napus*, involving transgenic HEAR lines with significant trierucin in their seed oil.

1. Introduction

Due to substantial progress in breeding and cultivation practice rapeseed and mustards - derived from several locally distributed members of the genus *Brassica* - have become one of the worldwide most important source of vegetable oil. Especially in several European countries with cool-temperate climates oilseed rape (*B. napus*) dominates field crop production. The primary demand is made up by food and animal feed industry furnished by rapeseed of 'double-low' or 'Canola' quality (Lühs & Friedt, 1994a). Modification of the fatty acid composition to make rapeseed oil more competitive in various segments of the food and industrial oil markets has been an important objective of plant breeding and molecular genetics in recent years. In order to use renewable resources growing interest exists to produce high-erucic acid rapeseed (HEAR) oils being directed to several industrial niche markets. The raw material is derived mainly from traditional and currently available, newly bred cultivars containing about 45-50% erucic acid (C22:1) in their seed oil (Lühs & Friedt, 1994b; Friedt & Lühs, 1998). Natural variation is limited in *B. napus* to a level of about 55-60%, while some accessions of related *Brassica* species possess an erucic acid content in their seed oil of 60% and even more (Lühs & Friedt, 1995a). In general, C22:1 content of *B. napus* varies with the allelic constitution of the genotype, differences in the ploidy level, the genetic background and environmental impact. Regarding the inheritance it is well known, that erucic acid content in *B. napus* (2n=4x=38; AACC genome) is controlled by two genes acting in an additive manner, which are derived from the progenitor species, viz. *B. rapa* (A genome; turnip rape, turnip, sarson, Chinese cabbage, etc.) and *B. oleracea* (C genome; cabbage, kohlrabi, cauliflower etc.), respectively. A series of alleles have been identified in *B. napus* and *B. rapa*, which makes it possible to breed strains containing almost any level of erucic acid from less than 1% to about 60% of total fatty acids. However, only in a few cases evidental data is available, which confirms the presence of true alleles residing in the A or C genome of *B. napus* (Jönsson, 1977; Chen & Heneen, 1989). Following *B. napus* resynthesis experiments we generated HEAR lines (RS lines) via interspecific hybridization by using *B. rapa* and *B. oleracea* genotypes with the most effective alleles for erucic acid synthesis as parents. We were able to show that C22:1 content can be modified by accumulating highly active alleles displaying an average additive contribution of 16-17% C22:1 per allele, which leads to a total C22:1 content of about 60% in the RS lines (Lühs & Friedt, 1995b).

In order to get a tailor-made raw material suited for industrial purposes, breeders, biochemists and genetic engineers are attempting to initiate the biosynthesis of trierucin (trierucoylglycerol) and to maximize the proportion of erucic acid in rapeseed oil. However, biochemical constraints are limiting fast success. At least two biosynthetic pathways emerge as the privileged targets for genetic modification: the capacity of erucic acid synthesis as such and the properties of the lysophosphatidic acid acyltransferase (LPAAT) being involved in the acylation reaction at the *sn*-2 position of the triacylglycerols (Frentzen, 1998; Friedt & Lühs, 1998).

2. Genetic engineering of rapeseed oil with regard to trierucin

In most members of the Brassicaceae and specifically in natural *B. napus*, erucic acid and other very-long chain fatty acids (VLCFA) are not incorporated into the central position of the glycerol backbone. This obstacle prevents the synthesis of trierucin and restricts the C22:1 content in rapeseed oil to a theoretical maximum of 66.7% (Lühs & Friedt, 1994c). To overcome the biochemical limitation of trierucin biosynthesis the strategy is to clone genes, which encode erucoyl-CoA compatible acyltransferase (LPAAT) enzymes and to insert these genes into appropriate *B.*

napus genotypes conferring rapeseed with the ability to esterify C22:1 into the central position of the TAG (Frentzen, 1998).

Different routes have been pursued in order to genetically modify the properties of the *sn*-2 acyltransferase (LPAAT) in *B. napus* and to increase the overall proportions of erucic acid in rapeseed oil. Following molecular breeding procedures, the erucoyl-compatible LPAAT of meadowfoam (*Limnanthes douglasii, L. alba*) or microorganisms, such as yeast or *Escherichia coli*, were functionally expressed in HEAR genotypes revealing significant alterations in both the stereochemical composition and the trierucin content of the transgenic seed oil. Keeping in mind that wild-type rapeseed has less than 1% erucic acid in the central position (*sn*-2 C22:1) it is clearly shown that the introduction of a foreign LPAAT has led to a drastic increase of erucic acid in the *sn*-2 position of the triacylglycerols (Table 1).

Table 1. Erucic acid content (%) of intact triacylglycerols (TAG) and of the *sn*-2 position detected in different transgenic *B. napus* seed oils containing trierucin.

Genotype	Source LPAAT	C22:1 TAG	C22:1 *sn*-2	Trierucin (C69:3)	Reference
'SLC1-1 Hero 8-6'	*Saccharomyces cerevisiae*	50.5	3.5	up to 0.4	Zou et al., 1997
'TN-E6'	*E. coli*	43.2	3.9	0.5	Weier et al., 1998
'7695-1'	*Limnanthes alba*	37.5	15.1	Significant peak	Lassner et al., 1995
'SCV144-2'	*L. douglasii*	32.1	28.3	2.8	Brough et al., 1996
'TR-E14'	*L. douglasii*	43.1	30.1	2.7	Weier et al., 1997
'T02-RS239'	*L. douglasii*	54.6	40.8	6.4	Weier et al., 1997
'pALM1-35'	*L. douglasii, Brassica napus*	49.5	49.3	5.3	Münster et al., 1998

LPAAT = lysophosphatic acid acyltransferase

The best results with ca. 40% *sn*-2 erucic acid and a corresponding trierucin content of about 6% were obtained by using the RS lines as recipient of the *L. douglasii* LPAAT (*Ld*-LPAAT) gene (Weier et al., 1997). Recently, the antisense blocking of the endogenous *B. napus* LPAAT in combination with the expression of the C22:1-CoA specific *Ld*-LPAAT has resulted in a further increase of *sn*-2 erucic acid content to about 50% (Münster et al., 1998).

Apart from these approaches, which have confirmed the feasibility of using chimeric LPAAT genes to alter the stereochemical composition of rapeseed oil, the fatty acid substrate specificity and selectivity of the LPAAT may also be changeable by site-directed mutagenesis as shown by Morand et al. (1998).

3. Development of genotypes with enhanced elongation capacity

Once the esterification of the *sn*-2 position with erucic acid had been accomplished it has become obvious, that the efficiency of erucic acid synthesis is a further target for metabolic engineering (Weier et al., 1997). The biosynthesis of very-long chain fatty acids (VLCFA) with chain lengths of C20 to C24 is catalyzed by acyl-CoA elongase(s) consisting of four enzyme activities being associated to a complex: condensing enzyme (β-ketoacyl-CoA synthase, KCS), reductase I, dehydratase and reductase II. The mechanismen of this microsomal pathway has been largely studied in various higher plants, such as *Arabidopsis thaliana*, jojoba (*Simmondsia chinensis*), leek (*Allium porrum*), honesty (*Lunaria annua*), and rapeseed (*B. napus*). It is generally admitted that *cis*-11-eicosenoic (C20:1), erucic and nervonic acid (C24:1), which are typical cruciferous VLCFA, are formed by successive 2-carbon chain extension of the precursor oleic acid (Cassagne et al., 1994). The activity of the KCS, representing the first and rate-limiting step of the microsomal fatty acid elongation reaction, determines the acyl chain length of the VLCFA produced by developing seeds (Millar & Kunst, 1997). In order to clarify the molecular basis of VLCFA biosynthesis the *FAE1* gene of *A. thaliana*, probably encoding a condensing enzyme, has been isolated by using direct transposon tagging (James et al., 1995). Subsequently, a couple of cDNAs encoding putative KCS enzymes have been isolated from different species, including jojoba, honesty as well as oilseed rape and other members of the genus *Brassica* (Table 2). Due to the high degree of both homology of nucleotide sequences and identity of the deduced amino acids there is strong evidence that C22:1 biosynthesis in rapeseed and other *Brassica* species is controlled through the expression and property of KCS enzymes being encoded by gene(s) homologous to the *FAE1* gene from *A. thaliana* (cf. Clemens & Kunst, 1997; Barret et al., 1998; Fourmann et al., 1998). The introduction of a KCS gene cloned from jojoba confirmed this assumption by showing the complementation of the Canola fatty acid elongation mutation (*fae*) leading to the restoration of erucic acid synthesis in transgenic rapeseed (Lassner et al., 1996). The isolation and expression of cDNAs encoding KCS from honesty (*L. annua*), a crucifer being rich in nervonic acid, led to alterations in the fatty acid composition of transgenic rapeseed, but did not dramatically increase the VLCFA content of the oil (Lassner, 1997).

Table 2. *FAE1* homologues (cDNAs) encoding microsomal condensing enzymes (KCS).

Species	**Genotype used as cDNA source**	**cDNA clone / Acc. No.**	**Reference**
Arabidopsis thaliana	strain 'WS'	Acc. U29142	James et al. (1995)
Simmondsia chinensis	not specified	Acc. U37088	Lassner et al. (1996)
Brassica napus	cv. 'Hokkaido'	Acc. U50771	Barret et al. (1998)
	cv. 'Golden'	Acc. AF009563	Clemens & Kunst (1997)
	strain 'Yudal'	Acc. AF054497	Fourmann et al. (1998)
	strain 'Yudal'	Acc. AF054498	Fourmann et al. (1998)
	cv. 'Askari'	clone KCSb5	Han et al. (1998)
B. rapa	cv. 'R500'	Acc. AF054499	Fourmann et al. (1998)
B. oleracea	'Rapid cycling capitata'	Acc. AF054500	Fourmann et al. (1998)
B. juncea	cv. 'Pusa Bold'	Acc. Y11007	Venkateswari et al. (unpubl.)

We were successful in isolating a *B. napus* cDNA clone (KCSb5) from a library derived from developing pods of the HEAR cultivar "Askari". For screening a probe was used being PCR amplified from the *FAE1* gene from *A. thaliana* (Han et al.,

1998). In further experiments KCSb5 was applied in a twofold way: 1) as a specific KCS probe in genomic Southern hybridization analyses of erucic acid genes/alleles in *Brassica* and 2) for the development of gene contructs used in the genetic transformation of rapeseed.

3.1 *Molecular characterization of erucic acid alleles*

Figure 1 illustrates the polymorphism we have obtained by digestion of genomic DNA with *Bam*HI and hybridization with the KCSb5 probe. The plant material being evaluated in the course of this Southern blot analyses consisted of different rapeseed and *Brassica* genotypes, which were known to possess a specific erucic acid content due to particular genetic constitution. The results manifest our hypothesis, that variation in erucic acid content is mainly induced by different alleles, which can be distinguished on a molecular level. Corresponding to the amphidiploid origin of *B. napus* and the digenic control of erucic acid content we observed in most of the cases two fragments (Figure 1).

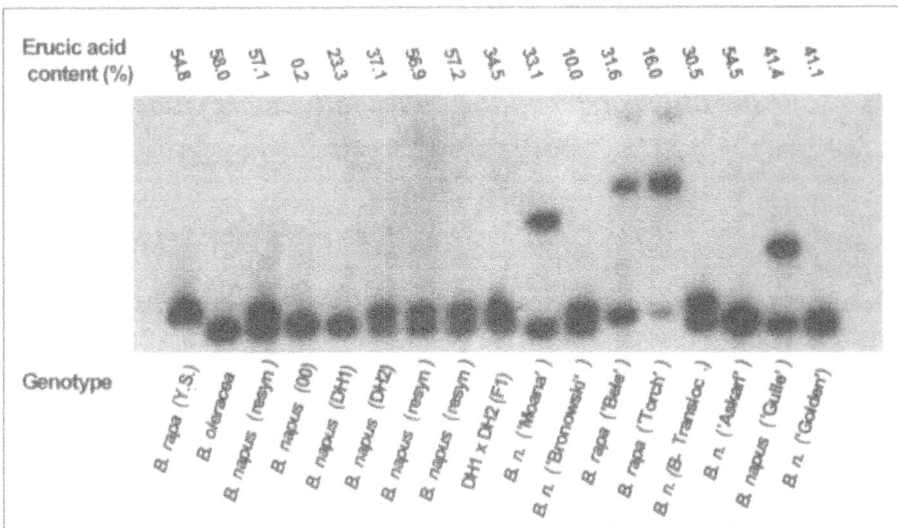

Figure 1. Molecular characterization of true *Brassica* erucic acid alleles performed by Southern analysis after digestion of genomic DNA with *Bam*HI and hybridization with a *FAE1* probe (KCSb5).

This is particularly the case for the RS lines (*B. napus* resyn, $E_AE_AE_CE_C$, ca. 57% C22:1) displaying a combination of the parental genomes, 'Yellow sarson' (*B. rapa* 'Y.S.', E_AE_A, 54.8% C22:1) and cauliflower (*B. oleracea*, E_CE_C, 58.0% C22:1), respectively. The erucic acid content (ca. 7-11%) of the spring rapeseed 'Bronowski' is referred to a genotype $e_Ae_AE_CE_C$, where the alleles contributing to the total erucic content (ca. 4% per allele) reside in the C genome and those for absence of C22:1 are present in the A genome (cf. Anand & Downey, 1981). Regarding *B. napus* 'Moana' also this kind of monogenic control of high erucic acid content was observed, since homozygosity of an "effective" allele at one locus gives about 33% C22:1 (cf. Jönsson, 1977). In our study the doubled haploid rapeseed lines DH1 ($e_Ae_AE_CE_C$, 23.1% C22:1) and DH2 ($E_AE_Ae_Ce_C$, 37.1% C22:1) were involved, which have been derived from intraspecific crosses between a double-low spring rapeseed (*B. napus* 00, $e_Ae_Ae_Ce_C$, 0.2% C22:1) and the RS lines mentioned above (cf. Lühs & Friedt, 1995b). The genetic constitution of DH1 and DH2, possessing one of the newly generated *B. napus* alleles, viz. E_A from *B. rapa* ('Yellow sarson') and E_C from *B. oleracea* (cauliflower), was confirmed by test crosses between these lines as well as to 'Bronowski' and 'Moana'. Due to intraspecific crosses involving 'Bronowski' the *B. nigra* erucic acid allele (E_B) of the *B. napus*/B genome translocation line ($E_A^BE_A^Be_Ce_C$, *B. napus* B-Transloc., 30.5% C22:1), originally developed at the University of Göttingen (cf. Struss et al., 1991), is supposed to reside at the A locus in homozygous condition.

Only for "Askari" ($E_AE_AE_CE_C$, 54.5% C22:1), used as a check in this study, the spring rapeseed 'Golden' ($E_AE_AE_CE_C$, 41.1% C22:1) and the low erucic acid variety ($e_Ae_Ae_Ce_C$) we obtained one fragment, indicating a high grade of homology between these erucic acid genes/alleles due to minor difference in the nucleotide sequence and the deduced amino acids of the KCS polypeptid (cf. Clemens & Kunst, 1997; Fourmann et al., 1998).

3.2 Restoration of very-long chain fatty acid synthesis

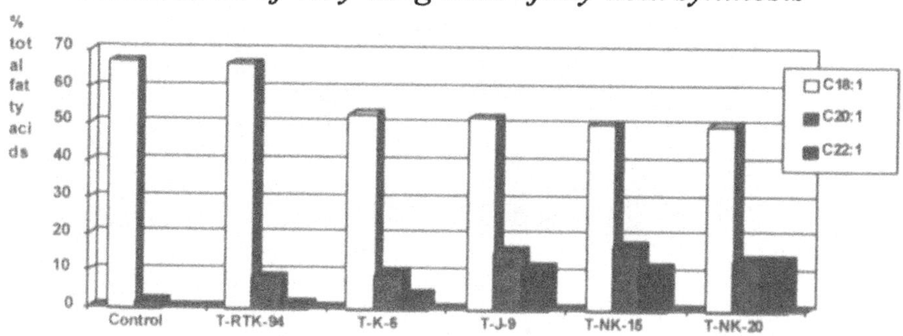

Figure 2. Modification of *(n-9)*-fatty acid content in the seed oil of transgenic *Brassica napus* cv. "Drakkar" expressing the *β*-ketoacyl-CoA-synthase (KCS) gene.

Following a genetic engineering approach, as described earlier by Calgene scientists (Lassner et al., 1996), the effect of specific KCS genes or alleles can be studied in a system that shows an almost complete abolishment of C22:1 synthesis. As shown in Figure 2 the integration of KCSb5, originating from the winter rapeseed "Askari", into the Canola variety "Drakkar" (00) has led to the restoration of VLCFA biosynthesis in *B. napus*. In the best transformants the content of cis-11-eicosenoic (C20:1) and erucic acid has been increased from about 2% to more than 25-30% of total fatty acids, which clearly demonstrates that KCS is essential within the fatty acid elongation process.

4. Concluding remarks

As the high erucic acid level has remained resistant to change until now, concerted research efforts in plant breeding and genetic engineering are still underway to maximize the proportion of C22:1 in rapeseed oil. The genetic alteration of the *sn*-2 acyltransferase revealed also that this approach was not a wholly sufficient route towards rapeseed oil with a very high erucic acid content, i.e. more than 70% C22:1. Since the number of elongase complexes as such might be limited further work aimes in the identification, molecular characterization and introduction of more effective alleles and/or additional genes for erucic acid synthesis in *B. napus*, in order to fulfill the opportunity being provided by transgenic HEAR lines with significant trierucin in their seed oil. To resume all the data available at present, nasturtium (*Tropaeolum majus*) is still the only species in plant kingdom producing seed oils with very high erucic acid (71-78%) content due to the occurrence of trierucin (42-52%), which has led to the hypothesis that this plant species must have developed a unique route for the biosynthesis of very high erucic acid seed oils (Löhden & Frentzen, 1992; Lühs & Friedt, 1994c).

Acknowledgements: The authors thank Dr. Karin Sonntag of the Bundesanstalt für Züchtungsforschung (BAZ) in Groß Lüsewitz as well as the Planta GmbH in Einbeck for developing transformants derived from cv. 'Drakkar'. Furthermore, we thank Dr. C. Möllers, Prof. Dr. H.C. Becker and Prof. Dr. G. Röbbelen, Institut für Pflanzenbau und Pflanzenzüchtung, University Göttingen, being involved during the initial stage of this programme. We are grateful for financial support granted by the Bundesministerium für Bildung, Wissenschaft, Forschung und Technologie, Bonn (BMBF), and German rapeseed breeding companies, namely Norddeutsche Pflanzenzucht, Hans-Georg Lembke KG, Hohenlieth (NPZ), Deutsche Saatveredelung Lippstadt-Bremen GmbH (DSV), and Kleinwanzlebener Saatzucht AG, Einbeck (KWS).

5. References

Anand, I.J. & Downey, R.K. (1981) A study of erucic acid alleles in digenomic rapeseed (*Brassica napus* L.), *Can. J. Plant Sci.* **61**, 199-203.

Barret, P., Delourme, R., Foisset, N., Renard, M., Domergue, F., Lessire, R., Delseny, M. & Roscoe, T. (1998) A rapeseed *FAE1* gene is linked to the E1 locus associated with variation in the content of erucic acid, *Theor. Appl. Genet.* **96**, 177-186.

Brough, C.L., Coventry, J.M., Christie, W.W., Kroon, J.T.M., Brown, A.P., Barsby, T.L. & Slabas, A.R. (1996) Towards the genetic engineering of triacylglycerols of defind fatty acid composition: major changes in erucic acid content at the *sn*-2 position affected by the introduction of a 1-acyl-*sn*-glycerol-3-phosphate acyltransferase from *Limnanthes douglasii* into oil seed rape, *Molecular Breeding* **2**, 133-142.

Cassagne, C., Lessire, R., Bessoule, J.J., Moreau, P., Creach, A., Schneider, F. & Sturbois, F. (1994) Biosynthesis of very long chain fatty acids in higher plants, *Prog. Lipid Res.* **33**, 55-69.

Chen, B.Y. & Heneen, W.K. (1989) Fatty acid composition of resynthesized *Brassica napus* L, *B. campestris* L. and *B. alboglabra* Bailey with special reference to the inheritance of erucic acid content, *Heredity* **63**, 309-314.

Clemens, S. & Kunst, L. (1997) Isolation of a *Brassica napus* cDNA (Accession No. AF009563) encoding 3-ketoacyl-CoA synthase, a condensing enzyme involved in the biosynthesis of very long chain fatty acids in seeds, *Plant Physiol.* **115**, 313-314.

Fourmann, M., Barret, P., Renard, M., Pelletier, G., Delourme, R. & Brunel, D. (1998) The two genes homologous to *Arabidopsis FAE1* co-segregate with the two loci govering erucic acid content in *Brassica napus, Theor. Appl. Genet.* 96, 852-858.

Frentzen, M. (1998) Acyltransferases from basic science to modified seed oils, *Fett/Lipid* 100, 161-166.

Friedt, W. & Lühs, W. (1998) Recent developments and perspectives of industrial rapeseed breeding, *Fett/Lipid* **100**, 219-226.

Han, J., Lühs, W., Sonntag, K., Borchardt, D.S., Frentzen, M. & Wolter, F.P. (1998) A *Brassica napus* cDNA restores the deficiency of canola fatty acid elongation at a high level. Proc. 13th Intern. Symp. on Plant Lipids, July 5-10, 1998, Sevilla, Spain. Publicaciones de la Universidad de Sevilla (in press).

James, D.W., Jr., Lim, E., Keller, J., Plooy, I., Ralston, E. & Dooner, H.K. (1995) Directed tagging of the *Arabidopsis* Fatty Acid Elongation 1 (*FAE1*) gene with the maize transposon activator, *Plant Cell* **7**, 309-319.

Jönsson, R. (1977) Erucic-acid heredity in rapeseed (*Brassica napus* L. and *Brassica campestris* L.), *Hereditas* **86**, 159-170.

Lassner, M. (1997) Transgenic oilseed crops: a transition from basic research to product development, *Lipid Technology* **9** (1), 5-9.

Lassner, M.W., Levering, C.K., Davies, H.M. & Knutzon, D.S. (1995) Lysophosphatidic acid acyltransferase from meadowfoam mediates insertion of erucic acid at the *sn*-2 position of triacylglycerol in transgenic rapeseed oil, *Plant Physiol.* **109**, 1389-1394.

Lassner, M.W., Lardizabal, K. & Metz, J.G. (1996) A jojoba β-ketoacyl-CoA synthase cDNA complements the canola fatty acid elongation mutation in transgenic plants, *Plant Cell* **8**, 281-292.

Löhden, I. & Frentzen, M. (1992) Triacylglycerol biosynthesis in developing seeds of *Tropaeolum majus* L. *and Limnanthes douglasii* R. Br., *Planta* **188**, 215-224.

Lühs, W. & Friedt, W. (1994a) Major oil crops, in D.J. Murphy (ed.), *Designer Oil Crops*, VCH Verlagsges., Weinheim, pp. 5-71.

Lühs, W. & Friedt, W. (1994b) Non-food uses of vegetable oils and fatty acids, in D.J. Murphy (ed.), *Designer Oil Crops*, VCH Verlagsges., Weinheim, pp. 73-130.

Lühs, W. & Friedt, W. (1994c) Stand und Perspektiven der Züchtung von Raps (*Brassica napus* L.) mit hohem Erucasäure-Gehalt im Öl für industrielle Nutzungszwecke, *Fat Sci. Technol.* **96**, 137-146.

Lühs, W. & Friedt, W. (1995a) Natural fatty acid variation in the genus *Brassica* and its exploitation through resynthesis, *Eucarpia Cruciferae Newsletter* **17**, 14-15.

Lühs, W. & Friedt, W. (1995b) Breeding high-erucic acid rapeseed by means of *Brassica napus* resynthesis. Proc. 9th Intern. Rapeseed Congr. (GCIRC), Cambridge, United Kingdom, Vol. 2, pp. 449-451.

Millar, A. & Kunst, L. (1997) Very long chain fatty acid biosynthesis is controlled through the expression and specificity of the condensing enzyme, *Plant J.* **12**, 121-131.

Morand, L.Z., Patil, S., Quasney, M. & German, J.B. (1998) Alteration of the fatty acid substrate specificity of lysophosphatidate acyltransferase by site-directed mutagenesis, *Biochem. Biophys. Res. Comm.* **244**, 79-84.

Münster, A., Gräfin zu, Lühs, W., Borchardt, D.S., Wolter, F.P. & Frentzen, M. (1998) Experiments to optimize the channelling of erucic acid into the *sn*-2-position of transgenic rapeseed oil. Proc. 13th Intern. Symp. on Plant Lipids, July 5-10, 1998, Sevilla, Spain. Publicaciones de la Universidad de Sevilla (in press).

Struss, D., Quiros, C.F. & Röbbelen, G. (1991) Construction of different B-genome addition lines of *Brassica napus* L. Proc. 8th Intern. Rapeseed Congress (GCIRC), Saskatoon, Sask., Canada, Vol. 2, pp. 358-363.

Weier, D., Hanke, C., Eickelkamp, A., Lühs, W., Dettendorfer, J., Schaffert, E., Möllers, C., Friedt, W., Wolter, F.P. & Frentzen, M. (1997) Trierucoylglycerol biosynthesis in transgenic plants of rapeseed (*Brassica napus* L.). Fett/Lipid **99**, 160-165.

Weier, D., Lühs, W., Dettendorfer, J. & Frentzen, M. (1998) sn-1-Acylglycerol-3-phosphate acyltransferase of *Escherichia coli* causes insertion of cis-11 eicosenoic acid into the sn-2 position of transgenic rapeseed oil, Molecular Breeding **4**, 39-46.

Zou, J.T., Katavic, V., Giblin, E.M., Barton, D.L., MacKenzie, S.L., Keller, W.A., Hu, X. & Taylor, D.C. (1997) Modification of seed oil content and acyl composition in the Brassicaceae by expression of a yeast sn-2 acyltransferase gene, Plant Cell **9**, 909-923.

G.T. Scarascia Mugnozza, E. Porceddu & M.A. Pagnotta (Eds.)
Genetics and Breeding for Crop Quality and Resistance, 331-338, 1999

Variability of seed storage proteins within the sunflower gene pool

I.N. Anisimova[1], Al.V. Konarev[2], V.T. Rozhkova[1], V.A. Gavrilova[1], R.J. Fido[3], A.S. Tatham[3] & P.R. Shewry[3]
[1]N.I. Vavilov Institute of Plant Industry, 42 Bolshaya Morskaya str., 190000 St.Petersburg, Russia; [2]All-Russian Institute of Plant Protection, 3 Shosse Podbelskogo, 189620 St.Petersburg, Russia; [3]IACR-Long Ashton Research Station, University of Bristol, Bristol BS18 9AF, UK.

Abstract: Sunflower seeds are distinguished from those of other oil crops by the lack of antinitritive and toxic components. Besides oil they accumulate significant amounts of storage proteins called helianthinin and 2S albumins. Due to that the sunflower seeds are considered as the most promising sources of plant proteins for food industry. Moreover recent finding of the low molecular weight methionine-rich albumin SFA8 renders the sunflower seed proteins both a target for breeding improvement and a source of a valuable agronomic character for genetic modifying of other plant crops. In the present study more than 30 thousands genotypes from sunflower gene pool representing inbred lines and interline hybrids, cultivars, wild species and interspecific hybrid progenies were screened and a variation in composition of major seed proteins was demonstrated. Analysis of segregation in F_2 and BC hybrid populations has showed helianthinin genes to be clustered in three loci. Two of them co-segregated with a recombination frequency of about 22%; the third locus showed independent inheritance. In the cotyledons of F_1 hybrids between distantly related sunflower species the helianthinin protein deficient in certain polypeptides was accumulated that probably explains inviability of seeds derived from interspecific crosses in *Helianthus*. An allelic variation for some albumin polypeptides was also observed. N-terminal amino acid sequencing of individual albumins and analysis of segregation in the F_2 of interline hybrid allowed to suggest that the albumin gene family consists of at least three subfamilies. The level of SFA8 protein varied from trace amounts (in perennial Helianthus species) up to 30% in total albumin fraction (interline F_1 hybrid and hybrid varieties). These data indicate a breeding potential for increasing content of sulfur essential amino acids in sunflower.

1. Introduction

In the past decades serious efforts were made to search for novel sources of proteins for animal nutrition and the food industry, among traditional and exotic crops. In this respect oilseeds have great potential because, in addition to lipids, they also accumulate significant amounts of storage proteins, accounting for a high proportion of the residue after oil extraction (Gassman, 1983). Sunflower seeds are distinct from some other oilseeds (for example, soybean) in that they lack antinitritional and toxic protein compounds. The major storage protein components of sunflower seed belong to the two groups, salt-soluble 11S globulins (helianthinin) and water soluble 2S albumins, in relative amounts of about 2:1. The protein content of sunflower seeds averages between 20% and 30% among different varieties (Chmeleva et al., 1981). Due to this the sunflower meal can be utilised as an alternative source of protein for animal food and human diet. Moreover, the recent identification of a low molecular weight single chain methionine-rich albumin SFA8 (Kortt & Caldwell, 1990) means the sunflower seed proteins are an attractive target for improvement by breeding and a source of a valuable agronomic character for the genetic modification of other agricultural crops. However, there is little information available on the genetic variability of sunflower seed storage proteins. The aim of our study was to describe the allelic variation in seed protein loci which was revealed by the analysis of selected samples from the 30,000 genotypes representing the sunflower gene pool.

2. Materials and Methods

2.1 Seed materials

The materials examined comprised 188 inbred lines, 10 interline hybrids (including F_2 segregating populations), 23 varieties, 20 ecotypes of wild *Helianthus* species and 13 interspecific hybrid progenies. They originated from the Sunflower World Collection (Vavilov Institute of Plant Industry, St.Petersburg, Russia), All-Russia Institute of Oil Crops (Krasnodar, Russia), Institute of Genetics (Sofia, Bulgaria), University of Birmingham (United Kingdom), University of Giessen (Germany).

2.2 Methods

Electrophoresis of helianthinin (SDS-PAGE) was carried out in a Tris/glycine system following the procedure described in our previous paper (Anisimova et al., 1991). The 2S albumins were fractionated in a Tris/tricine electrophoretic system or separated using high performance liquid chromotography (RP-HPLC) as described by Anisimova et al. (1995).

3. Results and Discussion

3.1 *Genetic variability of helianthinin*

Heianthinin is a legumin-like globulin with molecular mass of about 305 000 (Schwenke et al., 1979). It is composed of six subunits each consisting of disulfide-linked acidic (α,α') and basic (β,β') polypeptide chains. The charge and molecular mass heterogeneity of subunits and polypeptides which was initially demonstrated by Dalgalarrondo et al. (1984) suggests that helianthinin is the product of multigenic family. Two cDNA clones which probably represent two divergent helianthinin subfamilies have been isolated and characterized by Vonder Haar et al. (1988) and expressed in transgenic tobacco (Bogue et al., 1990). Genetic analysis carried out with using segregating hybrid populations has revealed three loci encoding helianthinins, *Hel*A, *Hel*B and *Hel*C. The *Hel*A locus includes two members (for subunits A1 and A2), while each of the loci *Hel*B and *Hel*C consists of three members (for subunits B1, B2, B3 and C1, C2, C3, respectively). The disulfide-linked polypeptide pairs constituting individual helianthinin subunits were identified as shown in Table 1. A series of polymorphic alleles was revealed which encoded variant polypeptides and subunits. Genotypes lacking certain polypeptides and subunits are likely to be homozygous for various null-alleles.

The helianthinin loci *Hel*B and *Hel*C were demonstrated to be linked with a recombination frequency of about 22% (Anisimova et al., 1996). The *Hel*A locus segregated independently of both the *Hel*B and *Hel*C loci. Linkage analysis indicated very loose linkage between the *Hel*B and *Est*1 (esterase) loci, *Hel*B and *Gpi*1 (glucose phosphate isomerase), *Hel*B and *Vs* (raised leaf veins) and *Hel*C and *Vs*. Independent inheritance was observed for the pairs of loci *Hel*B-*Ep* (striped pigmentation of epidermis) and *Hel*B-*P* (presence of armour layer in pericarp) (Anisimova et al., 1996). No linkage was found between the *Hel*B and *TIm* (main trypsin inhibitor) loci.

Table 1. Organisation of helianthinin gene family.

Locus	Subunits	Polypeptide pairs*	Polymorphic alleles
*Hel*A	A1	α33 β4(β7)	*hel* 34
	A2	α32 β7(β4)	*hel* 4^{nul}
*Hel*B	B1	α28 β10	*hel* 29
	B2	$\alpha 30^{a}$ β10	*hel* 30^{nul}, *hel* 10^{nul}
	B3	$\alpha 30^{b}$ β3'	
*Hel*C	C1	α'22 $\beta'12^{b}$	*hel* 9, *hel* 11
	C2	α'23 $\beta'12^{a}$	*hel* 12^{nul}
	C3	α'21 β'12	

*Polypeptide designations are given in Greek letters (Raymond et al., 1994) and figures (Anisimova et al., 1991)

All the varieties examined possessed unique combinations of polymorphic helianthinin alleles (Table 2). The *hel*34 allele (locus *Hel*A) was found to be characteristic of only the variety Armavirets and showed very limited distribution among various inbred lines (some originating from variety Armavirets). The *hel*29 allele (locus *Hel*B) occurred in the varieties Gigant 549 and Pervenets but was rather common among inbred lines. The *hel*9 and *hel*11 alleles (locus *Hel*C) showed wide distribution within the sunflower gene pool. They were present both in different inbred lines and in most varieties. The varieties VNIIMK1646, Krasnodarets and Start lacked these alleles but could be distinguished on the base of other polymorphic helianthinin variants (data not shown). The varieties Jenisei, Lider, Pervenets, Pochin, Skorospelyi and Yugo-Vostochnyi all possessed both alleles *hel*9 and *hel*11 (with the highest frequency of occurrence being in Jenisei) whereas other varieties had either *hel*9 (Peredovik, Kavkazets, Progress) or *hel*11 (Berezanskii, Voskhod, Zelenka368, Zenit, Lider, Nadeoyzhnyi).

Table 2. Distribution of some polymorphic helianthinin alleles among variety populations.

Variety	Frequency of occurrencies of alleles*						
	*hel*9	*hel*11	*hel*29	*hel*9	*hel*4^{nul}	*hel*10^{nul}	*Hel*12^{nul}
Peredovik	12	0	0	0	0	0	0
Gigant 649	0	0	68	0	100	0	0
Armavirets	0	0	0	4	0	0	0
Berezansky	0	6	0	0	0	0	0
VNIIMK 1646	0	0	0	0	0	0	0
Voskhod	0	4	0	0	0	0	0
Yenisei	9	19	0	0	0	0	0
Zarya	0	0	12	0	0	0	0
Zelenka 368	0	4	0	0	2	0	0
Zenit	0	24	0	0	0	0	0
Kavkazets	2	0	0	0	0	0	0
Krasnodarets	0	0	0	0	0	0	0
Lider	8	3	0	0	0	0	2
Lutch	0	0	0	0	0	0	2
Nadeozhnyi	0	22	0	0	0	0	0
Pervenets	3	1	48	0	4	0	1
Potchin	2	6	0	0	0	0	0
Progress	4	0	0	0	0	0	0
Salyut	0	0	0	0	0	2	1
Skorospelyi	2	4	0	0	0	0	0
Start	0	0	0	0	0	0	0
Yugo-Vostochnyi	4	8	4	0	0	0	0

*Data were obtained for random samples of 100 seeds.

A polymorphism for alleles *hel*9 and *hel*11 was observed within wild *H. annuus* ecotypes. Segregation analysis of crosses between wild *H. annuus* and several inbred lines has demonstrated that the helianthinin gene family of domesticated sunflower differed from that of wild ecotypes by a single allele (*hel*12) at the *Hel*C locus. This is consistent with molecular analysis data which reveal high identity of RAPD markers between wild and domesticated *H. annuus* (Arias & Riesberg, 1995). A number of facts indicate that «wild» allele *hel*9 is probably present in the suppressed state in the cultivated sunflower genome («sleeping» allele) and can be repressed, for example, as a result of interspecific crosses. Artificial hybridization of cultivated sunflower with wild perennial *Helianthus* is known to be a complex process which gives rise to a diversity of forms, some of which have valuable agronomic characters (Georgieva-Todorova, 1984). We examined a number of hybrid progenies between *H. annuus* inbred lines and perennial species which have been produced *via* «embryo rescue» or by conventional field technique. All the progenies, regardless of their origin, expressed helianthinin of the maternal («annual») phenotype. In certain progenies, however, the allele *hel*9 was expressed rather than the maternal type allele *hel*12. This fact can be explained by a possible de-suppression of an «ancient» *hel*9 allele present in the genotype of the maternal line in the «sleeping» state.

Eighty two inbred lines out of 188 analysed (~43%) possessed characteristic helianthinin polypeptides (data not shown). It should be noted that the *hel*29 allele was the most frequent among seven polymorphic alleles considered (Table 3). Several lines had two or three allelic variants, however, most inbreds expressed only one polymorphic allele. These data indicate a quite wide genetic base of domesticated sunflower that is consistent with data obtained with the use of RAPD markers (Lawson et al., 1994) and by RFLP-analysis (Berry et al., 1994).

Table 3. Distribution of some polymorphic helianthinin alleles within a set of 188 inbred lines.

Allele	Numbers of lines	
	total	%
hel 9	8	4.3
hel 11	6	3.2
hel 29	16	8.5
hel 34	4	2.1
hel 4^{nul}	5	2.7
hel 10^{nul}	3	1.6
hel 10^{nul}	3	1.6

3.2 Genetic variability of 2S albumins

The 2S albumins of sunflower seed exist as a heterogeneous mixture of single chain polypeptides with M_r 10,000-18,000, most, if not all, having intra-chain disulfide bonds (Kortt & Caldwell, 1990; Anisimova et al., 1995). Two methionine-rich albumins (SFA7 and SFA8) have nearly identical amino acid composition and were established to constitute of about 7% of the total protein in the seeds of the variety Hysun (Kortt & Caldwell, 1990). One of the 2S albumin genes which have been so far isolated and sequenced represented the single chain methionine-rich albumin SFA8 (Kortt et al., 1991). Another known gene, HaG5 (Allen et al., 1987; Brunel, 1994) was shown to be a member of small multigenic family consisting of at least two divergent genes. The precursor proteins encoded by these genes are presumed to undergo extensive proteolytic processing giving rise to two different polypeptides which are not disulfide linked but each of which contain disulfide bonds.

Up to 12 individual polypeptides have been identified in the 2S albumin fraction using SDS-PAGE and RP-HPLC analyses. According to their N-terminal sequences, the 2S albumin polypeptides of sunflower can be classified into three groups: the methionine-rich albumins SFA7 and SFA8 (M_r~10,000), and proteins derived from the N- and C-terminal parts of the HaG5 precursor protein. Variation in the presence or absence of minor albumin polypeptides was also observed among the genotypes analysed. A variant form of SFA8 with slightly slower mobility on SDS-PAGE compared to the normal type was identified. This variant was found in only five of 60 inbred lines examined, all of which were derived from the German accession k-2266 and the Polish line L1648/1. The two SFA8 forms clearly differed in a pI with the normal form having pI, of about 6.0 and the variant form having a pI of about 6.5. In F_1 seeds of a cross between the inbred lines VIR130 (variant SFA8) and VIR104 (normal SFA8) the two forms were expressed co-dominantly while the ratio of phenotypic classes in F_2 fitted 1:2:1 indicated that the normal and variant forms of SFA8 are encoded by allelic loci.

Significant variation in the level of SFA8 in the 2S albumin fractions was observed within the sunflower gene pool. SFA8 was absent from seeds of perennial *Helianthus* species, however, but present in both annual species and cultivated sunflower. The combined level of SFA7 and SFA8 and the relative proportions of these proteins varied significantly among different inbred lines (Figure 1). A considerable heterotic effect for the level of methionine-rich albumins (~50% prevalence over mean values of parental lines) was observed in F_1 hybrid seeds of interline crosses. However, no heterotic effect was noted when the F_2 seeds were analysed. Moreover, unexpectedly high variation in the amounts of SFA7 and SFA8 was observed among individual F_2 seeds. The data suggest that the levels of methionine-rich albumins in sunflower seed are probably under genetic control.

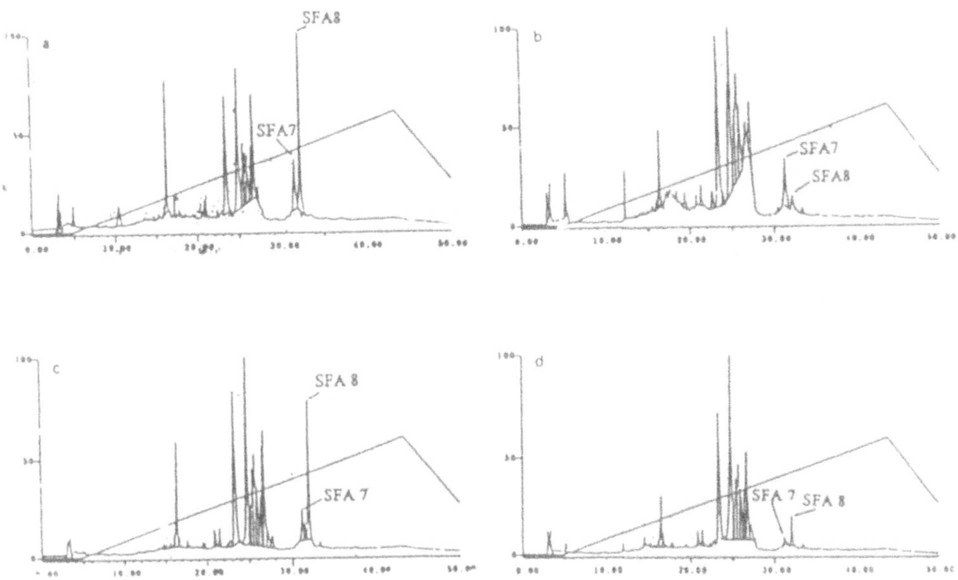

Figure 1. HPLC profiles of 2S albumin fractions from inbred lines VIR435 (a), VIR253 (b) and interspecific hybrid progenies *H. annuus* x (*H. an.* x *H. tuberosus*) (c) and *H. an.* x *H. tub.* (d); SFA7 and SFA8 are sulfur-rich polypeptides.

4. Conclusions

Allelic variation in the genetic loci encoding seed storage proteins of sunflower seed has been revealed. It can be used in linkage studies aimed to construction of genetic linkage map of sunflower. Variation in the amounts of individual polypeptides should be considered for when attempting to improve the seed protein composition by manipulation of sunflower and other agricultural crops.

Acknowledgements The authors wish to thank Prof. W. Friedt (University of Giessen) for providing seed samples of interspecific hybrid progenies. Part of this work was supported by a Joint Project (683072.P810) from the Royal Society of London. The authors are grateful to the EUCARPIA Organisation for providing a financial support of the visit to XV EUCARPIA General Congress. IACR receives grant- aided support from the BBSRC (UK).

5. References

Allen, R.D., Cohen, E.A., Vonder Haar, R.A., Adams, C.A, Ma, D.P., C.L.Nessler C.L. & Thomas, T.L. (1987) Sequence and expression of a gene encoding an albumin storage protein in sunflower, *Moecular and General Genetics* **210**, 211-218.

Anisimova, I.N., Gavrilyuk, I.P. & Konarev, V.G. (1991) Identification of sunflower lines and varieties by helianthinin electrophoresis, *Plant Varieties and Seeds* **4**, 133-141.

Anisimova, I.N., Gavrilova, V.A., Loskutov, A.V. & Tolmachev, V.V. (1996) Identification of gene loci encoding for 11S globulin of sunflower seed, in Proceedings of the 14th International Sunflower Conference (Beijing/Shenyang, China, 12-20 June 1996). **2**, pp. 1167-1170.

Anisimova, I.N., Fido, R.J., Tatham A.S. & Shewry, P.R. (1995) Genotypic variation and polymorphism of 2S albumins of sunflower, *Euphytica* **83**, 15023.

Arias, D.M. & Riesberg, L.H. (1995) Genetic relationships among domesticated and wild sunflowers (*Helianthus annuus, Asteraceae*), *Economic Botany* **49**, 239-248.

Berry, S.T., Allen R.J., Barnes, S.R. & Caligari, P.D.S. (1994) Molecular marker analysis of *Helianthus annuus* L. 1.Resriction fragment length polymorphism between inbred lines of cultivated sunflower., *Theoretical and Applied Genetics* **89**, 435-441.

Bogue, M.A., Vonder Haar, R.A., Nuccio, M.L., Griffing, L.R. & Thomas, T.L. (1990) Developmentally regulated expression of a sunflower 11S seed protein gene in transgenic tobacco, *Molecular and General Genetics* **222**, 49-57.

Brunel, D. (1994) Denaturing gradient gel electrophoresis (DGGE) and direct sequencing of PCR amplified genomic DNA: rapid and reliable identification of *Helianthus annuus* L. cultivars, *Seed Science and Technology* **22**, 185-194.

Chmeleva, Z.V., Zakharova, N.C. & Anashchenko, A.V. (1981) Characterisation of sunflower collection for the content of protein, essential amino acids and oil in the kernel. *Trudy Prikl. Bot.* **70**, 120-128 (in Russian).

Dalgalarrondo, M., Raymond, J. & Azanza, J.L. (1984) Sunflower seed proteins: characterization and subunit composition of the globulin fraction, *Journal of Experimental Botany* **35**, 1618-1628.

Gassman, B.C. (1983) Preparation and application of vegetable proteins, especially proteins from sunflower seed, for human consumption. An approach, *Die Nahrung* **27**, 351-369.

Georgieva-Todorova, J. (1984) Interspecific hybridization in the genus *Helianthus* L., *Zeitshrift fur Pflanzenzuchtung* **93**, 265-279.

Kortt, A.A. & Caldwell, J.B. (1990) Low molecular weight albumins from sunflower seed: identification of a methionine-rich albumin, *Phytochemistry* **29**, 2805-2810.

Kortt, A.A., Caldwell, J.B., Lilley, G.G. & Higgins, T.J.V. (1991) Amino acid and cDNA sequences of a methionine-rich 2S protein from sunflower seed (*Helianthus annuus* L.), *European Journal of Biochemistry* **195**, 329-334.

Lawson, W.R., Henry, R.J., Kochman, J.K. & Kong, G.A. (1994) Genetic diversity in sunflower (*Helianthus annuus* L.) as revealed by random amplified polymorphic DNA analysis, *Australian Journal of Agricultural Research* **45**, 1319-1327.

Raymond, J., Mimouni, B. & Azanza, J.-L. (1994) Variability in the 11S globulin fraction of seed storage proteins of *Helianthus* (*Asteraceae*), *Plant Systematics and Evolution* **193**, 69-79.

Schwenke, K.D., Pahtz, W., Linov, K., Raab, B. & Schultz, M. (1979) On seed proteins. Purification, chemical composition and some physico-chemical properties of the 11S globulin (helianthinin) in sunflower seed, *Die Nahrung* **23**, 241-254.

Vonder Haar, R.A., Allen, R.D., Cohen, E.A., Nessler, C.L. and Thomas, T.L. (1988) Organisation of the sunflower 11S storage protein gene family, *Gene* **74**, 433-443.

G.T. Scarascia Mugnozza, E. Porceddu & M.A. Pagnotta (Eds.)
Genetics and Breeding for Crop Quality and Resistance, 339-346, 1999

Sunflower breeding for different oil quality

D. Škoric, S. Jocic & N. Lecic
Institute of Field and Vegetable Crops, 21000 Novi Sad, M. Gorkog 30, Yugoslavia

Abstract: The sunflower is one of the four most important oil crops in the world. The common sunflower oil predominantly contains linoleic acid, followed by oleic, palmitic and stearic ones as well as some others that are found only in traces. As far as tocopherols are concerned, the predominant type are alpha tocopherols. Using sources for a high oleic acid content (Ol genes) and sources with genes tph_1 (50% alpha + 50% beta tocoph.), tph_2 (0-5% alpha + 95-100% gamma), and tph_1tph_2 (8-40% alpha + 0-25% beta + 25-84% gamma + 8-10% delta), the highly *Phompsis*-tolerant parental lines of the hybrids NS-H-45, NS-H-626, and NS-H-680 were translated into three isogenic forms. These have taken the following forms: a) Ol+tph_1; b) Ol+tph_2; and c) Ol+tph_1tph_2. In other words, three types of high-oleic oil with different tocopherol type and content have been obtained. Besides these isogenic lines, a number of oleic and linoleic B-lines with a different tocopherol composition have been developed as well. Also, we have developed a large number of high-oleic, high-oil restorer lines with different tocopherols. Some of these have a high level of tolerance to *Phomopsis*. The newly developed B- and Rf-lines can be used to develop new hybrids with different oil quality. Due to a synergy between Ol and tph genes, the oil oxistability of these new hybrids can be up to 20 times higher than in the common sunflower oil.

1. Introduction

Areas sown with the sunflower as one of the four most important oil crops in the world are constantly on the increase. According to Friedt et al. (1994), the sunflower oil is generally considered a premium edible oil because of its high linoleic and oleic acid contents (C 18:2 and C 18:1, respectively). These two acids comprise over 90% of the fatty acid content of the sunflower oil, which, in addition, contains the following acids: palmitic (C 16:0), stearic (C 18:0), minor amounts of myristic (14:0), myristoleic (C 14:1), palmitoleic (C 16:1), arachidic (C 20:0), behenic (C 22:0), as well as some others in traces. Many authors have found that the cultivated sunflower is characterized by variability in the composition and amount of fatty acids depending on the genotype and environmental factors. Seiler (1992) described the variability of wild *Helianthus* species with regard to oil quality and, especially, fatty acid composition. In Seiler (1998), the conclusion is that selected populations of *H. annuus*, *H. pauciflorus*, and *H. giganteus* have a relatively low (<70 kg^{-1}) content of the two major saturated fatty acids, palmitic and stearic ones. Based on his results, the author concludes that selection for lower levels of palmitic and stearic acids appears to be feasible.

Soldatov (1976) made a significant contribution to a change of the C 18:1 composition and developed a high-oleic sunflower genotype by means of induced mutation. Treatments of the variety VNIIMK8931 with dimethylsulfate (DMS) produced plants with more than 50% of oleic acid and further selections of 80-90% C 18:1 were used to derive the high-oleic variety Pervenets.

Friedt et al. (1994) gave an overview of literature data on altering the levels of palmitic (C 16:0), stearic (C 18:0), palmitoleic (C 16:1), and linoleic (C 18:2) acids by induced mutations (gamma-rays, X-rays, sodium azide, EMS).

The creation of sources with a high oleic acid content by Soldatov (1976) has enabled the development of high-oleic sunflower hybrids in a number of countries around the world. The mode of inheritance of a high oleic acid content and genes that control the expression of this trait has been studied by several authors (Urie, 1985; Fernandez-Martinez et al., 1989; Demurin & Škoric, 1996). The standard sunflower oil contains predominantly alpha tocopherol. A significant contribution to a change in sunflower oil tocopherol was made by Demurin (1993), who discovered genes tph_1 (line L6-15), tph_2 (line L6-17), and tph_1, tph_2 (line L6-24). Lines containing these genes enable us to develop hybrids with altered levels of alpha, beta, gamma, and delta tocopherols in the oil. Demurin et al. (1994), Demurin & Škoric (1995), and Škoric et al. (1996) point out that the existing genetic variability of sunflower offers great possibilities for altering its oil quality in terms of developing genotypes with different oleic acid contents as well as different levels of alpha, beta, gamma, and delta tocopherols.

This study was aimed at developing new B- and Rf-lines by crossing the parental lines of the hybrids NS-H-45, NS-H-626, and NS-H-680 with lines with a high oleic acid content and different alpha, beta, gamma, and delta tocopherol contents without losing resistance to *Phomopsis*.

2. Materials and Methods

Used in the study were the lines of the hybrids NS-H-45 (female: Ha-74, male RHA-SNRF-b), NS-H-626 (female: Ha-981, male: RHA-583), and NS-H-680 (female: CMS 3-8, male: RHA-583) as well as donors of resistance to *Phomopsis*. The above female lines were crossed with B-lines LG-21 (Ol + tph_1), LG-25 (tph_1 tph_2), LG-24 (tph_1 tph_2), and VK-OL-373 (Ol) (all developed at the VNIIMK, Krasnodar, Russia) as donors of Ol genes for different tocopherol content and composition. The restorer lines of NS-H-45, NS-H-626, and NS-H-680 were crossed with lines VK-66-1 (tph_1 tph_2), VK-66-2 (Ol + tph_1), and VK-66-3 (Ol + tph_2) as donors of different oil quality. In the summer of 1993, we made the aforementioned crosses between B-lines resistant to *Phomopsis* and B-lines with a different oil quality. At the same time, crosses were made between restorers resistant to *Phomopsis* and restorers with different oil quality.

The plants used as females were on a daily basis subjected to a manual removal of stamens in the early morning hours. Throughout the growing season, pollen from the donor lines was applied with the brush on a daly basis as well. During the autumn and winter of 1993 and 1994, F_1BC_{pl}, F_2 and F_3 combinations were produced in the greenhouse. In the course of 1994, new generations of breeding materials were produced in the field. After the harvest, the seed of individual plants was analysed for tocopherol composition and fatty acid content.

In the autumn of 1994, a selection of plants with desirable traits was sown in the greenhouse in order to create a new generation. After the harvest, a new selection was made and the plants were sown again in the greenhouse in January, 1995. After the harvest at the end of April, selected plants were sown in the field and there was yet another selection. Subsequent to harvesting, these plants were analysed in the laboratory for tocopherols and fatty acid composition.

During the autumn/winter of 1995/96, summer of 1996, autumn/winter of 1996/97, and autumn/winter of 1997/98, three generations of selection materials were produced per year in the field and greenhouse. During this time, we developed isogenic lines (three types) of the B-lines Ha-74, Ha-981, and CMS-3-8 and translated them into the CMS form. At the same time, a number of new B-lines was selected from the available B-materials and then translated into the CMS form as well.

In the restorers RHA-SNRF-b and RHA-583, three types of isogenic lines with different oil quality were developed by the end of 1996. In parallel with this, restorer lines with different oil quality were being developed from the newly created Rf-materials developed by crossing RHA-SNRF-b and RHA-583 with lines VK-66-1, VK-66-2, and VK-66-3.

Throughout the selection process, tocopherol composition was determined by means of thin-layer chromatography (TLC) followed by the Emmerie-Engel reaction and densitometer quantization (Popov & Aspiotis, 1991) and fatty acid composition by gas chromatography (GC) of methyl esters. Seed oil content was determined using an NMR-analyser, making it possible to directly use the analysed seed to develop new generations during the breeding process. The resistance of the breeding materials to *Phomopsis* was graded using a scale from 0 to 5 (0 -fully resistant, 5 - fully susceptible).

3. Results

As predicted by Škoric et al. (1996), during the selection cycle described herein the parental lines of the hybrids NS-H-45, NS-H-626, and NS-H-680 that are characterized by a high level of resistance to *Phomopsis* were translated into three new (isogenic) forms oil quality-wise. The standard sunflower oil is known to contain predominantly linoleic acid (>60%) and alpha tocopherols (over 95%). In the new isogenic lines, however, the situation is as follows: a) those containing the Ol+tph_1 genes are 85-87% oleic acid (C 18:1) and only about 6% linoleic acid (C 18:2), with 4-6% of palmitic (C 16:D) and 3-4% of stearic acid (C 18:0), such genotypes also contain equal parts of alpha and beta tocopherols (around 50%) instead of predominantly alpha ones (>95%); b) isogenic lines with the Ol+tph_2 genes have approximately the same proportions of the four basic fatty acids as those in a), but their oil has 95% (or sometimes even more) of gamma tocopherols and only 5% of alpha ones; c) the third form, which contains the Ol+tph_1 tph_2 genes, has a higher fatty acid composition similar to the above two, while its tocopherol composition is as follows: 8-40% alpha, 0-25% beta, 25-84% gamma, and 8-10% delta tocopherols. Obviously, there is a great variability of tocopherol composition. So why do the fatty acid content and tocopherol composition change both at the same time? The reason is quite simple - when an oil with a high oleic acid content and beta and gamma instead of alpha tocopherols, a kind of synergy occurs that gives it a high oxistability (Škoric et al., 1996).

Beside these isogenic lines with different oil quality, a number of new B- and Rf-lines with differing agronomic and other traits were developed during the study using F_1BC_1 and F_2 combinations. Particularly interesting are the newly developed B-lines that have a high oleic acid content and equal proportions of alpha and beta tocopherols. Some of them, such as UK-18 and UK-26, have a high seed oil content (over 52%) and are tolerant to *Phomopsis* (Table 1). It can be said with certainty that these new lines can be successfully used for the development of hybrids with different oil quality. Their combining abilities are currently being tested, so the first selection of promising hybrids with different oil quality can be expected as early as 1999.

Table 1. B-lines with increased oleic acid content and different tocopherol levels.

No.	Line	Acid content (%)		Tocopherols (%)		Oil content(%)	*Phomopsis* resistance (0-5 scale)
		oleic	linoleic	α	β		
1.	UK-4	88	3	50	50	43.86	4
2.	UK-8	85	5	100		46.90	3
3.	UK-9	86	3	50	50	48.36	4
4.	UK-18	86	4	40	60	52.72	2
5.	UK-26	88	3	50	50	53.05	2
6.	UK-29	86	3	50	50	41.64	2
7.	UK-56	88	2	50	50	42.80	3
8.	UK-57	89	2	50	50	47.96	3
9.	UK-58	86	4	50	50	43.92	2
10.	UK-87	88	3	100		47.10	2
11.	UK-88	86	4	50	50	/	2
12.	UK-89	87	3	50	50	/	2
13.	UK-90	88	3	50	50	40.90	2
14.	UK-91	88	2	50	50	40.17	2
15	UK-96	86	4	100		49.10	2

The development of high-linoleic B-lines with different types and levels of tocopherols has an important place in developing sunflower hybrids with different oil quality (Table 2). Unfortunately, the oil content of most of these lines is significantly lower than that of high-oleic ones. Nevertheless, some of these lines, most notably UK-20, UK-28, UK-19, UK-43, and some others, can still be used for developing hybrids with different oil quality. Also of interest from the point of view of breeding are lines that contain exclusively gamma tocopherols as well as those whose oil is equal parts gamma and delta tocopherols. Some of these lines also have a very high level of tolerance to *Phomopsis*.

Table 2. Linoleic B-lines with different tocopherol levels.

No.	Line	Tocopherols (%)				Oil content (%)	Phomopsi resistance (0-5 scale)
		α	β	γ	δ		
1.	UK-2	50		50		39.70	4
2.	UK-3	50		50		34.70	3
3.	UK-6	50	50			35.20	4
4.	UK-7	40		60		/	3
5.	UK-9	50	50			44.55	4
6.	UK-10			50	50	34.90	3
7.	UK-11	20	20	40	20	37.10	4
8.	UK-12	25	25	25	25	33.20	4
9.	UK-13	50	50			37.40	4
10.	UK-14			50	50	37.00	4
11.	UK-15	50	50			37.19	3
12.	UK-16			100		/	2
13.	UK-19	50	50			44.77	2
14.	UK-20	50	50			48.64	2
15	UK-22	50	50			42.60	2
16.	UK-23	50	50			43.40	2
17.	UK-27	50	50			43.60	4
18.	UK-28	50	50			45.05	2
19.	UK-30	50	50			46.55	3
20.	UK-31	50	50			44.40	2
21.	UK-33	50	50			43.10	3
22.	UK-35	50	50			39.77	1
23.	UK-36			100		38.90	3
24.	UK-37			50	50	38.50	4
25.	UK-38	20	20	40	20	33.40	3
26.	UK-41			100		39.60	2
27.	UK-42			100		37.19	2
28.	UK-43			100		43.30	2
30	UK-63	50		50		29.63	4
29.	UK-64	50	50			26.85	3
30.	UK-65			50	50	/	4
31.	UK-66	50	50			/	5
34.	UK-68	50	50			/	4
32.	UK-69	50	50			37.34	2
33.	UK-92	50	50			43.47	2
34.	UK-93	50	50			44.00	2
38.	UK-94	30	30	30	10	/	4

Significant results have also been achieved in developing restorer lines with a high seed oil content. The seed oil content of some of them exceeds 55%, and in one it is as high as 63.9%. These restorer lines can be divided into two groups. The first one is characterized by a high seed oil content as well as a high oleic acid content and the presence of the tph_1 gene, meaning that these restorers have equal amounts of alpha and beta tocopherols. Restorers from the second group have high seed oil and oleic acid contents but contain only alpha tocopherols. These newly developed restorer lines enable the development of oleic-type hybrids with a high oil content and either alpha or alpha-beta tocopherols. It should be noted that some among them have a high level of tolerance to *Phomopsis* as well (Table 3).

Table.3. High-oleic RHA-lines with different tocopherol levels.

No.	Lines	Acid content (%)		Tocopherols (%)		Oil content (%)	*Phomopsis* resistance (0-5)
		oleic	linoleic	α	β		
1.	Rus-rf-tph-ol-8	86	4	50	50	51.21	4
2.	Rus-rf-tph-ol-10	86	3	90	10	40.50	2
3.	Rus-rf-tph-ol-14	90	3	50	50	54.37	4
4.	Rus-rf-tph-ol-15	82	10	50	50	54.05	4
5.	Rus-rf-tph-ol-16	87	3	50	50	54.22	4
6.	Rus-rf-tph-ol-23	91	2	100	/	53.87	4
7.	Rus-rf-tph-ol-25	89	2	50	50	55.38	3
8.	Rus-rf-tph-ol-26	90	2	50	50	56.68	3
9.	Rus-rf-tph-ol-35	87	3	50	50	50.05	3
10.	Rus-rf-tph-ol-37	84	6	50	50	52.08	3
11.	Rus-rf-tph-ol-39	89	2	50	50	42.32	4
12.	Rus-rf-tph-ol-43	88	3	50	50	/	5
13.	Rus-rf-tph-ol-51	87	5	50	50	49.59	4
14.	Rus-rf-tph-ol-69	85	7	50	50	39.54	4
15	Rus-rf-tph-ol-81	83	5	50	50	55.92	4
16.	Rus-rf-tph-ol-83	86	6	50	50	49.93	4
17.	Rus-rf-tph-ol-87	81	9	50	50	51.70	4
18.	Rus-rf-tph-ol-93	84	4	50	50	51.40	4
19.	Rus-rf-tph-ol-96	87	4	50	50	33.73	4
20.	Rus-rf-ol-3	87	2	100		36.48	4
21.	Rus-rf-ol-11	86	4	100		30.28	3
22.	Rus-rf-ol-12	84	5	100		35.18	4
23.	Rus-rf-ol-14	85	4	100		38.07	4
24.	Rus-rf-ol-15	85	5	100		/	3
25.	Rus-rf-ol-27	85	4	100		47.68	4
26.	Rus-rf-ol-37	84	7	100		36.04	4
27.	Rus-rf-ol-38	89	4	100		54.70	4
28.	Rus-rf-ol-39	87	5	100		44.87	4
29.	Rus-rf-ol-40	88	4	100		37.70	4
30.	Rus-rf-ol-54	90	2	100		51.71	3
31.	Rus-rf-ol-67	89	2	100		49.41	4
32.	Rus-rf-ol-68	87	4	100		47.88	4
33.	Rus-rf-ol-70	85	5	100		43.71	4
34.	Rus-rf-ol-77	91	1	100		58.39	4
35.	Rus-rf-ol-78	88	2	100		56.00	4
36.	Rus-rf-ol-80	87	4	100		45.98	4
37.	Rus-rf-ol-91	88	3	100		55.26	3
38.	Rus-rf-ol-94	86	3	100		43.55	4
39.	Rus-rf-ol-134	85	6	100		43.46	4
40.	Rus-rf-ol-140	84	4	100		46.43	4
41.	Rus-rf-ol-142	88	2	100		59.13	3
42.	Rus-rf-ol-154	84	4	100		63.90	3
43	Rus-rf-ol-206	88	3	100		42.02	4
44.	Rus-rf-ol-207	90	2	100		53.46	4
45.	Rus-rf-ol-209	90	1	100		52.83	4
46.	Rus-rf-ol-222	90	2	100		52.57	4
47.	Rus-rf-ol-242	91	3	100		52.57	4

In addition to isogenic lines RHA-SNRFb and RHA-583 and the above restorers (Table 3), a lot of attention in the course of this study was payed to the development of high-oleic restorer lines that have the same levels of alpha and beta tocopherols and high tolerance to *Phomopsis* (Table 4). All of them have a high seed oil content as well (above 50%). Particularly interesting are those that have a consistently high oleic acid content (there are several such lines). These restorers should be used for developing high-oleic hybrids with a high level of tolerance to *Phomopsis*.

Table 4. Selected high-oleic restorer lines that are more than 50% oil and highly tolerant to *Phomopsis*.

No.	Line	Oleic acid content (%)					Tocopherols(%)	
		x	Max.	Min.	Std.Dev.	Std.Error	α	β
1.	Rf-kv-1	85.4	87	80	3.050	1.364	50	50
2.	Rf-kv-2	88.8	89	88	0.447	0.200	50	50
3.	Rf-kv-3	88.6	91	88	1.342	0.600	50	50
4.	Rf-kv-4	88.0	90	86	1.414	0.632	50	50
5.	Rf-kv-5	85.6	88	78	4.278	1.913	50	50
6.	Rf-kv-6	87.6	89	85	1.673	0.748	50	50
7.	Rf-kv-7	89.2	91	88	1.304	0.583	50	50
8.	Rf-kv-8	88.2	89	87	0.837	0.374	50	50
9.	Rf-kv-9	81.2	90	49	18.006	8.052	50	50
10.	Rf-kv-10	83.0	88	70	7.550	3.376	50	50
11.	Rf-kv-11	87.8	89	87	0.837	0.374	50	50
12.	Rf-kv-12	87.4	89	86	1.342	0.600	50	50
13.	Rf-kv-13	71.8	89	41	19.588	8.760	50	50
14.	Rf-kv-14	82.4	87	71	6.542	2.926	50	50
15.	Rf-kv-15	85.2	89	73	6.870	3.072	40	60
16.	Rf-kv-16	88.2	89	87	0.837	0.374	40	60

4. Discussion

Thanks to Soldatov (1976) and his variety Pervenets that was developed using induced mutations, high-oleic sunflower hybrids are now being developed in all major breeding centres around the world. In a number of cases, however, the high oleic acid content is not fully stable due to lack of knowledge of all the genes that control the expression of this trait. This is confirmed by the findings of Demurin & Škoric (1996), namely that the original Ol mutation has an ability to express itself in progenies of high-, intermediate-, or low-oleic phenotypic classes. The matter is further complicated in cases when the germ has a high oleic acid content and the cotyledons have a standard fatty acid content. Lines with a high but unstable oleic acid content were found in the present study as well.

The current level of knowledge cannot yet provide us with a definitive answer as to why a strong synergy occurs when a genotype has both the genes for a high oleic acid content and tph_1 and tph_2. According to Demurin et al. (1994) and Škoric et al. (1996), the oxistability of oleic-type oil is three times larger than that of the standard sunflower oil (in the case of Ol+tph_1 and Ol+tph_2 the difference can be up to 20 times). The new genotypes that have both the Ol and tph genes have a highly oxistable oil that can find various new uses in the food and pharmaceutical industries as well as in production of cosmetics and special machine oils.

The results of this study enable the development of hybrids with different oil quality and modified oleic or linoleic acid contents and alpha, beta, gamma, and delta tocopherol levels. The question of which type of oil should be given priority in future breeding work remains to be answered. Also, the mode of inheritance of *Ol* and *tph* genes in the F_1 and F_2 generations has not yet been sufficiently understood, and without this there can be no clear-cut hybrid development. Added to all of this are also problems related to the incorporation of resistance to dominant pathogens into genotypes with different oil quality.

5. Conclusions

Our study has led to the following conclusions:

1) We have developed isogenic lines from the parental lines of the hybrids NS-H-45, NS-H-626, and NS-H-680 with three new types of oil: Ol+$thph_1$, Ol+tph_2, and Ol+tph_1tph_2;
2) We have developed a number of high-oleic B-lines with equal levels of alpha and beta tocopherols and tolerance to *Phomopsis*;
3) We have developed a number of linoleic B-lines have been developed that have different levels of alpha, beta, gamma, and delta tocopherols and tolerance to *Phomopsis*;
4) We have developed a large number of restorer lines have been developed that are characterized by a high seed oil content, a high oleic acid content, and either 100% alpha or equal parts alpha and beta tocopherols (tph_1).
5) Particularly valuable are a group of new restorers that are highly tolerant to *Phomopsis* and have a seed oil content of more than 50% as well as equal proportions of alpha and beta tocopherols (tph_1).
6) The new isogenic lines and the newly developed lines with different oil quality will enable us to develop hybrids with different types of oil quality and high oxistability.

6. References

Demurin, Ya. (1993) Genetic variability of tocopherol composition in sunflower seeds, HELIA 16: 59-62.

Demurin, Ya., Škoric, D., Popov, P., Efimenko, S. & Bochkovoy, A. (1994) Tocopherol genetics in sunflower breeding for oil quality. Proc. of EUCARPIA - Symposium on Breeding of Oil and Protein Crops. Albena, Bulgaria. 22-24.09.1994. pp. 193-197.

Demurin Ya. & Škoric D. (1995) Genetic modification of sunflower seed oil. Proc. of Symposium: Breeding and cultivation of wheat, sunflower and legume crops in the Balkan countries. Albena, Bulgaria, 26-29/6/95. p. 55-9.

Demurin, Ya. & Škoric, D. (1996) Unstable expression of Ol genes for high oleic acid content in sunflower seeds. Proc. of 14th Int. Sunflower Conference I: Beiging/Shenyang, China, 12-20/6/96. pp. 145-151.

Fernandez-Martinez, J., Jimenez, A., Dominguez, J., Garcia, J.M., Garces, R. & Mancha, M. (1989) Genetic analysis of the high oleic acid content in cultivated sunflower-Euphytica 41: 39-51.

Friedt, W.,Ganssmann, M. & Korell, M. (1994) Improvement of sunflower oil quality. Proc. of EUCARPIA - Symposium on Breeding of Oil and Protein Crops. Albena, Bulgaria, 22-24.09.1994. pp. 1-30.

Popov, P.S. & Aspiotis E.H. (1991) Biochemical methods for breeding material estimation (In Russien). Biology, breeding and growing of sunflower Agropromizdat, Moscow. pp. 78-80.

Seiler, G.J. (1992) Utilization of wild sunflower species for the improvement of cultivated sunflower. In: Field Crops Research 30. Elsevier Sci. Publ., Amsterdam. pp. 195-230.

Seiler, G.J. (1998) The potential use of wild Helanthus species for selection of low saturated fatty acids im sunflower oil. Proc. of EUCARPIA - Symp. on Breeding of Protein and Oil Crops. Pontevedra, 1-4/4/98. pp. 109-110.

Soldatov, K. (1976) Chemical mutagenesis for sunflower breeding. 7th Int. Sunfl. Conf., Krasnodar, USSR,pp. 352-7.

Škoric, D., Demurin, Ya. & Jocic, S. (1996) Development of hybrids with various oil qualities. Poc. of 14th International Sunflower Conference I Beiging/Shenyang, China, 12-20 June 1996. pp. 54-60.

Urie, A.L. (1985) Inheritance of high oleic acid in sunflower. Crop.Sci. 24: 1113-1115.

G.T. Scarascia Mugnozza, E. Porceddu & M.A. Pagnotta (Eds.)
Genetics and Breeding for Crop Quality and Resistance, 347-355, 1999

Gluten agglomeration in batters as quality characteristic of wheat varieties for wafer making and starch production*)

D. Meyer, U. Hanneforth. & W. Bergthaller
Federal Centre for Cereal, Potato and Lipid Research, Detmold, P.O. Box 1354, D-32703 Detmold, Germany; E-mail: staerke.bagkf@t-online.de

1. Introduction

In the Federal Republic of Germany as well as in other EU-countries the production of varieties for bread making was the main objective in wheat breeding for many years. In recent years, an increasing demand for flours with special functional processing properties and for different uses is recognised. The cereal processing industry tried to overcome the lack of suitable raw material by applying special procedures like the air classification in mills. Therefore, specifications for wheat for different uses are needed. It is, for instance, necessary to find out the specific requirements for the production of unleavened pastries including flat wafers. Another subject is wheat starch production, as the role of wheat as substrate in relation to maize and potato is becoming more important. A statistic indicates a total of 7.0 million t for the European Union and a 25% share for wheat in 1996. In Germany, the use of wheat achieved a substantial increase from 0.29 million t in 1993 to 0.74 million t in 1996 (Bergthaller et al., 1997; Anonymous, 1998).

The processing procedures for wafer making and starch production are related now, as for both a batter of flour and water is prepared. Whereas for wafer making a low viscosity of the batter is prerequisite, a satisfying gluten agglomeration is required for the separation of starch.

*) Publ. No.: 6969 of the Federal Centre for Cereal, Potato and Lipid Research, Detmold and Münster

2. Material and Methods

2.1 *Material*

In the studies German and EU-varieties of different baking quality, dough properties and grain structure from different trial sites were tested. The influence of the protein content was evaluated with samples from trials with variable treatment levels for nitrogen fertilisation. For starch processing experiments in the pilot plant, pure commercial variety samples were used.

2.2 *Methods*

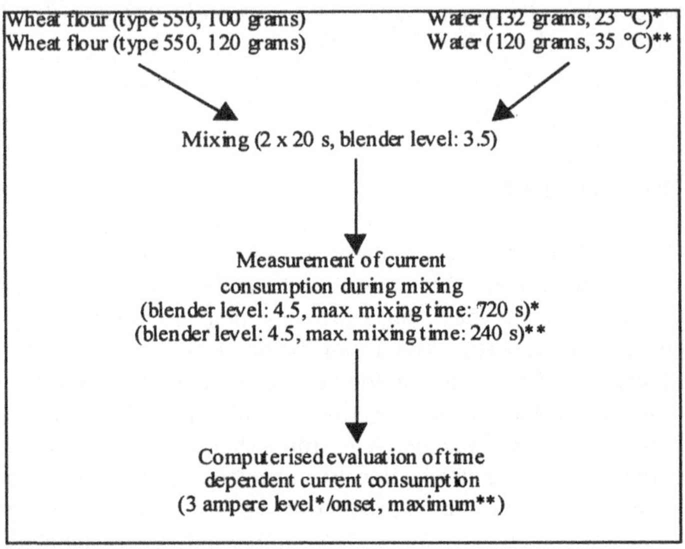

Figure 1. Schedule of the gluten agglomeration test (*test for wafer making, **test for starch production).

For measuring the viscosity of the batters, a gluten agglomeration test was introduced. A concentrated flour water suspension of 1:1.2 is prepared in a standard mixer blender, using conditions of batter formation according to wafer making processes (Figure 1). After formation of a homogenous suspension (20 s mixing at low speed) the energy consumption during high speed mixing is registered by an ammeter and evaluated by means of specific software. Data concerning time and/or current consumption are derived from diagrams (Figure 2). The time used until reaching 3 amperes is measured as agglomeration time (Hanneforth et al., 1997; Bergthaller et al., 1998) For the assessment of wheat for starch production, the method was modified according to industrial processing conditions. The flour/water ratio was increased to 1:1 and the temperature was raised to 35°C. For comparison wafer baking tests were made with a waffle-iron (Franz Haas Waffelmaschinen

Industrieges.mbH, Leobendorf, Austria). The wet separation of fractions of starch, gluten, fibres and solubles was investigated by applying the mixer method (Bergthaller et al., 1997, 1998).

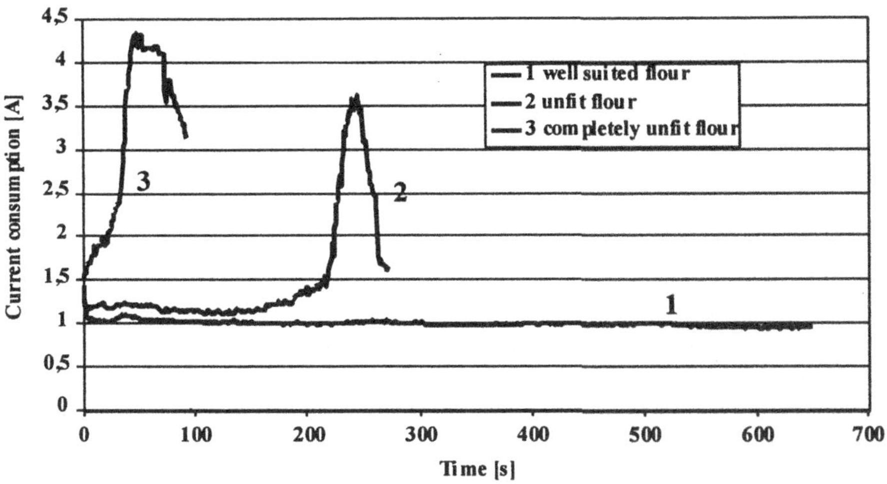

Figure 2. Diagrams for wheat flour with extreme differences in gluten agglomeration time applied to the wafer making test.

3. Wafer making

3.1 Quality characteristics and flour requirements

For wafers the main quality characteristics are a stable wafer sheet, no splintery break, a uniform structure and tender chewing properties (Table 1).

Table 1. Quality characteristics of wafers

Weight	constant
Surface	smooth
Structure	delicate, uniform
Break	not brittle
Chewing properties	tender
Sheet stability	stable

In order to meet these requirements, flour of poor breadmaking quality is generally recommended. Specifications of wheat for wafer making in industrial processes are shown in Table 2 (Seibel et al., 1978; Ludewig et al., 1980, 1991; Hanneforth et al., 1997). With respect to a uniform structure, flours with low protein and gluten contents are preferred. The structure of the gluten should be normal or short and not extensible. A poor gluten agglomeration is considered to produce wafers with tender chewing properties. Furthermore, wheat with soft grain structure and low water absorption, respectively is more suitable than hard structured wheat.

Table 2. Recommended flour characteristics for flat wafers.

Characteristic	Recommended value
Protein content (% d.b.)	<12.0
Wet gluten content (%)	<25
Wet gluten properties	not extensible, normal, little short or tough
Falling number (s)	>200
Water absorption capacity (%)	low
Gluten agglomeration time (s)	>720

For processability, a low viscosity of a batter, that does not plug the pipes and nozzles, is of great importance. In industrial procedures a batter with a low flour/water ratio is prepared with respect to a lower energy consumption. A more extended flour/water ratio decreases the viscosity of the batter, but on the other hand energy would be wasted for evaporation of additional water. Other procedures used in wafer making processes are the addition of starch or proteolytic enzymes. These measures are causing costs and, in addition, food produced with the aid of enzymes is not accepted by most of the consumers. A low viscosity of the batter has additional positive effects on the quality of the products like break stability, structure and chewing properties. As the viscosity of a batter at room temperature is mainly dependent on the gluten agglomeration besides damaged starch and pentosans, different wheat varieties have been tested for this property.

3.2 Results

The agglomeration time of the tested varieties could be found in the range from 100 to more than 720 s (Table 3). "Crousty", "Soissons" and "Contra" had agglomeration times of more than 720 s. The results from samples produced at different trials sites indicated, that agglomeration is effected to some degree by environmental conditions, but to a higher degree it is a property specific to the variety. Besides, the agglomeration was not correlated to general bread making quality. There were varieties of poor baking quality, like "Crousty" and "Contra", as well as varieties with a good baking performance, like "Soissons" and "Astron", which showed high agglomeration time values. On the other hand, the gluten of varieties of the C class, like "Caprimus", agglomerated rapidly. The agglomeration time did not correlate to most of the characteristics important for bread making. There was, to some degree, a relationship to the elasticity of the dough and the extensibility in the extensogram. In wafer baking tests, the samples with retarded agglomeration allowed an even distribution of the batters on the baking plate, a fundamental requirement for products of good quality.

Table 3. Gluten agglomeration time of wheat varieties with different baking quality.

Wheat variety	Quality class for bread making*	Agglomeration time (s)
Crousty	C	>1000
Soissons	A	900
Contra	C	800
Astron	A	800
Tambor	A	600
Batis	A	250
Ritmo	B	200
Caprinus	E	150
Bussard	C	100

*E=elite quality wheat, A=quality wheat, B=bread making wheat, C= wheat not suitable for bread making

Results of the samples from trials with different fertiliser levels indicated a relationship between agglomeration and the protein content. With rising protein content, the gluten agglomeration time was decreasing, in general. Some varieties like Contra showed even at protein contents of 13% a retarded agglomeration (Table 4).

Table 4. Agglomeration time of flours from wheat varieties - samples from fertilisation trials.

Contra		Astron		Batis	
Protein-content (% d.b.)	Agglom.-time (s)	Protein-content (% d.b.)	Agglom.-time (s)	Protein-content (% d.b.)	Agglom.-time (s)
8.7	-	9.3	-	9.2	-
11.3	-	12.8	-	12.2	180
11.3	-	12.6	-	12.4	225
12.1	-	13.6	-	12.5	180
12.6	820	15.0	380	13.0	150
13.1	760	15.4	350	13.3	230
13.8	600	16.2	320	13.7	150

4. Starch production

4.1 Quality characteristics

Quality characteristics of wheat and wheat flour, suitable for starch production, are not so precisely defined as for bread making. Specifications have been described by Witt (1990) and in more recent time by Zwingelberg & Lindhauer (1996, 1997). Protein levels of wheat for starch production should not pass 12 to 12.5% d.b. (nitrogen conversion factor 5.7) giving a flour protein content of approx. 11 to 11.5% d.b. (Table 5). Since starch and protein content are inversely correlated, a reduced grain protein content produces an increase in starch content, promising higher starch yields. Further wheat characteristics are low endosperm hardness and small shares of mechanically damaged starch granules. With respect to milling, wheat suitable to starch production, should have a soft to medium endosperm structure, low protein and pentosan concentrations and good starch quality, represented by high falling number and amylograph data. Furthermore, they should provide well milling ability, which means production of a high flour yield based on a mineral content 0.6% d.b. (in Germany: type 550).

Table 5. Specifications concerning wheat flour for starch production.

Characteristics	Limits
Max. moisture content	14.5%
Max. mineral content (d.b.)	0.62% (0.80%)
Min. raw protein content (d.b., Nitrogen conversion factor 6.25)	12.0 to 12.5%
Min. falling number	280 s
Min. amylograph consistency	500 AU
Portion of damaged starch	low

4.2 *Starch extraction procedures*

Although numerous attempts have been made to use whole grains as substrate for starch production, wheat flour remained the unique material. In particular, the economic interest in vital wheat gluten as an important side-stream product was at most responsible for retaining in wheat flour as starting material (Tegge, 1988)

Washing out starch from a concentrated flour-water system (1 part flour to 0.6 parts water), similar in consistency to bakers bread dough, dominated industrial processing for a rather long time. This procedure of wet extraction became popular as Martin process. After washing out starch, fibres and pentosans a gluten fraction remains, that is purified by a further washing treatment and finally carefully dried in order to receive vital gluten. The suspension separated by the washing procedure is then split off in a highly purified A starch fraction (= regular wheat starch), a protein-rich (up to 5%) B starch fraction and feed, consisting of fibres and insoluble pentosans (Tegge, 1988). A serious disadvantage of the Martin process is the high amount of water consumed in the extraction step, which is said to be approx. the 15 fold of the flour weight (Zwitserloot, 1989).

Modern wheat starch production utilises centrifugal separation as the more promising principle in separating starch and gluten. By the introduction of this technique the water consumption could be reduced to the 5-7 fold of the flour weight in case of the batter process and with further process improvements to less than the 5 fold. The new hydrocyclone based process allows for example a water flour relation of 4-5:1 and the decanter based process a relation of 4:1 (Zwitserloot, 1989). These developments were decisive milestones in modernising the wheat wet milling process irrespective of the applied centrifugal technique (Tegge, 1988; Zwitserloot, 1989; Meuser, 1989, 1997; Ludewig, 1991; Lindhauer, 1997). The density difference between A starch, gluten, B starch, fibres and pentosans allows an arrangement in separate layers in the gravity field (Zwitserloot, 1989; Meuser, 1994). In using hydrocyclones together with screens or decanters diluted flour/water mixtures (1:1.8) are separated after conditioning the concentrates (1:0.9). After refining a highly purified A starch as well as vital gluten are recovered. The fraction consisting of B starch, insoluble pentosans, and fibres is in general split off into B starch and feed.

The increasing importance of wheat as substrate in wet milling (Bergthaller, 1997; Anonymous, 1998), ongoing structural changes in the relevant industrial branch, and the replacement of the traditional Martin process (Tegge, 1988; Zwitserloot, 1989; Meuser, 1994, 1997) were reason for paying attention to grain and flour characteristics potentially connected with the wet separation of flour components (Witt, 1990; Bergthaller, 1997; Lindhauer, 1997). Further impulses were introduced by technologies, using centrifugal forces as an important separation principle. Requirements concerning the evaluation of the potential of wheat gluten to agglomerate induced investigations in a rapid method for measuring agglomeration (Bergthaller, 1998). This potential is compared with results of yield evaluation, for which satisfying gluten agglomeration is prerequisite.

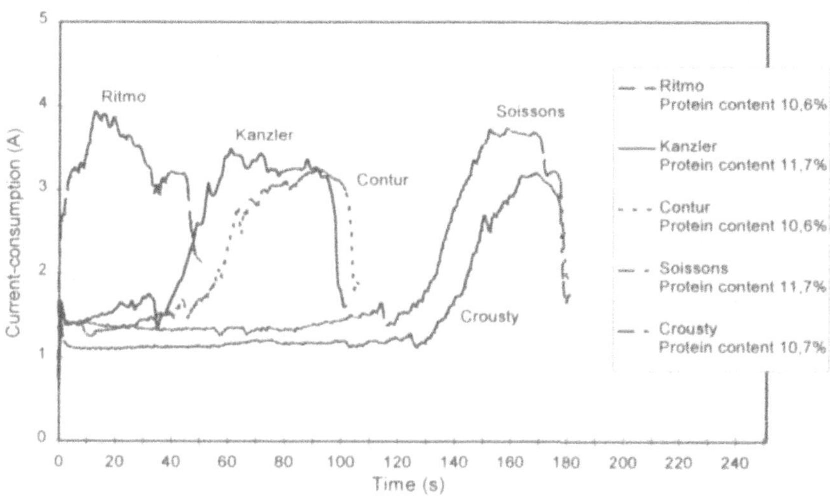

Figure 3. Gluten agglomeration test applied to flour samples of the varieties "Ritmo", "Kanzler", "Contur", "Soissons" and "Crousty" (flour/water ratio 0.9:1, water temperature 35°C).

4.3 Results

Results of some of the cultivars tested are given as representative example in Figure 3. Three groups of flour samples could be differentiated. In case of cv. "Ritmo" agglomeration occurred nearly instantaneously in batter formation and needed high current consumption. For the samples of cv. "Kanzler" and "Contur" agglomeration needed more time - respective agglomeration time values varied among 35 and 45 s. Flours of the third group (cv. "Soissons" and "Crousty") needed much longer (120-130 s) for agglomeration (Table 6).

Table 6. Selected results of the Mixer test of 6 flours (type 550) from soft winter wheat ordered according to increasing agglomeration time.

Characteristic	Ritmo	Kanzler	Contur	Soissons	Crousty
Agglomeration time (s)	0	35		120	130
Flour starch content (% d.b.)	81.7	80.6	82.4	80.9	81.6
Total starch yield (% d.b.)	80.8	81.3	83.0	81.0	82.4
A Starch (% d.b.)	66.6	73.8	75.0	69.7	70.8
B Starch (% d.b.)	14.2	7.5	8.0	11.3	11.6
Fibres (% d.b.)	1.1	1.1	1.2	1.8	1.6
Solubles (%)	6.3	4.0	5.9	6.7	5.5
Flour protein content (%)	10.6	11.6	10.5	11.6	10.8
Moist gluten (g)	25.9	29.5	25.8	29.4	26.3
Dry gluten yield (% d.b.)	9.7	11.0	10.0	10.9	10.1

The values for the starch content followed according to expectations inversely corresponding protein contents. With one exception (cv. "Ritmo") total starch yield followed flour starch content, but yield values were 0.1 to 1.6% higher as a result of impurities recovered with B starch. For cv. "Ritmo", however, a reduction (0.9%) was characteristic induced by a serious A starch loss, that could be found in B starch. These results in combination with the absence of agglomeration time need further investigations. Beside this observation, the discrimination between the analysed samples given by significant differences in agglomeration time (35 to 45 s versus 120 to 130 s) could be found by comparing yield values of B starch and fibres. Both samples needing more than two times the mixing time than cv. "Kanzler" and "Contur", showed higher yield levels for B starch and fibres. In contrast, A starch yield was reduced. With respect to industrial processing, the achieved results indicated a less promising processing behaviour of cv. "Soissons", but also cv. "Crousty" proved to give an undesired low A starch yield, when agglomeration took place slowly. For the yield of solubles found as dry matter of the process water the respective range was 4.0 to 6.7%. Looking on results of the mixer test for the same set of flours moist gluten weight as well as dry gluten yield followed well their protein content and met expectations.

5. Conclusions

The gluten agglomeration test proved as a useful method for the evaluation of wheat varieties for wafer making. A low viscosity was not only responsible for the improvement of the processability. It has had positive effects on quality characteristics of the wafers, too. The studies indicated that the gluten agglomeration was specific for a variety, independent on the quality for bread making. A relationship between agglomeration time and protein content could be observed. Suitable varieties, however, showed even at higher protein contents of more than 12.0% a retarded agglomeration.

For starch production, only sufficient gluten agglomeration, that could be evaluated on basis of the described agglomeration test, produced adequate conditions for separation of components of acceptable quality. This was in good accordance with expectations and requirements described in relevant literature (Witt, 1990; Lindhauer, 1997). In order to explain the observed differences in extraction rates, especially the loss of A starch of wheat with very fast agglomeration, further investigations have to be performed.

6. References

Anonymous (1998) LMC Analysis Starch & Fermentation, April issue, LMC International Ltd, Oxford.

Anonymous (1998) Zahlen und Fakten zur Särke-Industrie. Issue 1997. Fachverband der Stärke-Industrie e.V., Bonn.

Bergthaller, W., Lindhauer, M.G. & Zwingelberg, H. (1997) Variety testing of wheat for starch production using a small scale process, in Praznik, W., and Huber, A. (eds.) *Carbohydrates as Organic Raw Materials IV*, WUV-Universitätsverlag, Wien, pp. 163-175.

Bergthaller, W., Lindhauer, M.G., Zwingelberg, H. & Ehring, M. (1998) Laboratory scale evaluation of starch extractability of wheat varieties. *Unpublished.*

Hanneforth, U., Zwingelberg, H. & Gebhard, M. (1997) Mehle für besondere Verwendungszwecke. 3. Mitt. Untersuchungen zur Charakterisierung von Weizenmehlen zur Herstellung von Flachwaffeln (Flours for special use. 3rd part: Investigations to characterise wheat flours for the production of wafers), *Getreide Mehl und Brot* **51**, 227-231.

Lindhauer, M.G. & Zwingelberg, H. (1997) Weizen für besondere Verwendungszwecke - 2. Mitt.: Stärkegewinnung (Wheat flour for special processing. - 2. part: starch production), *Getreide Mehl und Brot* **51**, 67-70.

Ludewig, H.-G. (1980) Einfluß der Massenbeschaffenheit und der Lockerung auf die Blattqualität von Flachwaffeln, *Zucker- und Süßwarenwirtschaft* **33**, 374-377.

Ludewig, H.-G. (1991) Rezepturbestandteile und deren funktionelle Eigenschaften, in Seibel, W. (ed.), *Feine Backwaren,* Paul Parey, Berlin and Hamburg, pp. 47-96.

Meuser, F. (1994) Wheat utilization for the production of starch, gluten and extruded products, in Bushuk, W., and Rasper, V.F. (eds) *Wheat. Production, Properties and Quality,* Blackie Academic & Professional, London, Glasgow, Weinheim, New York, Tokyo, Melbourne, Madras, pp. 179-204.

Meuser, F. (1997) New developments in the production of wheat starch and gluten, in van Doren, H., and van Swaaij, N. (eds.) *Starch 96. The Book. Perspectives for a Versatile Raw Material on the Threshold of a New Millennium*, The Carbohydrate Research Foundation & Zestec bv, the Hague, pp.65-80.

Meuser, F., Althoff, F. & Huster, H. (1989) Developments in the extraction of starch and gluten from wheat flour and wheat kernels, in Pomeranz, Y. (ed), *Wheat is Unique*, American Association of Cereal Chemists, St. Paul, Minn., pp.479-499.

Seibel, W., Menger, Anita, Ludewig, H.-G., Seiler, K. & Bretschneider, F. (1978) Standardisierung eines Backversuches für Flachwaffeln, *Getreide Mehl und Brot* **32**, 188-193.

Tegge, G. (1988) *Stärke und Stärkederivate*, Behr's Verlag, Hamburg.

Witt, W. (1990) Anforderungsprofil an den Rohstoff Weizen, in Anonymous *Stärke im Nichtnahrungsbereich*, Landwirtschaftsverlag GmbH, Münster-Hiltrup, pp. 125- 141.

Zwingelberg, H. & Lindhauer, M.G. (1996) Weizensorten für die Stärkeherstellung (Wheat varieties for starch production), *Getreide Mehl und Brot* **50**, 303-306.

Zwitserloot, W.R.M. (1989) Production of wheat starch and gluten: Historical review and development into a new approach., in Pomeranz, Y. (ed.), *Wheat is Unique,* American Association of Cereal Chemists, St. Paul, Minn., pp. 509-519.

G.T. Scarascia Mugnozza, E. Porceddu & M.A. Pagnotta (Eds.)
Genetics and Breeding for Crop Quality and Resistance, 357-360, 1999

QTL for quality parameters for bread-making in a segregating wheat by spelt population

S. Zanetti[1], M. Keller[1], M. Winzeler[1], W. Saurer[1], B. Keller[2] & M. Messmer[1]
[1]Swiss Federal Research Station for Agroecology and Agriculture, FAL-Reckenholz, Reckenholzstr. 191, CH-8046 Zurich, Switzerland; [2]Institute for Plant Biology, University of Zurich, Zollikerstr. 107, CH-8008 Zurich, Switzerland.

Abstract: Breeding for quantitatively inherited traits, e.g. quality parameters, is difficult on a phenotypic basis. Therefore, markers linked with QTL (quantitative trait loci) represent a promising tool to make the breeding process more efficient. We aim at estimating the number of genes involved in the phenotypic expression of the quality traits, localising these genes on a genetic map and studying their effects. This investigation will contribute towards a better understanding of spelt and wheat quality and will provide tools for a more efficient selection for quality within spelt and wheat breeding programs. As mapping population we used 226 F_5 Recombinant Inbred Lines (RILs) of a *Triticum aestivum* cv Forno x *T. spelta* cv Oberkulmer cross. For the characterisation of the phenotype the population and the parental lines were grown at three different environments (Fal96, Es96, Ros97). The harvested material was analysed for protein content and Zeleny sedimentation which correlates with the quantity and quality of the gluten. Grit was used for the determination of falling number reflecting the resistance to pre-harvest sprouting. Averaged over environments six QTL were found for protein content explaining 49% of the phenotypic variation, while one QTL (partR^2=25%) located on 5A coincided with a QTL for Zeleny (part R^2=26%). At both QTL the positive allele was from the spelt parent. Over all three locations a total of 14 QTL were detected for Zeleny explaining 45% of the phenotypic variation. Out of the 14 QTL for Zeleny seven were in common with QTL for specific Zeleny (ratio between Zeleny and protein content) and may, therefore, mainly influence the gluten quality. QTL analysis for falling number was performed over single environments due to the weak correlations between environments (r= 0.65). We detected nine QTL for pre-harvest sprouting resistance in Fal96 and six QTL in the environment Ros97 where high falling numbers were observed. In both environments all QTL together explained over 60% of the phenotypic variation. The most predominant QTL for this trait (part R^2=45%) was situated on the long arm of chromosome 5A at the so called Q-locus. The high number of QTL for these three traits demonstrate the complex inheritance of baking quality and that several genes are responsible for the different baking quality of wheat and spelt. Therefore, we hope that bread-making quality can be improved in both crops by combining the positives alleles and that selection becomes more efficient by the use of markers linked with the QTL detected in this study.

1. Introduction

In the frame of the European project 'SESA' (spelt, a recovered crop for the future of sustainable agriculture in Europe) we want to elucidate the inheritance of spelt- and wheat-specific bread-making quality traits. Spelt (*Triticum spelta* L.), an interesting niche culture for marginal regions, has quite different bread-making quality in comparison to wheat (*Triticum aestivum* L.). Spelt has a higher protein content than wheat varieties with a good bread-making quality. Despite this high protein content, spelt flour has a low Zeleny sedimentation value (between 25 and 30), with a weak dough strength and a low loaf volume. Because of these differences we used a wheat/spelt cross to investigate the inheritance of these traits and to determine the genetic differences between wheat and spelt in regard to baking quality. Since the degree of polymorphism on the molecular level is larger between wheat and spelt than within wheat (Siedler et al., 1994), wheat/spelt crosses are suitable to develop molecular markers for quality traits. Such markers could be used for marker assisted selection which would make breeding for quality more efficient.

2. Materials and Methods

The investigated plant population consists of 226 F_5-Recombinant Inbred Lines (RILs) originating from a winter wheat/winter spelt cross (Forno x Oberkulmer) carried out in 1988. The parental varieties differ strongly in their bread-making quality as well as in resistance to pathogens and to lodging. The progenies of the cross were propagated as bulks up to the F_5-generation in which 226 RILs were randomly selected. The population was grown at two different locations in 1996 and at one location in 1997. The parental lines (each with three entries), eleven spelt and seven wheat varieties were integrated in the field trial, which was designed as α-lattice with 25 blocks consisting of 10 genotypes per incomplete block and two replications. Flour samples were analysed for protein content (Near Infrared Reflectance) and for Zeleny sedimentation (determined according to ICC-standards). The falling number (determined according to ICC-standards) was assessed with grit. The lattice analysis of single environments and the ANOVA over environments were performed with the program PLABSTAT (Utz, 1992).

For the construction of a genetic map, the DNA of the parental lines was digested by seven different restriction enzymes (*Eco*RI, *Hind*III, *Eco*RV, *Xba*I, *Dra*I, *Bam*HI, *Bgl*II) and tested for polymorphism with 310 DNA probes. Hundred seventy six markers showing polymorphism between parents were used for hybridisation of the 226 RILs. In addition, ten wheat microsatellites were used to complete the regions of chromosomes with an unsatisfactory RFLP-marker density. With the available marker data, the linkage analysis was performed with the program MAPMAKER (Lander et al., 1987) using the Haldane mapping function. Linkage groups were assigned to chromosomes according to published wheat maps (McGuire & Qualset, 1997) and nulli-tetrasomic data. The genetic map contains 231 loci with 23 linkage groups (2478 cM). QTL-analysis were performed by the software-package PLABQTL (Utz & Melchinger, 1995) based on the composite interval mapping (CIM) with 204 genotypes (excluding genotypes with more than 10% of the markers being heterozygous). Cofactors were assessed by the procedure cov SELECT. The threshold for the detection of the QTLs was fixed at a LOD value of 3.0.

3. Results and Discussion

Of the 183 probes tested 64% were polymorphic between "Forno" and "Oberkulmer" with at least one restriction enzyme. Of the 14 used wheat microsatellites ten were polymorphic between the parental varieties. 176 RFLP-probes and 10 microsatellites were tested on the 226 RILs. This resulted in 243 segregating loci (1.3 polymorphic loci per marker). The genetic map contains 231 loci (12 loci were not linked) assigned to 23 linkage groups. The spanning distance of the genetic map is 2,478 cM. This ensures a coverage of approximately two third of the genome. A detailed description of the map is given by Messmer et al. (submitted). One of the microsatellites analysed is located within the coding region of a low molecular weight glutenin on Chr. 1AS and one in the region of a γ-gliadin pseudogene on Chr. 1BS (Devos et al., 1995). Therefore, they are very interesting for the localisation of genes influencing baking quality. After the removal of the closely linked marked loci (<1 cM) the average marker density reached 13.6 cM for the QTL mapping.

For minimising sampling effects and environmental influences the investigated population was tested at three environments. Correlation between environments varied between 0.77 and 0.86 for protein content and Zeleny, which demonstrates the reliability of the phenotype data. Within the population protein content ranged from 9% to 20% and Zeleny from 16ml to 71ml. The two traits showed transgression and were weakly correlated to each other ($r = 0.37$; $P<0.01$). In the overall analysis we found six QTL for protein content explaining 49% of the phenotypic variation. The most predominant QTL ($partR^2$=25%) for protein content is located on 5A. At the same location a QTL for Zeleny was detected with a $partR^2$ of 26%. At this locus the spelt allele enhances Zeleny value and protein content (additive effect = 3.7ml and 0.7%, respectively). In the overall analysis, a total of 14 QTL were found for Zeleny value (R^2=45%). One QTL was found at the telomeric region of Chr. 1A at the location of the microsatellite marker for the coding region of low molecular weight glutenin. Since Zeleny reflects the quality and quantity of the gluten, QTL analysis was also performed for specific Zeleny (ratio between Zeleny and protein content) which is assumed to predict the gluten quality. We detected 9 QTL for specific Zeleny out of which seven were in common with QTL for Zeleny. These locations may, therefore, mainly influence the gluten quality.

The QTL analysis for falling number was performed over single environments. The humid conditions before harvest in 1996 permitted a good differentiation in the sprouting resistance within the segregating plant population (population mean=185s). Under the optimal weather conditions of 1997 (Ros97) the mean of the population was at 365s. In the environment Fal96 we detected nine QTL for pre-harvest sprouting resistance. In Ros97 six QTL were found, of which four were in common with QTL in Fal96. For both locations, all QTL together explained more than 60% of the phenotypic variation. On the long arm of chromosome 5A (Xpsr918b) one QTL with a major effect was found with the allele from the spelt parent. This locus seems to be identical to the, so called, Q-locus responsible for the long, lax and brittle ear of spelt. In the same chromosome segment also a gene for α-amylase activity (α-amy-3) was located (McGuire & Qualset, 1997). The falling number data were neither related to morphological traits relevant for lodging resistance (e.g. plant height, score of lodging) nor to earliness (e.g. heading, flowering). The presented results illustrate the complex inheritance of quality traits. Markers linked to the QTL detected in this study can be used in the selection process to improve the baking quality within the wheat and spelt genpool and to introgress desirable quality traits from wheat to spelt and *vice versa.*

Acknowledgements: In the frame of an European project (SESA; FAIR 3-CT 96-1569) this investigation is supported by an grant from the Swiss Federal Office for Education and Sciences (BBW, No. 96.0174-2). The lab work was supported by a grant of the Priority Program Biotechnology (Module 6) of the Swiss National Science Foundation (SPP, No. 5002-38967).

4. References

Devos, K.M., Bryan, G.J., Collins, A.J., Stephenson, P. & Gale, M.D. (1995) Application of two microsatellite sequences in wheat storage proteins as molecular markers, *Theor Appl Genet* **90**, 247-252.

Lander, E.S., Green, P., Abrahamson, J., Barlow, A., Daly, M.J., Lincoln, S.E. & Newburg, L. (1987) MAPMAKER: an interactive computer package for constructing primary genetic linkage maps of experimental and natural populations, *Genomics* **1**, 174-181.

McGuire, P.E. & Qualset, C.O. (1997) Progress in genome mapping of wheat and related species. Joint proceedings 5th and 6th public workshops of the International Triticeae Mapping Initiative. Report n° 18. University of California Genetic Resources Conservation Program, Davis, CA, USA.

Messmer, M.M., Keller, M., Zanetti, S. & Keller, B. (submitted) Genetic linkage map of a wheat x spelt cross.

Siedler, H., Messmer, M.M., Schachermayr, G.M., Winzeler, H., Winzeler, M. & Keller, B. (1994) Genetic diversity in European wheat and spelt breeding material based on RFLP data, *Theor Appl Genet* **88**, 994-1003.

Utz, H.F. (1992) PLABSTAT- Ein Computerprogramm zur statistischen Analyse von pflanzenzüchterischen Experimenten. Version 2G vom 30.6.1992. Institute of Plant Breeding, Seed Science and Population Genetics, University of Hohenheim, Stuttgart, Germany.

Utz, H.F. & Melchinger, A.E. (1995) PLABQTL. A computer program to map QTL. Institute of Plant Breeding, Seed Science and Population Genetics, University of Hohenheim, Stuttgart, Germany.

G.T. Scarascia Mugnozza, E. Porceddu & M.A. Pagnotta (Eds.)
Genetics and Breeding for Crop Quality and Resistance, 361-366, 1999

Detection of QTL for bread-making quality in wheat using molecular markers

P. Sourdille[1], M.R. Perretant[1], G. Charmet[1], T. Cadalen[2], M.H. Tixier[1], P. Joudrier[3], M.F. Gautier[3], G. Branlard[1], S. Bernard[1], C. Boeuf[1] & M. Bernard[1]
[1]INRA, Station d'Amélioration des Plantes, 234, Av. du Brézet, 63039 Clermont-Ferrand Cedex 2, France; [2]West Africa Rice Development Association, 01 BP 2551, 01 Bouake, Côte d'Ivoire; [3]INRA, Unité de Biochimie et Biologie Moléculaire des Céréales, 2, Place Viala, 34060 Montpellier Cedex 01, France

Abstract: Bread-making quality in wheat (*Triticum aestivum*) is a complex trait influenced at once by genetic factors and environmental conditions. The development of molecular biology technics together with intervarietal genetic maps provide a useful tool for the understanding of complex agronomic traits such as quality. The aim of our study was to identify QTL for bread-making quality using a doubled-haploid lines population issued from a cross between Courtot and Chinese Spring. Three hundred and eighty loci were mapped using a subset of 106 lines and 187 lines were genotyped for the anchor markers of this map. Between 144 and 172 DH lines were grown under field conditions for three consecutive years (1994 and 1995: one replication; 1996 two replications) and between 106 and 163 were submitted to several technological tests. We were thus able to approach the genetic control of quality traits (strength (W), grain hardness and protein content) that were highly contrasted between the two parents. Despite the fact that the D genome was under represented because of a lack of polymorphism, it appeared that an important part of the variation was explained by the QTL located at different loci than those of the storage proteins (glutenins and gliadins), which seemed to play a minor role compared to other regions of the genome. We are now planning to study some of these regions to identify the genes that could be involved in the baking process.

1. Introduction

Wheat quality could be defined as the ability of a wheat cultivar to meet the industrial requirements and to satisfy the consumers demand. This complex trait depends on multiple factors including (i) the plant genotype which is largely submitted to environmental conditions (nutrients, water, diseases, temperature), (ii) the diversity and the quantity of the grain components like storage proteins, mainly glutenins and gliadins, lipids, carbohydrates, or enzymes, but also (iii) the multiple interactions that take place through the technological processes in order to obtain the final product (bread, biscuit, cookies). The quality evaluation remains a problem especially during breeding, a process where only a small quantity of grain is available. In order to assess the quality, many indirect tests have been elaborated like alveograph, SDS tests, Pelshenke. Nevertheless, most of these tests are strongly correlated to each other.

Numerous studies have also shown the influence of storage proteins on the dough quality (Payne et al., 1981, 1987; Branlard & Dardevet, 1985). But despite the high level of polymorphism revealed by these proteins, they do not represent the globality of the genome and thus, they account for only a part of the total quality variation observed among all the genotypes. One way to overcome this problem would be to use a saturated genetic linkage map of molecular markers (including glutenins and gliadins).

The construction of such a map comes up against the difficulty due to the size (16,000 Mb) and the complexity of the wheat genome (relationships existing between the three homoeologous chromosomes of each of the seven groups) but also due to the poor level of molecular polymorphism often observed in wheat (Chao et al., 1989; Cadalen et al., 1997). Thus, the choice of the parents for the production of the segregating populations depends on the aim to reach. Molecular mapping and elaboration of reference linkage maps requires to work with progenies issued from an interspecific context in order to maximize the polymorphism while populations used for QTL detection need to be constructed using parents with an agronomical interest i.e. to work in an intervarietal context. In the work presented here, we have chosen an intermediate situation that means that we have done an intervarietal cross by using two wheat cultivars, Courtot and Chinese Spring that were sufficiently geneticaly distant from each other to have enough molecular polymorphism.

2. Material and Methods

Courtot (Ct) is a semi-dwarf variety with a good productivity and a good bread-making quality while Chinese Spring (CS) is of bad productivity and bread-making quality but represents the international reference in wheat genetics. Nevertheless, these two cultivars presented a good level of polymorphism (60%; Cadalen et al., 1997). In addition the Ct and the CS monosomic lines were available for each of the 21 chromosomes. A lot of differences for several agronomic traits exist between Ct and CS: plant height (Sourdille et al., 1998), wheat/rye crossability (Tixier et al., 1998), spike morphology, etc, and bread-making quality.

The investigated plant population consisted in 217 doubled-haploid (DH) lines and was produced by anther culture from Ct x CS F_1-hybrids. One hundred and six lines were genotyped for all the markers and 187 DH lines were genotyped only for anchor loci. The segregating population was grown at Clermont-Ferrand for 3 years (1994, 1995 (one replication) and 1996 (two replications)) together with the parents. Three rows (1.5 m) of 144 to 172 entries were sown in the nursery. Plants were grown under normal field conditions with fungicide application to control rusts and powdery mildew. According to the harvesting year, between 106 and 163 DH lines were tested for their technological properties of the dough

The kernels from each DH line were milled and 50g of flour of each sample were used for the micro-alveograph test. This test allows measurement of the rheological caracteristics of the dough: tenacity as maximum overpressure (P), extensibility as average absissa at rupture of the bubble (L), curve configuration ratio (P/L), and strength (W) as required energy for dough deformation. Proteins content (Prot) and hardness (H) were evaluated by near infrared reflectance spectroscopy (NIR). For protein content, the NIR instrument was calibrated with a set of 60 cultivars and with two sets of hard and soft samples for kernel hardness.

For construction of the genetic map, 380 RFLP, microsatellites and AFLP markers were used. The map covered approximately 2,900 cM and most of the 21 chromosomes were well saturated excepted chromosomes from the D genome and especially chromosomes 1D and 4D. Linkage analysis was performed using Mapmaker/exp 3.06 (Lander et al., 1987) and linkage groups were assigned to chromosomes using monosomic hybrids obtained from the cross between reciprocal monosomic lines of Ct and CS.

QTL analyses were performed using either one-way and two-way ANOVA (threshold for significance $P<0.001$) or regression mapping (Kearsey & Hyne, 1994). Estimates of the locations of the QTL, additive value of each QTL and origin of the positive alleles were evaluated using the marker regression method.

3. Results

A principal component analysis (PCA) showed clearly that the strength (W) explained the most of the variation of the first axis while the P/L ratio explained most of the variation of the second axis. The strength was strongly correlated with grain hardness (H: $r^2=0.56$). The results presented hereafter will only concern W, H and also the protein content (Prot.). The distribution of these 3 traits indicated clearly the presence of a major gene for grain hardness (known to be Ha on chromosome 5D: bimodal distribution) while the distribution for the strength showed also influence of this same gene and the distribution of protein content fitted normality.

All the QTLs found for the three traits are summarized in Table 1. Analysis of variance and regression analysis revealed a strong QTL on chromosome 5D for grain hardness. The closest locus corresponded to the puroindoline-b gene (Xmta10) and explained up to 80% of the variation for this trait. This result was in accordance to those obtained by Sourdille et al. (1996) who also found a QTL for grain hardness close to the puroindoline-a locus (Xmta9), a neighbouring locus of Xmta10 on chromosome 5D.

Concerning protein content, two QTL on chromosomes 4B and 6A explained respectively 16 and 22% of the variation while four others on chromosomes 1B, 4B, and 7B (2 QTL) explained between 8 and 10% of the remaining variation. The alleles of Chinese Spring were favourable in most of the cases (4/6) which was consistent with the fact that CS had a higher protein content than Ct.

Two QTL were detected for strength on chromosome 1A and 1B close to the locus for Glu-A1 and Glu-B3 respectively suggesting that he storage proteins may greatly influence this technological parameter. Nevertheless, these QTL explained only 7.7 and 3.6% of the variation respectively while two others on chromosomes 3B and mainly 5D explained 7.6 and 16.9% respectively. In this latter case, this QTL arose close to Xmta10, the same locus that influenced grain hardness. This result was certainly due to the strong correlation existing between W and H and this QTL was the same as the one detected for hardness.

Table 1. Markers significantly associated with strength (W), protein content (Prot.) and grain hardness (H). Add is the additive value of the QTL and R2 the percentage of additive variance explained by the QTL. - indicates that positive alleles come from Chinese Spring.

		W		Prot.		H	
Chrom	Markers	Add	R2	Add	R2	Add	R2
1A	Xfba92	13.5	7.7				
	XksuH9					6.4	9.5
	XksuG34					-3.6	3.0
1B	Xmta14	9.1	3.6				
	XksuG34			-0.35	9.2		
	Xcdo393			0.16	1.8		
2D	Xgwm382			-0.18	2.3		
3B	XksuE3	13.3	7.6				
4B	Xfba1			-0.47	16.1		
	Xbcd15			0.34	8.7		
5A	Xglk407	8.9	3.4				
	Xtam75	-8.8	3.3				
5D	Xmta10	19.8	16.9			18.8	82.1
	Xgwm358					-4.1	3.9
	Xfbb278	4.4	1.0				
6A	XksuG8			0.14	1.5		
	XE3M3			-0.54	21.8		
6B	Xfbb250	-8.3	3.0				
6D	Xgwm55					-4.0	3.7
7A	Xtam51					-2.0	1.0
7B	Xfbb53			0.32	7.6		
	Xfba382			-0.36	9.6		

4. Discussion

Genetic dissection of complex traits such as quality through the use of molecular markers, is an attractive goal for cereal breeders. The population used in the present study had two main advantages: (i) a large range of variation for several quality tests as well as the presence of transgressive lines indicating the influence of numerous genetic factors. (ii) Courtot and Chinese Spring displayed a polymorphism for only one out of the three high-molecular-weight glutenin locus (Glu-A1) and a previous study including Glu-A1 and the polymorphic loci for gliadins (Gli-A1, Gli-A2, GliA5) and low-molecular-weight glutenins (Glu-B3, Glu-D3) did not reveal any relationship between these loci and indirect tests (Felix, 1996).

It is well known that correlations exist between all the quality parameters. This situation was illustrated for the locus Xmta10 coding for puroindoline-b on the short arm of chromosome 5D. This locus explained most of the variation for grain hardness (>80%) but also the major part of the variation for strength (W). This result is easily explained if the strong correlation (r^2=0.56) existing between these two traits is considered.

The genetic bases of grain hardness is well established and one major gene named *ha* is assigned on the short arm of chromosome 5D. The closest marker to *ha* that we found was the gene for puroindoline-b at the end of chromosome arm 5DS confirming thus the results obtained by Sourdille et al. (1996) using a different segregating population. This also confirmed the results of Giroux & Morris (1998) who indicated that grain hardness resulted from highly conserved mutations in puroindoline a and b. However, Sourdille et al. (1996) found additional regions influencing grain hardness on chromosome arms 2AL, 2DL, 5BL, 6DS while in our study, we revealed regions on chromosome arms 1AL, 6DL and 7AL (Table 1). This may indicate either different genes or genotype/environment interactions in the two crosses.

The properties of the dough are known to be influenced by the quality of storage proteins. Here, we revealed two important QTL on chromosome 1A close to the XGlu-A1 locus and on chromosome 1B close to the XGlu-B3 locus confirming thus the preponderance of these proteins in the explaination of the strength (W). Nevertheless, additional loci were found, especially one on chromosome 3B (Table 1), that had never been described before. This locus had a similar effect than those of the storage proteins indicating that W was under a complex polygenic control.

Most of the positive alleles came from the best parent. Nevertheless, some positive alleles were also found in the other parent. This explained the fact that we observed transgressive lines for all of the three traits. But this also reinforced the necessity to incorporate molecular markers in wheat breeding schemes in order to make marker assisted selection. Subsequent efforts also need to be enhanced on genetic mapping of expressed sequences (EST) to try to determine the genes that are involved in the QTL expression.

5. References

Branlard, G. & Dardevet, M. (1985) Diversity of grain proteins and bread wheat quality. I- Correlation between gliadin bands and flour quality characteristics. J. Cereal Sci. 3: 329-343.

Cadalen, T., Boeuf, C., Bernard, S. & Bernard, M. (1997) An intervarietal molecular marker map in *Triticum aestivum* L. em Thell and comparison with a map from a wide cross. Theor. Appl. Genet. 94: 367-377.

Chao, S., Sharp, P.J., Worland, A.J., Warham, E.J., Koebner, R.M.D. & Gale, M.D. (1989) RFLP-based genetic maps of wheat homoeologous group 7 chromosomes. Theor. Appl. Genet. 78: 495-504.

Félix, I. (1996) Etude de la diversité allélique des protéines de réserve (gluténines et gliadines) et relations avec des tests de technologie appréciant la valeur d'utilisation du blé tendre (*Triticum aestivum* L.). PhD Thesis, University of Clermont II. pp146.

Giroux, M.J. & Morris, C.F. (1998) Wheat grain hardness results from highly conserved mutations in the friabilin components puroindoline a and b. Proc. Natl. Acad. Sci. USA 95: 6262-6266.

Kearsey, M.J. & Hyne, V. (1994) QTL analysis: a simple 'marker-regression approach'. Theor. Appl. Genet. 89: 698-702.

Lander, E.S., Green, P., Abrahamson, J., Barlow, A., Daly, M., Lincoln, S.E. & Newburg. L. (1987). MAPMAKER: an interactive computer package for constructing primary genetic linkage maps of expermimental and natural populations. Genomics, 1: 174-181.

Payne, P.I., Corfield, K.G., Holt, L.M. & Blackman, J.A. (1981) Correlations between the inheritance of certain high-molecular-weight subunits of glutenin and bread-making quality in progenies of six crosses of bread wheat. J. Sci. Food Agric. 32: 51-60.

Payne, P.I., Seekinggs, J.A., Worland, A.J., Jarvis, M.G. & Holt, L.M. (1987) Allelic variation of glutenin subunits and gliadins and its effect on bread making quality in wheat: analysis of F5 progeny from Chinese Spring x Chinese Spring (Hope 1A). J. Cereal Sci. 6: 103-118.

Sourdille, P., Perretant, M.R., Charme, G., Leroy, P., Gautier, M.F., Joudrier, P., Nelson, J.C., Sorrells, M.E. & Bernard, M. (1996) Linkage between RFLP markers and genes affecting kernel hardness in wheat. Theor. Appl. Genet. 93: 580-586.

Sourdille, P., Cadalen, T., Charmet, G., Tixier. M.H., Gay, G., Boeuf, C., Bernard, S., Leroy, P. & Bernard, M. (1998) Molecular marker linked to genes affecting plant height in wheat using a doubled-haploid population. Theor. Appl. Genet. 96: 933-940.

Tixier, M.H., Sourdille,P., Charmet, G., Gay, G., Jaby, C., Cadalen, T., Bernard, S., Nicolas, P. & Bernard, M. (1998) Detection of QTL for crossability in wheat using a doubled-haploid population. Theor. Appl. Genet. 97: 1076-1082.

G.T. Scarascia Mugnozza, E. Porceddu & M.A. Pagnotta (Eds.)
Genetics and Breeding for Crop Quality and Resistance, 367-374, 1999

Indirect selection for total digestible dry matter yield in forage maize, using stem diameter

F. Casañas[1], L. Bosch[1], E. Sánchez[1], A. Almirall[1] & F. Nuez[2]

[1]Escola Superior d'Agricultura de Barcelona. Urgell, 187. 08036 Barcelona. Spain. Tel. 34 93 4304207 Fax. 34 93 4192601 e-mail casanas@esab.upc.es; [2]Escuela Técnica Superior de Ingenieros Agrónomos de Valencia. Camino de Vera. 46022 Valencia. Spain. Tel 34 96 3877421

Abstract: Breeding for increased Total Digestible dry matter yield (TDdmy) in forage maize presents problems due to trait complexity, low heritabilities of its components, and difficulty of some analysis. Consequently the use of indirect selection would be an alternative procedure in order to advance in this character. Previous studies on several morphological traits pointed to the diameter of the first stem elongated internode as a good trait to select. This diameter has an acceptable heritability (h^2=0.37), is positively correlated with yield (r=0.52), is not negatively correlated with nutritive quality, and can be measured before flowering, thus allowing to make crosses among the selected plants. To check this in a practical manner five cycles of mass selection were undertaken in Lancaster variety. The comparative study of the five selected generations and the base population yielded the following results: a) a constant increase in stem diameter of about 4% per generation, b) a constant increase of stover yield, and an increase of ear yield interrupted at the third cycle of selection, c) no changes in stover digestibility but a 24% increase in TDdmy at the end of the process, due to the increase in biomass yield, d) a 4.4% increase in days to flowering at the end of the process. Although the correlated response of the ear yield stopped at the third cycle of selection it is concluded that stem diameter is a good trait to select in order to improve TDdmy.

1. Introduction

Improvement of forage maize entails modifying one or more of the components involved in Total Digestible dry matter yield (TDdmy), i.e., ear yield, ear digestibility, stover yield, and stover digestibility. Assuming ear digestibility to be very constant (Demarquilly, 1969; Aerts et al., 1976; White & Winter, 1980), attempts at improving TDdmy must be focused on the other three traits. Unfortunately all of them are measured after harvest, so, selection of the male parent is impossible and the selection process is less effective. In this context it would be desirable to find high-heritability traits well correlated with TDdmy or one of its components that could be measured before crossing. This would enable efficient indirect selection.

Our team has been studying this topic for several years, using a wide range of germplasm (Ferret et al., 1991; Bosch et al., 1992, 1994; Almirall et al., 1996; Mas, 1997). We have deduced that one of the most suitable traits to select in the improvement of forage maize is the diameter of the first elongated internode of the stalk. This trait has a relatively high heritability (h^2=0.37, Almirall et al., 1996; h^2=0.29, Mas, 1997). Moreover, several studies have found positive correlations between this trait and production variables such as ear yield (Pé et al., 1982; Bosch et al., 1992, 1994; Almirall et al., 1996), stover yield (Paramathma & Balasubramanian, 1986; Bosch et al., 1994; Almirall et al., 1996; Mas, 1997) and dry matter yield and/or TDdmy (Gallais et al., 1975; Bosch et al., 1992, 1994; Almirall et al., 1996; Mas, 1997). On the other hand, stalk thickness seems to have no influence on the digestibility of the stover (Dolstra et al., 1987; Bosch et al., 1992, 1994; Argillier et al., 1995; Almirall et al., 1996). As stalk diameter shows little change after flowering, both parents can be selected before pollination.

Mass selection was performed within the Lancaster Surecrop variety to determine whether the theoretical predictions derived from the genetic study of the same population (Almirall et al., 1996) would be fulfilled. After five generations of selection, a comparative study was carried out between the resultant populations plus the initial population with the following objectives: i) to study the response to selection for stalk thickness and its variation throughout the different generations, and ii) to assess the correlated effects in total digestible dry matter yield and its components.

2. Material and Methods

2.1 *Selection process*

Lancaster Surecrop was used as the starting point for the selection program. Two thousand competitive plants, with an effective density of 83,300 pl ha^{-1} were used in each selection cycle. The smallest diameter of the central part of the first elongated internode of each plant was measured immediately before the earliest plants shed, and a 5% selection pressure was applied, which corresponds to a selection intensity of i=2, as normal distributions were assessed in our populations (Falconer, 1996).

2.2 *Comparative study of the different generations obtained in the selection process*

The comparative study of the five generations of selection plus the initial population was performed from July through October 1996 under irrigation at Caldes de Montbui (location 1), Cardedeu (location 2) and La Roca del Vallès (location 3), all in Northeast Spain.

An experimental design of six randomized blocks was used in each location. Each block consisted of 6 rows (one for each generation of selection, plus one for the initial generation). Each row (plot) consisted of 42 competitive plants (density 83,300 pl ha^{-1}), harvested at the optimum time for forage.

2.3 Traits studied

Days to flowering (ps) were recorded when 50% of plants in a plot shed pollen. Each generation was harvested at each location when the average of its plots reached the late-dough stage in its grain in the location which was at different time for generation and location. This resulted in an overall percentage of dry matter at harvest of 29.96±0.35 for the whole plant, within the optimum interval for forage harvest. At that moment the following traits were recorded:

- -Stalk diameter in millimeters (d). The thinnest diameter in the central part of the first elongated internode of the stalk.
- -Dry ear yield (ey), Dry stover yield (sy) and Total dry matter yield (Tdmy) in g/plant.

A representative sample of the stover of each half-plot was chopped as if for silage, dried at 60°C in a forced-air oven and ground to pass a 1-mm screen of a mill. With this material, duplicate determinations were performed on the following:

- -Ash content expressed as a percentage (as).
- -Neutral Detergent Fibre of the stover expressed as a percentage (NDF)(Van Soest & Wine, 1967).
- -Stover dry matter digestibility expressed as a percentage (Ddm)(Aufrère, 1982).
- -Cell wall digestibility expressed as a percentage (Dcwc)(Struik, 1983). Cell wall digestibility = 100 - (100 - (digestible organic matter + 9)) x (100 - ash content)/NDF; digestible organic matter = 0.875 x stover digestibility + 9.19, (Aufrère & Demarquilly, 1989).
- -Total Digestible dry matter yield in g $plant^{-1}$ (TDdmy). Calculated from the stover and ear yields and their respective digestibilities according to the expression TDdmy = ey x 0.83 + sy x Ddm. The ear was assumed to have a constant digestibility value (Deinum & Bakker, 1981), since a constant value of ear digestibility has been reported for a wide range of dry matter content (Demarquilly, 1969; Aerts et al., 1976; White & Winter, 1980).

2.4 Statistical treatment of data

Data were analysed according to the linear equation

$$x_{ijkl} = \mu + g_i + l_j + b_{k(j)} + gl_{ij} + e_{l(ijk)}$$

where,

g_i = generation effect

l_j = location effect

b_k = block effect within location

$e_{l(ijk)}$ = residual effect (plot within generation, block and location).

Responses to selection (R) per unit of applied intensity of selection (i) were calculated according the expression $R/i = h^2\sigma_P$ (Falconer, 1996). The realised heritability was calculated from the same formula. In addition, the correlated response to selection in the trait Y (CR_Y) per unit of applied intensity of selection in the trait X (i) was estimated according to the expression $CR_Y/i = h_X r_A \sigma_{AY}$, being h_X the square root of the heritability of the trait X, r_A the additive genetic correlation between the involved traits (Y and X), and σ_{AY} the additive standard deviation of the trait Y (Falconer, 1996). Genetic parameters to compare estimated advance in selection with real accomplishments were taken from a previous study done in the same Lancaster variety by Almirall et al. (1996).

3. Results and Discussion

3.1 Evolution of the stalk diameter throughout the selection process

Significant differences in stalk diameter were found between all generations (Table 1). No significant generation of selection x location interactions were found ($p \leq 0.05$).

Table 1. Mean values for stalk diameter in each generation. Observed responses to selection between each pair of successive generations. Realised heritability in each generation of selection (h^2).

generation	diameter*	response to selection	h^2**
S0	2.25a		
S1	2.30b	0.05	0.07
S2	2.38c	0.08	0.08
S3	2.47d	0.09	0.12
S4	2.58e	0.11	0.11
S5	2.70f	0.12	0.15
Mean	2.44±0.07	0.090±0.010	0.10

* Values of diameter column followed by the same letter are not significantly different according to the Newman-Keuls test ($p \leq 0.05$).
** All values for heritability differ significantly from 0 ($p \leq 0.05$).

The correlation between diameter and generation of selection were very high (r=0.99; see Figure 1).

Based on data obtained by Almirall et al. (1996) from a study of sibs and half-sibs in the same Lancaster variety, the expected increase in stalk diameter per unit of intensity of applied selection was 4.62%. As we applied a selection pressure of 5%, corresponding to an intensity of selection i=2 (Falconer, 1996) the expected mean response would be 0.209 cm, i.e., twice that observed in the present experiment (Table 1). The mean realised heritability was 0.10 while the estimation based on Almirall et al. (1996) was 0.37. These differences in expected and observed results are probably due to differences in plant density: Almirall et al. (1996) used 62,500 pl ha[-1] while we used 83,300 pl ha[-1] (the usual planting density in materials of this earliness under irrigation) in this study of selection and comparison of generations. Stuber & Moll (1977) pointed out that predictions based on studies carried out using a particular density can be affected by changes in density.

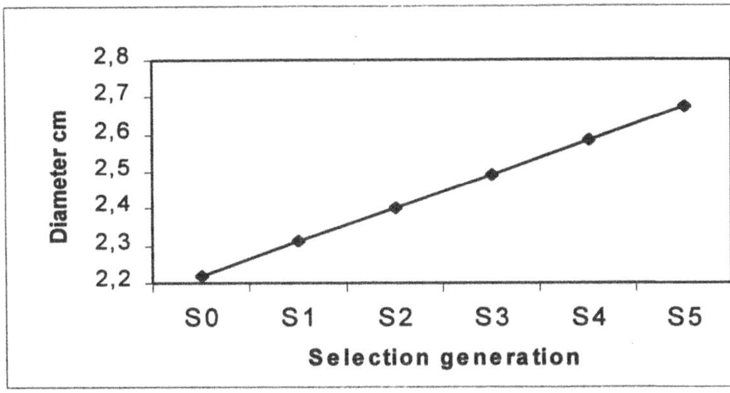

Figure 1. Regression line for stalk diameter over generation of selection.

3.2 Evolution of the other traits. Indirect selection

Significant differences were found between some selection generations for all traits except Dcwc (Table 2). These differences are atributable to the selection process, given the additive genetic correlations reported by Almirall et al. (1996) for the same Lancaster variety between stalk thickness and ear yield, stover yield, and TDdmy. None of the traits showed significant interaction generation x location ($p \leq 0.05$).

Table 2. Mean values of the measured traits. Values of a column followed by the same letter are not significantly different according to the Newman-Keuls test ($p \leq 0.05$).

	ps	ey	sy	as	Ddm	NDF	Dcwc	TDdmy
S0	75.6a*	90a	135a	8.0ab	51.9b	52.5a	36.2a	144a
S1	76.2b	103b	139ab	8.0ab	52.0b	51.9a	35.5a	158b
S2	76.2b	108b	152bc	8.1ab	51.1ab	53.2ab	36.7a	167bc
S3	76.7b	109b	153bc	8.9b	50.8ab	53.4ab	36.3a	168bc
S4	77.9c	105b	164c	7.2a	50.2a	55.9c	36.9a	169bc
S5	78.9d	105b	178d	7.1a	51.1ab	54.8bc	37.4a	178d

The selection process increased yields (Table 2). It also lengthened time to maturity (later flowering –Table 2) suggesting a closer additive genetic correlation between diameter and earliness than that found by Almirall et al. (1996). These differences may be due to the above-mentioned difference in planting density. Whatever the reason may be, lengthening the maturity range can be a drawback when dealing with large modifications.

The mean indirect responses achieved by selection for stalk thickness are similar to those predicted by calculations from the data of Amirall et al. (1996) for the traits stover yield, percentage of ash, digestibility of the organic matter, NDF and Dcwc (Table 3).

Table 3. Observed and expected responses to diameter selection in the rest of the traits studied for each generation of selection.

	Observed responses						Mean expected
Trait	S1	S2	S3	S4	S5	mean	response
ps	0.6	0	0.5	1.2	1	0.66±0.18	ns*
ey	13	5	1	-4	0	3.0±2.58	14.0
sy	4	13	1	11	14	8.7±3.85	9.03
as	0	0.1	0.8	-1.7	-0.1	-0.18±0.37	ns
Ddm	0.1	-0.9	-0.3	-0.6	0.9	-0.16±0.28	ns
NDF	-0.6	1.3	0.2	2.5	-1.1	0.46±0.58	ns
Dcwc	-0.7	1.2	-0.4	0.6	0.5	0.24±0.31	ns
TDdmy	14	9	1	1	9	6.8±2.27	14.8

*ns= additive genetic correlation between diameter and the corresponding trait not significant $p \leq 0.05$ according to data from Almirall et al. (1996).

In relation to ear yield, the response observed in the first selection cycle is in line with those expected, but for the rest of the cycles this is no longer true. Beginning with the second selection cycle, increases in grain yield tend to zero with consequent repercussions in the mean values. Indeed, grain yield responds irregularly to mass selection and there are reports in the literature of direct mass selection with null responses (Sevilla-Panizo & Quevedo-Willis, 1975; Vega & Agudelo, 1972), with responses only in the first selection generations (Subandi et al., 1980), or with responses only in the central cycles of the selection programme (Josephson & Kincer, 1976). Gene fixation and/or linkage disequilibrium could be responsible of this behaviour but, in the case of indirect selection, interpretation is even more difficult.

Total digestible dry matter yield suffered from decreased ear yield from the second generation onward, but its values continued to rise because of the constant increase in stover yield and the constancy of the Ddm (in spite of the increased NDF)(Tables 2 and 3).

4. Conclusions

The response to selection for increased diameter was constant throughout the five selection cycles, with a mean increase of approximately 4% per generation. These gains, however, were not as great as those expected from the results of previous studies. Increases in yield were also achieved; these were constant for stover yield but fell toward zero for ear yield after the first generation of selection.

No substantial changes in the nutritional traits of the stover were observed but a slight increase in NDF; therefore, the increases in the total digestible dry matter yield were due to the increase in biomass production. This improvement in yield, which reached 24% in TDdmy, was accompanied by an overall 3.3 day delay in the flowering time, which represents a 4.4% increase, so it seems that the trait stalk diameter at the first elongated internode is a suitable trait for selection as it drags other traits along with it in turn increasing TDdmy.

Acknowledgements: We would like to thank Dr. A.R. Hallauer for supplying the initial Lancaster seeds. This study has been carried out with a grant CICYT AGF94-0405.

5. References

Aerts J.V., D.L. De Brabander, B.G. Cottyn, C.V. Boucque & F.X. Buysse (1976) Evolution de la composition de la digestibilité et du rendement du maïs en fonction du stade de maturité. *Rev. Agric.* (Brussels) **29**: 379-430.

Almirall A., F. Casañas, L. Bosch, E. Sánchez, A. Perez & F. Nuez (1996). Genetic study of the forage nutritive value in the Lancaster variety of maize. *Maydica* **41**: 227-234.

Argillier, O., Y. Hebert & Y. Barrière, 1995 Relationship between biomass yield, grain production, lodging susceptibility and feeding value in silage maize. *Maydica* **40**: 125-136.

Aufrère, J & C. Demarquilly, 1989 Predicting organic matter digestibility of forage by two pepsin-cellulase methods. pp. 878-878. In: *Proc. 16th Int. Grassl. Congr.*, Nice, Oct. 4-11, 1989. The French Grassland Society, Versailles.

Aufrère, J., 1982 Étude de la prévision de la digetibilité des fourrages par un méthode enzymatique. *Ann. Zootech.* **31**: 111-130.

Bosch L., F. Muñoz, F. Casañas, E. Sànchez & F. Nuez, 1992 Valoración forrajera de 24 híbridos comerciales de maíz de ciclo largo: parámetros de producción de biomasa y de calidad nutritiva. *Investigación Agraria. Prod. Prot. Veg.* **7**: 129-142.

Bosch L., F. Casañas, A. Ferret & E. Sànchez, F. Nuez, 1994 Screening tropical maize populations to obtain semiexotic forage hybrids. *Crops Science* **34**: 1089-1096.

Deinum B. & J.J. Bakker, 1981 Genetic differences in digetibility of forage maize hybrids. *Neth. J. Agric. Sci.* **29**: 93-98.

Demarquilly C., 1969 Valeur alimentaire du maïs fourrage. I. Composition chimique et digestibilité du maïs sur pied. *Ann. Zootech.* **18**: 17-32.

Dolstra O., M.A. Jongmans & A.W. De Jong, 1987 Genetic variation for digestibility of cell-wall constituents in the stalks and its relation to feeding value and various stalk traits in maize (Zea mays L.). In *Proc. of the 14th Congress of the Maize and Sorghum Section of Eucarpia*, Nitra, Slovakia. Pp. 394-402

Falconer D.S., 1996 Introduction to quantitative genetics. 3rd ed. Longman Inc., New York. pp.438

Ferret A., F. Casañas, A.M. Verdù & L. Bosch, F. Nuez, 1991 Breeding for yield and nutritive value in forage maize: An easy criterion for stover quality, and genetic analysis of Lancaster variety. *Euphytica* **53**: 61-66.

Gallaris A., M. Pollacsek & L. Huguet, 1975 Possibilités de sélection du maïs en tant que plante fourragère. Proc. Eucarpia Maize-Sorghum Section, Versailles, France. *Ann. Amèlior. Plantes*, **26**: 591-605.

Josephson L.M. & H.C. Kincer, 1976 Mass selection for yield in corn. *Agronomy Abstracts*. American Society of Agronomy. n.54.

Mas M.T., 1997 Estudio de la base genética de caracteres agronómicos cuantitativos de interés forrajero en poblaciones semiexóticas de maíz. Memoria de Tesis Doctoral. Universidad de Barcelona.

Paramathma M. & M. Balasubramanian, 1986 Correlations and path-coefficient analysis in forage maize (Zea mays L.) *Madras Agric. Jour.* **73**: 6-10.

Pé E., E.Ottaviano & A. Camussi, 1982 Structural analysis of relationships beween ear and plant traits in maize. *Maydica* **27**: 41-53.

Sevilla-Panizo, R. & S. Quevedo-Willis, 1975 Respuesta a la selección masal en tres poblaciones de maíz de la Sierra del Perú. *Investigaciones Agropecuarias* **5**: 109-121.

Struik P.C., 1983 The effects of switches in photoperiod on crop morphology, production pattern and quality of forage maize (Zea mays L.) under field conditions. *Meded. Landbouwhogesch.*, Wageningen **83**: 1-27.

Stuber C.W. & R.H. Moll, 1977 Genetic variances and hybrid predictions of maize at two plant density. *Crop Sci.* **17**: 503-506.

Subandi A., A. Sudjana & M.M. Dahlan, 1980 Mass selection in two varieties of corn (Zea mays L.). *Contributions, Central Research Institute for Agriculture*, Bogor. n.56, 12 pp.

Van Soest P.J. & R.H. Wine, 1967 Use of detergents in the analysis of fibrous feeds. IV. Determination of plant cell-wall constituents. *J. Assoc. Off. Anal. Chem.* **50**: 50-55.

Vega U. & L.C. Agudelo, 1972 Selección masal estratificada para rendimiento en dos variedades de maíz. *Agronomia Tropical* **22**: 159-168.

White R.P. & K.A. Winter, 1980 Effect of harvest date on yield, dry matter content, plant nutrient content and in-vivo digestibility of various parts of forage maize plants in a short season environment, in E.S. Bunting (ed.), *Production and utilization of the maize crop*, Proc. 1st European Maize Congr. Euromaïs 79, Cambridge, Sept. 3-7, 1979. Martinus Nijhoff, The Hague, pp. 65-69.

SESSION 7

OVERVIEWS

G.T. Scarascia Mugnozza, E. Porceddu & M.A. Pagnotta (Eds.)
Genetics and Breeding for Crop Quality and Resistance, 377-396, 1999

Agricultural production and natural resources

E. Porceddu
Department of Agrobiology and Agrochemistry, University of Tuscia, Via S.C. de Lellis, 01100 Viterbo, Italy

1. Introduction

Over twelve thousand years ago mankind had to face one of the greatest environmental crises ever known. Vegetation was changing as result of the decline of the Wurm and, in many areas, natural resources were no longer capable of providing a sufficient quantity of food to sustain the Earth's human population. Humans began to domesticate animals and plants, to breed and grow them: that domestication was the beginning of agriculture, the great Neolithic revolution that made human civilisation possible.

Today, scientists are able to modify, construct and/or transfer genes, regulate their expression, modifying a metabolic pathway to obtain particular end products, as well as protect plants from viruses, bacteria, fungi, nematodes and insects. These aspects will receive proper attention in the presentations and discussions during this Congress. They are some of the technical advances achieved through an unprecedented development of scientific research during this century.

These advances represent potential opportunities that usher in a future characterised by optimism and uncertainty. Though human beings began to make technological use of natural resources a long ago, giving rise to agriculture, it took them thousands of years to fine-tune the process, domesticating and setting aside species on the basis of the experience that they gradually gained. Changes were extremely gradual and slow. Until the present century, people were able to witness only a limited part of a change, so that no given change was ever experienced in its entirety or appreciated in all its implications. Today scientists modify the most intimate nature of organisms and changes have at times become so rapid that a person may experience a great number of changes in the course of his/her life. Many societies have attained an affluence such as not to have any very pressing concern for today and can therefore concentrate their attention on the future. What causes concern is no longer the quantity of natural resources available to day, but rather what will be left after present use, its quantity and quality for future generations.

2. Natural resources and sustainability

Present-day concerns on sustainability of natural resources are not new. Ruttan (1989) listed at least three main waves of concerns during the last 50 or so years. The first appeared immediately after World War II, when the old Malthusian question of the quantitative relationships between available natural resources - energy, land, water, etc. - and economic growth was raised. The response to this concern was technological innovation. New knowledge was acquired, innovations were set up and adopted. New and high yielding crop varieties and adequate agronomic practices, including irrigation, soil preparation and fertilisation, and crop protection made possible the great production improvements that Gaud (1968) called "green revolution".

A second wave of problems emerged in the 'sixties and the 'seventies, and concerned the environment's capacity to assimilate the multiple forms of pollution generated by economic growth. Problems had been anticipated, when Swaminathan (1968) put it as follows: "Exploitive agriculture offers great possibilities if carried out in a scientific way, but poses great dangers if carried out with only an immediate profit or productive motive. Intensive cultivation of land without conservation of soil fertility and soil structure would lead ultimately to the springing up of deserts. Indiscriminate use of pesticides, fungicides and herbicides could cause adverse changes in biological balance as well as lead to an increase in the incidence of cancer and other diseases, through the toxic residues present in the grain or other edible parts".

In fact, two opposing demands for environmental services acquired prominence. One was the elimination of the residues generated by increased production, such as pesticide residues in food, spreading of pollution in the natural habitats, and workers' safety, whereas the other concerned the request for amenities and services, generated by a rapid growth in *pro capite* income. New technologies and new products, characterised by some dematerialization, were developed. For example, 20 g of pirethroids replaced 2 kg of DDT in protecting one hectare of crop from insects, and 15 g of sulphonyl urea replaced 3kg ha^{-1} of dinitrocreosole in weed control (Pasquon, 1995); high yielding varieties allowed the production of 0.95 Kg of wheat grain per m^3 of water instead of previous 0.48 kg (FAO Yearbooks, several years).

These issues had not yet been resolved when, in the mid-'eighties, a third wave of concerns emerged, centred on the implications of a series of changes on trans-national scale, such as climatic change, and the erosion of bio-diversity, for the quality of the environment, adequate food production and human health.

To some extent each wave of concerns has recycled the concerns of the previous wave, that through time had gradually been rendered less acute by innovation.

Today concerns on natural resources can be related to three aspects of the production process (Quadrio Curzio & Zoboli, 1995). The first is the deterioration of natural resources, especially those - like the climate and the genetic resources - that cannot readily be valued and have an undefined ownership, even though they constitute essential inputs for every production process. The second is the growth in the demand for components of multifunctional natural resources, such as the forest ecosystems and agricultural systems, which have to provide both environmental protection and support to production. The third is the environmental impact of the use of natural resources, such as the pollution resulting from agro-chemicals or the possible spreading of undesirable genes from Genetically Modified Organisms (GMO) to the environment.

However, while developed countries find themselves in the midst of the above concerns, their developing country counterparts are experiencing the acute problem of the quantitative depletion of their productive resources: forests slush down, soil erosion, loss of bio-diversity, water pollution, and carbon dioxide and methane emissions, which are seriously endangering air quality.

Thus the natural resources concerns have acquired a global dimension, and are no longer limited to few aspects, but involve all natural resources and their vital relations with the biosphere. The present response to the new dimension of the problem consists in overcoming the concepts of zero growth and/or limited growth, to conceive the idea of sustainable development. The latter has been defined as "what meets the needs of the present without compromising the ability of future generations to meet their own" (WCED, 1987). As a consequence, sustainable development is not that fixed state of economic, ecological, agro-ecological harmony, discussed by several authors and/or organisations, but rather "a balanced and adaptive process of change" (Nijkcamp et al., 1991), with a different solution any time there is an important change in one or more key variables. This concept takes into account the capability of biological systems and the ability of social systems to adapt and to evolve, so that a better future does not necessarily imply a sacrifice today. This requires an increase in knowledge, an enhancement in the ability of dealing with, and solving, problems, a higher efficiency in the use of natural resources, and more numerous options for future generations, even at the expenses of some natural resources, because, as Brooks (1992) put it, "Knowledge.....is a resource which is not depleted by more intensive utilisation for human benefit today".

So conceived, the concept of sustainability acquires a new dimension, puts in a different light the productive use of natural resources and modifies the traditional relationships between science, technology and economic, cultural and social development, underlining how more complex things are, compared with the previous vision. It also opens a whole series of questions, including the role of uncertainty in knowledge on resources, the interactions between resources and technology, the globalisation in the resources and development relations, the implications of human choices (Quadrio Curzio & Zoboli, 1995). The relevance of these questions urges to explore some insights and to hypothesise some possible answers.

3. Questions

3.1 Scientific uncertainties

Uncertainty is one of the characteristics of science. A theory, a result has to be validated, and sooner or later limits and exceptions are identified. But in the case of natural resources, uncertainty assumes a wider dimension, because it involves aspects that are not well understood yet or altogether unknown.

A great number of national and international research programmes on natural resources have been promoted and carried out in recent years. They concern, among others, systems for surveying and monitoring the available resources, new approaches to ecosystems analysis, detection of pollution, definition of the critical parameters to be monitored, etc.

Unfortunately, the results are as yet incomplete and far from conferring certainty upon the various hypotheses regarding the future of resources. Equally uncertain is the understanding of the consequences of human action on them (Parry, 1994). Particularly stimulating, and matter of anxiety, is the case of the climate, the factors of which - solar radiation, air, water, wind, etc., - constitute natural resources of unquestioned importance, essential to agricultural production (Gommes & Fresco, 1998). The climate complex, *i. e.* the set of variables which behave coherently as a result of atmospheric physics and dynamics (Sombroeck & Gommes, 1996), underlies, in fact, the various agro-ecological zones, promotes certain soil features, determines the bio-diversity spectrum and contributes to the general production potential. On the other hand, climate variations from year to year are responsible for annual fluctuations in crop production, as documented by losses associated with drought, storms, diseases, and pest attacks. Many agronomic practices have been devised essentially to mitigate direct or indirect consequences of climatic variability on crops, and have evolved acquiring location-specific features in response to the climate complex in the concerned area.

It is now a well established evidence that climate is changing, but future scenarios are uncertain, especially with regard to the timing, the magnitude and the regional patterns of climate change. The various global models insist on CO_2 and temperature, leaving aside factors also relevant for agriculture, such as potential evapo-transpiration, water balance, climate variability (Katz & Brown, 1992). All models agree that the clouds will be higher if carbon dioxide and temperature were to increase, but they do not, by any means, agree as to clouds quantity, thickness, and reflectivity. An increase in their quantity and reflectivity could considerably reduce and even cancel the global warming. Also aerosols can absorb and reflect radiation; changes in aerosol concentration could modify clouds' reflectivity and reduce the earth's temperature (Charlson, 1995); but aerosols are short-lived and their effect on global warming could well prove to be transient.

There are two points on which scientists agree: CO_2 will rise (Figure 1), resulting in higher temperatures that, however, will be not uniformly distributed. The implications for agriculture and natural resources will be important (Reilly, 1996), and more critical at low than at high latitudes, especially if they are coupled with changes in precipitation (Rosenzweig & Parry, 1994; Fischer et al., 1996).

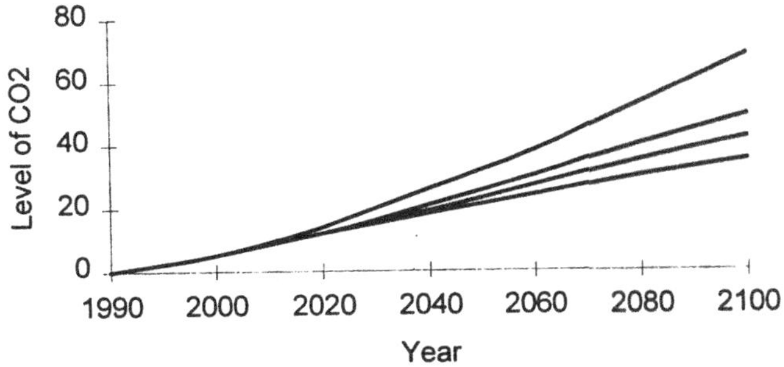

Figure 1. IPCC scenarios (Upper-line = Business as usual) for CO_2 levels in the atmosphere (Houghton et al., 1990).

Global impact on agricultural production would include, among others (Parry & Carter, 1998):

- modification of the agricultural areas as a result of the possibility of utilising areas currently under cold climates;
- increase in soil erosion as a result of a loss in organic matter;
- longer growing seasons in some areas;
- shorter crop cycles;
- zonal migration of species;
- changes in biodiversity organisation and distribution.

Very limited is the knowledge on the geographical distribution of the more complex impacts, which Fisher et al. (1996) indicate as generally positive in industrialised countries and negative elsewhere. Besides macroscopic effects, carbon dioxide change has in fact direct effects on plants (Bazzaz & Sombroeck, 1996). An increase in atmospheric CO_2 would lead to:

- an initial increase in photosynthesis per leaf area unit in C_3 plants;
- a decrease in evapo-transpiration and, consequently, greater water use efficiency;
- an increase in plant growth and therefore a greater bio-mass;
- an increase in the dry matter allocated in the roots;
- a decrease in stress resistance, especially cold resistance.

All these changes will have great implications for agriculture. It has been estimated that crop yield may increase by 10-35%, provided soil fertility will be maintained, water will be abundant, and weeds, insects, and diseases will be kept under control (Idso & Kimball, 1992; Kimball, 1993).

But this scenario is uncertain, giving rise to other questions (Bazzaz, 1995; Bazzaz & Sombroeck, 1996). For example, the plant tissue quality will change. A rise in the CO_2 content would lead to a downturn in the N/C ratio and therefore a reduction in the protein value, with relevant implications for animal and human nutrition. Reduced N/C ratio could also have important effects on primary production as a consequence of the increased consumption of plants by insects and other animals. The poorer food nutrition value could lead insects to eat more, consume more, and cause additional crop yield reduction; but it could also lead to slower insect growth and cause less damage. These facts may promote changes in selection pressure leading to a change in insect populations, which would, ultimately, cause a need for new insecticides, new strategies of control, new sources of resistance genes.

Lower N/C ratios would reduce the rate at which soil bacteria and fungi decompose organic matter, which is critical for maintaining a nutritive supply to crop plants, while an increase of plant residues could hinder some soil mechanical operations. This change may imply the need for greater use of fertilisers, especially nitrogen. But, as temperature rises and the soil retains sufficient moisture, decomposition rates could also increase, reducing the effects of the lower nitrogen amount in the litter.

Species present in an ecosystem may differ in their responsiveness to carbon dioxide. This difference is important in natural ecosystems and in weed system composition. It is expected that CO_2-responsive species will become dominant, suppressing the less responsive ones. As result, there may be a change in the wild flora of cultivated fields and in plants' diversity. A similar change may occur in soil micro-organisms and ultimately in total biological diversity. The relationships between plants and fungi, bacteria, insects, viruses, etc., may change; new races of pathogens and pests may emerge, resulting in the need for a research devoted to face up these new situations.

More modern technologies will be needed to deal with the changes. There will be a stronger need for plant breeding efforts producing plants:

- better suited to resist to abiotic stresses,
- with higher photosynthetic efficiency,
- with more specific pest and disease resistance.

Biotechnology promises to make research more efficient, but there are concerns and uncertainties about the spread of genetically modified plants. These concerns introduce a second set of questions.

3.2 *Technology dynamics and resource efficiency*

A critical point, currently under debate, is whether technological change, *i. e.* changes in the way resources are utilised, has had beneficial or harmful effects on natural resources in the past and what these effects are today and will be in the future.

Research on technological change is scanty and relevant case histories are neither numerous nor well documented, so that the answers to the relevant questions are perceptions and impressions often contrasting and not readily reconcilable. In fact, this topic can be subdivided into two parts depending upon one's perspective. One concerns the stimuli that the state of the resources has exerted towards the development and adoption of technologies; the other takes into consideration the effects that technologies exert on resources.

As far as the first is concerned, results of research have shown that, in developed countries, efficiency of productive systems is improving, both in terms of use of energy and materials and emission of pollutants per unit of output.

Output has increased both per unit of land and per unit of labour (Porceddu & Rabbinge, 1997); output has also increased per unit of solar energy absorbed, thanks to a greater crop leaf area index and a greater trans-location and accumulation of photosynthetates in the grain.

It has been previously mentioned that the use of insecticides and herbicides has diminished during the last two decades, thanks to their greater efficiency; they are also more easily and completely degraded, with a consequent decrease in their residues in soils. The trend was stimulated by the evidence that pesticide residues damage the fauna, and that a continuous application promotes resistance in insects and weeds (Georghiu, 1990).

Particular attention has been paid to control of insects, a field in which – over and above the previously recalled evolution of chemicals – there has been a change in the use of the natural sources of resistance. The use of genetic sources of plant protection commenced as early as 1831, when apple plants resistant to *Eriosoma lanigerum* were identified (Painter, 1951), making possible to continue the cultivation of this species in areas where *Eriosoma* was present, rather than having to change species, as had happened only a few decades earlier, when in Ceylon the coffee plantations, affected by rust, were replaced by tea.

Today progress in molecular biology and increasing knowledge in ecology are expanding the options for improving the efficiency in resources use. For example, the Bt toxin, which had been used as insecticide for many years, is today being synthesised directly in plants into which the coding gene has been inserted. Engineered plants are resistant to Lepidopters and Coleopters, thus saving enormous quantities of energy and materials and providing additional ecological benefits. Results from the USA indicate that Bt potatoes offer a season-long control against the Colorado potato beetle, which is considered the most devastating insect pest in North America, and that the protein Bt affects only the Colorado beetle and not the beneficial insects that can be important elements of integrated pest management strategies and biological control programmes (James, 1997).

Since the Colorado beetle is a devastating pest in most of the 18 million ha of potato growing areas in the world, and potato crop is one of the heaviest users of pesticides among the field and vegetable crops, it has been estimated that Bt potatoes could avoid the use of a relevant part of the over 360 million US$ worth of insecticides (Mackenzie, 1997). But it is interesting to notice that, apart from expenditure, the protection of potato crops requires about 500 t of insecticides as well as 1500 t of inert materials for over 2000 t of commercial product, and that the preparation of these insecticides absorbs 1600 t of raw materials and 30 t of oil, generating 1000 t of waste. The transport and distribution of the 2000 t of product, carried in 180,000 separate containers, requires about 600 t of fuel. In the end, only 5% of the active ingredient actually affects the target, while the remaining 95% are dispersed in the environment (Harvard Business Review, 1997). All this can be saved by *Bt* genes. In addition, to be efficiently expressed in plants, the *Bt* gene had to be greatly modified, making its re-expression in bacteria unlikely. Tissue-specific gene expression avoids damage to non-target insects and slows the emergence of resistance to the toxin in the target insects. The development of resistance can be hindered and slowed by strategies that range from a reduction of selection pressure on insects – for example: by limiting the transgene expression, that makes possible a rapid reversal of the resistance phenomenon (Tobashnik et al., 1994) – to the adoption of multiple resistances, the search for a wider range of genes, like those involved in the production of secondary metabolites, or the use of innovative strategies, devised upon same recent results in molecular biology. In any case, an increase in knowledge can lead to a reduction of material, energy and residues that exert a negative effect on resources.

The use of insecticides can be lowered also by the use of virus-resistant genetically modified crops, since insecticides are used to control the virus-transmitting aphids. In China, the use of CMV and TMV resistant tobacco has halved the number of insecticide applications (James & Krattiger, 1996).

The use of genetic resistance for controlling and/or for avoiding fungal and bacterial plant diseases is a more traditional breeding activity, notwithstanding the complexity and diversity of these systems, which lead many pathogens to have more than a single attack strategy and to easily evolve to overcome the genetic resistance mechanisms. Sources of resistance are good, even though many of them do not last long, and chemical protection is restricted to rather specific situations. New resistance strategies are now being developed. They include, among others, the use of chitinase genes, as in resistance to *Rhizoctonia solani* (Broglie et al., 1991), the manipulation of the phytoalexine synthesis (Hain et al., 1990, 1993), the introduction of antifungal toxin genes (Logemann et al., 1992), and the detoxification of the pathogen toxins, by transferring to plants a bacterial gene encoding for acetyl-tranferase (Anzai et al., 1989).

Considerable effort has been put on identifying the mechanisms by which plants are naturally, or become, resistant to herbicides, and on breeding herbicide-tolerant crop plants. Herbicide tolerance has been obtained through an overproduction of the herbicide target site, reduced absorption of the active ingredient, degradation and lesser affinity to the herbicide (Holt, 1996). Herbicide-tolerant crop plants allow the use of herbicides such as glyphosate, that are more readily degraded in the soil, can be used in smaller doses per unit of land, and allow a shift of treatments from pre-emergence to post-emergence (Darmency, 1996). Results indicate that 80% of glyphosate-tolerant canola crops in Canada required only one herbicide application and that the amount of herbicide applied decreased from 1400 to 400 g ha^{-1}. Similarly, 75% of the herbicide-tolerant soybean crops in USA required only one herbicide application (James, 1997). This approach has a fourfold benefit: (i) smaller costs for the farmer, (ii) presence in crop fields of a population of weeds that, without affecting yield, are useful in preventing or delaying the development of resistance in weed populations, and (iii) reduction of environmental impact by lowering the amount of ingredients and their breakdowns in soil, thus avoiding groundwater contamination and accumulation in the food chain; (iv) provision of significant management advantages, *i.e.* increased flexibility, better yield dependability, improved soil and moisture conservation, compatibility with tillage conservation that reduces soil erosion, which is a critically important factor in some soils.

Remedial technology is therefore shifting towards a more environment-friendly technology. The new technologies render also possible the utilisation of difficult environments. Examples include the introduction of genes for metallothioneins that improve cadmium tolerance (Pan et al., 1994), and the manipulation of genes affecting compatible solutes, for improving salinity tolerance (Weretilnyk & Hanson, 1990).

Economic stimuli are promoting an unprecedented technological creativity. Institutions, concerned for natural resources and consequent environmental policies, are gearing this creativity towards a more sustainable use of natural resources. However, these stimuli and concerns represent only a partial explanation of the positive developments mentioned above. They represent a change in the commitment of research organisations to solve problems but are also the result a particular momentum without which many problems could not have been solved, since, as Mokyr (1995) stated, " it is not the steering wheel but the engine that makes the car go forward".

Developed countries are today experiencing a transition phase in which technological revolution is inducing a qualitative change (Losoure & Barré, 1991). But the new technological paradigm, that changes products and processes into a new coherent system (De Liso & Metcalfe, 1994), may not be fully utilised, because there is a lack of appropriate social and organisational conditions.

Genetic improvement advances have led to a formidable progress in production, both directly, by increasing yield per unit of land and labour, and indirectly, by supporting mechanisation and fertiliser adoption. However, advanced biotechnology remains a developing revolution. This is not due to a lack of results, since the range of products is widening every day and transgenic crops cover this year more than 27 millions hectares of land (James, 1998), but rather because their development may be hindered by the concerns for GMO.

Surprisingly, the debate on biotechnology products is centred on the possible implications for their use on natural resources and food, rather than on the consideration of technical and production of data for assessing the possibility that those events may occur, namely that trans-genes may pass into the natural vegetation or into non-transgenic crop plants. Results from research on pollen diffusion, isolation distances, seed dormancy, pollen-style compatibility, relative time of flowering, presence of wild relatives, degree of crop plant domestication, cross-ability of hybrids, embryo-endosperm relationships in the developing seeds, hormonal equilibrium in hybrid plants, etc. would all produce the knowledge bases for assessing risks, and for designing and adopting risk-avoidance strategies.

Worries are related to the possible development of new pests, as feral populations of cultivated plants, to the enhancement of weediness in wild relatives through the acquisition of trans-genes, to the disruptive effects on biotic communities, adverse effects on ecosystem processes, etc.

Since the release of genetically modified organisms has been limited by regulations, information derive from a small number of *ad hoc* experiments and from indirect evidence, including a long history of cultivation and breeding, studies on natural and agricultural populations ecology, definition of weediness, knowledge of the transgene and its possible phenotypic effects.

It is thus possible to indicate that:

- the widespread use of herbicides has led to an accumulation of herbicide resistance in weeds. Pollen dispersal from plants genetically modified for herbicide tolerance could exacerbate this problem, particularly if crosses between transgenic and normal plants were to became widespread. However it has to be noticed that the development of herbicide tolerant crops by traditional methods does not appear to have been accompanied by an increase in problems due to herbicide-tolerant weed species (The Royal Society, 1998);
- gene transfer could increase weediness in areas where wild crop relatives are present, even though it is unlikely that a small number of trans-genes can alter the weediness of a plant in an agricultural context. Experiments carried out on 73 releases failed to show an increase in weediness (Crawley et al., 1993). Gene flow between crops and weeds has probably occurred since the beginning of agriculture and many weeds may have co-evolved with crop plants (Bartsch et al., 1993). Common wheat originated in this way, and, as it shows, it is easier for pollen from wild plants to pollinate crop plants than vice-versa. Thus the addition of a few, new genes into a plant seems unlikely to present new problems;
- domestication has deprived crop plants of a series of traits typical of wild plants, such as seed shattering, stem length, seed dormancy, etc.. Thus possible hybrids would suffer, in nature, from some lack of fitness (Ellstrand & Hoffman, 1990);
- genetic modifications *per se* do not alter fitness (Holt, 1996); however fitness of genetically modified plants will depend on the effect of the trans-gene at particular stages of development. For example, modified seed oils may affect seed longevity and seedling establishment (Schmitt & Linder, 1994);
- new species are more likely to invade disturbed habitats than stable communities; the same could happen for genetically modified plants (Crawley et al., 1993);
- the effects of insecticidal proteins on useful insects, such as bees, can be avoided by using tissue specific promoters (Fuchs et al., 1992) or by silencing the trans-genes with anti-sense sequences (Wilkinson et al., 1994) or targeting an antibody that inactivates the transgenic product in the non-target tissues;
- another important factor is the variability in the expression level and the instability of the transgene, as it has been highlighted in tests with lucerne in which the expression of the transgene was lost after 10 days at temperatures above 37°C (Broer et al., 1992; Walter et al., 1992).

More intense and focused research effort is needed to clarify these and other aspects and to gain insight so that the risk's nature and magnitude may be assessed and appropriate actions designed.

Some people are opposing technological innovation simply because it implies a risk, or for fear of unknown. This leads them to ask that unless science can guarantee that a maximum level of an ingredient is safe, the ingredient's level should be reduced to zero (Lewis, 1989). It means giving to science a dogmatic character that – by nature – it can never have; it means denying that future research can clarify something that is unknown today. It should be remembered that physical and biological research has shown that many past technologies, that were considered to be good and safe, are in fact inappropriate and harmful. It should also be noticed that research progress, in addition to technological innovation, has provided also the instruments for detecting harmful effects associated therewith.

Every innovation implies some risk. In agriculture risks concern natural resources. If protection of natural resources implies modifications to existing agricultural technologies, the resulting dis-equilibrium will generate a constant technological evolution and continuous progress. That does not mean to be trapped into a vicious circle, characterised by the use of increasingly harmful products, as some one states, but rather entering a virtuous circle in which humans face some problems at the best of their possibilities, leaving others to future generations, together with a greatly improved basis of knowledge, that will enable them to solve those problems successfully. This approach is the paradigm that humans established when they invented agriculture.

Rejecting technological innovation for fear of risks or the unknown, means foregoing this paradigm, and renouncing the effort to build a better future, the very characteristic that distinguishes man from all other creatures.

3.3 Globalisation

The dynamics described thus far are positive for the developed countries, but this optimism rapidly diminishes when they are considered in a global perspective. It is not so much a question of content, but rather of the gap existing between the developed countries and their developing counterparts. Developed countries have an overall pollution and natural resources consumption levels that – in absolute terms – are far greater than developing countries. However, the opposite is true when the pollution and consumption levels are referred to GDP and sometimes even on a *pro capite* basis. The greenhouse gases are a good case in point: the developing countries record a gas emission, per unit of GDP, five times greater than that in the developed countries. The emission level depends on both the economic growth rate and technology efficiency. For example a production of 240 t of milk a year can be obtained either from 60 cows yielding 4 t each or from 24 cows yielding 10 t each. The difference is that in the first case the cows will emit 6.6 t of methane, whereas in the second case only 3.5 t of the same will be released in the atmosphere (Piva, 1995). The point is relevant because methane cattle emissions account for 15% of the methane emissions at global level. The methane emissions from rice paddies follow a similar pattern, and they account for 20% of the total methane emissions in the atmosphere (Table 1).

Table 1. Agricultural contribution to total global gas emissions (%).

Contributor	Methane	Nitrous oxide	Carbon dioxide	Ammonia
Paddy rice	21			?
Livestock and wastes	15	?	-	80-90
Fertilizers	-	5-20	-	<5
Biomass burning	8	5-20	20-30	?
Total	44	10-25	20-30	90

From: G. Conway, 1997.

The developing countries are in a transition phase characterised by a high environmental inefficiency with little chances for improvements, there by limiting their sustainability, unless there is economic growth, promoting a more efficient use of the natural resources. The implications in terms of quality of the natural resources, atmosphere, climate, biodiversity, etc., are not limited to these countries: they will be felt on a global scale. The need follows for a convergence of the development paths of developing and developed countries, there by assuring a more diffuse welfare and a greater environmental efficiency.

The specifics of the problems may be summarised in the following points.

Human population growth rate has never been greater than in the present century. However, in the developed countries, where the transition to lower mortality and birth rate levels commenced with the industrial revolution, the annual growth rate has now diminished to 0.5%, a pre-industrial level. In the developing countries, instead, the transition commenced in the middle of the present century, and the young age of most of their populations has triggered a phase of considerable energy that the downturn in the birth rate, now underway, cannot neutralise (Livi Bacci, 1995). Forecasts suggest that the world population will be approximately 7.7 billion people in 2020 and that 84% of them will be in the developing countries (UN, 1996). Food production will have to be increased tremendously (Figure 2), requiring additional natural resources and more specifically land to be cropped.

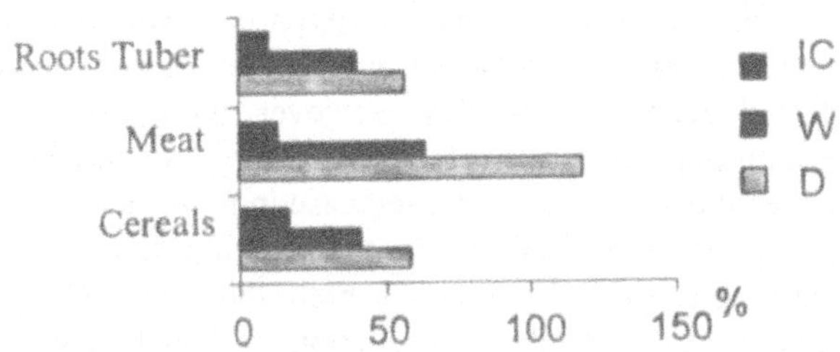

Figure 2. Increase in total demand for cereals, meat, and roots and tubers, 1993-2020. (From. Pinstrup-Andersen et al., 1997).

The population's growth rate is particularly high in the urban areas. Urbanisation will produce changes in food preferences: consumption of proteins, animal fats and fruit will increase, whereas the consumption of the traditional foods, whose increase heralds the improvement of earnings in the rural areas, will go down. The sources of carbohydrates will change, from the traditional foods to finely ground cereals and other processed foods (Hossain, 1988). These facts will increase the need for natural resources. The production of one Kg of chicken meat requires, in fact, at least 3 Kg of grain equivalent, and one Kg of beef requires at least 8 Kg of grain equivalents (Fresco & Rabbinge, 1997).

A third point concerns the specific role played by agriculture in the developing countries. In addition to food production, it generates employment, income and growth in the rest of the economy (Hazel & Haggblade, 1993). In most of the developing countries agriculture accounts for more than 80% of the labour force and more than 50% of the GDP. Thus, even modest growth rates have a considerable multiplier effect, increasing rural income, which in turn generates consumption and therefore growth also in the non-agricultural sectors (Conway, 1997). As a matter of fact only a very limited number of countries have experienced economic growth without either a previous or a contemporaneous growth in agriculture (Mellor, 1995).

The satisfaction of these needs will require doubling or tripling the yield per hectare. The amount of land suitable for cultivation is limited and most is already in use. Estimates indicate that about 900 million out of 1.5 billion hectares of currently cultivated land have high or average productivity characteristics, and that the same may be said of 500 million ha of pastures and 400 million ha of (Figure 3).

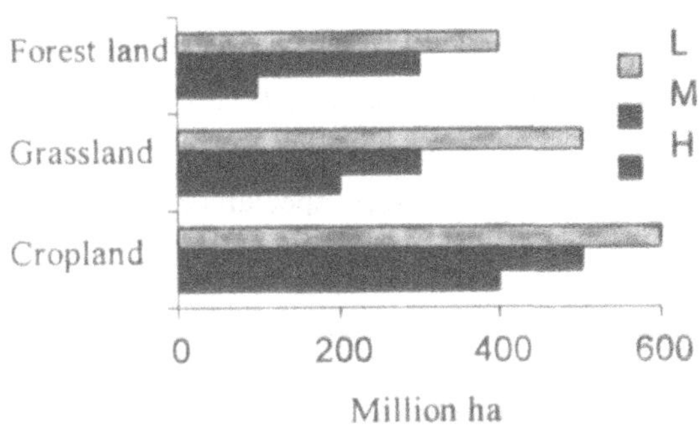

Figure 3. Land in various uses classified by potential for crop production (Source: Buring & Dudal, 1987)

Cropping of present pastures and forests is highly problematic. Apart from the fact that the greater part of the cultivable lands is located in Africa and Latin America (Figure 4) and is therefore far from Asia, home of the greater part of the population of the developing countries, and away from the potential markets, the greatest concern is related to the transformation cost in terms of natural resources. Clearing of forests would reduce the services they render, the biological diversity they host, a more marginal role in the control of the greenhouse and temperature effects, etc. Likewise, farming of the land presently under pastures would compromise livestock breeding and the complementary benefits that animals bring to the fertility of the land. Clearing these lands would also clash with the commitment to conserve nature that was solemnly underwritten by more than 150 heads of state at the UNCED Rio Conference in 1992.

It should also be noted that tropical lands have their organic matter – a key element of fertility – concentrated in the upper 5-10 cm of soil and that this matter would disappear within a few years of cropping, oxidised at a rate some four times faster than in the temperate areas (Lal, 1984). Experience also indicates that an improper expansion of agriculture into fragile lands is one of the causes of drought crises and failure of production, and more generally of that set of negative side effects on agricultural production known as unsustainable spiral.

The implication of this situation complex is that the increased demand for agricultural production has to be met by a greater yield per unit of land. The less land is required for agricultural production, the greater area can be retained to provide other desirable services, such as the maintenance of the biological diversity.

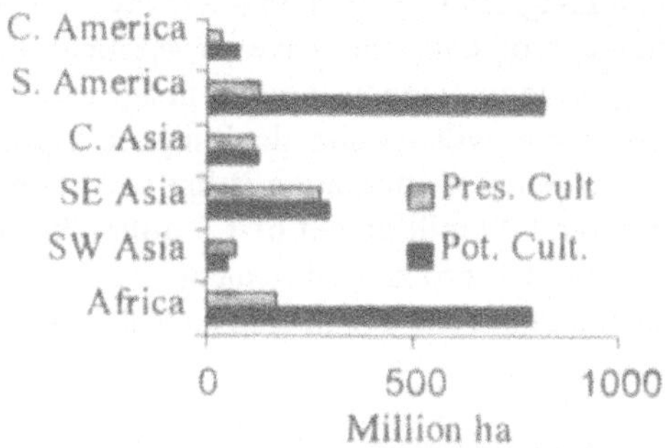

Figure 4. Potential cropland (Present *vs* Potential) in different areas (Source: Buoring & Dudal, 1987).

Biological diversity represents a very special case of globalisation. Genetic resources, the basic material for plant improvement are mainly a resource from the developing countries, while the knowledge of the most advanced breeding technologies is mainly an asset of developed countries. The need for co-operation and the diverging interests between the developed and the developing countries were highlighted at the UNCED Rio Conference. In fact, the problem of bio-diversity is far more complex than those experienced in connection with energy and mineral resources (Perrings et al., 1992), where the main problem was the international redistribution of the income and the value added deriving from the exploitation of the resources.

Genetic resources are part of a complex circularity, with components having profound interdependence (Quadrio Curzio & Zoboli, 1995). Future agricultural development of the developing countries depends on the innovations controlled by the developed countries, who in turn need the genetic resources of the developing countries. Furthermore, if food production and economic growth in the developing countries is obtained by the expansion of the cropped land, this will jeopardise the conservation of genetic resources and generate greater uncertainties on breeding, biotechnological innovations and their future developmet.

On the other hand, if the view is expanded to include the political choices that govern the evolution of agricultural technologies in the developed countries, some additional contrasts will emerge between technologies and natural resources. In fact, a paradox exists in connection with the extensivation of agriculture in Europe. Conceived to reduce the agricultural surplus and the environmental impact, the rules and provisions which aimed at, for example, the return to extensive livestock breeding will increase methane emissions, one of the most important greenhouse-effect gases. Many European regions must farm their lands, because the clovers readily spreading on un-tilled soils, if not harvested, will eventually release the nitrogen, they absorbed from the atmosphere, into the underlying water tables (Rockenbauer, *in litteris*).

In this situation, the primary aim of research should be the planning of agricultural systems different from place to place, but having in common an optimisation of the input use efficiency and a minimisation of emissions that may damage natural resources.

The adoption of such a strategy requires an approach that ranges from a proper planning of land use, pointing out where agriculture can serve more goals than the classical one of crop production, to management of agricultural practices, from the choice of the most suited varieties to a differentiation of cropping practices.

Geneticists and breeders should carefully understand the needs of different environments and breed varieties with high potential production, high and stable resistance to pests and diseases, high efficiency in use of inputs, stability against climatic vagaries, etc.

All this implies, once more, an urgent need for greater scientific knowledge.

3.4 Choices

The points so far considered stimulate some considerations on the evolution of research for agriculture and its resources. The first point concerns the need for an intensification of research. That has already furnished important results and has triggered a process of increasing knowledge which, together with the discovery of new knowledge gaps, has played a crucial role in calling for the attention on natural resources and finding solutions.

The aim of a modern agricultural research is to make agriculture produce greater harvests while protecting the natural resource base through the application of science and technology.

The key players – economists, social scientists, biologists, agronomists, geneticists, plant breeders, etc. – play different roles, but they all require policy makers able to understand the realities of science and the ways it operates and to create an environment that stimulates innovative processes and provides novel solutions. The present situation is however rather controversial from this point of view:

- the need for better and sustainable harvests is universally recognised, but the support for agricultural research, which has rarely been high, is now in most cases falling down;
- molecular biology and biotechnology have made a revolution in science, in the way it approaches problems, but they have also stimulated the interests of many basic scientists, who are now devoting increased attention to real problems and to their solution. The research community for agriculture is no more restricted to agriculturists but it includes expertises that range from atmospheric physics to molecular biology;
- paradoxically, people who have never suffered from food shortages are seeking to block advanced research programmes, that can make food available at low cost for every one in a sustainable way. Those who are striving for a reduction in use of chemical fungicides and pesticides are against resistance genes, which have been the traditional way breeding has fought epidemics, with full respect of natural resources. Those who make large use of DNA, in vaccines made from

human pathogens, are afraid of the infinitesimal possibility that a disease resistance gene inserted in a crop plant may pass through the intestinal barriers and enter human cells.

This sort of controversy has to be cleared.

Research on disease resistance and quality, to which this congress is devoted, is complex and needs to be imaginative, forward looking, and exploring different avenues. This requires the pursuit of long term goals, adequate planning, and quality control.

The second point, complementary to the first, is, in fact, that of identifying principles, instruments and decision mechanisms on which to define action and future research directions. This has to include proper consideration for uncertainties. That is to say that programmes for the utilisation of resources have to be devised allowing for a combination of certainties and uncertainties, assigning to natural resources a value that underscores the risk of going beyond the point of no-return. Present action should not compromise the future, as the very definition of sustainability holds it.

Experience indicates that success may smile on even when residual uncertainty is present, provided it is coupled with increasing knowledge. The appropriate mix for attaining significant results would seem to be that of securing an evolving interaction between growing scientific knowledge, institutional development and favourable socio-economic factors. Very often success has come when policies have set in motion specific economic interests, namely, when technology had an environmental content that conferred a competitive advance to it. Good examples are provided by the cases of the CFCs and the lead-free gasoline. The situation of agricultural research is rather dynamic and delicate from this point of view.

Thanks to biotechnology, private sector is becoming a major actor in global agricultural research. Its interest has increased because of expanding market opportunities throughout the value chain, from agricultural inputs to consumer goods. The market potential of genetically engineered crops is high; it is expected to be 2 billion $ per year world wide by the year 2000, 6 billion $ by 2005, and 20 billion $ by 2010 (Tasker & Goldberg, 1996).

Thanks to biotechnology, private sector is becoming a major player also in the basic research end of the basic-strategic-applied-adaptive research continuum, in contrast with its traditional role of user of the basic and strategic research findings of public sector institutions.

An issue is specifically critical on this point, that of intellectual property rights. Patenting is no more restricted to the final product but to the technology to set it up. Technology for identifying, studying, preparing, and modifying genes that encode for the protein responsible for specific traits are patented, as the technologies related to the techniques for inserting the gene in cells through plant transfection systems, selectable markers, and gene expression techniques (Seghal, 1996).

This string of protections associated with germplasm is leading to the emergence of a complex set of intellectual property right claims and imposition of legal barriers that may limit the resources freedom to operate. Any IPR holder of even one marginal element could block the release of a new line or variety. Freedom to operate will become even more complicated as companies seek to bundle traits to gain competitive advantage.

In such a situation, the public sector may have no access to the enabling technologies that are controlled by the private sector and that are so important for addressing the problems of production sustainability, and natural resources preservation.

Special arrangements are urgently needed.

4. Conclusions

In summary the above discussion would suggest the following considerations.

Uncertainties still characterise scientific knowledge about natural resources dynamics. More research is needed to remove these uncertainties, and time is needed to dispel doubts surrounding some crucial issues.

Indeed, recent experience indicates that technological innovations have played an important role in improving the impact of economic growth on natural resources and in some cases resulted in a downturn of their consumption and pollution.

The scenario is potentially favourable, but there are questions due to the uncertainty: some technological solutions may well result in some negative surprises as scientific knowledge becomes more comprehensive.

One of the basic problems is the extension of favourable trends to developing countries, thus rendering them global. Given the expanding population dynamics in those countries, development implies an enormous need for economic growth at the global level. Thus the central question is what limits the capacity of globalising the most advanced technologies and what enables developing countries to rapidly adopt technological systems so that they can be closer to those in the advanced countries.

This is particularly important with regard to the mix represented by the use of natural resources, agricultural production and economic growth in developing countries. From this point of view the great demand for technological innovation remains unsatisfied.

The world situation of the natural resources is therefore at a crucial point, because it is impossible to isolate the problem of the natural resources from other issues, from inter-country situations to the amount of resources devoted to agricultural research, to public perceptions, to the issues related to the protection of intellectual property rights.

It is essential that the developed countries shoulder their responsibilities and progress towards a rationale of optimised strategy of sustainable use of natural resources. Science and technology are showing the path to be followed in this endeavour.

5. References

Anzai H., K. Yoneyama & I. Yamaguchi (1989) Transgenic tobacco resistant to bacterial disease by the detoxification of pathogenic toxin. *Molecular and General Genetics*. 219, 492-494.

Bartsch D., H. Sukopp & U. Sukopp (1993) Introduction of plants with special regard to cultigens running wild. In: Wohrmann K. & J. Tomiuk eds. Transgenic organisms: risk assessment of deliberate release. Basel: Birkhauser Verlag. 135-151.

Bazzaz F.A. (1995) Global climate change and ecosystem: implications for the future of technology. In: Quadrio Curzio, A. & R. Zoboli (Eds). *Scienza, economia e tecnologia per l'ambiente.* Milano, Fondazione Cariplo per la Ricerca Scientifica. Quaderni, 125-139.

Bazzaz F. & W. Sombroeck (Eds) (1996) Global climate change and agricultural production. Direct and indirect effects of changing hydrological, pedological and plant physiological processes. FAO and John Wiley & Sons. 353 pp.

Broer I., W. Droge, D. Hillemann, K. Neumann, C. Walter & A. Puhler (1992) Instability of herbicide resistance in transgenic suspension cultures and plants. In: Casper R. & J. Landsmann eds. *Proceedings of the second international symposium on the biosafety results of field tests of genetically modified plants and microorganisms*, Goslar. Braunschweig: BBA, 230-236.

Broglie K., I. Chet, M. Holliday, R. Cressman, P. Biddle, S. Knowlton, C.J. Mauvais & R. Broglie (1991) Transgenic plants with enhanced resistance to the fungal pathogen *Rhizoctonia solani. Science* 254: 1194-1197.

Brooks H. (1992) Sustainability and technology. In: *Science and Sustainability.* Selected papers in IIASA 20th Anniversary. Laxembury IIASA.

Buring P. & R. Dudal (1987) Agricultural land use in time and space. In: M. G. Wolman & F. G. A. Fournier (eds), *Land transformation in agriculture.* Wiley, Chiechester, 9-43.

Charlson R.J. (1995) The vanishing climate role of dymethyl sulfide. In: Woodwell GM and Mackenzie FT (Eds), *Biotic Feedbacks in the Global Climate System.* Oxford, Oxford University Press.

Conway G. (1997) The double green revolution. Food for All in the Twenty first Century. Penguin Books.

Crawley M. J., R. S. Hails, M. Rees, D. Kohn & J. Buxton (1993) Ecology of transgenic oilseed rape in natural habitats. *Nature* 363, 620-622.

Darmency H. (1996) Potential advantages of herbicide-resistant crops in weed resistance management. In: *Proc. II nd IWCC* 427-433.

De Liso M. & J. S. Metcalfe (1994) Towards a New Notion of the Technological System. ISDE-CNR. Quaderni n.3/94.

Ellstrand N. C. & C. A. Hoffman (1990) Hybridization as an avenue of escape for engineered genes. *BioScience* 40, 438-442.

Fischer G., K. Frohberg, M.L. Parry & C. Rosenzweig (1996) Impacts of potential climate change an global and regional food production and vulnerability. In: *Climate change and world Food Security* (T.E. Downing ed.) NATo ASI Series, vol. 37. Berlin: Springer Verlag.

Fresco L. O. & R. Rabbinge (1997) Keeping World Food Security on the Agenda: Implications for the United Nations and the CGIAR. *Issues in Agriculture*, CGIAR, Washington DC.

Fuchs R. L., S. A. Berberich & F. S. Serdy (1992) The biosafety aspects of commercialization: insect resistant cotton as a case study. In: Casper R. & J. Landsmann eds. *Proceedings of the second international symposium on the biosafety results of field tests of genetically modified plants and microorganisms.* Goslar. Braunschweig: BBA, 230-236.

Gaud W. (1968) Quoted by Conway, 1997.

Georghiou G. P. (1990) Overview of insecticide resistance. In: Green MB., *et al.*, eds. *Managing resistance to agrochemicals: from fundamental research to practical strategies.* Washington DC: American Chemical Society, 18-41.

Gommes R. & L. O. Fresco (1998) Everybody camplanis about climate. What can agricultural Science and the CGIAR do about it?. *Mobilizing Science for Global Food security.* CGIAR mid–term meeting 1998. May 25-29 Brasilia, Brazil (Mimelographed).

Hain R., B. Biesler, H. Kindl., G. Schroder & R. Stoker (1990) Expression of a stilbene synthase gene in *Nicotiana tabacum* results in synthesis of the phytoalexin resveratrol. *Plant Molecular Biology* 15, 325-335.

Hain R. H., J. Reif, E. Krause, R. Langebartels, H. Kindl, B. Vormam, W. Wiese, E. Schmeizer, P. H. Shreier, R. H. Stocker & K Strenzel (1993) Disease resistance results from foreign phytoalexin expression in a novel plant. *Nature.* 361: 153-156.

Harvard Business Review (1997) Growth through Global sustainability: an interview with Monsanto's CEO Robert Shapiro, By Robert Magretta. January-February 1997.

Hazel P. & S. Haggblade (1993) Farm-non farm linkages and the welfare of the poor. In: Lipton M. & J. van der Gaag (Eds.). *Including the poor*. Washington DC World Bank.

Holt J.S. (1996) Ecological fitness of herbicide-resistant weed. In: *Proc. II nd IWCC* 387-392.

Hossain M. (1988) Credit Alleviation of Rural Poverty: the Grameen Bank in Bangladesh. Washington DC. IFPRI. Research Report 65.

Houghton J.T., G. J. Jenkins & J. J: Ephraums (Eds.) (1990) Climate change. The IPCC scientific assessment. Cambridge, Cambridge University Press.

Idso B.S. & B.A. Kimball (1992) Effects of atmospheric carbon dioxide enrichment on photosynthesis and respiration, and growth of sour orange trees. *Plant Physiology* n.99: 341-343.

James C. (1997) Global status of transgenic Crops in 1997. ISAAA, Briefs N.5. ISAAA, Ithaca, NY. USA. pp. 30.

James C. (1998) Global Review of Commercialized Transgenic Crops: 1998. ISAAA Briefs No. 8. ISAAA: Ithaca, NY.

James C. & A. F. Krattiger (1996) Global review of the Field testing and Commercialization of trangenic Plants, 1986 to 1996: the first decade of Crop Biotechnology. ISAAA Series 1. ISAAA: Ithaca, NY. USA pp.31.

Katz R. W. & B. G. Brown (1992) Extreme events in a changing climate: variability is more important than averages. *Clim. Change*, 21:289-302.

Kimball B. A. (1993) Response of cotton to varying carbon dioxide, irrigation and nitrogen, yield growth. *Agronomy Journal* n.85: 706-712.

Lal R. (1984) Productivity assessment of tropical soils and the effects on erosion. In: Rijsberman F. & M. Wolman (eds). Quantification of effect of erosion on soil productivity in an international context. Delf. Hydraulics Laboratory, The Netherlands.

Lesoure J. & R. Barrè (1991) On the emergence of a new techno-economic system, in OECD, Technology and Productivity. The Challenge for Economic Policy, Paris.

Lewis H.W. (1989) Techonological risk. W.W. Norton, NewYork.

Livi Bacci M. (1995) Population and environment. In: Quadrio Curzio A. & R. Zoboli (Eds). *Scienza, economia e tecnologia per l'ambiente*, Milano, Fondazione Cariplo per la Ricerca Scientifica. Quaderni, 75-105.

Logemann J., G. Jach, H. Tommerup, J. Mundy & J. Schell (1992) Expression of a berley ribosome-inactivating protein leads to increased fungal protection in transgenic tobacco plants. *Bio/Technology* 10, 305-308.

Mackenzie W. (1997) Quoted by James 1997.

Mellor J.W. (1995) Introduction. In: J.W. Mellor (Ed.). Agriculture on the Road to Industrialization. Baltimore Md. John Hopkins University Press. 1-22.

Mokyr J. (1995) Environmental crisis and technological change. In: Quadrio Curzio A. & R. Zoboli (Eds). *Scienza, economia e tecnologia per l'ambiente*, Milano, Fondazione Cariplo per la Ricerca Scientifica. 223-243

Nijkamp P., C. J. M. van den Begh & F. J. Soeteman (1991) Regional sustainable development and natural resources use. In: Proc. *World Bank Animal Conference and Development Economics*. Washington D.C. World Bank.

Painter R.H. (1951) Insect resistance in crop plants. New York, McMillan. Co.

Pan A., M. Yang, F. Tie, L. Li., Z. Chen & B. Ru (1994) Expression of mouse metallothionein-I gene confers cadmium resistance in transgenic tobacco plants. *Plant Molecular Biology* 24, 341-351.

Parry M. (1994) climate change and the Future of Agriculture. In: A.Quadrio Curzio et. al. (Eds). *Innovation, Research and Economic Growth*. Berlin. Springer Verlag.

Parry M. & T. Carter (1998) Climate impact and adoptation assessment: a guide to the IPCC approach. Earthscan. London. 166 pp.

Pasquon I. (1995) Industria chimica, tecnologia ed ambiente. In: Quadrio Curzio A. & R. Zoboli (Eds). *Scienza, economia e tecnologia per l'ambiente*, Milano, Fondazione Cariplo per la Ricerca Scientifica. 162-170

Pinstrup-Andersen P., R. Pandya-Lorch & R. W. Rosegrant (1997) The world food situation: Recent developments, Emerging isuues, and long term Prospects. CGIAR, ICW, Washington DC October 27, 1997.

Perrings C., C. Folke & C.G. Maler (1992) The ecology and economics of Biodiversity loss. The research Agenda. *Ambio* vol. 21 N.3.

Piva G. (1995) Considerazioni sull'impatto ambientale dell'attività agricola e sulle possibili linee di intervento. In: Quadrio Curzio A. & R. Zoboli (Eds). *Scienza, economia e tecnologia per l'ambiente*, Milano, Fondazione Cariplo per la Ricerca Scientifica. 171-187

Porceddu E. & R. Rabbinge (1997) Role of research and education in the development of agriculture in Europe. *European Journal of Agronomy*, 7: 1-13.

Quadrio Curzio A. & R. Zoboli (1995) A preliminary survey of the Conference issues from the point of view of Economics. In: Quadrio Curzio A. & R. Zoboli (Eds.). *Scienza, economia e tecnologia per l'ambiente*, Milano, Fondazione Cariplo per la Ricerca Scientifica. 3-10.

Reilly J. (1996) Climate change. Global agriculture and regional vulnerability. In. F. Bazzaz & W. Soembroeck (Eds.) 237-265.

Rosenzweig C. & M. L. Parry (1994) Potential impact of climate change on world food. *Nature* 367, 133-8.

Ruttan V.W. (1989) Challanges to Agriculture Research in the 21st Century. In: Pardey, P.G., J. Roseboom & J.R. Anderson (Eds) *Agricultural Research Policy International Quantitative Perspective*. Cambridge University Press. Cambridge 399-412.

Schmitt J. & C. R. Linder (1994) Will escaped transgenes lead to ecological release? *Molecular Ecology* 3, 71-74.

Seghal S. (1996) IPR driven restructuring of the seed industry. *Biotechnology and Development Monitor*. No.29, December.

Sombroeck W. S. & R. Gommes (1996) The climate change-agricultural conundrum. In F. Bazzaz & W. Sombroeck (Eds.). Global climate change and agricultural production. Direct and indirect effects of changing hydrological, pedological and plant physiological processes. FAO and John Wiley & Sons. 1-14.

Swaminathan M.S. (1968) The age of Algeny, genetic destruction of yield barriers and agricultural transformation. Presidential Address. Agricultural Science Section. Fifty fifth Indian Science Congress. January 1968. *Proc. Indian Science Congress.* Varanasi. India.

Tasker C.A. & R.A. Goldberg (1996) Delta &Pine Land: measuring the value of Transgenic Cotton. Harvard Business School no. 597-1005.

The Royal Society (1998) Genetically modified Plants for Food use. pp.29.

Tobashnik BE., N. Finson, FR. Groeters, WJ. Moar, MW. Johnson, K. Luo & MJ. Adang (1994) Reversal of resistance to *Bacillus thuringiensis* in *Plutella xylostella. Proceedings of the National Academy of Sciences*. USA: 91: 4120-4124.

UN (United Nations) (1996) World population prospects. The 1996 revisions, New York. UN.

Walter C., I. Broer, D. Hillemann & A. Puhler (1992) High frequency, heat treatment-induced inactivation of the phosphinothricin resistance gene in transgenic single cell-suspension cultures of *Medicago sativa. Molecular and General Genetics* 235, 189-196.

WCED (World Commission an Environment and Development) (1987) Our Common Future. Oxford, Oxford University Press.

Weretilnyk E. A. & A. D. Hanson (1990) Molecular cloning of a plant betaine-aldehyde dehydrogenase, an enzyme implicated in adaptation to salinity and drought, *Proceedings of the National Academy of Sciences*, USA 87, 2745-2749.

Wilkinson J., C. Eady, D. Twell & K. Lindsey (1994) Pollen as a vector in the dispersal of transgene products. *Journal of Experimental Botany* 45, Supplement 42.

G.T. Scarascia Mugnozza, E. Porceddu & M.A. Pagnotta (Eds.)
Genetics and Breeding for Crop Quality and Resistance, 397-417, 1999

Where do we go from this point

F. Salamini
Max-Planck-Institut für Züchtungsforschung Köln, Carl-von-Linné-Weg 10, D-50829 Köln, Germany

1. Introduction

With the rediscovery of Mendel's laws in 1900, radical changes took place in plant breeding methods. The new science of genetics contributed new concepts: the gene, the role of chromosomes as gene carriers, gene linkage, the Mendelian basis of continuous variation, heterosis, maternal inheritance, experimental mutagenesis, polyploidy and gene-enzyme relationships. These scientific discoveries rapidly permeated breeding theory, to the point that plant breeding became synonymous with applied genetics. It should, nevertheless, be emphasized that the science of plant breeding has received and continues to assimilate relevant contributions from cytology, systematics, physiology, pathology, entomology, chemistry, statistics, and, more recently, from molecular biology.

At the end of the 5th decade of this century, plant breeding was a mature technology capable of guiding the evolution of modern agricultural systems. As such, in those years its methods were applied to the solution of emerging problems of worldwide relevance. These were mainly related to the need to mechanize agriculture in developed countries and, in the developing parts of the world, to provide more food for the growing population. Both targets were achieved. In the last 40 years, in the Western world an impressive shift to crop mechanization has been made possible by breeding varieties with modified plant shape, resistance to diseases, simultaneous maturation of produce, resistance of plant parts to mechanical harvest. In developing countries, the intensive application of plant breeding by the CGIAR International Institute led to the success of the Green Revolution which attenuated, and in places eliminated, their recurrent crises in food supply.

2. Where do we go

The use of the existing technology in agricultural research may prove difficult to match to world needs of the year 2010. This is due to unprecedented increments in world population and because problems of resource degradation and mismanagement are emerging. Sustainability of the technological paradigms on which the expansion of food production has depended in the last decades is, in fact, questioned. The main challenge is to expand again agricultural production. But this must be done under conditions of low expansion of arable land. Simultaneously, those agricultural practices which affect negatively agricultural sustainability must be replaced by benign technologies.

Plant breeding remains a resource. In plant breeding, care is taken to have available sufficient genetic variability in the population to be selected and to reduce errors inherent to the phenotypic choice. Seen in a modern context, to the first point belong all conventional and single gene transfer-related procedures, which add to the genetic variability supporting a trait. Plant transformation is here particularly efficient: genes can be moved across species and kingdom borders, allowing a radical expansion of the genetical basis of selection. To the second point belong the choice of selection schemes considering the level of heritability of the trait to be selected. Heritability can be made efficient by the use of molecular markers linked to the genes supporting that trait, including those contributing to quantitative variation (QTLs). Thus, in the breeding process, the adoption of marker assisted selection allows to select single traits with heritability values approaching 1.

2.1 Transgenic crops

The possibility of transforming crop plants become realistic more than 20 years ago (Walden & Wingender, 1995). The first procedures, based on vectors derived from *Agrobacteria*, were followed by other more species-specific methods and at present almost all crop plants can be transformed. The technology has revealed powerful. One example is sufficient: the Bt toxin, in the different versions active against Lepidoptera, Diptera and Coleoptera, when expressed *in planta* has the capacity to control insect pests (Fischhoff et al., 1987; Perlack et al., 1993; Armstrong et al., 1995; Wünn et al., 1996). According to Krattiger (1997) the *Bt* genes have the potential to substitute almost one third of the 8,100 million US dollars necessary to chemically control the insect pests. A large part of the first wave of transgenic crops, currently cultivated, are Bt insect resistant. Their success demonstrates that agricultural productivity will be improved in the future in a context of more safe and sustainable farming systems (James, 1997). Other toxins are now offered as alternatives to the Bt protein. These are chitinase, lectins, and α-amylase-inhibitors (Gatehouse et al., 1996; Ishimoto et al., 1996; Schroeder et al., 1995), protease-inhibitor II (Duan et al., 1996; Thomas et al., 1995), cystatin (Irie et al., 1996), and cholesterol-oxidase (Purcell et al., 1993). The interest in their study and use, reveals both the economical relevance and the potential for the future of genetically modified organisms (GMOs.)

2.1.1 Transgenic plants as feed and food producers

Improvement of feed utilisation; protection of health. A new cellulase from a nematode can be used in transgenics to improve colour and feel of cellulose fibers and as feed additive to improve nutrient utilisation (Shots et al., 1998). Also a new *Aspergillus* enzyme with cellulase activity can be expressed in transgenic plants to improve digestibility (ex. of barley; Madrid et al., 1996). Other enzymes like xylanase are reported to increase xylan degradation and foodstuff value when expressed in transgenics (Fagerstrom et al., 1997). An animal feed containing a recombinant phospholipase, produced by transgenic plants, has an improved lipid digestibility and promote food efficiency utilisation, particularly for non ruminants (Beudeker et al., 1996).

Tissues from transgenic plants containing a fungal phytase and phytic acid-hydrolysing phosphatase (the phosphatase being more thermostable than the phytase) can hydrolyse phytic acid. This improves feedstuff digestion, releasing inorganic phosphorus and inositol, thereby reducing phosphorus excretion in excrements (Avene, 1994). A similar gene from *Aspergillus ficuum* expressed *in planta* allow to use transgenic plant organs in the production of inositol or inositol phosphate. The engineered crop used as food promotes growth and, as cited, reduces phosphorus content in excrement (Brocades, 1991). Phytase genes can be derived from *Thermomyces lanuginosus* and from *Bacillus* sp. (Berka et al., 1997; Oh et al., 1997). An *Aspergillus niger* phytase gene was expressed in tobacco leaves at the level of 1% of soluble proteins and the enzyme can be now added to feedstuff (Verwoerd et al., 1995).

Animal care can include transgenic foods. The subunit LT-B of the heat labile *E. coli* enterotoxin has been expressed in tobacco and potato. Mice immunised orally produced anti LT-B immunoglobulins. Similarly, oral vaccine against toxin B cholera subunit is shown to induce specific antibodies in mice. Immunization was done by oral boosters of transgenic potatoes expressing cholera toxin B subunit (Arakawa et al., 1998). A therapy against lactose intolerance in mammals is based on the oral administration of transgenic plants containing a gene coding for beta-galactosidase. Transgenic tissues may be added to the diet or used as food. Transformed plants are maize, soybean, sunflower, rape (Howard et al., 1997). An antiparasitic protein (against helmith or protozoon) can be produced from maize or rice. The protein has cysteine protease-inhibiting activity. The gene is expressed in transgenic plants, which allows the production of a dietary crop for the animal host (Atkinson et al., 1995).

Decrease of feed-food toxicity. Reduction of feed-food toxicity concern: i) a thioredoxin protein gene from hard or soft wheat which can improve food quality or suppress the antinutritional quality of leguminous plants. It can be expressed in yeast to be used directly but also in transgenics (Gauthier et al., 1996); ii) an aflatoxin-degrading enzyme from *Bacillus licheniformis* used against pathogenic fungi in cereal and fruits. The enzyme can increase fattening of domestic animals. The corresponding gene improves quality of transgenic cereals (Iizuka, 1993); iii) fumosin degrading enzymes which can detoxify seeds from fumosin or related mycotoxins. The genes are from *Exophiala spinifera* or *Rhinocladiella atrovirens*. Maize seeds are treated pre- or post-harvest (Duvick & Rood, 1996).

Improved content of specific nutrients. Special efforts have been dedicated in the past to the improvement of the content of limiting aminoacids in seed crops, like lysine in cereals and methionine in legumes. Falco et al. (1996) report that in transgenic canola the methionine content in seeds is increased by 100% through the expression of a maize protein. The same group has expressed, in tobacco, maize and soybean, a synthetic protein having a content of 31% of lysine and 22% of methionine. Level of free-L-lysine can be increased in maize seeds through the expression of a gene encoding dihydropicolinate synthase (DHPS), resistant to feed-back-inhibition by lysine. The transgene is active in the chloroplast (Gengenbach et al., 1996). Mutant *E. coli* genes encoding a lysine feed-back insensitive form of aspartate kinase (AS) and DHPS, when transformed into barley, increase 14x free lysine and 8x methionine in leaves. In seeds of DHPS plants, lysine, arginine and asparagine were increased 2x (Brinch-Pedersen et al., 1996). Overproduction of threonine, lysine and methionine in maize kernels used for nutrition can also be achieved by inducing mutation during *in vitro* culture (Green, 1983).

Bread quality; other industrial applications. The high molecular weight subunits of wheat glutenins are the determinants of the elastic properties of gluten. They have qualitative and quantitative effects on bread quality. Novel subunits conferring high quality to gluten were expressed in wheat at higher (Barro et al., 1997), or similar level (Blechl & Anderson, 1996) than endogenous subunits. The transgenic dough has an increased elasticity. Peptides and proteins with capacity to substitute carbohydrate sweeteners may have important future applications. Among these, cyclamate, glycyrrhizin, aspartame, saccharine, monellin and thaumatin have high sweetness capacity (Zemanek & Wasserman, 1995). Thaumatin is from *Thaumatococcus danielli*, a plant used as sweetener in cooking. The protein has a weak antifungal activity and is produced *in planta* as a preprothaumatin. The protein contributes no calories to food and has been used to enhance flavour and aroma, as a sweetener, to mask bitterness, as animal feed additive. Thaumatin is produced commercially from ripe fruits, but is difficult to be produced in *E. coli*. In yeast, a mature thaumatin is synthesised; yield can be improved with a chemically synthesised gene and arrives to 20% of total yeast proteins. With 140 copies of the gene, yield can reach 100,000 molecules per cell. In transgenic potatoes the protein is processed and transformed tissues are sweet with protein levels superior by 10x to the threshold level of sweetness (Zemanek & Wasserman, 1995).

Oils. Töpfer et al. (1995) have summarised the possibility of oil plant modification using genetic engineering. Seeds of canola where the stearyl-ATP desaturase is inhibited, increase their content of stearic acid by 2-40% (Knutzon et al., 1992). Also the increase of mono-unsaturated fatty acids is possible through the expression of desaturase encoding genes (Grayburn et al., 1992). Thioesterases have been largely utilised: one of these genes isolated from *Cuphea hookeriana* can increase short chain triglycerids (Dehesh et al., 1996). A transgenic canola producing up to 40% of laurate is already in cultivation. Soybean oil containing more oleic acids has higher heat and oxidation stability. The oil can be used as spray to improve appearance, gloss and mouthfeel of bakery products. The C14 fatty acid myristate present in plant seed triglycerides is produced by transgenic *Brassica* sp. (up to 5% myristate in total fatty acids). The gene added encode a *Cuphea palustris* acyl-ACP-thioesterase hydrolyzing c14:0 acyl-acyl-carrier protein substrate. The product of the transgenic *Brassica* can be used in the surfactant industry, but also in the food industry (Dehesh et al., 1996).

Starches and others polysaccharides. Special starches with variable ratio between amylopectin and amylose can be produced in several transgenic crop species (Visser & Jacobsen, 1993; Shewmaker et al., 1994). The amylopectin potato offer new opportunities: higher viscosity, more stability, application in dairy, soup and sauce products and starch coatings (de Vries, 1997). The amylopectine type of starch obtained from transgenic potatoes can be used as filling agent in food preservation (Stahl, 1997). A potato starch with 35-66% amylose and with a specific viscosity is produced by transgenic plants. The gene used in antisense is a class-A starch branching enzyme. The application of the patent has been requested for 10 plant species (Cooke et al., 1996). A bacterial ADP-glucose pyrophosphorylase gene can increase level of starch (Stark et al., 1992), but contrasting results are also reported (Sweetlove et al., 1996). The expression in transgenics of the gene encoding the bacterial enzyme glucose-1-phospate-adenylyltransferase in a form less sensitive to metabolic regulation enhances tuber and fruit dry matter content (+20-30% under field conditions; high starch potato; high solid tomato; high starch rape; Stark et al., 1996). The expression of *E. coli* branching enzyme in tubers of amylose-free transgenic potato leads to an increased branching degree of the amylopectin. Fructans are polyfructose molecules that function as non-structural storage carbohydrates. By using microbial (*B. subtilis* or *S. aureus*) genes, transgenics have been obtained which accumulate high molecular mass (> $5x10^5$) fructan molecules (1-30% dm in leaves; 1 to 70% dm in microtubers; Van der Meer et al., 1994). Also in transgenic maize endosperm expressing a *B. amyloliquefaciens* fructosyltransferase, fructans accumulate, although in presence of low crop yield (Caimi et al., 1996).

2.1.2 Transgenic horticultural crops

Herbicide resistance. Herbicide resistance in horticultural crop is easily achieved by the methods used for industrial and feed and food crops. Two recent examples are the generation of asulam-resistant (Surov et al., 1998) and bromoxynil-resistant potatoes (Eberlein et al., 1998). The asulam-resistant potatoes may facilitate the eradication of the parasitic broomrapes (*Orobanche* spp.).

Virus resistance. Kavanagh & Spillane (1995) and Shah et al. (1995) have summarized the available strategies for the improvement, by transformation, of levels of resistance of crop species to viruses. The most successful approach to this problem is the expression of genes encoding viral functions, like capsid proteins, replicates, transport proteins or viral satellite RNAs. *Solanaceae* have been frequently considered and potato is a good example. Recent activities concern potato plants transformed with cDNA sequences encoding coat protein of potato leafroll luteovirus (PLRV). Resistance was observed over 5 years of field evaluation

(Kawchuk et al., 1997). Similar results were obtained by Tacke et al. (1996) using a PLRV transport protein gene. In some of these transformants, resistance to PVY and PVX was also observed. A PVY coat protein gene introduced into potato results in drastic reductions in the accumulation of virus in some lines (Song et al., 1996). Transgenic potato lines expressing a yeast-derived double-stranded RNA gene, suppressed potato spindle tuber viroid (PSTV) infection and accumulation. All of the progeny potato tubers were also free of PSTV (Sano et al., 1997). Transgenic hot pepper plants that express cucumber mosaic virus (CMV) satellite RNA show attenuation of symptoms (Kim et al., 1997). A tomato yellow leaf curl bigeminivirus (TYLCV) gene encoding the capsid protein was transformed into tomato. Some plants showed expression of delayed disease symptoms and recovered from the disease with increased resistance (Kunik et al., 1997); from the same virus, a truncated version of the gene encoding a replication protein was also introduced into tomato. The progeny expressed high levels of the protein and was resistant to TYLCV (Brunetti et al., 1997). Transformation of a tomato line with the tomato spotted wilt virus (TSWV) nucleoprotein gene resulted in high levels of resistance to TSWV, which were maintained in hybrids derived from the parental tomato line (Ultzen et al., 1995).

Problems related to viral infections are relevant in perennial plants, particularly when these are reproduced vegetatively. Papaya plants, transgenic for coat protein genes, have become resistant to papaya-ringspot virus (PRV; Yeh et al., 1997; Gonsalves et al., 1997; Lius et al., 1997). In *Prunus* species, which have very low levels of natural resistance to the Sharka virus, the transgenic approach may be the only available counter-measure: transgenic plum trees (*Prunus domestica*) containing the plum pox potyvirus coat protein (PPV-CP) gene were inoculated with PPV. Based on analyses over 3 years one out of five transgenic clones was found to be resistant to infection. In fact, PPV could not be detected in inoculated plants of this clone (Ravelonandro et al., 1997, Scorza et al., 1995).

Modification of fruit maturation. The genetic manipulation of the metabolic routes leading to ethylene synthesis and degradation has a well demonstrated capacity to modify plant organ maturation. This allows, for example, the production and harvest of fruits which show resistance to postmaturation and which can be transported without refrigeration. The resistance to maturation is obtained either by inhibiting, via antisense, the enzyme amino acylpropane carboxylic acid (ACC) oxidase, or overexpressing ACC deaminase (Oeller et al., 1991; Hamilton et al., 1990; Klee 1993; Botella & Sanewski, 1997; Theologis & Sato, 1998), S-adenosylmethione decarboxylase (Grierson et al., 1995; Monsanto, 1992), or S-adenosylmethionine hydrolase (Good et al., 1994). In melon, the enzyme ACC oxidase can be inhibited by antisense RNA (Ayub et al., 1996) and the transgenic plants produce only 1% of normal ethylene levels; ripening of transgenic melons is thus blocked. Treatment with exogenous ethylene reverses the ripening-inhibited phenotype of these fruits. The transgenic fruits display extended shelf life and improved quality (Ayub et al., 1996). Transformation of tomato with a truncated *Acc2* gene results in fruits that reach the pink stage 2-3 weeks after breaker stage. When fruits are picked at breaker stage and stored at 15°C, they do not reach the red stage in the absence of ethylene application (Bedbrook et al., 1997).

Modification of the ethylene response based on the engineering of the hormone receptor is a further possible mean of interfering with fruit maturation. A plant ethylene response ETR element transformed into tomato plants induces a decrease in response to ethylene permitting e.g. controlled fruit ripening, delayed floral senescence and abscission during growth, and improved flexibility in transport and storage (Ecker & Kieber, 1997; Meyerowitz et al., 1995).

Antisense sequences complementary to part of the mRNAs that encode softening enzymes, e.g. polygalacturonase or pectinesterase, can also be used to inhibit fruit maturation. This technology reduces expression of fruit softening enzymes and thus delays fruit softening (ICI, 1993; Picton et al., 1995). Transgenic tomatoes with reduced levels of pectin methylesterase produce juice and ketchup with increased total solids (+5%), soluble solids (+6%) and viscosity (+8%). In these GMOs, the time of harvest has no effect on juice quality (Thakur et al., 1996). Similar transgenics showed an increase in fruit number and yield compared to wild-type. Transgenic fruits had higher soluble and total solids and higher pH than control fruits. Thus, the introduction of the antisense gene does not adversely affect fruit yield or vegetative growth of plants (Tieman et al., 1995).

Resistance to fungi, bacteria and nematodes. Proteins tested *in planta* and shown to be active against pathogenic fungi include chitinases, glucanases and ribosome-inactivating proteins (summarized in Shah et al., 1995). The expression of more than one of these proteins leads to a further increase in resistance (Zhu et al., 1994; Jach et al., 1995; Jongedijk et al., 1995). Several other proteins, particularly some pathogenesis-related proteins like protein IA, osmotin and protein 2 (Alexander et al., 1993; Liu et al., 1994; Terras et al., 1995), are also active, as is the expression *in planta* of glucose oxidase, which leads to accumulation of H_2O_2 (Wu et al., 1995). H_2O_2-mediated disease resistance in transgenic potato plants is effective against a broad range of plant pathogens. The constitutively elevated levels of H_2O_2 enhance the accumulation of total salicylic acid severalfold in the leaf tissue of transgenic plants. Several potential antibacterial proteins, like lysozyme, L-thionine and attacin E, have been tested *in planta* (Düring et al., 1993; Anzai et al., 1989; Norelli et al., 1993, 1994; Zhu et al., 1996). A critical consideration of the results obtained using the strategy outlined above, does not lead to optimistic conclusions.

New ways to engineer resistance against fungi, bacteria and nematodes are emerging from the cloning of plant resistance genes. These genes mediate the gene-for-gene interaction -- the recognition of a pathogen avirulence gene product by its receptor encoded by the resistance gene R. The cloning of a number of avirulence (*Avr*) and *R* genes provides new insights into the molecular basis of resistance and opens new perspectives for the design of improved resistances. The cloned *R* genes (now more than a dozen) can be assigned to two major classes (Dangl, 1995; Hammond Kosack & Jones, 1997; Staskawicz et al., 1995). The *Cf* genes, *Xa21* and *Pto* belong to a group of structurally related genes that code for an N-terminal leucine-rich repeat (LRR) and/or a serine-threonine kinase domain. Other members of this group with no apparent resistance gene function are the receptor-like kinases

(Li & Chory, 1997; Walker, 1993; Wang et al., 1996). *N*, *Rps2*, *Rpp5*, *I2*, *Xa1*, *M*, *L6* form a more abundant class of *R* genes containing a nucleotide binding site (NBS) and a C-terminal LRR. Interestingly, theses structural features have also been found in genes of *Caenorhabditis elegans* and of human that function during programmed cell death (van der Biezen & Jones, 1998). The strong conservation of several short motifs in the NBS-LRR class has facilitated PCR-based isolation of a large number of homologous genes, allowing the study of the genomic organization of resistance genes (Kanazin et al., 1996; Leister et al., 1996, 1998; Spielmeyer et al., 1998; Yu et al., 1996). It has been shown that NBS-LRR and *Cf* genes are organized in clusters of copies linked in head-to-tail orientation, which reflects a particular mode of R gene evolution (Leister et al., 1998; Parniske et al., 1997). The introduction of the tomato *R* genes *Pto* and *Cf9* and of the tobacco *R* gene *N* into other species (Hammond-Kosack et al., 1998; Thilmony et al., 1995; Whitham et al., 1996) shows that resistance genes can be functional in transgenics, a finding which opens the prospect of using the cloned R genes in horticultural practice. A second attractive possibility involves the generation, and simultaneous expression, of resistance and avirulence gene products in plants. This can lead to necrosis and cell death, as has been shown for *Cf9* and *avr9* of tomato and *Cladosporium fulvum*, respectively (Hammond-Kosack et al., 1998; Honee et al., 1995). The fine-tuning of this reaction may prove to be a powerful means for designing improved resistances.

Control of plant form, flowering time, flower characteristics and fertility. Plant form can be altered through the use of hormone genes from microbial sources, like those encoding triptophan monoxygenase, indolacetamide hydrolase, isopentenyl transferase, IAA-lysine synthetase, or the *rolA*, *B*, *C* genes from *Agrobacterium rhizogenes*. Plant form can also be altered through the use of phytochrome genes from plant sources (Robson & Smith, 1997). When the phytochrome genes are overexpressed, the result is dwarfing through reduction of internode length, increased leaf coloration, increased branching and delayed foliar senescence (McCown, 1997).

Several strategies are now available that allow one to modify flowering time by gene transfer. A gene for nucleoside diphosphate kinase, when over-expressed, alters flowering time by rendering the plant hypersensitive to red light. An inactive form may also be over-expressed to render the plant hyposensitive to red light (Mitsui-Seiyaku, 1997). A gene encoding a cauliflower CAL protein converts a shoot meristem to a floral meristem in an angiosperm. The gene is a part of a kit for promoting early flowering in angiosperms (Yanofsky, 1997). The CENTRORADIALIS (CEN) protein of *Antirrhinum majus* influences flowering characteristics in transgenic plants. These include the switching of apical meristems to a floral fate, which can be promoted or inhibited, and the control of time of flowering (Bradley et al., 1997). The sequence encoding the *Arabidopsis thaliana* protein FCA may be used in the production of transgenic plants to modify flowering characteristics, i.e. advance or delay flowering (Dean et al., 1996). Transgenics, tobacco or aspen, that express *LEAFY* or *APETALA1* homologs show modified flower meristem development. A dominant-negative mutant or a corresponding antisense RNA, when expressed in transgenic angiosperm, gymnosperm, monocotyledon or dicotyledon plants alters the time of flower induction (Colasanti & Sundaresan, 1996). A gene encoding a late elongated hypocotyl protein (LHY) influences flowering characteristics. Overexpression delays flowering in transgenics. The *LHY* promoter can be used to confer oscillatory expression on a target sequence in a circadian manner, for products which are only required or desirable at certain times of the day (Coupland & Schaffer, 1997).

Flower architecture depends on the position and identity of floral organ primordia. Flower organ identity is regulated by specific transcription factors, the MADS-box type of transcription function (Schwarz-Sommer et al., 1990). When these genes are repressed, the identity of the floral organs is modified (Mandel et al., 1992; Tsuchimoto et al., 1993; Halfter et al., 1994; Van der Krol & Vorst, 1997). An example of application of this knowledge is a sequence encoding a protein required for normal flower development in a woody perennial of the genus *Eucalyptus*. This sequence can be used to produce sterile transgenic *Eucalyptus* trees useful for establishing wood-lot plantations in reforestation projects (Harcourt et al., 1996). The same results can be obtained with a DNA encoding the Tasselseed-2 protein of maize, a product which results a in plant producing predominantly male flowers (Dellaporta, 1995).

Transgenic plants can be made to produce seedless fruits and vegetables by the use of a seed-specific promoter, preferably a maternal tissue promoter, coupled to genes which encode a methylase, or the genes for indole acetamide hydrolase (IamH) and indole acetamide synthase (IamS). IamS is capable of converting a substance endogenous to a plant cell into a non-toxic substance which is rendered cytotoxic by IamH. The progeny will therefore be seedless. An extension of the method leads to the production of hybrid seeds which will produce a plant with seedless fruits (Tomes et al., 1997). A similar procedure leading to parthenocarpic tobacco, melon and eggplants is based on the indole acetic acid monoxigenase gene (Rotino et al., 1997).

Flower colour. In breeding of ornamentals, one of the most important traits is flower color. Davies & Schwinn (1997) have recently summarized the state of the art in this field. They conclude that the prospects for genetic modifications of flavonoid-based colors are good. Genes have been isolated for the majority of the biosynthetic enzymes involved. Color has been modified in transgenic plants of *Petunia* (Meyer et al., 1987; van der Krol et al., 1988; Holton, 1995, 1996; Brugliera et al., 1994; Oud et al., 1995), *Gerbera* (Elomaa et al., 1993), *Eustoma* (Delores et al., 1995), *Nicotiana* (van der Krol et al., 1988), *Rosa* (Courtney-Gutterson, 1994) and *Dendranthema* (Courtney-Gutterson et al., 1994). Transgenic carnation, rose, gerbera, or chrysanthemum can be produced by transformation with a gene for flavonoid-3'5'-hydrolase dihydrokaempferol-4-reductase. This method may be used to produce flower colors in the range covering lilac, violet, purple and blue (Holton, 1996). A petunia gene which regulates vacuolar pH and has a helix-loop-helix motif, used in sense or antisense orientation, alters anthocyanin pigment accumulation in transgenics, producing bluish flowers and fading flower colour with aging (Chuck et al., 1994). DNA encoding a protein with flavonoid-3',5'-hydroxylase activity, transformed into cells of ornamentals such as rose, petunia, tobacco and carnation, gives transgenic plants having petals with enhanced blue colour (Kyowa-Hakko, 1993). Chalcone synthase DNA isolated from gentian may be introduced into the same species or a different species to diversify flower colour (Iwate-Ken, 1996). A gene encoding a plant protein aromatic acyltransferase, when transformed into plants, results in acylation of pigments by the enzyme, generating altered or novel colours (Ashikari et al., 1996).

2.1.3 Transgenic acceptance

In the coming years, several plant transgenic products are expected to be commercialised. They will affect both consumer behaviour and market-economy of food and feed. The current problems of transgenic acceptance will be likely overcame, as soon as more data will be made available on the security state of cultivation and use of GMOs. For example, already the United Kingdom Advisory Commitee on Novel foods and Processes (1995) concluded that the tomato paste and products derived from it were safe for use in food, and compositionally similar to paste of conventional tomatoes. Safety of transgenic tomatoes expressing *Bt* genes has been assessed; the food is safe, lacking toxicity in man and other mammals. However, it is not qualitatively different from standard tomato products (Noteborn et al., 1995). An analysis of gene transfer to natural ecosystems was undertaken using 391 European field trails. Crops were classified having minimal, low or high risk, depending on their capacity for cross-pollination with wild relatives. Engineered traits were also classified according to their potential capacity to enhance competitiveness of wild relatives, if transferred. Trial crops studied include oilseeds, cereals, root crops and horticultural produce. Results showed that 91% of trails were likely to have a minimal, if any, potential environmental impact, whilst the remaining 9% of trails had a low potential environmental impact (Ahl Goy & Duesing, 1996). More studies like those cited are necessary in Europe. They will contribute to reduce the present relevance given to perceived risk in comparison to actual risk of GMOs. It is in any case already possible to predict, with high degree of probability, that transformed plant crops will be in the future largely cultivated also in Europe.

2.2 Improving plants by marker assisted breeding

A different contribution of molecular biology to plant breeding is the development of molecular techniques for use as diagnostic tools to facilitate the breeding process. Standard breeding procedures utilize the genetic variability present within the available gene pools of crop species to synthesize new cultivars. Most characters selected have a complex inheritance pattern. The availability of genetic markers diagnostic for superior expression of a trait greatly facilitates the selection process. Such markers are also helpful in reducing the handling of plant pathogens necessary for testing the inheritance of resistance genes.

Molecular methods have been developed to identify and use several types of probe that reveal polymorphisms in genomic DNAs. These probes are known and described under the acronyms RFLP, RAPD, AFLP, SCAR, etc. They have been used to generate a wealth of molecular data on the genomic organization of crop species. Dense molecular linkage maps have been constructed and several genes contributing to agronomically relevant traits mapped. Summaries of this area can be found in several books. For example, an overview of the state of the art up to 1994 (Phillips & Vasil, 1994) already included maps for 15 crops. An imcomplete review of papers published on the subject in 1997 and 1998 concerning horticultural crop would have to include: mapping of important genes for disease resistance in apple, bean, coffee, tomato and citrus (King et al., 1998; Geoffrey et al., 1998; Agwanda et al., 1997; Mestre et al., 1997; Roche et al., 1997); molecular maps of almond, banana, chicory, apple and avocado (Socias & Company, 1998; Kaemmer et al., 1997; De Simone et al., 1997; Conner et al., 1997; Sharon et al., 1997); molecular maps and genomic fingerprinting of organisms harmful to horticultural crops like *Colletotrichum lindemunthianum* (Mesquita et al., 1998; Sicard et al., 1997); germplasm characterization in bean, pistachio, onion, mango (Skroch et al., 1998; Hormaza et al., 1998; Le Thierry et al., 1997; Lopez-Valenzuela et al., 1997); markers for special traits like pollen fertility in garlic, acidity in citrus, sex-related genes in asparagus and hop, flower type in citrus (Hong et al., 1997; Fang et al., 1997; Jiang & Sink, 1997; Polley et al., 1997; Scovel et al., 1998); QTL analysis and cloning in tomato and other plants (Fulton et al., 1997; Paran et al., 1997; Alpert & Tanksley, 1996; Young, 1996).

Current plant breeding programs that already exploit molecular markers are few. However, before the potential of the new varieties can be fully evaluated, present methods are destined to be substituted by a more integrated approach. This will be based on the capacity to direct the evolution of crop genomes at the level of the structure, function, location and expression of a large number of genes. EST (expressed sequence tag) sequencing will provide extensive catalogs of DNA sequences that are expressed in various tissues. The EST data can be "fed into" gene knock-out systems leading to the association of sequence with function, and used to establish hybridization technologies for the assessment of gene expression at the genome level (Marshall & Hodgson, 1998). The integration of the two approaches has evident potential: for example, the metabolic reaction of a plant to biotic or abiotic stresses can be described based on a large set of genes activated or repressed. The relevance of these genes to the phenomenon under study – and ultimately to crop yield – can be established by the use of mutants in which single genes are knocked out. EST assignment to linkage and physical maps raises the possibility of adopting a candidate gene approach to the cloning of genes with mutant phenotypes, included QTL alleles of genes influencing yield, heterosis, disease resistance, etc. Mapped ESTs, moreover, contribute to the available array of molecular markers, linked to genes of interest. Markers based on micro-satellite variation detectable by PCR, for example, can be generated based on EST sequencing and are extremely useful in DNA fingerprinting, in marker-assisted selection, and gene cloning. Large-insert DNA libraries, and the physical alignment of chromosomal DNA are necessary steps toward complete genome descriptions. They allow the localization of ESTs on the physical maps, the cloning of genes based on their positions, the integration of genetic, molecular, functional and physical maps, and the assignment, in heterologous species, of putative gene functions based on sintenic gene relationships.

It can be concluded that the capacity to reveal any type of DNA variation and use it in application-oriented projects will become the central issue in plant breeding. The genetic control of quantitative traits involves interactions with the environment. Because of this, the definition of the genetic components contributing to plant traits requires that a large number of markers be applied to large populations of segregating individuals (Schafer & Hawkins, 1998). At the moment, attention is focused on the possibility of using single-nucleotide polymorphisms (SNPs) which are plentiful. SNPs can be identified based on several techniques (Grompe, 1993; Mashal & Sklar, 1996; Hawkins, 1997; Cotton, 1997; Orita et al., 1989; White et al., 1992; Fischer & Lerman, 1983; Riesner et al., 1989). Fluorescence technology potentially has a high throughput capacity when large populations have to be scored for SNPs (see for example Tyagi et al., 1998); technologies based on micro-arrays ("genotypers") may offer a more radical solution for the future (Ramsay, 1998).

3. The future: a need of change in breeding targets

Intensive agricultural practices damage the productivity of land, in some cases severely. The consequence can be water- and wind-induced erosion, salination, compaction, waterlogging, overgrazing, and other problems. Erosion, in particular, has made a billion hectares of arable soil unusable, with Asia having the highest percentage (nearly 30 percent of arable surface). It is thus clear that current agricultural production is unsustainable. The consequence is that new farming systems should be developed: intensive but more environmentally friendly; based more on biological than on chemical inputs; sustainable, also with respect to the delicate nature of the agricultural soils in the tropics.

Modern plant breeding has the potential to offer varieties more adapted to be integrated into new farming systems. The emerging concept is to negate the agronomic principle of searching and eliminating those environmental constraints which limit crop productivity. The alternative view is to modify the plant genome such that the plant can adapt to the environment, included biotic stresses. These modifications are feasible, particularly when moving genes across species barriers or fully exploiting the potential of crop genomes to evolve. The plant breeding community, in this context, has the urgent task to redirect its future targets. I personally see as necessary the need of reconsidering perennialism in crop plants as one of the major challenges.

4. References

Agwanda, C.O., Lashermes, P., Trouslot, P., Combes, M.C. & Charrier, A. (1997) Identification of RAPD markers for resistance to coffee berry disease, *Colletotrichum kahawae*, in Arabica coffee. *Euphytica* 97, 241-248.

Ahl Goy, P. & Duesing, J.H. (1996) Assessing the environmental impact of gene transfer to wild relatives. *BioTechnology* 14, 39-40.

Alexander, D., Goodman, R.M., Gut-Rella, M., Glascock, C., Weymann, K., Friedrich, L., Maddox, D., Ahl-Goy, P. & Luntz, T. (1993) Increased tolerance to two oomycete pathogens in transgenic tobacco expressing pathogenesis-related protein 1A. *Proc. Natl. Acad. Sci. USA* **90**, 7327-7331.

Alpert, K.B. & Tanksley, S.D. (1996) High-resolution mapping and isolation of a yeast artificial chromosome contig containing fw2.2: A major fruit weight quantitative trait locus in tomato. *Proceedings of the National Academy of Sciences of the United States of America* **93**,15503-15507.

Anzai, H., Yoneyama, K. & Yamaguchi, I. (1989). Transgenic tobacco resistant to a bacterial disease by detoxification of a pathogenic toxin. *Mol. Gen. Genet.* **219**, 492-494.

Arakawa, T., Chong, D.K.X. & Langridge, W.H.R. (1998) Efficacy of a food plant-based oral cholera toxin B subunit vaccine. *Nature Biotechnology* **16**, 292-297.

Armstrong, C.L., Marker, G.B. & Pershing, J.C. (1995) Field evaluation of European corn borer control in progeny of 173 transgenic corn events expressing an insecticidal protein from *Bacillus thuringiensis*. *Crop Science* **35**, 550-557.

Ashikari, T., Tanaka, Y., Fujiwara, H., Masahiro, N., Fukui, Y., Yonekura, K., Mizutani, M. & Kusumi, T. (1996) DNA coding for aromatic-acyltransferase activity; from *Gentiana triflora* var. *japonica*, *Petunia hybrida*, *Perilla ocimoides*, *Senecio cruentus* or lavender, for transgenic plant production with altered flower color. Patent WO 9625500.

Atkinson, H.J., Koritas, V.M., Lee, D.L., Macgregor, A.N., Smith, J.E., Univ. Leeds (1995) Patent No. WO 9523229.

Avene (1994) Synergistic enzyme composition for hydrolyzing phytic acid in feed; Aspergillus spp. phytase and acid phosphatase gene expression in transgenic plant; thermostable enzyme composition application in phytic acid hydrolysis for improved feedstuff digestibility. Patent No. EP 619369.

Ayub, R., Guis, M., Ben-Amor, M., Gillot, L., Roustan, J-P., Latche, A., Bouzayen, M. & Pech, J.C. (1996) Expression of ACC oxidase antisense gene inhibits ripening of cantaloupe melon fruits. Nature Biotechnology 14, 862-866

Barro, F., Rooke, L., Békés, F., Gras, P., Tatham, A.S., Fido, R., Lazzeri, P.A., Shewry & P.R. Barceloó, P. (1997) Transformation of wheat with high molecular weight subunit genes results in improved functional properties. *Nature Biotechnology* **15**, 1295-1299.

Bedbrook, J.R., Dunsmuir, P., Howie, W.J., Joe, L. & Lee, K.Y. (1997) Tomato plant comprising genetic locus conferring delayed fruit ripening phenotype; truncated ethene biosynthesis 1-aminocyclopropane-1-carboxylate-synthase gene transfer and expression in a transgenic plant. Patent WO 9701952.

Berka, M., Ray, M.W., Klotz, A.V. Novo-Nordis-Biotech (1997) New 3,6-phytase from *Thermomyces lanuginosus* and related DNA, vectors and transformed cells; vector plansmid expression in host cell for application in the food industry. Davis, USA, Patent No. WO 9735017.

Beudeker, R.F., Kies, A.K. & Brocades (1996), Patent No. EP 743017, Delft, The Netherlands.

Blechl, A.E. and Anderson, O.D. (1996) Expression of a novel high-molecular-weight glutenin subunit gene in transgenic wheat. *Nature Biotechnology* **14**, 875-879.

Botella, J. & Sanewski, G. (1997) New isolated ACC-synthase genes from pineapples; 1-aminocyclopropane-1-carboxylate-synthase sense or antisense gene transfer and expression in a pineapple transgenic plant, for flowering initiation inhibition. Patent AU 9719963.

Bradley, D.J, Carpenter, R. & Coen, E.S. (1997) *Antirrhinum majus* centroradialis (CEN) gene and *Arabidopsis* homolog, Tfl1; gene expression in transgenic plant for flowering modification. Patent WO 9710339.

Brinch-Pedersen, H., Galili, G., Knudsen, S. & Holm, P.B. (1996) Engineering of the aspartate family biosynthetic pathway in barley (*Hordeum vulgare* L.) by transformation with heterologous genes encoding feed-back-insensitive aspartate-kinase and dihydrodipicolinate-synthease; crop improvement by *Escherichia coli* enzyme gene expression in transgenic plant. *Plant Mol Biol* **32**, 611-620.

Brocades, Mogen-Int. (1991) Patent No. EP 449375.

Brugliera, F., Holton, T., Stevenson, T.W., Farcy, E., Lu, C.-Y. & Cornish, E.C. (1994) Isolation and characterization of a cDNA clone corresponding to the Rt locus of Petunia hybrida. The Plant Journal 5, 81-92.

Brunetti, A., Tavazza, M., Noris, E., Tavazza, R., Caciagli, P., Ancora, G., Crespi, S. & Accotto, G.P. (1997) High expression of truncated viral rep protein confers resistance to tomato yellow leaf curl virus in transgenic tomato plants. *Molecular Plant-Microbe Interactions* **10**, 571-579.

Caimi, P.G., McCole, L.M., Klein, T.M. & Kerr, P.S. (1996) Fructan accumulation and sucrose metabolism in transgenic maize endosperm expressing a *Bacillus amyloliquefaciens SacB* gene. *Plant Physiol.* **110**, 355-363.

Chuck, G.S., Dooner, H.K., Courtney-Gutterson, N., Keller, J., Nijjar, C.S. & Ralston, E.J. (1994) Petunia *Ph* gene and constructs containing it; petunia, rose or tomato transgenic plant production with e.g. altered flower color, by *Ph* gene sense or antisense DNA expression Patent WO 9423561.

Colasanti, J.J. & Sundaresan V. (1996) New isolated plant Id gene; zinc finger, dominant-negative mutant, antisense RNA or ribozyme expression in a maize or sorghum transgenic plant for accelerated or delayed flowering. Patent WO 9634088.

Conner, P.J., Brown, S.K. & Weeden, N.F. (1997) Randomly amplified polymorphic DNA-based genetic linkage maps of apple cultivars. *Journal of the American Society for Horticultural Science* **122**, 350-359.

Cooke, D., Debet, M., Gidley, M.J., Jobling, S.A., Safford, R., Sidebottom, C.M. & Westcott, R.J. (1996) New potato plant starch having high amylose content; and starch branching enzyme to alter viscosity for using in the food, biodegradable product, adhesive and textile industry. Nat. Starch-Chem Investment, Wilmington, USA. Patent No. WO 9634968.

Cotton, R.G.H. (1997) Mutation detection. Oxford University Press.

Coupland, G.M. & Schaffer, R.J. (1997) New isolated late elongated hypocotyl gene; from *Arabidopsis thaliana*, for use in delaying or advancing flowering in transgenic plant. Patent WO 9749811.

Courtney-Gutterson, N. (1994) The biologist's palette: genetic engineering of anthocyanin biosynthesis and flower colour, in: B.E. Ellis, G.W. Kuroki and H. Stafford (eds.), Genetic Engineering of Plant Secondary Metabolism, *Recent Advances in Phytochemistry 28*, Plenum Press New York, pp. 93-124.

Courtney-Gutterson, N., Napoli, C., Lemieux, C., Morgan, A., Firoozabady, E. & Robinson, K.E.P. (1994) Modification of flower color in florist's chrysanthemum: production of a white flowering variety through molecular genetics. *Bio/Technology* **12**, 268-271.

Dangl, J.L. (1995) Piece de Resistance: novel classes of plant disease resistance genes. *Cell* **80**, 363-366.

Davis, K.M. & Schwinn, K.E. (1997) Flower colour, in: R.L. Geneve, J.E. Preece, and S.A Merkle (eds.), Biotechnology of ornamental plants. CAB International AL, pp 259-294.

De Simone, M., Morgante, M., Lucchini, M., Parrini, P.& Marocco, A. (1997) A first linkage map of *Cichorium intybus* L. using a one-way pseudo-testcross and PCR-derived markers. *Molecular Breeding* **3**, 415-425.

de Vries, J.A. (1997) New opportunities with amylopectin potato starch. *Food Marketing & Technology* **11**, 12, 14.

Dean, C., MacKnight, R.C., Bancroft, I. & Lister, C.K. (1996) Methods of influencing flowering characteristics of plants; *Arabidopsis thaliana* FCA gene cloning, expression in transgenic plant using *Agrobacterium tumefaciens* and propagation for flowering modification. Patent WO 9638560.

Dehesh, K., Voelker, T.A. & Howkins, D. (1996) New recombinant production of myristate in plant cells; myristic acid production using a vector encoding *Cuphea palustris* acy-ACP-thioesterase gene, for expression in *Brassica* sp. seed, for application in the surfactant and food industry. Calgene, Davis, USA. Patent No. WO 9623892.

Dellaporta, S.L. (1995) *Ts2* gene and promoter and expression systems for transforming plants; propagation of plants with predominantly single-sex flowers Patent WO 9505732.

Delores S.C., Bradley J.M., Davies K.M., Schwinn K.E. & Manson, D.G. (1995) Generation of novel patterns in lisianthus flowers using an antisense chalcone synthase gene, in Proceedings of the XVIIIthe Eucarpia Symposium, Ornamental Section, Tel Aviv, Israel.

Duan, X., Li, X., Xue, Q., Abo-El-Saad, M., Xu, D. & Wu. R. (1996). Transgenic rice plants harboring and introduced potato proteinase inhibitor II gene are insect resistant. *Bio/Technology* **14**, 494-498.

Düring, K., Porsch, P., Fladung, M. & Lörz, H. (1993) Transgenic potato plants resistant to the phytopathogenic bacterium *Erwinia carotovora. Plant J.* **3**, 587-598.

Duvick, J. & Rood, T.A. (1996) Detoxifying fumonisin and related mycotoxin compounds; Exophiala spinifera, Rhinocladielle atrovirens and bacterium enzymes production and expression in maize transgenic plant, for application as a feedstuff. Pioneer-Hi-Bred Des Moines, USA, Patent No. WO 9606175.

Eberlein, C.V., Guttieri, M.J. & Steffen-Campbell, J. (1998) Bromoxynil resistance in transgenic potato clones expressing the bxn gene. *Weed Science* **46**, 150-157.

Ecker, J.R. & Kieber, J.J. (1997) Plants overexpressing constitutive triple response genes; *Arabidopsis thaliana* ethene insensitivity gene transfer and expression in a transgenic plant, for altered fruit ripening, flower and leaf senescence and leaf abscission. Patent US 5602322.

Elomaa, P., Honkanen, J., Puska, R., Seppänen, P., Helariutta, Y., Mehto, M., Hotilainen, M., Nevalainen, L. & Teeri, T.H. (1993) *Agrobacterium*-mediated transfer of antisense chalcone synthase cDNA to *Gerbera hybrida* inhibits flower pigmentation. *Bio/Technology* **11**, 508-511.

Fagerstrom, R., Palo-Heimo, M., Lantto, R., Lahtinen, T. & Suominen, P. (1997) Patent No. WO 9722691, Primalco, Rajamaki, Finland.

Falco, S.C., Beaman, C. & Chui, C.F. (1996) Transgenic crops with improved amino acid composition. 2nd. Colmar Symp. Biol. Sci. Plant Biology. Colmar, Programme abstracts. pp. 25-27.

Fang, D.Q., Federici, C.T. & Roose, M.L. (1997b) Development of molecular markers linked to a gene controlling fruit acidity in citrus. *Genome* **40**, 841-849.

Fischer, S.G. & Lerman, L.S. (1983) DNA fragments differing by single base pair substitutions are separated in denaturing gradient gels: correspondence with melting theory. *Proc. Natl. Acad. Sci. USA* **80**, 1579-1583.

Fischhoff, D.A., Bowdish, K.S. & Perlak, F.J. (1987) Insect tolerant transgenic tomato plants. *Bio/Technology* **5**, 807-813.

Fulton, T.M., Beck-Bunn, T., Emmatty, D., Eshed, Y., Lopez, J., Petiard, V., Uhlig, J., Zamir, D. & Tanksley, S.D. (1997) QTL analysis of an advanced backcross of *Lycopersicon peruvianum* to the cultivated tomato and comparisons with QTLs found in other wild species. *Theoretical and Applied Genetics* **95**, 881-894.

Gatehouse, A.M.R., Davison, G.M., Newell, C.A., Merryweather, A., Hamilton, W.D.O., Burgess, E.P.J., Gilbert, R.J.C. & Gatehouse, J.A. (1997) Transgenic potato plants with enhanced resistance to the tomato moth, *Lacanobia oleracea*: Growth room trials. *Molecular Breeding* **3**, 49-63.

Gatehouse, A.M.R., Down, R.E., Powell, K.S., Sauvion, N., Rahbe, Y. & Newell, C.A. (1996) Transgenic potato plants with enhanced resistance to the peach-potato aphid Myzus persicae. Entomology Experimentalis et Applicata 79, 295-307.

Gauthier, M.F., Lullien-Pellerin, V., de Lamotte, F. & Joudrier, P. INRA (1996) Patent No. WO 9603505, Paris, France.

Gengenbach, B.G., Somers, D.A., Bittel, D.C., Shaver, J.M. & Sellner, J.M. (1996) DNA encoding maize dihydropicolinic-acid-synthase mutant; dihydropicolinate-synthase mutant resistant to feedback inhibition by lysine expression in maize transgenic plant for nutritional value enhancement and crop improvement. Univ. Minnesota, USA. Patent No. US 5545545.

Geoffrey, V., Creusot, F., Falquet, J., Sevignac, M., Adam-Blondon, A.F., Bannerot, H., Gepts, P. & Dron, M. (1998) A family of LRR sequences in the vicinity of the CO2 locus for anthracnose resistance in *Phaseolus vulgaris* and its potential use in marker assisted selection. *Theoretical and Applied Genetics* **96**, 494-502.

Gonsalves, C., Cai, W., Tennant, P. & Gonsalves, D. (1997) Efficient production of virus resistant transgenic papaya plants containing the untranslatable coat protein of the papaya-ringspot virus. American Phytopathological Society, 1997 Annual Meeting, Rochester, NY, 9-13 August, 1997. *Phytopathology* **87**, Suppl., p34.

Good, X., Kellog, J.A., Wagoner, W., Langhoff, D., Matsumara, W. & Bestwick, R.K. (1994) Reduced ethylene synthesis by transgenic tomatoes expressing S-adenosyl methionine hydrolase. *Plant Mol. Biol.* **26**, 781-790.

Grayburn, W.S., Collins, G.B. & Hildebrand, D.F. (1992) Fatty acid alteration by a Δ9 desaturase in transgenic tobacco tissue. *Bio/Technology* **10**, 675-678.

Green, C.E. (1983) New developments in plant tissue culture and plant regeneration; application to improved animal feedstuff production etc. (conference paper). *Basic Life Sci.* **25**, 195-209.

Grierson, D., Fray, R.G. & Wallace, A.D. (1995). New S-adenosylmethionine-decarboxylase DNA; expression modulatin in transgenic plant for senescence or fruit ripening modification. Patent WO 9514092.

Grompe, M. (1993) The rapid detection of unknown mutations in nucleic acids. *Nat. Genet.* **5**, 111-117.

Halfter, U., Ali, N., Stockhaus, J., Ren, L. & Chua, N.-H. (1994) Ectopic expression of a single homeotic gene, the *Petunia* gene *green petal*, is sufficient to convert sepals to petaloid organs. *EMBO J.* **13**, 1443-1449.

Hamilton, A.J., Lycett, H.W. & Grierson, D. (1990) Antisense gene that inhibits synthesis of the hormone ethylene in transgenic plants. *Nature* **346**, 284-287.

Hammond Kosack, K. E. & Jones, J.D.G. (1997) Plant disease resistance genes. *Annual Review of Plant Physiology and Plant Molecular Biology* 48, 575-607.

Hammond-Kosack, K.E., Tang, S., Harrison, K. & Jones, J.D.G. (1998) The tomato Cf-9 disease resistance gene functions in tobacco and potato to confer responsiveness to the fungal avirulence gene product Avr9. *Plant Cell* in press.

Harcourt, R.L., Llewellyn, D., Kyozuka, J., Peacock, W.J., Southerton, S. & Dennis, E.S. (1996) Isolated *Eucalyptus* reproductive genes; antisense, ribozyme or cosuppression or overexpression-inducing copies of the normal lower development genes used in the construction of a sterile transgenic plant. Patent AU 9539013.

Hawkins, J.R. (1997) Finding mutations. IRL Press at Oxford University Press.

Holton, T.A. (1995) Modification of flower colour via manipulation of P450 gene expression in transgenic plants. Drug Metabolism and Drug Interaction 12, 359-368.

Holton, T.A. (1996) New transgenic plants; carnation, rose, gerbera or chrysanthemum transgenic plant construction by flavonoid-3'5'-hydroxylase and dihydrokaempferol-4-reductase gene transfer for flower color alteration. Patent WO 9636716.

Honee, G., Melchers, L.S., Vleeshouwers, V.G.A.A., Van Roeckel, J.S.C. & De Wit, P.J.G.M. (1995) Production of the AVR9 elicitor from the fungal pathogen *Cladosporium fulvum* in transgenic tobacco and tomato plants. *Plant Molecular Biology* 29, 909-920.

Hong, C.J., Etoh, T., Landry, B. & Matsuzoe, N. (1997) RAPD markers related to pollen fertility in garlic (*allium sativum* L.), *Breeding Science* 47, 359-362.

Hormaza, J.I., Pinney, K. & Polito, V.S. (1998) Genetic diversity of pistachio (*Pistacia vera*, anacardiaceae) germplasm based on randomly amplified polymorphic DNA (RAPD) markers. *Economic Botany* 52, 78-87.

Howard, J.A., Culver, K.W., Pioneer-Hi-Bred-Int. Hum. Gene-Ther. Res. Inst. Des-Moines (1997) Patent no. WO 9748810. Des Moines, USA.

ICI (1993). DNA generating antisense RNA to fruit softening enzyme sequence; vector with polygalacturonase or pectinesterase antisense DNA for use in tomato transgenic plant crop improvement. Patent EP 532060.

Iizuka T (1993) Patent No. JP 05146289.

Irie, K., Hosoyama, H., Takeuchi, T., Iwabuchi, K., Watanabe, H., Abe, M., Abe, K. & Arai, S. (1996) Transgenic rice established to express corn cystatin exhibits strong inhibitory activity against insect gut proteinases. *Plant Molecular Biology* 30, 149-157.

Ishimoto, M., Sato, T., Chrispeels, M..J. & Kitamura, K. (1996) Bruchid resistance of transgenic azuki bean expressing seed alpha-amylase inhibitor of common bean. *Entomologia Experimentalis et Applicata* 79, 309-315.

Iwate-Ken (1996) Novel chalcone-synthase gene; Gentiana sp. gene cloning for flower color modification in transgenic plant. Patent JP 08089251.

Jach, G., Gornhardt, B., Mundy, J., Logemann, J., Pinsdorf, E., Leah, R., Schell, J. & Maas, C. (1995) Enhanced quantitative resistance against fungal disease by combinatorial expression of different barley antifungal proteins in transgenic tobacco. *Plant Journal* 8, 97-109.

James, C. (1997) Global status of transgenic crops in 1997. *ISAAA Briefs* No. 5. ISAAA, Ithaca, NY, pp. 31.

Jiang, C. & Sink, K.C. (1997) RAPD and SCAR markers linked to the sex expresson locus M in asparagus. *Euphytica* 94, 329-333.

Jongedijk, E., Tigelaar, H., Van Roekel, J.S.C., Bres-Vloemans, S.A., Dekker, I., Van Den Elzen, P.J.M., Cornelissen, B.J.C. & Melchers, L.S. (1995) Synergistic activity of chitinases and beta-1,3-glucanases enhances fungal resistance in transgenic tomato plants. *Euphytica* 85, 173-180.

Kaemmer, D., Fischer, D., Jarret, R.L., Baurens, F.C., Grapin, A., Dambier, D., Noyer, J.L., Lanaud, C., Kahl, G. & Lagoda, P.J.L. (1997) Molecular breeding in the genus Musa: A strong case for STMS marker technology. *Euphytica* 96, 49-63.

Kanazin, V., Marek, L.F. & Shoemaker, R.C. (1996) Resistance gene analogs are conserved and clustered in soybean. *Proc. Natl. Acad. Sci. USA* 93, 11746-11750.

Kavanagh, T.A. & Spillane, C. (1995) Strategies for engineering virus resistance in transgenic plants. *Euphytica* 85, 149-158.

Kawchuk, L.M., Lynch, D.R., Martin, R.R., Kozub, G.C. & Farries, B. (1997) Field resistance to the potato leafroll luteovirus in transgenic and somaclone potato plants reduces tuber disease symptoms. *Canadian Journal of Plant Pathology* 19, 260-266.

Kim, S.J., Lee, S.J., Kim, B.D. & Paek, K.H. (1997) Satellite-RNA-mediated resistance to cucumber mosaic virus in transgenic plants of hot pepper (*Capsicum annuum* vs. Golden Tower). *Plant Cell Reports* **16**, 825-830.

King, G.J., Alston, F.H., Brown, L.M., Chevreau, E., Evans, K.M., Dunemann, F., Janse, J., Laurens, F., Lynn, J.R., Maliepaard, C., Manganaris, A.G., Roche, P., Schmidt, H., Tartarini, S., Verhaegh, J. & Vrieling, R. (1998) Multiple field and glasshouse assessments increase the reliability of linkage mapping of the Vf source of scab resistance in apple. *Theoretical and Applied Genetics* **96**, 699-708.

Klee, H.J. (1993) Ripening physiology of fruit from transgenic tomato (*Lycopersicon esculentum*) plants with reduced ethylene synthesis. *Plant Physiology* **102**, 911-916.

Knutzon, D.S., Thompson, G.A., Radke, S.E., Johnson, J., Knauf, V.C. & Kridl, J.C. (1992) Modification of *Brassica* seed oil by antisense expression of a stearoyl-acyl-carrier protein desaturase gene. *Proc. Natl. Acad. Sci. USA* **89**, 2624-2628.

Krattiger, A.I. (1997) Insect resistance in crops: a case study of *Bacillus thuringiensis* (*Bt*) and its transfer in developing countries. *ISAAA Briefs No.* **2**. ISAAA, Ithaca, NY pp. 42.

Kunik, T., Gafni, Y., Czosnek, H. & Citovsky, V. (1997) A. Altman and M. Ziv (eds.) *Transgenic tomato plants expressing TYLCV capsid protein are resistant to the virus: the role of the nuclear localization signal (NLS) in the resistance.* Proceedings of the third international ISHS symposium on in vitro culture and horticultural breeding, Jerusalem, Israel, 16-21 June 1996. *Acta Horticulturae* **447**, 387-391.

Kyowa-Hakko (1993) Gene encoding flavonoid-3',5'-hydroxylase of Petunia petal; useful for blue flower color imparting to ornamental transgenic plant, e.g. rose or tobacco. Patent WO 9318155.

Le Thierry, D'ennequin, M., Panaud, O., Robert, T. & Richroch, A. (1997) Assessment of genetic relationships among sexual and asexual forms of *Allium cepa* using morphological traits and RAPD markers. *Heredity* **78**, 403-409.

Leister, D., Ballvora, A., Salamini, F. & Gebhard, C. (1996) A PCR-based approach for isolating pathogen resistance genes from potato with potential for wide application in plants. *Nature Genetics* **14**, 421-429.

Leister, D., Kurth, J., Laurie, D.A., Yano, M., Sasaki, T., Devos, K., Graner, A. & Schulze Lefer,t P. (1998) Rapid reorganization of resistance gene homologues in cereal genomes. *Proceedings of the National Academy of Sciences of the United States of America* **95**, 370-375.

Li, J. & Chory, J. (1997) A putative leucine-rich repeat receptor kinase involved in brassinosteroid signal transduction. *Cell* **90**, 929-938.

Liu, D., Raghothama, K.G., Hasegawa, P.M. & Bressan, R.A. (1994) Osmotin overexpression in potato delays development of disease symptoms. Proc. Natl. Acad. Sci. USA **91**, 1888-1892.

Lius, S., Manshardt, R.M., Fitch, M.M.M., Slightom, J.L., Sanford, J.C. & Gonsalves, D. (1997) Pathogen-derived resistance provides papaya with effective protection against papaya ringspot virus. *Molecular Breeding* **3**, 161-168.

Lopez-Valenzuela, J.A., Martinez, O. & Paredes-Lopez, O. (1997) Geographic differentiation and embryo type identification in *Magnifera indica* L. cultivars using RAPD markers. *HortScience* **32**, 1105-1108.

Madrid, S.M., Rasmussen, P. & Baruch, A. (1996) New endo-beta-1,4-glucanase from *Aspergillus*; new recombinant cellulase production; signal peptide, promoter and terminator DNA sequence; expression in fungus and transgenic plant for use in the brewing industry and feedstuff. Danisco, Copenhagen, DK. Patent No. WO 9629415.

Mandel, M.A., Bowman, J.L., Kempin, S.A., Ma, H., Meyerowitz, E.M. & Yanowski, M.F. (1992) Manipulation of flower structure in transgenic tobacco. *Cell* **71**, 133-143.

Marshall, A. & Hodgson, J. (1998) DNA chips: An array of possibilities. *Nature Biotechnology* **16**, 27-31.

Mashal, RD & Sklar, J. (1996) Practical methods of mutation detection. *Current Opinion in Genet. Dev.* **6**, 275-280.

McCown, B.H. (1997). Approaches to modify plant form, in: R.L. Geneve, J.E. Preece and S.A. Merkle (eds.) Biotechnology of ornamental plants.*CAB International AL*, pp 199-214.

Mesquita, A.G.G., Faleiro, F.G., Paula, T.J.D. Jr., Ragagnini, V.A., Moreira, M.A. & Barros, E.G.D. (1998) Use of molecular markers to differentiate *Colletotrichum lindemuthianum* races 89 and 69. *Fitopatologia Brasileira* **23**, 58-61.

Mestre, P.F., Asins, M.J., Pina, J.A., Carbonell, E.A. & Navarro, L. (1997) Molecular markers flanking citrus tristeza virus resistance gene from *Poncirus trifoliata* L. Raf. *Theoretical and Applied Genetics* 94,458-464.

Meyer, P., Heidmann, I., Forkmann, G. & Saedler, H. (1987). A new petunia flower colour generated by transformation of a mutant with a maize gene. *Nature* **330**, 677-678.

Meyerowitz, E.M., Chang, C. & Bleecker, A.B. (1995). Modified ethene response nucleic acid; tomato transgenic plant with modified response to ethene for controlled fruit ripening, floral senescence, etc. Patent WO 9501439.

Mitsui-Seiyaku (1997). Protein, its gene and use; plant nucleoside-diphosphate-kinase expression in a transgenic plant, to alter flowering time. Patent JP 09252781.

Monsanto (1992). Delaying fruit ripening and senescence in plant; by controlling ethene production, preferably by expression of 1-aminocyclopropane-1-carboxylate-deaminase in transgenic plant. Patent WO 9212249.

Norelli, J., Aldwinckle, H., Destefano-Beltran, L. & Jaynes, J. (1993) Transgenic apple plants containing lytic proteins have increased resistance to Erwinia amylovora. Phytopathology 83 (12), 1395.

Norelli, J.L., Aldwinckle, H.S., Destefano-Beltran, L. & Jaynes, J.M. (1994). Transgenic "Malling 26" apple expressing the attacin E gene has increased resistance to *Erwinia amylovora. Euphytica* **77**, 123-128.

Noteborn, H.P.J.M., Bienenmann-Ploum, M.E. & Kuiper, H.A. (1995) Product safety of transgenic tomatoes with *Bacillus thuringiensis* genes. Voedingmiddelentechnolgie **28**, 31-34.

Oeller, P.W., Min-Wong, L., Taylor, L.P., Pike, D.A. & Theologis, A. (1991) Reversible inhibition of tomato fruit senescence by antisense RNA. *Science* **254**, 437-439.

Oh, T.K., Kim, H.K., Bae, K.S., Park, Y.S., Kim, Y.O., Che, Y.U. & Lee, D.K. (1997) Novel *Bacillus* sp. DS11 and novel phytase produced thereby; for use as a feed-additive. Korea Inst. Sci. Technol., Seoul, Korea, Patent No. WO 9733976.

Orita, M., Iwahana, H., Kanazawa, H., Hayashi, K. & Sekiya, T. (1989) Detection of polymorphisms of human DNA by gel electrophoresis as single-strand conformation polymorphisms. *Proc. Natl. Acad. Sci. USA* **86**, 2766-2770.

Oud, J.S.N., Schneiders, H., Kool, A.J. & van Grinsven, M.Q.J.M. (1995) Breeding of transgenic orange *Petunia hybrida* varieties. *Euphytica* **84**,175-181.

Paran, I., Goldman, I. & Zamir, D. (1997) QTL analysis of mophological traits in a tomato recombinant inbred line population. *Genome* **40**, 242-248.

Parniske, M., Hammond-Kosack, K.E., Golstein, C., Thomas, C.M., Jones, D.A. Harrison, K., Wulff, B.B. & Jones, J D. (1997) Novel disease resistance specificities result from sequence exchange between tandemly repeated genes at the Cf-4/9 locus of tomato. *Cell* **91**, 821-32.

Perlak, F.J., Stome, T.B., Muskopf, Y.M. & Petersen, L.J. (1993) Genetically improved potatoes: protection from damage by Colorado potato beetles. *Plant Mol. Biol.* **22**, 313-321.

Phillips, R.L. & Vasil, I.K. (Eds.) (1994) DNA-based markers in Plants. Kluwer Academic Publishers, Dordrecht, Boston, London.

Picton, S., Gray, J.E. & Grierson, D. (1995) The manipulation and modification of tomato fruit ripening by expression of antisense RNA in transgenic plants. *Euphytica* **85**, 193-202.

Polley, A., Seigner, E. & Ganal, M.W. (1997) Identification of sex in hop (*Humulus lupulus*) using molecular markers. *Genome* **40**, 357-361.

Purcell, J.P., Greenplate, J.T. & Jennings, M.G. (1993) Cholesterol oxidase: a patent insecticidal protein active against Boll weefil larvae. Biochem. *Biophys. Res. Comm.* **196**: 1406-1413.

Ramsay, G. (1998) DNA chips: State-of-the art. *Nature Biotechnology* **16**, 40-44.

Ravelonandro, M., Scorza, R., Bachelier, J.C., Labonne, G., Levy, L., Damsteegt, V., Callahan, A.M. & Dunez, J. (1997) Resistance of transgenic *Prunus domestica* to plum pox virus infection. *Plant Disease* **81**, 1231-1235.

Riesner, D., Steger, G., Zimmat, R., Owens, R.A., Wagenhöfer, M. & Hillen, W. (1989) Temperature-gradient gel electrophoresis of nucleic acids: analysis of conformational transitions, sequence variations, and protein-nucleic acid interactions. *Electrophoresis* **10**, 377-389.

Robson, P.H.R. & Smith, H. (1997) Fundamental and biochemical applications of phytochrome transgenes. *Plant, Cell and Environment* **20**, 831-839.

Roche, P., Alston, F.H., Maliepaard, C., Evans, K.M., Vrielink, R., Dunemann, F., Markussen, T., Tartarini, S., Brown, L.M., Ryder, C. & King, G.J. (1997) RFLP and RAPD markers linked to the rosy leaf curling aphid resistance gene (Sd-1) in apple. *Theoretical and Applied Genetics* **94**, 528-533.

Rotino, G.L., Perri, E., Zottini, M., Sommer, H. & Spena, A. (1997) Genetic engineering of parthenocarpic plants. *Nature Biotechnology* **15**, 1398-1401.

Sano, T., Nagayama, A., Ogawa, T., Ishida, I. & Okada, Y. (1997) Transgenic potato expressing a double-stranded RNA-specific ribonuclease is resistant to potato spindle tuber viroid. *Nature Biotechnology* **15**, 1290-1294.

Schafer, A.J. & Hawkins, J.R. (1998) DNA variation and the future of human genetics. *Nature Biotechnology* **16**, 33-39.

Schroeder, H.E., Gollasch, S., Moore, A., Tabe, L.M., Craig, D. & Higgins, T.J.V. (1995) Bean alpha-amylase inhibitor confers resistance to the pea weevil (*Bruchus pisorum*) in transgenic peas (*Pisum sativum* L.). *Plant Physiology* **107**, 1233-1239.

Schwarz-Sommer, Z., Huijser, P., Nacken, W., Saedler, H. & Sommer, H. (1990) Genetic control of flower development by homeotic genes in *Antirrhinum majus*. *Science* **250**, 931-936.

Scorza, R., Levy, L., Damsteegt, V., Yepes, L.M., Cordts, J., Hadidi, A., Slightom, J. & Gonsalves, D. (1995) Transformation of plum with the papaya ringspot virus coat protein gene and reaction of transgenic plants to plum pox virus. *Journal of the American Society for Horticultural Science* **120**, 943-952.

Scovel, G., Bein-Meir, H., Ovadis, M., Itzhaki, H. & Vainstein, A. (1998) RAPD and RFLP markers tightly linked to the locus controlling carnation (*Dianthus caryophyllus*) flower type. *Theoretical and Applied Genetics* **96**, 117-122.

Shah, D.M., Rommens, C.M.T. & Beachy, R.N. (1995) Resistance to diseases and insects in transgenic plants: progress and applications to agriculture. *Trends Biotechnology* **13**, 362-368.

Sharon, D., Cregan, P.B., Mhameed, S., Kusharska, M., Hillel, J., Lahav, E. & Lavi, U. (1997) An integrated genetic linkage map of avocado. *Theoretical and Applied Genetics* **95**, 911-921.

Shewmaker, C.K., Boyer, C.D., Wiesenborn, D.P., Thompson, D.B., Boersig, M.R., Oakes, J.V. & Stalker, D.M. (1994) Expression of *Escherichia coli* glycogen synthase in the tubers of transgenic potatoes (*Solanum tuberosum*) results in highly branched starch. *Plant Physiology* **104**, 1159-1166.

Shots, A., Bakker, J., Helder, J., Gommers, F., Stiekema, W.J., Roosien, J., Goverse, A., Shouten, A., Smant, G., de Boer, J.M., Stokkermans, J.P.W., Abad, P. & Rosso, M.N.F. (1998) New cellulases from nematodes and related nucleic acid, vectors, transformed cells and antibodies recombinant cellulase production for use in the food, clothing and waste industry and transgenic plant construction with nematode resistance. Patent no. WO 9801569, Wageningen, The Netherlands.

Sicard, D., Michalakis, Y., Dron, M. & Neema, C. (1997) Genetic diversity and pathogenic variation of *Colletotrichum lindemuthianum* in the three centers of diversity of its host, *Phaseolus vulgaris*. *Phytopathology* **87**, 807-813.

Skroch, P.W., Nienhuis, J., Bebee, S., Tohme, J. & Pedrata, F. (1998) Comparison of Mexican common bean (*Phaseolus vulgaris* L.) core and reserve germplasm collections. *Crop Science* **38**, 488-496.

Socias, I. & Company, R. (1998) Fruit tree genetics at a turning point: The almond example. *Theoretical and Applied Genetics* **96**, 588-601.

Song, Y.-R., and Ma, Q.-H., Hou, L.-L., Zhang, L.-Z., Yang, W.-Y., Peng, X.-X. & Wang, H.-Y. (1996) Transgenic potato with PVY coat protein gene and its small-scale field test. *Acta Botanica Sinica* **38**, 711-718.

Spielmeyer, W., Robertson, M., Collins, N., Leister, D., Schulze-Lefert P., Seah, S., Moullet, O. & Lagudah, E.S. (1998) A superfamily of disease resistance gene analogs is located on all homeologous chromosome groups of wheat (*Triticum aestivum*). *Genome*, in press.

Stahl, A. (1997) A filling agent obtained from amylopectin-type starch; amylopectin-type potato transgenic plant starch for use in the food industry. Sveriges-Starkelseproducenter-Forening, Karlshamn, Sweden. Patent No. WO 9703573.

Stark, D., Timmermann, K.-P., Barry, G.F., Preiss, J. & Kishore, G.M. (1992) Regulation of the amount of starch in plant tissues by ADP glucose pyrophosphorylase. *Science* **258**, 287-292.

Stark, D.M., Barry, G.F. & Kishore, G.M. (1996) Improvement of food quality traits through enhancement of starch biosynthesis; potato, tomato and rape transgenic plant construction with e.g. bacterium glucose-1-phosphate-adenylyltransferase expression for crop improvement. *Ann. N.Y. Acad. Sci.* **792**, 26-36.

Staskawicz, B.J., Ausubel, F.M., Baker, B.J., Ellis, J.G. & Jones, J.D. (1995) Molecular genetics of plant disease resistance. *Science* **268**, 661-667.

Surov, T., Aviv, D., Aly, R., Joel, D.M., Goldman-Guez, T. & Gressel, J. (1998) Generation of transgenic asulam-resistant potatoes to facilitate eradication of parasitic broomrapes (*Orobanche* spp.), with the sul gene as the selectable marker. *Theoretical and Applied Genetics* **96**, 132-137.

Sweetlove, L.J., Burrell, M.M. & ap Rees, T. (1996) Starch metabolism in tubers of transgenic potato (*Solanum tuberosum*) with increased ADPglucose pyrophosphorylase. *Biochem J.* **320**, 493-498.

Tacke, E., Salamini, F. & Rohde W. (1996) Genetic engineering of potato for broad-spectrum protection against virus infection. *Nature Biotechnology* **14**, 1597-1601.

Terras, F.R., Eggermont, K., Kovaleva, V., Raikhel, N.V., Osborn, R.W., Kester, A., Rees, S.B., Torrekens, S., Van Leuven, F. & Vanderleyden, J. (1995) Small cysteine-rich antifungal proteins from radish: their role in host defense. *Plant Cell* **7**, 573-588.

Thakur, B.R., Singh, R.K., Tieman, D.M. & Handa, A.K. (1996) Tomato product quality from transgenic fruits with reduced pectin methylesterase. Journal of Food *Science* **61**, 85-87.

Theologis, A. & Sato, T. (1998) DNA encoding antisense RNA blocking plant ACC-synthase expression; e.g. tomato transgenic plant construction by vector plasmid-mediated antisense 1-aminocyclopropane-1-carboxylate-synthase gene expression, used to delay fruit ripening Patent US 5723766.

Thilmony, R.L., Chen, Z., Bressan, R.A. & Martin, G.B. (1995) Expression of the tomato *Pto* gene in tobacco enhances resistance to *Pseudomonas syringae* pv *tabaci* expressing avrPto. *Plant Cell* **7**, 1529-1536.

Thomas, J.C., Adams, D.G., Keppenne, V.D., Wasmann, C.C., Brown, J.K., Kanost, M.R. & Bohnert, H.J. (1995) Protease inhibitors of *Manduca sexta* expressed in transgenic cotton. *Plant Cell Reports* **14**, 758-762.

Tieman, D.M., Kausch, K.D., Serra, D.M. & Handa, A.K. (1995) Field performance of transgenic tomato with reduced pectin methylesterase activity. *Journal of the American Society for Horticultural Science* **120**, 765-770.

Tomes, D.T., Huang, B. & Miller, P.D. (1997) Production of seedless fruit and vegetables; vector-mediated indole-acetamide-hydrolase, indole-acetamide-synthase and DAM-methylase gene transfer for melon transgenic plant construction. Patent WO 9742333.

Töpfer, R., Martini, N. & Schell, J. (1995) Modification of plant lipid synthesis. *Science* **268**, 681-686.

Tsuchimoto, S., van der Krol, A.R. & Chua, N.-H. (1993) Ectopic Expression of *pMADS3* in transgenic Petunia phenocopies the Petunia *blind* mutant. *Plant Cell* **5**, 843-853.

Tyagi, S., Bratu, D.P. & Kramer, F.R. (1998) Multicolor mulecular beacons for allele discrimination. *Nature Biotechnology* **16**, 49-53.

Ultzen, T., Gielen, J., Venema, F., Westerbroek, A., De Haan, P., Tan, M.L., Schram, A., Van Grinsven, M. & Goldbach, R. (1995) Resistance to tomato spotted wilt virus in transgenic tomato hybrids. *Euphytica* **85**, 159-168.

van der Biezen, E.A. & Jones, J.D. (1998) The NB-ARC domain: a novel signalling motif shared by plant resistance gene products and regulators of cell death in animals. *Current Biology* **8**, R226-227.

van der Krol, A.R. & Vorst, O. (1997) Manipulation of flower shape. in: R.L. Geneve, J.E. Preece, and S.A. Merkle (eds.), Biotechnology of ornamental plants. *CAB International AL*, pp 237-258.

van der Krol, A.R., Lenting, P.E., Veenstra, J.G., van der Meer, I.M., Loes, R.E., Gerats, A.G.M., Mol, J.N.M. & Stuitje, A.R. (1988) An antisense chalcone synthase gene in transgenic plants inhibits flower pigmentation. *Nature* **333**, 866-869.

Van der Meer, I.M., Ebskamp, M.J.M., Bisser, R.G.F., Weisbeek, P.J. & Smeekens, S.C.M. (1994) Fructan as a new carbohydrate sink in transgenic potato plant. *The Plant Cell* **6**, 561-570.

Verwoerd, T.C., van Paridon, P.A., van Ooyen, A.J.J., van Len,t J.W.M., Hoekema, A. & Pen, J (1995) Stable accumulation of *Adpergillus niger* phytase in transgenic tobacco leaves. *Plant Physiol* **109**, 1199-1205.

Visser, R.G.F. & Jacobsen, E. (1993) Towards modifying plants for altered starch content and composition. *Trends in Biotechnology* **11**, 63-68.

Walden, R. & Wingender, R. (1995) Gene transfer and plant-regeneration techniques. *Trends in Biotechnology* **13**, 324-331.

Walker, J.C. (1993) Receptor-like protein kinase genes of *Arabidopsis thaliana. Plant Journal* **3**, 451-456.

Wang, X., Zafian, P., Choudhary, M. & Lawton, M. (1996) The PR5K receptor protein kinase from *Arabidopsis thaliana* is structurally related to a family of plant defense proteins. *Proc Natl Acad Sci USA* **93**, 2598-2602.

White, M.B., Carvalho, M., Derse, D., O'Brien, S.J. & Dean, M. (1992) Detecting single base substitutions as heteroduplex polymorphisms, *Genomics* **12**, 301-306.

Whitham, S., McCormick, S. & Baker, B. (1996) The N gene of tobacco confers resistance to tobacco mosaic virus in transgenic tomato. *Proceedings of the National Academy of Sciences of the United States of America* **93**, 8776-8781.

Wu, G., Shortt, B.J., Lawrence, E.B., Levine, E.B., Fitzsimmons, K.C. & Shah, D.M. (1995) Disease resistance conferred by expression of a gene encoding H2O2 generating glucose oxidase in transgenic potato plants. Plant Cell 7, 1357-1368.

Wünn, J., Klöti, A., Burkhardt, P.K., Biswas, G.C.G., Launis, K., Iglisias, V.A. & Potrykus, I. (1996) Transgenic indica rice breeding line IR58 expressing a synthetic cryIA(b) gene from Bacillus Thuringiensis provides effective insect pest control. *Bio/Technology* **14**, 171-176.

Yanofsky, M.F. (1997) New isolated cauliflower floral meristem identity genes transgenic plant construction with early flowering Patent WO 9727287.

Yeh, S.D., Cheng, Y.H., Bau, H.J., Yu, T.A. & Yang J.S. (1997) Coat protein transgenic papaya immune or highly resistant to different strains of papaya-ringspot-poty virus. American Phytopathological Society, 1997 Annual Meeting, Rochester, NY, 9-13 August, 1997. *Phytopathology* **87**, Suppl., p. 107.

Young, N.D. (1996) QTL mapping and quantitative disease resistance in plants. *Annual Review of Phytopathology* **34**, 479-501.

Yu, Y.G., Buss, G.R. & Maroof, M.A. (1996) Isolation of a superfamily of candidate disease-resistance genes in soybean based on a conserved nucleotide-binding site. *Proc Natl Acad Sci USA* **93**, 11751-11756.

Zemanek, E.C. & Wasserman, B.P. (1995) Issues and advances in the use of transgenic organisms for the production of thaumatin, the intensely sweet protein from Thaumatococcus danielli. *Critical Reviews in Food Science and Nutrition* **35**, 455-466.

Zhu, B., Chen, T.H.H. & Li, P.H. (1996) Analysis of late-blight disease resistance and freezing tolerance in transgenic potato plants expressing sense and antisense genes for an osmotin-like protein. *Planta* **198**, 70-77.

Zhu, Q., Maher, E.A., Masoud, S., Dixon, R.A. & Lamb, C.J. (1994) Enhanced protection against fungal attack by constitutive co-expression of chitinase and glucanase genes in transgenic tobacco. *Bio/Technology* **12**, 807-812.

